Special Functions: Fractional Calculus and the Pathway for Entropy

Special Issue Editor

Hans J. Haubold

MDPI • Basel • Beijing • Wuhan • Barcelona • Belgrade

Special Issue Editor
Hans J. Haubold
Office for Outer Space Affairs, United Nations
Austria

Editorial Office
MDPI AG
St. Alban-Anlage 66
Basel, Switzerland

This edition is a reprint of the Special Issue published online in the open access journal *Axioms* (ISSN 2075-1680) from 2015–2017 (available at: http://www.mdpi.com/journal/axioms/special_issues/special_functions-fractional_calculus).

For citation purposes, cite each article independently as indicated on the article page online and as indicated below:

Author 1; Author 2. Article title. *Journal Name*. **Year**. Article number/page range.

First Edition 2018

ISBN 978-3-03842-665-3 (Pbk)
ISBN 978-3-03842-664-6 (PDF)

Dedicated to
Professor Dr. A.M. Mathai
on the Occasion of His 80th Birthday

Table of Contents

About the Special Issue Editor

Hans J. Haubold is a Professor of Theoretical Astrophysics. He has published more than 200 research papers and more than 10 books on topics of physics, astrophysics, and the development of basic space science worldwide. In cooperation with A.M. Mathai he was one of the initiators of the United Nations Basic Space Science Initiative (UN BSSI) for the worldwide development of astronomy, physics, and mathematics that was implemented in 1991–2015. He has contributed to A.M. Mathai' research program on mathematics and statistics for physics since the 1970's focusing on entropy, probability, and dynamics. This research pursues, among other themes, the interpretation of solar neutrino detection in terms of new physics by utilizing fractional calculus.

Preface to "Special Functions: Fractional Calculus and the Pathway for Entropy"

To commemorate the eightieth birthday of A. M. Mathai, we invited colleagues and former students to prepare special papers that meet Mathai's academic interests as well as research and teaching achievements in one way or another. The papers of this Special Issue of Axioms bear evidence to Mathai's everlasting and fundamental contributions to the development of mathematics, statistics, and their applications in natural sciences.

We wish to thank all contributors to this Special Issue of Axioms for the care they exercised in the composition of their papers. With the broad spectrum from fractional calculus and special functions of mathematical physics to Mathai's entropic pathway and solar neutrinos, we hope that this Festschrift will be useful and exciting for fellow colleagues and future students working in the eld related to entropy, probability, and fractional dynamics as provided by mathematics, physics, and statistics. None of this would have been possible without the help and support of Axioms sta, particularly Qiang Liu and Luna Shen. We would like to thank all of them for their professional editing and handling of the papers and this volume as a whole, as well as their patience in long-term and long-distance international cooperation that also has been made possible by mechanisms of the United Nations.

A.M. Mathai (Figure 1) was born on 28 April 1935 in Arnakulam, near Palai, in the Idukki district of Kerala, India, as the eldest son of Aley and Arakaparampil Mathai. After completing his high school education in 1953 at St. Thomas High School, Palai, he joined St. Thomas College, Palai, with record marks and obtained his B.Sc. degree in mathematics in 1957. In 1959 he completed his Master's degree in statistics at the University of Kerala, Thiruvananthapuram, Kerala, India; he achieved the university degree First Class, First Rank and Gold Medal. Then he joined St. Thomas College, Palai, University of Kerala, as a Lecturer in Statistics and served there until 1961. He obtained a Canadian Commonwealth scholarship in 1961 and went to the University of Toronto, Canada to complete his M.A. degree in mathematics in 1962. He was awarded a Ph.D. from the University of Toronto, Canada, in 1964. Mathai joined McGill University, Montreal, Canada, as an Assistant Professor until 1968. From 1968 to 1978 he was an Associate Professor there. He became a Full Professor at McGill University in 1979 (at this occasion also contributing to the anniversary of Albert Einstein's birthday) and served the Department of Mathematics and Statistics until 2000. Mathai is the founder of the Canadian Journal of Statistics and the Statistical Society of Canada. As of this date, A.M. Mathai is an Emeritus Professor of Mathematics and Statistics at McGill University, Canada, and Director of the Centre for Mathematical and Statistical Sciences, India. He has published over 400 research papers and more than 47 books on topics in mathematics, statistics, (astro)physics, chemistry, and biology. He is a Fellow of the Institute of Mathematical Statistics, National Academy of Sciences of India, served as President of the Mathematical Society of India, and a Member of the International Statistical Institute. At several occasions he has been honored by the United Nations for his services to the international scientific community in terms of education and research in mathematics, statistics, and natural sciences.

Figure 1. Professor A.M. Mathai, Director of the Centre for Mathematical and Statistical Sciences in a typical pose when engaging visitors in technical discussions at the Centre.

Figure 2. The famous painting of Evert Collier (1640-1708) showing 'The Wise Scholar' in research spirit engaged in thinking, reading, and scrippling, more than 350 years before Mathai appeared in a similar situation as shown in Figure 1 (Private Collection Haubold, Vienna and New York).

Outline of Mathai's Long-Term Research Programme: From Neutrinos, Entropy, and Probability to Fractional Dynamics (reaction and diffusion)

This Festschrift is a collection of independent essays illustrating elements of Mathai's research programme in mathematics and statistics applied to selected problems in physics, particularly the relations between entropy, probability, and fractional dynamics as they appeared in solar neutrino astrophysics since the 1970's. The very original research programme was published in three monographs (Mathai and Rathie 1975, Mathai and Pederzoli 1977, Mathai and Saxena 1978). An update of Mathai's research programme and selected results achieved since the 1970's is contained in Mathai et al. (1988, 2010).

Boltzmann's derivation of the second law of thermodynamics was based on mechanics arguments. In his paper of 1872, Boltzmann considered the dynamics of binary collisions and stated that "One has therefore rigorously proved that, whatever the distribution of the kinetic energy at the initial time might have been, it will, after a very long time, always necessarily approach that found by Maxwell" (Boltzmann 1872). Boltzmann's Stosszahlansatz, i.e. the assumption of molecular chaos used in his equation, was a statistical assumption which had no dynamical basis. His equally famous relation between entropy and probability, $S \sim logW$, in his paper "On the relation between the second law of the mechanical theory of heat and probability theory with respect to the laws of thermal equilibrium" (Boltzmann 1877) was not based on dynamics either. At that time Boltzmann's Stosszahlansatz was heavily criticized by Loschmidt's reversibility paradox (Boltzmann 1877) and Zermelo's recurrence paradox (Boltzmann 1896, 1897).

In the remarkable year 1900 for physics, Planck elaborated on the connection between entropy and probability based on the universality of the second law of thermodynamics and the established laws of probability and put in writing the final form of the relation between entropy S and permutability $P \sim W$ in its definitive form $S = klogW$. He called k Boltzmann's constant and came to the conclusion that in every finite region of phase space the thermodynamic probability has a finite magnitude limited by h, representing Planck's constant. At this point Planck introduced his quantum hypothesis (Schoepf 1978). Concerning Planck's hypothesis of light quanta he strictly preserved Maxwell's theory in vacuum and applied the quantum hypothesis only to matter that interacts with radiation (Planck 1907).

In 1911 at the first Solvay Conference, Einstein literally put it as an requirement that one needs a fundamental theory of dynamics to make sense of Boltzmann's connection between entropy and probability, even in the case of Planck's use of Boltzmann's formula in the process of discovery of the quantum of action. Einstein's immediate reaction to Planck's extensive report at the first Solvay Congress was (Eucken 1914):

> "What I find strange about the way Mr. Planck applies Boltzmann's equation is that he introduces a state probability W without giving this quantity a physical definition. If one proceeds in such a way, then, to begin with, Boltzmann's equation does not have a physical meaning. The circumstance that W is equated to the number of complexions belonging to a state does not change anything here; for there is no indication of what is supposed to be meant by the statement that two complexions are equally probable. Even if it were possible to define the complexions in such a manner that the S obtained from Boltzmann's equation agrees with experience, it seems to me that with this conception of Boltzmann's principle it is not possible to draw any conclusions about the admissibility of any fundamental theory whatsoever on the basis of the empirically known thermodynamic properties of a system."

Recently, Brush (2015) commented on the above Boltzmann-Planck-Einstein dispute from a historical point of view on how the interaction of theory and experiment in physics with available applicable mathematics and statistics lead to established theories and subsequently to predictions and explanations of natural phenomena. He perceives Planck's derivation of an equation for black-body radiation that this equation, when explored with Boltzmann's formula for entropy, implied that radiation is composed of particles. Planck, as a strong supporter of the wave theory of electromagnetic radiation, could not believe what the mathematics was telling him. Similarly, Kuhn (1978) pointed out that Planck did not propose a physical quantum theory but he used quantization only as a convenient method of approximation.

Following the above reasoning of Boltzmann, Planck, and Einstein, the Mathai programme turned to neutrino radiation and utilized the statistical methodology developed by Scafetta (2010) by evaluating the scaling exponent of the probability density function, through Boltzmann's entropy, of the diffusion process generated by complex fluctuations in the measurements of the solar neutrino flux in the Super-Kamiokande experiment (Yoo et al. 2003, Cravens et al. 2008, Sakurai 2014, Haubold et al. 2014). This turn was justified by earlier explorations of possible solutions to the so-called solar neutrino problem, established in Davis' Homestake experiment (Treder 1974, Haubold and John 1978, Haubold and Gerth 1985). Scafetta's method does focus on the scaling properties of the Super-Kamiokande time series (see Figure 3) generated by a supposedly unknown complex dynamical phenomenon. By summing the terms of such a time series one gets a trajectory and this trajectory can be used to generate a diffusion process. The method is thus based upon the evaluation of the Boltzmann entropy of the probability density function of a diffusion process. The numerical result of diffusion entropy analysis of the solar neutrino data from Super-Kamiokande is shown in Figure 4.

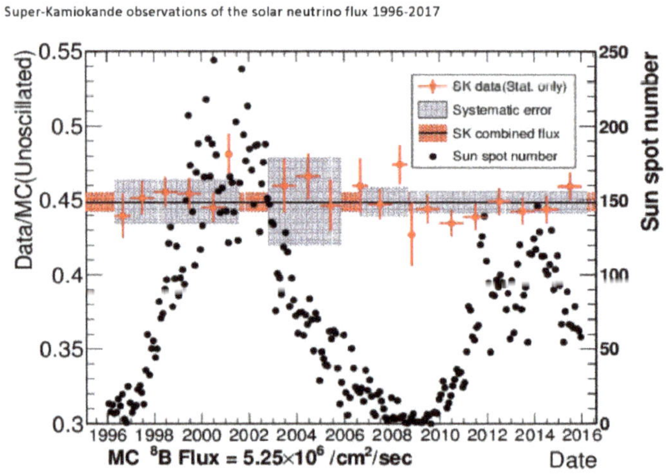

Figure 3. Super-Kamiokande I (1996–2001: 1496 days), II (2002-2005: 791 days), III, and IV solar neutrino data (Y. Takeuchi 2017), http://vietnam.in2p3.fr/2017/neutrinos/program.php.

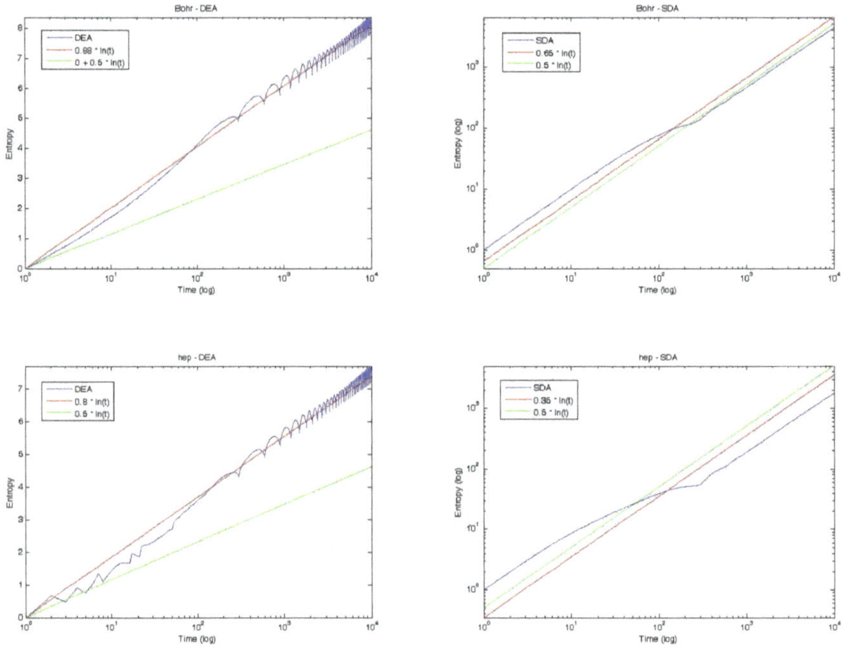

Figure 4. Diffusion Entropy Analysis and Standard Deviation Analysis of the Super-Kamiokande I and II solar neutrino data (Haubold et al. 2014).

In principle, one can perceive the graphical result in Figure 4 of the diffusion entropy analysis (and standard deviation analysis for comparison) of solar neutrino radiation similar to Planck's analysis of black body radiation. What physical meaning this carries remains to be seen. Assuming that the solar neutrino signal is governed by a probability density function (pdf) with scaling given by the asymptotic time evolution of a pdf, obeying the property:

$$p(x,t) = \frac{1}{t^{\delta}} F(\frac{x}{t^{\delta}}),$$

where δ denotes the scaling exponent of the pdf. The scaling exponent δ can be expressed in terms of respective parameters introduced in generalized entropies (Tsallis 2009, Mathai et al. 2010).

Todays perception of the quantum mechanics of neutrino flavour oscillations can be analyzd in a variety of ways in physics. There are treatments of this oscillation phenomenon based on plane waves, on wave packets, and on quantum field theory. These treatments have yielded the standard expression for the probability of oscillations. Neutrinos have been detected in three distinct flavours which interact in particular ways with electrons, muons, and tau leptons, respectively. Flavour oscillations occur because the flavour states are distinct from the neutrino mass states. In particular, a given flavour state may be represented as a coherent superposition of different mass states. In the recent MINOS experiment it was discovered that the phenomenon of neutrino oscillations violates the Leggett-Garg inequality, an analogue of Bell's inequality, involving correlations of measurements on neutrino oscillations at different times (Formaggio et al. 2017). The MINOS experiment analysis did show a violation of the classical limits imposed by the Leggett-Garg inequality. This provided evidence for the existence of the quantum effect of entanglement between the mass eigenstates which make up a flavour state. The entropy of entanglement (Liu et al. 2017) is an entanglement measure for a many-body quantum state and the question arises if the results shown in Figure 4 may find an interpretation in terms of the evolution of an entanglement entropy over time.

Back to Figure 4, it shows a phenomenon that follows certain scaling laws. This Diffusion Entropy Analysis (DEA) measures the correlated variations in the Super-Kamiokande solar neutrino time series. The analysis is based on the diffusion process generated by the time series and measures the time evolution of the Boltzmann entropy of the probability density function of this diffusion process, possibly a quantum diffusion phenomenon. Similar to Brownian motion trajectories, the value of a time series is intepreted as the steps of a diffusion process. The trajectories of this process are defined by the cumulative sum of these steps and obtain a different trajectory for each value of the time series over the full period of time of measurements. Subsequently the probability density function $p(x,t)$ is evaluated that describes the probability that a given trajectory has a displacement of x after t steps. For every particular t the temporal Boltzmann entropy of the probability density function $p(x,t)$ at time t is evaluated by $S(t) = \delta logt$, where δ is the diffusion exponent. For a random uncorrelated diffusion process with finite variance, the $p(x,t)$ will converge according to the Central Limit Theorem to a Gaussian pdf which exhibits $\delta = 1/2$. Figure 4 shows clearly that all δ's are different from the value $\delta = 1/2$. These diffusion exponents are non-Gaussian and exhibit diffusive fluctuations that cannot be modeled by random Gaussian diffusion processes.

To evaluate the Boltzmann entropy of the diffusion process at time t, Scafetta (2010) defined $S(t)$ as:

$$S(t) = -\int_{-\infty}^{+\infty} dx\, p(x,t) \ln\, p(x,t)$$

and with the previous $p(x,t)$, one has:

$$S(t) = A + \delta \ln(t), \quad A = -\int_{-\infty}^{+\infty} dy F(y) \ln F(y).$$

The scaling exponent, δ, is the slope of the entropy against the logarithmic time scale. The slope is visible in Figures 4 for the Super-Kamiokande data I and II measured for the solar neutrino fluxes generated in ^8B and *hep* nuclear reactions in the gravitationally stabilized solar fusion reactor. The

Hurst exponents of the Standard Deviation Analysis (SDA) of the same time series are $H = 0.66$ and $H = 0.36$ for ^8B and *hep*, respectively, shown in Figure 4. The pdf scaling exponents for DEA are $\delta = 0.88$ and $\delta = 0.80$ for ^8B and *hep*, respectively. The values for both SDA and DEA indicate a deviation from Gaussian behavior, which would require that $H = \delta = 1/2$

One of the well known random walk models is the Continuous Time Random Walk (CTRW) introduced by Montroll and Weiss (Oppenheim et al. 1977). It describes a large class of random walks, both normal and anomalous, and can be described as follows. Suppose a particle performs a random walk in such a way that the individual jump x in space is governed by a probability density function and that all jumps are independent and identically distributed. The characteristic function of the position of the particle relative to the origin after n jumps is $f^n(k)$, where $f^*(k)$ is the Fourier transform of $f(x)$. Unlike discrete time random walks, the CTRW describes a situation where the waiting time t between jumps is not a constant. Rather, the waiting time is governed by the pdf $\psi(t)$ and all waiting times are mutually independent and identically distributed. Thus, the number of jumps n is a random variable. Let $p(x,t)$ be the Green function of the CTRW, the Montroll–Weiss equation yields this function in Fourier–Laplace (k,u) space:

$$p\,(k,u) = \frac{1-\psi(u)}{u}\frac{1}{1-f^*\,(k)\,\psi\,(u)}.$$

All along the above the convention was used that the arguments in the parenthesis define the space we are working in, thus $\psi(u)$ is the Laplace transform of $\psi(t)$. Properties of $p(x,t)$ based on the Fourier–Laplace inversion of the previous equation are well investigated, see Mainardi et al. (2001). In particular, it is well known that the asymptotic behavior of $p\,(x,t)$ depends on the long time behavior of $\psi(t)$. An important assumption made in the derivation of the previous equation is that the random walk begun at time $t = 0$. More precisely, it is assumed that the pdf of the first waiting time, i.e., the time elapsing between start of the process at $t = 0$ and the first jump event is $\psi(t)$. Thus the Montroll-Weiss CTRW approach describes a particular choice of initial conditions, called non-equilibrium initial conditions.

The following diffusion model utilizes fractional-order spatial and fractional-order temporal derivatives (Naik and Haubold 2016)

$${}_0D_t^{\beta}p(x,t) = \eta\;{}_xD_{\theta}^{\alpha}p(x,t),$$

with the initial conditions ${}_0D_t^{\beta-1}p(x,0) = \sigma(x), 0 \le \beta \le 1, \lim_{x\to\pm\infty} p(x,t) = 0$, where η is a diffusion constant; $\eta, t > 0, x \in R; \alpha, \theta, \beta$ are real parameters with the constraints $0 < \alpha \le 2, |\theta| \le min(\alpha, 2-\alpha)$, and $\delta(x)$ is the Dirac-delta function. Then for the fundamental solution of the previous fractional differential equation with initial conditions, there holds the formula

$$p(x,t) = \frac{t^{\beta-1}}{\alpha|x|}H_{3,3}^{2,1}\left[\frac{|x|}{(\eta t^{\beta})^{1/\alpha}}\left|\begin{matrix}(1,1/\alpha),(\beta,\beta/\alpha),(1,\rho)\\(1,1/\alpha),(1,1),(1,\rho)\end{matrix}\right.\right], \alpha > 0$$

where $\rho = \frac{\alpha-\theta}{2\alpha}$, in terms of Fox's H-function. The following special cases of the previous fractional differential equation are of special interest for fractional diffusion models:

(i) For $\alpha = \beta$, the corresponding solution of the fractional differential equation, denoted by p_{α}^{θ}, can be expressed in terms of the H-function and can be defined for $x > 0$:
Non-diffusion: $0 < \alpha = \beta < 2; \theta \le \min\{\alpha, 2-\alpha\}$,

$$p_{\alpha}^{\theta}(x,t) = \frac{t^{\alpha-1}}{\alpha|x|}H_{3,3}^{2,1}\left[\frac{|x|}{t\eta^{1/\alpha}}\left|\begin{matrix}(1,1/\alpha),(\alpha,1),(1,\rho)\\(1,1/\alpha),(1,1),(1,\rho)\end{matrix}\right.\right], \; \rho = \frac{\alpha-\theta}{2\alpha}.$$

(ii) When $\beta = 1, 0 < \alpha \leq 2; \theta \leq \min\{\alpha, 2-\alpha\}$, then the previous fractional differential equation reduces to the space-fractional diffusion equation, which is the fundamental solution of the following space-time fractional diffusion model:

$$\frac{\partial p(x,t)}{\partial t} = \eta\, {}_x D_\theta^\alpha p(x,t), \eta > 0, x \in R,$$

with the initial conditions $p(x, t = 0) = \sigma(x), \lim_{x \to \pm\infty} p(x,t) = 0$, where η is a diffusion constant and $\sigma(x)$ is the Dirac-delta function. Hence for the solution of the previous fractional differential equation there holds the formula

$$p_\alpha^\theta(x,t) = \frac{1}{\alpha(\eta t)^{1/\alpha}} H_{2,2}^{1,1}\left[\frac{(\eta t)^{1/\alpha}}{|x|}\middle|\begin{matrix}(1,1),(\rho,\rho)\\(\frac{1}{\alpha},\frac{1}{\alpha}),(\rho,\rho)\end{matrix}\right], 0 < \alpha < 1, |\theta| \leq \alpha,$$

where $\rho = \frac{\alpha-\theta}{2\alpha}$. The density represented by the above expression is known as α-stable Lévy density. Another form of this density is given by

$$p_\alpha^\theta(x,t) = \frac{1}{\alpha(\eta t)^{1/\alpha}} H_{2,2}^{1,1}\left[\frac{|x|}{(\eta t)^{1/\alpha}}\middle|\begin{matrix}(1-\frac{1}{\alpha},\frac{1}{\alpha}),(1-\rho,\rho)\\(0,1),(1-\rho,\rho)\end{matrix}\right], 1 < \alpha < 2, |\theta| \leq 2-\alpha.$$

(iii) If one takes $\alpha = 2, 0 < \beta < 2; \theta = 0$, then one obtains the time-fractional diffusion, which is governed by the following time-fractional diffusion model:

$$\frac{\partial^\beta p(x,t)}{\partial t^\beta} = \eta \frac{\partial^2}{\partial x^2} p(x,t), \eta > 0, x \in R,\ 0 < \beta \leq 2,$$

with the initial conditions
${}_0D_t^{\beta-1} p(x,0) = \sigma(x), {}_0D_t^{\beta-2} p(x,0) = 0$, for $x \in r, \lim_{x\to\pm\infty} p(x,t) = 0$, where η is a diffusion constant and $\sigma(x)$ is the Dirac-delta function, whose fundamental solution is given by the equation

$$p(x,t) = \frac{t^{\beta-1}}{2|x|} H_{1,1}^{1,0}\left[\frac{|x|}{(\eta t^\beta)^{1/2}}\middle|\begin{matrix}(\beta,\beta/2)\\(1,1)\end{matrix}\right].$$

(iv) If one sets $\alpha = 2, \beta = 1$ and $\theta \to 0$, then for the fundamental solution of the standard diffusion equation

$$\frac{\partial}{\partial t} p(x,t) = \eta \frac{\partial^2}{\partial x^2} p(x,t),$$

with initial condition $p(x, t = 0) = \sigma(x), \lim_{x\to\pm\infty} p(x,t) = 0$, there holds the formula

$$p(x,t) = \frac{1}{2|x|} H_{1,1}^{1,0}\left[\frac{|x|}{\eta^{1/2} t^{1/2}}\middle|\begin{matrix}(1,1/2)\\(1,1)\end{matrix}\right] = (4\pi\eta t)^{-1/2} \exp[-\frac{|x|^2}{4\eta t}],$$

which is the classical Gaussian density.

In a different way the above fractional differential equations for $p(x,t)$ can also be written (Pagnini 2012)

$$\frac{\partial p(x,t)}{\partial t} = \frac{2H}{\beta} t^{2H-1} \mathcal{D}_{2H/\beta}^{\beta-1,1-\beta} \frac{\partial^2 p(x,t)}{\partial x^2},$$

where $\mathcal{D}_\eta^{\xi,\mu}$ is the Erdélyi–Kober fractional derivative with respect to t and then the process was also referred to as *Erdélyi–Kober fractional diffusion*. Special cases of the previous equation are: the classical diffusion ($\beta = 2H = 1$), the fractional Brownian motion master equation ($\beta = 1$), and the time-fractional diffusion equation ($\beta = 2H$). A similar approach can be developed in the framework of the space-time fractional diffusion equation, which includes all its special cases. Propagation of neutrino radiation may put forward a new class of phenomena that nonequiilibrium quantum systems may exhibit as shown in Figure 2. This could be an Erdélyi-Kober fractional diffusion operator, a mathematical operator that describes the evolution of the probability density function of the quantum system, and the partition function which describes the statitiscal properties of the system in thermal nonequilibrium with the environment. This will be worked out in future research.

History has seen a great relation between mathematics and statistics and their impact on physics: Mathematical structures entered the development of theoretical physics or, vice versa, problems aising in physics influenced strongly developments in mathematics and statistics. Famous nineteenth-century and twentieth-century examples are Boltzmann's statistical mechanics and the mathematical concept of entropy, the role of Riemannian geometry in general relativity, and the influence of quantum mechanics in the development of functional analysis. Einstein finalized general relativity in 1915 and quantum feld theory has been an open problem since its foundation in 1927 by Dirac. Today there are three fundamental theories in twenty-first century physics: statistical mechanics, general relativity, and quantum field theory. These theories describe the same natural world on very different scales. General relativity describes gravitation on an astronomical scale, quantum field theory describes the interaction of elementary particles through electromagnetic, strong, and weak forces, and statistical mechanics starts from appropriate microscopic laws (classical, relativistic, quantum) and by adequately using probability theory, to ultimately arrive to the thermodynamical relations and laws. The unification of such theories is pursued by mathematicians and physicists so far with no great success. Einstein invented general reativity to resolve an inconsistency between special relativity and Newtonian gravity. Quantum field theory was invented to reconcile Maxwell's electromagnetism and special relativity with nonrelativistic quantum mechanics. Einstein's thought experiments guided the discovery of general relativity based on the mathematics of Riemannian geometry. For quantum field theory experimental results played the important role with no a priori mathematical model available. Boltzmann-Gibbs entropy works perfectly but only within certain limits and if the physical system is out of equilibrium or its component states depend strongly on one another a generalized entropy should be used. Witten (1987) summarized this situation by saying that

> "Experiment is not likely to provide detailed guidance about reconciliation of general relativity with quantum field theory. One might, therefore, believe that the only hope is to emulate the history of general relativity, inventing by sheer thought a new mathematical framework which will generalize Riemannian geometry and will be capable of encompassing quantum field theory. Many ambitious theoretical physicists have aspiredto do such a thing, but little has come of such efforts."

In the above sense, Mathai's research programme is analysing data of solar neutrino experiments to better understand the theory of 'entropy, probability, and fractional dynamics'.

Hans J. Haubold
Special Issue Editor
Office for Outer Space Affairs, United Nations, Austria
Centre for Mathematical and Statistical Sciences, India

References

1. Boltzmann, L. Weitere Studien ueber das Waermegleichgewicht unter Gasmolekuelen. *Wiener Berichte* **1872,** *66,* 275–370, Wissenschaftliche Abhandlungen, Band I, 316–402; English translation: Further studies on the thermal equilibrium of gas molecules, in Kinetic Theory 2, S.G. Brush, Editor, Pergamon, Oxford, 1966, pp. 88–174.
2. Boltzmann, L. Ueber die Beziehung zwischen dem zweiten Hauptsatz der mechanischen Waermetheorie und der Wahrscheinlichkeitsrechnung respektive den Saetzen ueber das Waermegleichgewicht. *Wiener Berichte* **1877,** *76,* 373–435; Wissenschaftliche Abhandlungen, Band II, 164–223.
3. Boltzmann, L. Bemerkungen ueber einige Probleme der mechanischen Waermetheorie. *Wiener Berichte* **1877**, *75,* 62–100; Wissenschaftliche Abhandlungen, Band II, 112–150.
4. Boltzmann, L. Entgegnung auf die waermetheoretischen Betrachtungen des Hrn. E. Zermelo. *Wiedener Annalen* **1896**, *57,* 773–784; Wissenschaftliche Abhandlungen, Band III, 567–578.
5. Boltzmann, L. Zu Hrn. Zermelos Abhandlung ueber die mechanische Erklaerung irreversibler Vorgaenge. *Wiedener Annalen* **1897**, *60,* 392–398; Wissenschaftliche Abhandlungen, Band III, 579–586.
6. Boltzmann, L. Ueber einen mechanischen Satz Poincares. *Wiener Berichte* **1897**, *106,* 12–20; Wissenschaftliche Abhandlungen, Band III, 587–595.
7. Brush, S.G.; Segal, A. *Making 20th Century Science: How Theories Became Knowledge*; Oxford University Press: Oxford, UK, 2015.
8. Cravens, J.P.; Abe, K.; Iida, T.; Ishihara, K.; Kameda, J.; Koshio, Y.; Nakahata, M. Solar neutrino measurements in Super-Kamiokande-II. *Phys. Rev. D* **2008**, *78,* 032002.
9. Eucken, A. Die Theorie der Strahlung und der Quanten, Verhandlungen auf einer von E. Solvay einberufenen Zusammenkunft (30. Oktober bis 3. November 1911). Mit einem Anhang ueber die Entwicklung der Quantentheorie vom Herbst 1911 bis zum Sommer 1913, Druck und Verlag von Wilhelm Knapp, Halle 1914, 95p.
10. Formaggio, J.A.; Kaiser, D.I.; Murskyj, M.M.; Weiss, T.E. Violation of the Leggett-Garg inequality in neutrino oscillations. *Phys. Rev. Lett.* **2017,** *117,* 050402.
11. Haubold, H.J.; John, R.W. On the evaluation of an integral connected with the thermonuclear reaction rate in closed form. *Astronomische Nachrichten* **1978**, *299,* 225–232.
12. Haubold, H.J.; Gerth, E. The search for possible time variations in Davis' measurements of the argon production rate in the solar neutrino experiment. *Astronomische Nachrichten* **1985**, *306,* 203–211.
13. Haubold, H.J.; Mathai, A.M.; Saxena, R.K. Analysis of solar neutrino data from Super-Kamiokande I and II. *Entropy* **2014**, *16,* 1414–1425.
14. Kuhn, T.S. *Black-Body Theory and the Quantum Discontinuity 1894–1912*; Clarendon Press: Oxford, UK, 1978.
15. Liu, Z.-W.; Lloyd, S.; Zhu, E.Y.; Zhu, H. Generalized entanglement entropies of quantum design. arXiv:1709.04313.
16. Mainardi, F.; Luchko, Y.; Pagnini, G. The fundamental solution of the spacetime fractional diffusion equation. *Fract. Calc. Appl. Anal.* **2001**, 4, 153–192.
17. Mathai, A.M.; Rathie, P.N. *Basic Concepts in Information Theory and Statistics: Axiomatic Foundations and Applications*; John Wiley: New York, NY, USA, 1975.
18. Mathai, A.M.; Pederzoli, G. *Characterizations of the Normal Probability Law*; John Wiley: New York, NY, USA, 1977.
19. Mathai, A.M.; Saxena, R.K. *The H-funtion with Applications in Statistics and Other Disciplines*; John Wiley: New York, NY, USA, 1978.
20. Mathai, A.M.; Haubold, H.J. *Modern Problems in Nuclear and Neutrino Astrophysics*; Akademie-Verlag: Berlin, Germany, 1988.

22. Mathai, A.M.; Saxena, R.K.; Haubold, H.J. *The H-Function: Theory and Applications*; Springer: New York, NY, USA, 2010.
23. Naik, S.; Haubold, H.J. On the q-Laplace Transform and Related Special Functions. *Axioms* **2016**, *5*, 24, doi:10.3390/axioms5030024.
24. Oppenheim, I.; Shuler, K.E.; Weiss, G.H. *Stochastic Processes in Chemical Physics: The Master Equation*; MIT Press: Cambridge, MA, USA, 1977.
25. Pagnini, G. Erdelyi-Kober fractional diffusion. *Frac. Calc. Appl. Anal* **2012**, *15*, 117–127.
26. Planck, M. *July 1907 Letter from Planck to Einstein, The Collected Papers of Albert Einstein, Volume 5: The Swiss Years, Correspondence 1902–1914*;, Document 47; Klein, M.J., Kox, A.J., Schulmann, R., Eds.; Princeton University Press: Princeton, NJ, USA, 1995.
27. Sakurai, K. *Solar Neutrino Problems: How They Were Solved*; TERRAPUB: Tokyo, Japan, 2014.
28. Scafetta, N. Fractal and Diffusion Entropy Analysis of Time Series: Theory, concepts, applications and computer codes for studying fractal noises and Levy walk signals, VDM Verlag: Saarbruecken, Germany, 2010.
29. Schoepf, H.-G. *Von Kirchhoff bis Planck, Theorie der Waermestrahlung in Historisch-Kritischer Darstellung*; Akademie-Verlag: Berlin, Germany, 1978, pp. 105–127.
30. Treder, H.-J. Gravitation und weitreichende schwache Wechselwirkungen bei Neutrino-Feldern (Gedanken zu einer Theorie der solaren Neutrinos). *Astronomische Nachrichten* **1974**, *295*, 169–184.
31. Tsallis, C. Introduction to Nonextensive Statistical Mechnics: Approaching a Complex World; Springer: New York, NY, USA, 2009.
32. Witten, E. Physics and Geometry. In Proceedings of the International Congress of Mathematicians, Berkeley, CA, USA, 1986, American Mathematical Society, Providence, RI, 1987, 267–303.
33. Yoo, J.; Ashie, Y.; Fukuda, S.; Fukuda, Y.; Ishihara, K.; Itow, Y.; Nakahata, M. Search for periodic modulations of the solar neutrino fl in Super-Kamiokande-I. *Phys. Rev. D* **2003**, *68*,092002.

axioms

MDPI

Review

Approach of Complexity in Nature: Entropic Nonuniqueness

Constantino Tsallis [1,2]

[1] Centro Brasileiro de Pesquisas Fisicas and National Institute for Science and Technology for Complex Systems, Rua Xavier Sigaud 150, Rio de Janeiro-RJ 22290-180, Brazil; tsallis@cbpf.br; Tel.: +55-21-21417190

[2] Santa Fe Institute, 1399 Hyde Park Road, Santa Fe, NM 87501, USA

Academic Editor: Hans J. Haubold

Received: 8 July 2016 ; Accepted: 8 August 2016; Published: 12 August 2016

Abstract: Boltzmann introduced in the 1870s a logarithmic measure for the connection between the thermodynamical entropy and the probabilities of the microscopic configurations of the system. His celebrated entropic functional for classical systems was then extended by Gibbs to the entire phase space of a many-body system and by von Neumann in order to cover quantum systems, as well. Finally, it was used by Shannon within the theory of information. The simplest expression of this functional corresponds to a discrete set of W microscopic possibilities and is given by $S_{BG} = -k\sum_{i=1}^{W} p_i \ln p_i$ (k is a positive universal constant; BG stands for Boltzmann–Gibbs). This relation enables the construction of BGstatistical mechanics, which, together with the Maxwell equations and classical, quantum and relativistic mechanics, constitutes one of the pillars of contemporary physics. The BG theory has provided uncountable important applications in physics, chemistry, computational sciences, economics, biology, networks and others. As argued in the textbooks, its application in physical systems is legitimate whenever the hypothesis of ergodicity is satisfied, i.e., when ensemble and time averages coincide. However, what can we do when ergodicity and similar simple hypotheses are violated, which indeed happens in very many natural, artificial and social complex systems. The possibility of generalizing BG statistical mechanics through a family of non-additive entropies was advanced in 1988, namely $S_q = k\frac{1-\sum_{i=1}^{W} p_i^q}{q-1}$, which recovers the additive S_{BG} entropy in the $q \to 1$ limit. The index q is to be determined from mechanical first principles, corresponding to complexity universality classes. Along three decades, this idea intensively evolved world-wide (see the Bibliography in http://tsallis.cat.cbpf.br/biblio.htm) and led to a plethora of predictions, verifications and applications in physical systems and elsewhere. As expected, whenever a paradigm shift is explored, some controversy naturally emerged, as well, in the community. The present status of the general picture is here described, starting from its dynamical and thermodynamical foundations and ending with its most recent physical applications.

Keywords: complex systems; statistical mechanics; non-additive entropies; ergodicity breakdown

1. Introduction

In light of contemporary physics, the qualitative and quantitative study of nature may be done at various levels, which here we refer to as microcosmos, mesocosmos and macrocosmos. At the macroscopic level, we have thermodynamics; at the microscopic level, we have mechanics (classical, quantum, relativistic mechanics, quantum chromodynamics) and the laws of electromagnetism, which enable in principle the full description of all of the degrees of freedom of the system; at the mesoscopic level, we focus on the degrees of freedom of a typical particle, representing, in one way or another, the behavior of most of the degrees of freedom of the system. The laws that govern the microcosmos together with theory of probabilities are the basic constituents of statistical mechanics, a theory, which then establishes the connections between these three levels of description of nature. At the microscopic level, we typically address classical or quantum equations of evolution with time, trajectories in phase space, Hamiltonians, Lagrangians, among other mathematical objects. At the mesoscopic level, we address Langevin-like, master-like and Fokker–Planck-like equations. Finally, at the macroscopic level, we address the laws of thermodynamics with its concomitant Legendre transformations between the appropriate variables.

In all of these theoretical approaches, the thermodynamical entropy S, introduced by Clausius in 1865 [1] and its corresponding entropic functional $S(\{p_i\})$ play a central role. In a stroke of genius, the first adequate entropic functional was introduced (for what we nowadays call classical systems) by Boltzmann in the 1870s [2,3] for a one-body phase space and was later on extended by Gibbs [4] to the entire many-body phase space. Half a century later, in 1932, von Neumann [5] extended the Boltzmann–Gibbs (BG) entropic functional to quantum systems. Finally, in 1942, Shannon showed [6] the crucial role that this functional plays in the theory of communication. The simplest expression of this functional is that corresponding to a single discrete random variable admitting W possibilities with nonvanishing probabilities $\{p_i\}$, namely:

$$S_{BG} = -k\sum_{i=1}^{W} p_i \ln p_i \quad \left(\sum_{i=1}^{W} p_i = 1\right) \tag{1}$$

where k is a conventional positive constant (in physics, typically taken to be the Boltzmann constant k_B). This expression enables, as is well known, the construction of what is usually referred to as (BG) statistical mechanics, a theory that is notoriously consistent with thermodynamics. To be more precise, what is well established is that the BG thermostatistics is sufficient for satisfying the principles and structure of thermodynamics. Whether it is or not also necessary is a most important question that we shall address later on in the present paper. This crucial issue and its interconnections with the Boltzmann and the Einstein viewpoints have been emphatically addressed by E.G.D. Cohen in his acceptance lecture of the 2004 Boltzmann Award [7].

On various occasions, generalizations of the expression (1) have been advanced and studied in the realm of information theory. In 1988, [8] (see also [9,10]) the generalization of the BG statistical mechanics itself was proposed through the expression:

$$S_q = k\frac{1-\sum_{i=1}^{W} p_i^q}{q-1} = k\sum_{i=1}^{W} p_i \ln_q \frac{1}{p_i} \quad \left(\sum_{i=1}^{W} p_i = 1; q \in \mathbb{R}; S_1 = S_{BG}\right) \tag{2}$$

where the q-logarithmic function is defined through $\ln_q z \equiv \frac{z^{1-q}-1}{1-q}$ ($\ln_1 z = \ln z$). Its inverse function is defined as $e_q^z \equiv [1+(1-q)z]^{\frac{1}{1-q}}$ ($e_1^z = e^z$). Various predecessors of S_q, q-exponentials and q-Gaussians abound in the literature within specific historical contexts (see, for instance, [11] for a list with brief comments).

2. Additive Entropy versus Extensive Entropy

2.1. Definitions

An entropic functional $S(\{p_i\})$ is said to be additive (we are adopting Oliver Penrose's definition [12]) if, for any two probabilistically independent systems A and B (i.e., $p_{i,j}^{A+B} = p_i^A p_j^B,\ \forall(i,j)$),

$$S(A+B) = S(A) + S(B) \quad [S(A+B) \equiv S(\{p_{i,j}^{A+B}\});\ S(A) \equiv S(\{p_i^A\}); S(B) \equiv S(\{p_j^B\})] \tag{3}$$

It can be straightforwardly proven that S_q satisfies:

$$\frac{S_q(A+B)}{k} = \frac{S_q(A)}{k} + \frac{S_q(B)}{k} + (1-q)\frac{S_q(A)}{k}\frac{S_q(B)}{k} \tag{4}$$

Consequently, $S_{BG} = S_1$ is additive, whereas S_q is non-additive for $q \neq 1$.

The definition of extensivity is much more subtle and follows thermodynamics. A specific entropic functional $S(\{p_i\})$ of a specific system (or a specific class of systems, with its N elements with their corresponding correlations) is said to be extensive if:

$$0 < \lim_{N\to\infty} \frac{S(N)}{N} < \infty \tag{5}$$

i.e., if $S(N)$ grows like N for $N >> 1$, where $N \propto L^d$, d being the integer or fractal dimension of the system, and L its linear size.

Let us emphasize that determining whether an entropic functional is additive is a very simple mathematical task (due to the hypothesis of independence), whereas determining if it is extensive for a specific system can be a very heavy one, sometimes even intractable.

2.2. Probabilistic Illustrations

If all nonzero-probability events of a system constituted by N elements are equally probable, we have $p_i = 1/W(N), \forall i$.

In that case, $S_{BG}(N) = k \ln W(N)$ and $S_q(N) = k \ln_q W(N)$.

Therefore, if the system satisfies $W(N) \propto \mu^N$ ($\mu > 1$; $N \to \infty$) (e.g., for independent coins, we have $W(N) = 2^N$), referred to as the exponential class, we have that the additive entropy S_{BG} is also extensive. Indeed, $S_{BG}(N) \propto N$. For all other values of $q \neq 1$, we have that the non-additive entropy S_q is nonextensive.

However, if we have instead a system such that $W(N) \propto N^\rho$ ($\rho > 0$; $N \to \infty$), referred to as the power-law class, we have that the non-additive entropy S_q is extensive for:

$$q = 1 - \frac{1}{\rho} \quad (\rho > 0) \tag{6}$$

Indeed, $S_{1-1/\rho}(N) \propto N$. For all other values of q (including $q = 1$), we have that S_q is nonextensive for this class; the extensive entropy corresponding to the limit $\rho \to \infty$ precisely is the additive S_{BG}.

Let us now mention another, more subtle, case where the nonzero probabilities are not equal [13]. We consider a triangle of N ($N = 2, 3, 4, ...$) correlated binary random variables, say n heads and $(N-n)$ tails ($n = 0, 1, 2, ..., N$). The probabilities $p_{N,n}$ ($\sum_{n=0}^{N} p_{N,n} = 1, \forall N$) are different from zero only within a strip of width d (more precisely, for $n = 0, 1, 2, ..., d$)) and vanish everywhere else. This specific probabilistic model is asymptotically scale-invariant (i.e., it satisfies the so-called Leibniz triangle rule for $N \to \infty$): see [13] for full details. For this strongly-correlated model, the non-additive entropy S_q is extensive for a unique value of q, namely:

$$q = 1 - \frac{1}{d} \quad (d = 1, 2, 3, ...) \tag{7}$$

We see that the extensive entropy corresponding to the limit $d \to \infty$ precisely is the additive S_{BG}.

These examples transparently show the important difference between entropic additivity and entropic extensivity. What has historically occurred is that, during 140 years, most physicists have been focusing on systems that belong to the exponential class, typically either non-interacting systems (ideal gas, ideal paramagnet) or short-range-interacting ones (e.g., d-dimensional Ising, XY and Heisenberg ferromagnets with first-neighbor interactions). Since for this class, but not so for many others, the additive BG entropic functional is also extensive, a frequent confusion has emerged in the understanding of very many people and textbooks, which has led, and is unfortunately still leading, to somehow considering additive and extensive as synonyms, which is definitively false (this error is so easy to make, such that, by inadvertence, the book [14] by Gell-Mann and myself was entitled Nonextensive Entropy, whereas it should have been entitled Non-additive Entropy; obviously, we definitively regret this misnomer).

Further classes of systems do exist, for example the stretched exponential one, for which other entropic functionals (e.g., S_δ [15]) are necessary in order to achieve extensivity. Indeed, no value of q exists such that $S_q(N) \propto N$ for this class. In fact, a plethora of entropic functionals are now available in the information-theory literature (see, for instance, [16–29]).

2.3. Physical Illustrations

The entropic index q is to be determined from first principles, namely from the time evolution (in phase space, Hilbert space and analogous) of the state of the full system. This typically is an analytically hard task. Nevertheless, this task has been accomplished in some few cases. Let us briefly review some of them:

1. The logistic map at its Feigenbaum point;
2. The entropy of a subsystem of a $(1+1)$-dimensional system characterized by a central charge c at its quantum critical point;
3. The entropy of a subsystem of a $(1+1)$-dimensional generalized isotropic Lipkin–Meshkov–Glick model at its quantum critical point.

For the logistic map $x_{t+1} = 1 - ax_t^2$ $(0 < a < 2;\ t = 0,1,2,...;\ x_t \in [-1,1]$, we have that a value of q exists, such that S_q asymptotically increases linearly with time, where the value of q is dictated by the Lyapunov exponent being positive or zero, which in turn depends on the value of the external parameter a. To be more precise, we assume the interval $[-1,1]$ of x divided into W tiny intervals (identified with $i = 1,2,...,W$); we then place in one of those intervals many M initial conditions (with $M >> W$); and finally, we iterate the map for each of these initial conditions. The number of points $M_i(t)$ that are located at the i-th interval satisfy $\sum_{i=1}^{W} M_i(t) = M, \forall t$. We define next the probabilities $p_i(t) \equiv M_i(t)/M$, which enable the evaluation of the entropy $S_q(t)/k = \frac{1-\sum_{i=1}^{W}[p_i(t)]^q}{q-1}$. It can be shown that a unique value of q exists such that $K_q \equiv \lim_{t\to\infty} \lim_{W\to\infty} \lim_{M\to\infty} \frac{S_q(t)/k}{t}$ is finite. For any value of q above this special one, the ratio K_q vanishes, and for any value of q below this special one, the ratio K_q diverges.

For all values of a such that the Lyapunov exponent λ_1 is positive (i.e., in the presence of strong chaos, where the sensitivity to the initial conditions $\xi \equiv \lim_{\Delta x(0)\to 0} \frac{\Delta x(t)}{\Delta x(0)}$ increases exponentially with time, $\xi = e^{\lambda_1 t}$), we have that $q = 1$, and the ratio precisely equals the Lyapunov exponent (i.e., $K_1 = \lambda_1$; Pesin-like identity).

In contrast, at the edge of chaos, i.e., for the value of a where successive bifurcations accumulate (sometimes referred to as the Feigenbaum point), i.e., $a = 1.401155189092...$, we have that the Lyapunov exponent vanishes, and consistently [30,31],

$$q = 0.2444877013412820661 98... \tag{8}$$

(in fact, 1018 exact digits are numerically known nowadays [32]; see [11] for full details). At such special values of a, we verify that $\xi = e_q^{\lambda_q t}$, where a q-generalized version of the Pesin-like identity has been rigorously established [31]. The edge of chaos of logistic-like maps provides a remarkable connection of q-statistics with multifractals [30]. This is particularly welcome because the postulate of the entropy S_q in order to have a basis for generalizing BG statistics was inspired precisely by the structure of multifractals. The present status of our knowledge strongly suggests that a BG system typically "lives" in a smoothly-occupied phase-space, whereas the systems obeying q-statistics "live" in hierarchically-occupied phase-spaces.

Let us now address the entropy of an L-sized block of an N-sized quantum system at its quantum critical point, belonging to the universality class, which is characterized by a central charge c (e.g., the universality classes of the short-range Ising and the short-range isotropic XY ferromagnets correspond respectively to $c = 1/2$ and $c = 1$). It has been shown [33] that S_q is extensive for:

$$q = \frac{\sqrt{9+c^2}-3}{c} \tag{9}$$

We verify that $c \to \infty$ yields $q = 1$ (BG).

Finally, let us address the generalized isotropic Lipkin–Meshkov–Glick model [34], characterized by (m,k), where m is the number of states of the model (e.g., if the system is constituted by s-sized spins, we have $m = 2s$, $s = 1/2, 1, 3/2, ...$), and k ($k = 0, 1, 2, ...$) is the number of vanishing magnon densities. The entropy S_q is extensive for:

$$q = 1 - \frac{2}{m-k} = 1 - \frac{2}{2s-k} \quad (m-k = 2s-k \geq 3;\, q \geq 1/3) \tag{10}$$

Notice that, in the limit $s \to \infty$, $q = 1$ (BG).

Numerical results are available as well in the literature. For example, for a random antiferromagnet with s-sized spins, we have [35]:

$$q \simeq 1 - \frac{1.67}{\ln(2s+1)} \tag{11}$$

Before we proceed with analyzing thermodynamical aspects, let us stress that we have addressed here two different types of linearities, the thermodynamical one (i.e., $S_q(N) \propto N$) and the dynamical one (i.e., $S_q(t) \propto t$). Although the nature of these linearities is different and even the values of q, which guarantee them, may be different (although possibly related), there are reasons to expect both to be satisfied on similar grounds: this question was in fact (preliminarily) addressed in [36] and elsewhere.

2.4. Renyi Entropy versus q-Entropy

Let us address here a question that frequently appears in the literature, generating some degree of confusion. We refer to the discussion of Renyi entropy versus q-entropy on thermodynamical and dynamical grounds. The Renyi entropy [16] is defined as:

$$S_q^R \equiv \frac{\ln \sum_{i=1}^{W} p_i^q}{1-q} \quad \Big(\sum_{i=1}^{W} p_i = 1; q \in \mathbb{R}; S_1^R = S_{BG}\Big) \tag{12}$$

hence:

$$S_q^R = \frac{\ln[1+(1-q)S_q/k]}{1-q} \tag{13}$$

It is straightforward to verify that $S_q^R(S_q)$ is a monotonic function of S_q, $\forall q$. Consequently, under the same constraints, the extremization of S_q^R yields precisely the same distribution as the extremization of S_q (in total analogy with the trivial fact that maximizing, under the same constraints, S_{BG} or say $[S_{BG}]^3$ yields one and the same BG exponential weight). This mathematical triviality is at the basis of sensible confusion in the minds of some members of the community. Thermodynamics and statistical mechanics is much more than a mere probability distribution, and the reader has surely never seen, and this for more than one good reason, constructing a successful theory such as thermodynamics by using say $[S_{BG}]^3$ instead of S_{BG}.

To make things more precise, let us list now several important differences between S_q and S_q^R (see, for instance, [11] and the references therein).

(i) Additivity: If A and B are two arbitrary probabilistically-independent systems, S_q^R is additive, $\forall q$, whereas S_q satisfies the non-additive property in Equation (4).
(ii) Concavity: $S_q(\{p_i\})$ is concave for all $q > 0$, whereas $S_q^R(\{p_i\})$ is concave only for $0 < q \leq 1$. Both S_q and S_q^R are convex for $q < 0$. These properties have consequences for characterizing the thermodynamic stability of the system.
(iii) Lesche stability: S_q is Lesche-stable $\forall q > 0$, whereas S_q^R is Lesche-stable only for $q = 1$. Lesche stability characterizes the experimental reproducibility of the entropy of a system.
(iv) Pesin-like identity: For many physically important low-dimensional conservative or dissipative nonlinear dynamical systems with zero Lyapunov exponent, it is verified that, in the $t \to \infty$ limit, $S_q(t) \propto t$ for a unique special value of $q \neq 1$. This linearity property for $t >> 1$ is lost for $S_q^R(t)$; indeed, for those systems, it can be easily verified that $S_q^R(t) \propto \ln t \; (\forall q)$. No dynamical systems are yet known for which $S_q^R(t)$ is linear for $q \neq 1$. This linearity enables, $\forall q$, a natural connection with the coefficient (Lyapunov exponent for the $q = 1$ systems), which characterizes the dynamically meaningful sensitivity to the initial conditions.
(v) Thermodynamical extensivity: For various N-sized quantum systems, it can be shown that a fixed value of $q \neq 1$ exists, such that, in the $N \to \infty$ limit, $S_q(N) \propto N$, thus satisfying the necessary thermodynamic extensivity for the entropy. For those systems, $S_q^R(N) \propto \ln N \; (\forall q)$, which violates thermodynamics. For this statement, we have of course assumed that a (physically meaningful) limit $q \neq 1$ exists in the $N \to \infty$ limit. Various papers exist in the literature that focus on situations such that a phenomenological index q can be defined, which depends on N (see, for instance, [37,38] and the references therein), but they remain out of the present scope, since their $N \to \infty$ limit yields $q = 1$.
(vi) The likelihood function that satisfies Einstein's requirement of factorizability coincides with the function, which extremizes the entropic functional of the system (currently, the inverse function of the generalized logarithm, which characterizes that precise entropic functional: For $q = 1$ systems, the factorizable likelihood function is well known to be $\mathcal{W} \propto e^{S_{BG}/k}$, the exponential function being the inverse of $S_{BG}/k = \ln W$ (for equal probabilities), and for appropriate constraints, it maximizes the entropy S_{BG}. For $q \neq 1$, we have [39] $\mathcal{W} \propto e_q^{S_q/k}$, where the q-exponential function precisely is the inverse of $S_q/k = \ln_q W$ (for equal probabilities), and for appropriate constraints, it extremizes the entropy S_q. In contrast with this property, the factorizable likelihood function for the Renyi entropy is $e^{S_q^R}$, where the exponential function is the inverse of $S_q^R = \ln W$ (for equal probabilities), but it differs from the q-exponential function, which is the one that extremizes S_q^R. These properties plausibly have consequences for the large deviation theory of these systems (see the discussion about this theory below).

3. Why Must the Entropic Extensivity Be Preserved in All Circumstances?

Since we are ready to permit the entropic functional to be non-additive, should we not also allow for possible entropic nonextensivity? This question surely is a most interesting one, but to the best of our understanding, the answer is no. Indeed, there exist at least two important reasons for always demanding the physical (thermodynamical) entropy of a given system to be extensive. One of them is based on the Legendre transformations structure of thermodynamics; the other one is so suggested by the large deviations in some anomalous probabilistic models where the limiting distributions are q-Gaussians.

3.1. Thermodynamics

This argument has been developed in [11] and more recently in [15] (which we follow now). We briefly review this argument here. Let us first write a general Legendre transformation form of a thermodynamical energy G of a generic d-dimensional system (d being an integer or fractal dimension):

$$G(V,T,p,\mu,H,\ldots) = U(V,T,p,\mu,H,\ldots) - TS(V,T,p,\mu,H,\ldots) \tag{14}$$
$$+pV - \mu N(V,T,p,\mu,H,\ldots) - HM(V,T,p,\mu,H,\ldots) - \cdots \tag{15}$$

where T,p,μ,H are the temperature, pressure, chemical potential and external magnetic field and U,S,V,N,M are the internal energy, entropy, volume, number of particles and magnetization. We may identify three types of variables, namely: (i) those that are expected to always be extensive $(S,V,N,M,\ldots)$, i.e., scaling with $V \propto L^d$, where L is a characteristic linear dimension of the system (notice the presence of N itself within this class); (ii) those that characterize the external conditions under which the system is placed $(T,p,\mu,H,\ldots)$, scaling with L^θ; and (iii) those that represent energies (G,U), scaling with L^ϵ. Ordinary thermodynamical systems are those with $\theta = 0$ and $\epsilon = d$; therefore, both the energies and the generically extensive variables scale with L^d, and there is no difference between Type (i) and (iii) variables, all of them being extensive in this case. There are, however, physical systems where $\epsilon = \theta + d$ with $\theta \neq 0$. Let us divide Equation (15) by $L^{\theta+d}$, namely,

$$\frac{G}{L^{\theta+d}} = \frac{U}{L^{\theta+d}} - \frac{T}{L^\theta}\frac{S}{L^d} + \frac{p}{L^\theta}\frac{V}{L^d} - \frac{\mu}{L^\theta}\frac{N}{L^d} - \frac{H}{L^\theta}\frac{M}{L^d} - \cdots \tag{16}$$

If we consider now the thermodynamical $L \to \infty$ limit, we obtain:

$$\tilde{g} = \tilde{u} - \tilde{T}s + \tilde{p}v - \tilde{\mu}\,n - \tilde{H}m - \cdots \tag{17}$$

where, using a compact notation, $(\tilde{g},\tilde{u}) \equiv \lim_{L\to\infty}(G,U)/L^{\theta+d}$ represent the energies, $(s,v,n,m) \equiv \lim_{L\to\infty}(S,V,N,M)/L^d$ represent the usual extensive variables and $(\tilde{T},\tilde{p},\tilde{\mu},\tilde{H}) \equiv \lim_{L\to\infty}(T,p,\mu,H)/L^\theta$ correspond to the usually intensive ones. For a standard thermodynamical system (e.g., a real gas ruled by a Lennard–Jones short-ranged potential, a simple metal, etc.) we have $\theta = 0$ (hence, $(\tilde{T},\tilde{p},\tilde{\mu},\tilde{H}) = (T,p,\mu,H)$, i.e., the usual intensive variables), and $\epsilon = d$ (hence, $(\tilde{g},\tilde{u}) = (g,u)$, i.e., the usual extensive variables); this is of course the case found in the textbooks of thermodynamics.

The thermodynamic relations (15) and (16) put on an equal footing the entropy S, the volume V and the number of elements N, and the extensivity of the latter two variables is guaranteed by definition. In fact, a similar analysis can be performed using N instead of V since $V \propto N$.

An example of a nonstandard system with $\theta \neq 0$ is the classical Hamiltonian discussed in what follows. We consider two-body interactions decaying with distance r like $1/r^\alpha$ $(\alpha \geq 0)$. For this system, we have $\theta = d - \alpha$ whenever $0 \leq \alpha < d$ (see, for example, Figure 1 of [40]). This peculiar scaling occurs because the potential is not integrable, i.e., the integral $\int_{\text{constant}}^{\infty} dr\, r^{d-1}\, r^{-\alpha}$ diverges for $0 \leq \alpha \leq d$; therefore, the Boltzmann–Gibbs canonical partition function itself diverges. Gibbs was aware of this kind of problem and has pointed out [4] that whenever the partition function diverges, the BG theory cannot be used because, in his words, "the law of distribution becomes illusory". The divergence of the total potential energy occurs for $\alpha \leq d$, which is referred to as long-range interactions. If $\alpha > d$, which is the case of the $d = 3$ Lennard–Jones potential, whose attractive part corresponds to $\alpha = 6$, the integral does not diverge, and we recover the standard behavior of short-range-interacting systems with the $\theta = 0$ scaling. Nevertheless, it is worth recalling that nonstandard thermodynamical behavior is not necessarily associated with long-range interactions in the classical sense just discussed. A meaningful description would then be long-range correlations (spatial or temporal), because for strongly quantum-entangled systems, correlations are not necessarily connected with the interaction range. However, the picture of long- versus short-range interactions in the classical sense, directly related to the distance r, has the advantage of illustrating clearly the thermodynamic relations (15) and (16) for the different scaling regimes, as shown in Figure 1.

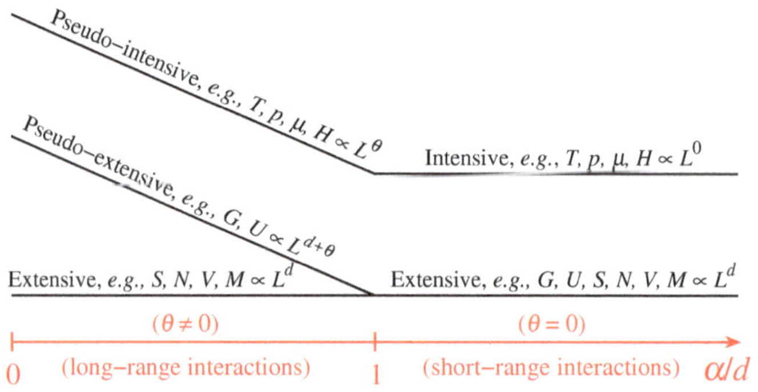

Figure 1. Representation of the different scaling regimes of Equation (16) for classical d-dimensional systems. For attractive long-range interactions (i.e., $0 \leq \alpha/d \leq 1$, α characterizing the interaction range in a potential with the form $1/r^{\alpha}$), we may distinguish three classes of thermodynamic variables, namely, those scaling with L^{θ}, named pseudo-intensive (L is a characteristic linear length; θ is a system-dependent parameter), those scaling with $L^{d+\theta}$, the pseudo-extensive ones (the energies), and those scaling with L^d (which are always extensive). For short-range interactions (i.e., $\alpha > d$), we have $\theta = 0$, and the energies recover their standard L^d extensive scaling, falling in the same class of S, N, V, etc., whereas the previous pseudo-intensive variables become truly intensive ones (independent of L); this is the region with two classes of variables that is covered by the traditional textbooks of thermodynamics. From [15].

To summarize this crucial subsection, we may insist that what is thermodynamically relevant is that the entropy of a given system must be extensive, not that the entropic functional ought to be additive. This is consistent with the fact that Einstein's principle for the factorizability of the likelihood function is satisfied not only for the additive BG entropic functional, but also for nonadditive ones [39,41].

3.2. Large Deviation Theory

The so-called large deviation theory (LDT) [42] constitutes the mathematical counterpart of the heart of BG statistical mechanics, namely the famous canonical-ensemble BG factor $e^{-\beta\mathcal{H}(N)} = e^{-N[\beta h(N)]}$ with $h(N) \equiv \mathcal{H}(N)/N$. Since, for short-range interactions, $\beta h(N)$ is a thermodynamically-intensive quantity in the limit $N \to \infty$, we see that the BG weight represents an exponential decay with N. This exponential dependence is to be associated [42–46] with the LDT probability $P(N;x) \simeq e^{-N r_1(x)}$, where Subindex 1 in the rate function $r_1(x)$ will soon become clear. Since $r_1(x)$ is directly related to a relative entropy per particle (see, for instance, [43]), the quantity $N r_1(x)$ plays the role of an extensive entropy.

If we focus now on, say, a d-dimensional classical system involving two- body interactions whose potential asymptotically decays at long distance r like $-A/r^{\alpha}$ ($A > 0; \alpha \geq 0$), the canonical BG partition function converges whenever the potential is integrable, i.e., for $\alpha/d > 1$ (short-range interactions), and diverges whenever it is non-integrable, i.e., for $0 \leq \alpha/d \leq 1$ (long-range interactions). The use of the BG weight becomes unjustified ("illusory" in Gibbs words [4] for, say, Newtonian gravitation, which in the present notation corresponds to $(\alpha, d) = (1,3)$; hence, $\alpha/d = 1/3$) in the later case because of the divergence of the BG partition function. We might therefore expect the emergence of some function $f(\mathcal{H}_N)$ different from the exponential one, in order to describe some specific stationary (or quasi-stationary) states differing from thermal equilibrium. The Hamiltonian $\mathcal{H}_N$ generically scales like $N\tilde{N}$ with $\tilde{N} \equiv \frac{N^{1-\alpha/d}-1}{1-\alpha/d} \equiv \ln_{\alpha/d} N$ (with the q-logarithmic function defined as $\ln_q z \equiv \frac{z^{1-q}-1}{1-q}$; $z > 0$; $\ln_1 z = \ln z$). Notice that ($N \to \infty$) $\tilde{N} \sim N^{1-\alpha/d}/(1-\alpha/d)$ for $1 \leq \alpha/d < 1$, $\tilde{N} \sim \ln N$ for $\alpha/d = 1$ and $\tilde{N} \sim 1/(\alpha/d - 1)$ for $\alpha/d > 1$. The particular case $\alpha = 0$ yields $\tilde{N} \sim N$, thus recovering the usual prefactor of mean field theories. The quantity $\beta\mathcal{H}_N$ can be rewritten as $[(\beta\tilde{N})\mathcal{H}_N/(N\tilde{N})]N = [\tilde{\beta}\mathcal{H}_N/(N\tilde{N})]N$, where $\tilde{\beta} \equiv \beta\tilde{N} \equiv 1/k_B\tilde{T} = \tilde{N}/k_BT$ plays the role of an intensive variable. The correctness of all of these scalings has been profusely verified in various kinds of thermal, diffusive and geometrical (percolation) systems (see [11,45]). We see that, not only for the usual case of short-range interactions, but also for long-range ones, $[\tilde{\beta}\mathcal{H}_N/(N\tilde{N})]$ plays a role analogous to an intensive variable. The q-exponential function $e_q^z \equiv [1+(1-q)z]^{\frac{1}{1-q}}$ ($e_1^z = e^z$) (and its associated q-Gaussian) has already emerged, in a considerable amount of nonextensive and similar systems, as the appropriate generalization of the exponential one (and its associated Gaussian). Therefore, it appears as rather natural to conjecture that, in some sense that remains to be precisely defined, the LDT expression $e^{-r_1 N}$ becomes generalized into something close to $e_q^{-r_q N}$ ($q \in \mathcal{R}$), where the generalized rate function r_q is expected to be some generalized entropic quantity per particle. As shown in Figures 2 and 3 (see the details in [45]), it is precisely this $e_q^{-r_q N}$ behavior that emerges in a strongly correlated nontrivial model [43,45]. Since, as for the $q = 1$ case, $r_q N$ appears to play the role of a total entropy, this specific illustration is consistent with an extensive entropy.

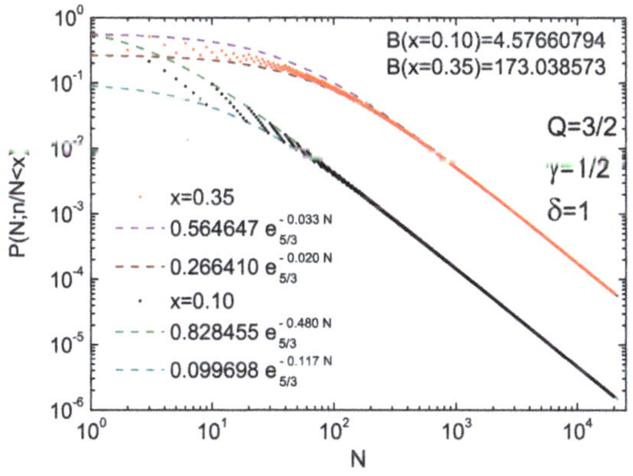

Figure 2. Comparison of the numerical data (dots) of [45] with $a(x)e_q^{-r_q N}$, where $(a(x), r_q(x))$ are positive quantities. From [45].

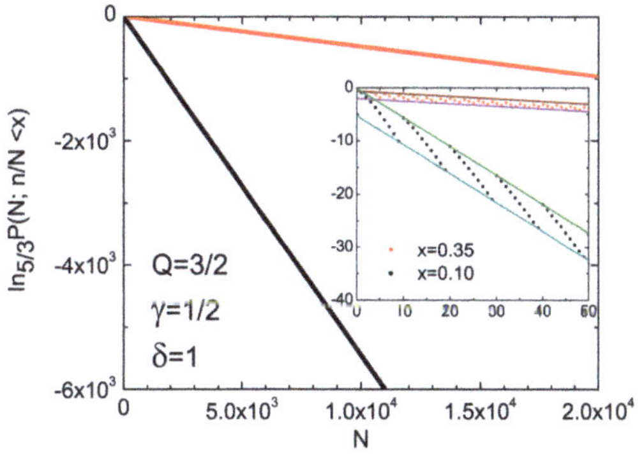

Figure 3. The same data of Figure 2 in (q-log)-linear representation. Let us stress that the unique asymptotically-power-law function, which provides straight lines at all scales of a (q-log)-linear representation, is the q-exponential function. The inset shows the results corresponding to N up to 50. From [45].

4. Further Applications and Final Words

A regularly-updated bibliography on the present subject can be found at [47]. At the same site, a selected list of theoretical, experimental, observational and computational papers can be found, as well. From these very many papers, let us briefly mention here a few recent ones.

For those systems that may be well described by a specific class of nonlinear homogeneous $d = 1$ Fokker–Planck equations, a prediction was advanced [48] in 1966, namely the scaling $\mu = 2/(3-q)$, where μ is the exponent that characterizes the scaling between space and time (specifically the fact that x^2 scales like t^μ) and q is the index of the q-Gaussian, which describes the paradigmatic solution of the equation. Notice that $q = 1$ yields the well-known Einstein 1905 result $\mu = 1$ for Brownian motion. The prediction was experimentally verified (within a 2% precision along an entire experimental decade), in 2015 [49], for confined granular material. It would be surely interesting to also verify for higher-dimension confined granular matter, the d-dimensional generalization of that scaling, namely $\mu = \frac{2}{2+d(1-q)}$ [50]; hence, once again $\mu = 1$ for $q = 1$.

For an area-preserving two-dimensional map, namely the standard map, it was neatly shown [51] how q-statistics, or BG statistics, or even a combination of both emerges as a function of the unique external parameter (K) of the map. This and various other emergencies of q-Gaussian and q-exponential distributions in many natural, artificial and social complex systems are most probably connected with q-generalizations of the central limit theorem (see, for instance, [52–63]).

Another q-statistical connection that certainly is interesting is the one with the so-called (asymptotically) scale-free networks. Indeed, their degree distribution has been shown in many cases to be given by $p(k) \propto e_q^{-k/\kappa}$ (k being the number of links joining a given node), which plays the role of the Boltzmann–Gibbs factor for short-range-interacting Hamiltonian systems. This connection was already established in the literature since one decade ago (see, for instance, [64,65]). Moreover, it has been recently shown [66] that neither q nor κ depend independently on the dimensionality d and from the exponent α characterizing the range of the interaction, but, interestingly enough, only depend on the ratio α/d. Very many papers focus on the degree distributions of (asymptotically) scale-free networks from a variety of standpoints. For example, an interesting exactly solvable master-equation approach is available in [67]. The novelty that we remind about in this mini-review is that the q-exponential degree distribution is here obtained from a simple entropic variational principle (under a constraint where the average degree plays the role of the internal energy in statistical mechanics).

High-energy physics has also been a field of many applications of q-statistics and related approaches, such as Beck–Cohen superstatistics [68] and Mathai's pathways (see [69–73] and the references therein). For example, a focus on the solar neutrino problem started long ago by Quarati and collaborators [74–77] and has been revisited in several occasions, even recently [78,79]. In the area of particle high-energy collisions, an intensive activity is currently in progress. It usually concerns experiments performed at LHC/CERN (ALICE, ATLAS, CMS and LHCb Collaborations) and RHIC/Brookhaven (STAR and PHENIX Collaborations). As typical illustrations of such measurements and their possible theoretical interpretations, let us mention [80–98]. A rich discussion about the thermodynamical admissibility of the possible constraints under which the entropic functional can be optimized is also present in the literature (see, for instance, [10,11,83,99–101]).

Many other systems (e.g., related to those mentioned in [102–105]) are awaiting for approaches along the above and similar lines. They would be very welcome. Even so, we may say that the present status of the theory described herein is at a reasonably satisfactory stage of physical and mathematical understanding.

Acknowledgments: I warmly dedicate this review to Arak M. Mathai on his 80th anniversary. I am deeply indebted to Hans J. Haubold, whose insights and encouragements have made this overview possible, and also to the anonymous referees who, through highly constructive remarks, made possible an improved version of the present manuscript. Finally, I acknowledge partial financial support from CNPq and Faperj (Brazilian agencies) and from the John Templeton Foundation (USA).

Conflicts of Interest: The author declares no conflict of interest.

References

1. Clausius, R. Über verschiedene für die Anwendung bequeme Formen der Hauptgleichungen der mechanischen Wärmetheorie. *Ann. Phys.* **1865**, *125*, 353.
2. Boltzmann, L. Weitere Studien über das Wärmegleichgewicht unter Gas molekülen. *Wien Ber.* **1872**, *66*, 275.
3. Boltzmann, L. *Über die Beziehung eines Allgemeine Mechanischen Satzes zum Zweiten Haupsatze der Wärmetheorie*; Sitzungsberichte, K., Ed.; Akademie der Wissenschaften in Wien, Math.-Naturwissenschaften: Wien, Austria, 1877; Volume 75, pp. 67–73.
4. Gibbs, J.W. *Elementary Principles in Statistical Mechanics—Developed with Especial Reference to the Rational Foundation of Thermodynamics*; C. Scribner's Sons: New York, NY, USA, 1902.
5. Von Neumann, J. *Mathematische Grundlagen der Quantenmechanik*; Springer: Berlin, Germany, 1932.
6. Shannon, C.E. A Mathematical Theory of Communication. *Bell Syst. Tech. J.* **1948**, *27*, 379–423; 623–656.
7. Cohen, E.G.D. Boltzmann and Einstein: Statistics and dynamics—An unsolved problem, Boltzmann Award Lecture at Statphys-Bangalore-2004. *Pramana* **2005**, *64*, 635–643.
8. Tsallis, C. Possible generalization of Boltzmann-Gibbs statistics *J. Stat. Phys.* **1988**, *52*, 479–487.
9. Curado, E.M.F.; Tsallis, C. Generalized statistical mechanics: Connection with thermodynamics. *J. Phys. A Math. Gen.* **1991**, *24*, L69–L72.
10. Tsallis, C.; Mendes, R.S.; Plastino, A.R. The role of constraints within generalized nonextensive statistics. *Phys. A* **1998**, *261*, 534–554.
11. Tsallis, C. *Introduction to Nonextensive Statistical Mechanics—Approaching a Complex World*; Springer: New York, NY, USA, 2009.
12. Penrose, O. *Foundations of Statistical Mechanics: A Deductive Treatment*; Pergamon: Oxford, UK, 1970; p. 167.
13. Tsallis, C.; Gell-Mann, M.; Sato, Y. Asymptotically scale-invariant occupancy of phase space makes the entropy Sq extensive. *Proc. Natl. Acad. Sci. USA* **2005**, *102*, 15377–15382.
14. Gell-Mann, M.; Tsallis, C. *Nonextensive Entropy—Interdisciplinary Applications*; Oxford University Press: New York, NY, USA, 2004.
15. Tsallis, C.; Cirto, L.J.L. Black hole thermodynamical entropy. *Eur. Phys. J. C* **2013**, *73*, 2487.
16. Renyi, A. On measures of information and entropy. In Proceedings of the Fourth Berkeley Symposium, Los Angeles, CA, USA, 20 June–30 July 1961; Volume 1, p. 547.
17. Varma, R.S. Generalizations of Renyi's entropy of order α. *J. Math. Sci.* **1966**, *1*, 34–48.
18. Aczel, J.; Daroczy, Z. *Mathematics in Science and Engineering*; Bellman, R., Ed.; Academic Press: New York, NY, USA, 1975.
19. Sharma, B.D.; Mittal, D.P. New non-additive measures of entropy for discrete probability distributions. *J. Math. Sci.* **1975**, *10*, 28–40.
20. Sharma, B.D.; Taneja, I.J. Entropy of type (α, β) and other generalized measures in information theory. *Metrika* **1975**, *22*, 205–215.
21. Sharma, B.D.; Taneja, I.J. Three generalized additive measures of entropy. *Elect. Infor. Kybern.* **1977**,*13* , 419–433.
22. Landsberg, P.T.; Vedral, V. Distributions and channel capacities in generalized statistical mechanics. *Phys. Lett. A* **1998**, *247*, 211–217.
23. Landsberg, P.T. Entropies galore! Nonextensive Statistical Mechanics and Thermodynamics. *Braz. J. Phys.* **1999**, *29*, 46–49.
24. Curado, E.M.F. General aspects of the thermodynamical formalism. *Braz. J. Phys.* **1999**, *29*, 36–45.
25. Anteneodo, C.; Plastino, A.R. Maximum entropy approach to stretched exponential probability distributions. *J. Phys. A* **1999**, *32*, 1089–1097.
26. Curado, E.M.F.; Nobre, F.D. On the stability of analytic entropic forms. *Phys. A* **2004**, *335*, 94–106.
27. Zaripov, R.G. Geometric representation of the group of entropy vectors in non extensive statistical mechanics. *Russ. Phys. J.* **2014**, *57*, 861–869.
28. Hanel, R.;Thurner, S. A comprehensive classification of complex statistical systems and an axiomatic derivation of their entropy and distribution functions. *EPL* **2011**, *93*, 20006.
29. Hanel, R.; Thurner, S. When do generalised entropies apply? How phase space volume determines entropy. *EPL* **2011**, *96*, 50003.

30. Lyra, M.L.; Tsallis, C. Nonextensivity and multifractality in low-dimensional dissipative systems. *Phys. Rev. Lett.* **1998**, *80*, 53–56.
31. Baldovin, F.; Robledo, A. Universal renormalization-group dynamics at the onset of chaos in logistic maps and nonextensive statistical mechanics. *Phys. Rev. E* **2002**, *66*, R045104.
32. Broadhurst, D. Available online: http://pi.lacim.uqam.ca/piDATA/feigenbaum.txt (accessed on 8 July 2016).
33. Caruso, F.; Tsallis, C. Non-additive entropy reconciles the area law in quantum systems with classical thermodynamics. *Phys. Rev. E* **2008**, *78*, 021102.
34. Carrasco, J.A.; Finkel, F.; Gonzalez-Lopez, A.; Rodriguez, M.A.; Tempesta, P. Generalized isotropic Lipkin-Meshkov-Glick models: Ground state entanglement and quantum entropies. *J. Stat. Mech.* **2016**, *3*, 033114.
35. Saguia, A.; Sarandy, M.S. Nonadditive entropy for random quantum spin-*S* chains. *Phys. Lett. A* **2010**, *374*, 3384–3388.
36. Tsallis, C.; Gell-Mann, M.; Sato, Y. Extensivity and entropy production. *Europhys. News Spec. Europhys. News* **2005**, *36*, 186–189.
37. Parvan, A.S.; Biro, T.S. Extensive Renyi statistics from non-extensive entropy. *Phys. Lett. A* **2005**, *340*, 375–387.
38. Parvan, A.S.; Biro, T.S. Renyi statistics in equilibrium statistical mechanics. *Phys. Lett. A* **2010**, *374*, 1951–1957.
39. Tsallis, C.; Haubold, H.J. Boltzmann-Gibbs entropy is sufficient but not necessary for the likelihood factorization required by Einstein. *EPL* **2015**, *110*, 30005.
40. Tamarit, F.; Anteneodo, C. Long-range interacting rotators: Connection with the mean-field approximation. *Phys. Rev. Lett.* **2000**, *84*, 208–211.
41. Sicuro, G.; Tempesta, P. Groups, information theory and Einstein' s likelihood principle. *Phys. Rev. E* **2016**, *93*, 040101.
42. Touchette, H. The large deviation approach to statistical mechanics. *Phys. Rep.* **2009**, *478*, 1–69.
43. Ruiz, G.; Tsallis, C. Towards a large deviation theory for strongly correlated systems. *Phys. Lett. A* **2012**, *376*, 2451–2454.
44. Touchette, H. Comment on "Towards a large deviation theory for strongly correlated systems". *Phys. Lett. A* **2013**, *377*, 436–438.
45. Ruiz, G.; Tsallis, C. Reply to Comment on "Towards a large deviation theory for strongly correlated systems". *Phys. Lett. A* **2013**, *377*, 491–495.
46. Ruiz, G.; Tsallis, C. Emergence of q-statistical functions in a generalized binomial distribution with strong correlations. *J. Math. Phys.* **2015**, *56*, 053301.
47. Tsallis, C. Nonextensive Statistical Mechanics and Thermodynamics. Available online: http://tsallis.cat.cbpf.br/biblio.htm (accessed on 8 July 2016).
48. Tsallis, C.; Bukman, D.J. Anomalous diffusion in the presence of external forces: exact time-dependent solutions and their thermostatistical basis. *Phys. Rev. E* **1996**, *54*, 2197–2200.
49. Combe, G.; Richefeu, V.; Stasiak, M.; Atman, A.P.F. Experimental validation of nonextensive scaling law in confined granular media. *Phys. Rev. Lett.* **2015**, *115*, 238301.
50. Malacarne, L.C.; Mendes, R.S.; Pedron, I.T.; Lenzi, E.K. Nonlinear equation for anomalous diffusion: Unified power-law and stretched exponential exact solution. *Phys. Rev. E* **2001**, *63*, R030101.
51. Tirnakli, U.; Borges, E.P. The standard map: From Boltzmann-Gibbs statistics to Tsallis statistics. *Sci. Rep.* **2016**, *6*, 23644.
52. Moyano, L.G.; Tsallis, C.; Gell-Mann, M. Numerical indications of a q-generalised central limit theorem. *Europhys. Lett.* **2006**, *73*, 813–819.
53. Hilhorst, H.J.; Schehr, G. A note on q-Gaussians and non-Gaussians in statistical mechanics. *J. Stat. Mech.* **2007**, *6*, P06003.
54. Umarov, S.; Tsallis, C.; Steinberg, S. On a q-central limit theorem consistent with nonextensive statistical mechanics. *Milan J. Math.* **2008**, *76*, 307–328.
55. Umarov, S.; Tsallis, C.; Gell-Mann, M.; Steinberg, S. Generalization of symmetric α-stable Lévy distributions for $q > 1$. *J. Math. Phys.* **2010**, *51*, 033502.
56. Hilhorst, H.J. Note on a q-modified central limit theorem. *J. Stat. Mech.* **2010**, *10*, P10023.
57. Jauregui, M.; Tsallis, C. q-generalization of the inverse Fourier transform. *Phys. Lett. A* **2011**, *375*, 2085–2088.

58. Jauregui, M.; Tsallis, C.; Curado, E.M.F. q-moments remove the degeneracy associated with the inversion of the q-Fourier transform. *J. Stat. Mech.* **2011**, *10*, P10016.
59. Hahn, M.G.; Jiang, X.X.; Umarov, S. On q-Gaussians and exchangeability. *J. Phys. A* **2010**, *43*, 165208.
60. Jiang, X.X.; Hahn, M.G.; Umarov, S. On generalized Leibniz triangles and q-Gaussians. *Phys. Lett. A* **2012**, *376*, 2447–2450.
61. Plastino, A.; Rocca, M.C. Inversion of Umarov-Tsallis-Steinberg q-Fourier Transform and the complex-plane generalization. *Phys. A* **2012**, *391*, 4740–4747.
62. Plastino, A.; Rocca, M.C. q-Fourier Transform and its inversion-problem. *Milan J. Math.* **2012**, *80*, 243–249.
63. Budini, A.A. Central limit theorem for a class of globally correlated random variables. *Phys. Rev. E* **2016**, *93*, 062114.
64. Soares, D.J.B.; Tsallis, C.; Mariz, A.M.; da Silva, L.R. Preferential attachment growth model and nonextensive statistical mechanics. *Europhys. Lett.* **2005**, *70*, 70–76.
65. Thurner, S.; Tsallis, C. Nonextensive aspects of self-organized scale-free gas-like networks. *Europhys. Lett.* **2005**, *72*, 197–203.
66. Brito, S.G.A.; da Silva, L.R.; Tsallis, C. Role of dimensionality in complex networks. *Sci. Rep.* **2016**, *6*, 27992.
67. Kullmann, L.; Kertesz, J. Preferential growth: Exact solution of the time-dependent distributions. *Phys. Rev. E* **2001**, *63*, 051112.
68. Beck, C.; Cohen, E.G.D. Superstatistics. *Phys. A* **2003**, *322*, 267–275.
69. Mathai, A.M. A pathway to matrix-variate gamma and normal densities. *Linear Algebra Its Appl.* **2005**, *396*, 317–328.
70. Mathai, A.M.; Haubold, H.J. Pathway model, superstatistics, Tsallis statistics, and a generalized measure of entropy. *Phys. A* **2007**, *375*, 110–122.
71. Mathai, A.M.; Haubold, H.J. On generalized entropy measures and pathways. *Phys. A* **2007**, *385*, 493–500.
72. Mathai, A.M.; Haubold, H.J. Pathway parameter and thermonuclear functions. *Phys. A* **2008**, *387*, 2462–2470.
73. Mathai, A.M.; Haubold, H.J. On generalized distributions and pathways. *Phys. Lett. A* **2008**, *372*, 2109–2113.
74. Kaniadakis, G.; Lavagno, A.; Quarati, P. Generalized statistics and solar neutrinos. *Phys. Lett. B* **1996**, *369*, 308–312.
75. Quarati, P.; Carbone, A.; Gervino, G.; Kaniadakis, G.; Lavagno, A.; Miraldi, E. Constraints for solar neutrinos fluxes. *Nucl. Phys. A* **1997**, *621*, 345c–348c.
76. Kaniadakis, G.; Lavagno, A.; Quarati, P. Non-extensive statistics and solar neutrinos. *Astrophys. Space Sci.* **1998**, *258*, 145–155.
77. Coraddu, M.; Kaniadakis, G.; Lavagno, A.; Lissia, M.; Mezzorani, G.; Quarati, P. Thermal distributions in stellar plasmas, nuclear reactions and solar neutrinos. *Braz. J. Phys.* **1999**, *29*, 153–168.
78. Mathai, A.M.; Haubold, H.J. On a generalized entropy measure leading to the pathway model with a preliminary application to solar neutrino data. *Entropy* **2013**, *15*, 4011–4025.
79. Haubold, H.J.; Mathai, A.M.; Saxena, R.K. Analysis of solar neutrino data from Super-Kamiokande I and II. *Entropy* **2014**, *16*, 1414–1425.
80. Biro, T.S.; Purcsel, G.; Urmossy, K. Non-extensive approach to quark matter. Statistical Power-Law Tails in High Energy Phenomena. *Eur. Phys. J. A* **2009**, *40*, 325–340.
81. Cleymans, J.; Hamar, G.; Levai, P.; Wheaton, S. Near-thermal equilibrium with Tsallis distributions in heavy ion collisions. *J. Phys. G* **2009**, *36*, 064018.
82. Cleymans, J. Recent developments around chemical equilibrium. *J. Phys. G* **2010**, *37*, 094015.
83. Biro, T.S. Is there a temperature?—Conceptual challenges at high energy, acceleration and complexity. In *Fundamental Theories in Physics*; Springer: Berlin/Heidelberg, Germany, 2011; Volume 171.
84. Biro, T.S.; Van, P. Zeroth law compatibility of non-additive thermodynamics. *Phys. Rev. E* **2011**, *83*, 061147.
85. Wong, C.Y.; Wilk, G. Tsallis fits to pt spectra for pp collisions at LHC. *Acta Phys. Pol. B* **2012**, *43*, 2047–2054.
86. Wong, C.Y.; Wilk, G. Tsallis fits to p_T spectra and relativistic hard scattering in pp collisions at LHC. *Phys. Rev. D* **2013**, *87*, 114007.
87. Marques, L.; Andrade, E., II; Deppman, A. Nonextensivity of hadronic systems. *Phys. Rev. D* **2013**, *87*, 114022.
88. Biro, T.S.; Van, P.; Barnafoldi, G.G.; Urmossy, K. Statistical power law due to reservoir fluctuations and the universal thermostat independence principle. *Entropy* **2014**, *16*, 6497–6514.
89. Wilk, G.; Wlodarczyk, Z. Tsallis distribution with complex nonextensivity parameter q. *Phys. A* **2014**, *413*, 53–58.

90. Deppman, A. Properties of hadronic systems according to the nonextensive self-consistent thermodynamics. *J. Phys. G Nucl. Part. Phys.* **2014**, *41*, 055108.
91. Wilk, G.; Wlodarczyk, Z. Quasi-power laws in multiparticle production processes. *Chaos Solitons Fractals* **2015**, *81*, 487–496.
92. Rybczynski, M.; Wilk, G.; Wlodarczyk, Z. System size dependence of the log-periodic oscillations of transverse momentum spectra. *EPJ Web Conf.* **2015**, *90*, 01002.
93. Wilk, G.; Wlodarczyk, Z. Tsallis distribution decorated with log-periodic oscillation. *Entropy* **2015**, *17*, 384–400.
94. Wong, C.Y.; Wilk, G.; Cirto, L.J.L.; Tsallis, C. Possible implication of a single nonextensive p_T distribution for hadron production in high-energy pp collisions. *EPJ Web Conf.* **2015**, *90*, 04002.
95. Wong, C.Y.; Wilk, G.; Cirto, L.J.L.; Tsallis, C. From QCD-based hard-scattering to nonextensive statistical mechanical descriptions of transverse momentum spectra in high-energy pp and $p\bar{p}$ collisions. *Phys. Rev. D* **2015**, *91*, 114027.
96. Marques, L.; Cleymans, J.; Deppman, A. Description of high-energy pp collisions using Tsallis thermodynamics: Transverse momentum and rapidity distributions. *Phys. Rev. D* **2015**, *91*, 054025.
97. Deppman, A.; Marques, L.; Cleymans, J. Longitudinal properties of high energy collisions. *J. Phys. Conf. Ser.* **2015**, *623*, 012009.
98. Deppman, A. Thermodynamics with fractal structure, Tsallis statistics, and hadrons. *Phys. Rev. D* **2016**, *93*, 054001.
99. Ferri, G.L.; Martinez, S.; Plastino, A. Equivalence of the four versions of Tsallis' statistics. *JSTAT J. Stat. Mech. Theory Exp.* **2005**, *4*, PO4009.
100. Thistleton, W.; Marsh, J.A.; Nelson, K.; Tsallis, C. Generalized Box-Muller method for generating q-Gaussian random deviates. *IEEE Trans. Inf. Theory* **2007**, *53*, 4805–4810.
101. Abe, S. Generalized molecular chaos hypothesis and H-theorem: Problem of constraints and amendment of nonextensive statistical mechanics. *Phys. Rev. E* **2009**, *79*, 041116.
102. Caride, A.O.; Tsallis, C.; Zanette, S.I. Criticality of the anisotropic quantum Heisenberg-model on a self-dual hierarchical lattice. *Phys. Rev. Lett.* **1983**, *51*, 145–147.
103. Tirnakli, U.; Tsallis, C.; Lyra, M.L. Circular-like maps: Sensitivity to the initial conditions, multifractality and nonextensivity. *Eur. Phys. J. B* **1999**, *11*, 309-314.
104. Tsallis, C.; Lenzi, E.K. Anomalous diffusion: Nonlinear fractional Fokker-Planck equation. *Chem. Phys.* **2002**, *84*, 341–347.
105. Tsallis, C. Non-additive entropy: The concept and its use. *Eur. Phys. J. A* **2009**, *40*, 257–266.

axioms

MDPI

Article

On the Fractional Poisson Process and the Discretized Stable Subordinator

Rudolf Gorenflo [1] and Francesco Mainardi [2,*]

[1] Department of Mathematics & Computer Science, Free University Berlin, Berlin 14195, Germany; gorenflo@mi.fu-berlin.de

[2] Department of Physics & Astronomy, University of Bologna, and INFN, Bologna 40126, Italy

* Author to whom correspondence should be addressed; francesco.mainardi@unibo.it or francesco.mainardi@bo.infn.it; Tel.: +39-051-2091098.

Academic Editor: Hans J. Haubold

Received: 20 June 2015; Accepted: 28 July 2015; Published: 4 August 2015

Abstract: We consider the *renewal counting number process* $N = N(t)$ as a forward march over the non-negative integers with independent identically distributed waiting times. We embed the values of the counting numbers N in a "pseudo-spatial" non-negative half-line $x \geq 0$ and observe that for physical time likewise we have $t \geq 0$. Thus we apply the Laplace transform with respect to both variables x and t. Applying then a modification of the Montroll-Weiss-Cox formalism of continuous time random walk we obtain the essential characteristics of a renewal process in the transform domain and, if we are lucky, also in the physical domain. The process $t = t(N)$ of accumulation of waiting times is inverse to the counting number process, in honour of the Danish mathematician and telecommunication engineer A.K. Erlang we call it the *Erlang process*. It yields the probability of exactly n renewal events in the interval $(0, t]$. We apply our Laplace-Laplace formalism to the fractional Poisson process whose waiting times are of Mittag-Leffler type and to a renewal process whose waiting times are of Wright type. The process of Mittag-Leffler type includes as a limiting case the classical Poisson process, the process of Wright type represents the discretized stable subordinator and a re-scaled version of it was used in our method of parametric subordination of time-space fractional diffusion processes. Properly rescaling the counting number process $N(t)$ and the Erlang process $t(N)$ yields as diffusion limits the inverse stable and the stable subordinator, respectively.

Keywords: renewal process; Continuous Time Random Walk; erlang process; Mittag-Leffler function; wright function; fractional Poisson process; stable distributions; stable and inverse stable subordinator; diffusion limit

---

1. Introduction

Serious studies of the fractional generalization of the Poisson process—replacement of the exponential waiting time distribution by a distribution given via a Mittag-Leffler function with modified argument—have been started around the turn of the millenium, and since then many papers on its various aspects have appeared. There are in the literature many papers on this generalization where the authors have outlined a number of aspects and definitions, see e.g., Repin and Saichev (2000) [1], Wang *et al.* (2003,2006) [2,3], Laskin (2003,2009) [4,5], Mainardi *et al.* (2004) [6], Uchaikin *et al.* (2008) [7], Beghin and Orsingher (2009) [8], Cahoy *et al.* (2010) [9], Meerschaert *et al.* (2011) [10], Politi *et al.* (2011) [11], Kochubei (2012) [12], so that it seems impossible to list them all exhaustively. However, in effect this generalization was used already in 1995: Hilfer and Anton [13] (without saying it in our words) showed that the Fractional Kolmogorov-Feller equation (replacement of the first order time derivative by a fractional derivative of order between 0 and 1) requires the underlying random walk to be subordinated to a renewal process with Mittag-Leffler waiting time.

Here we will present our formalism for obtaining the essential characteristics of a generic renewal process and apply it to get those of the fractional Poisson counting process and its inverse, the fractional Erlang process. Both of these comprise as limiting cases the corresponding well-known non-fractional processes that are based on exponential waiting time. Then we will analyze an alternative renewal process, that we call the "Wright process", investigated by Mainardi *et al.* (2000,2005,2007) [14–16], a process arising by discretization of the stable subordinator. In it the so-called M-Wright function plays the essential role. A scaled version of this process has been used by Barkai (2002) [17] for approximating the time-fractional diffusion process directly by a random walk subordinated to it (executing this scaled version in natural time), and he has found rather poor convergence in refinement. In Gorenflo *et al.* (2007) [18] we have modified the way of using this discretized stable subordinator. By appropriate discretization of the relevant spatial stable process we have then obtained a simulation method equivalent to the solution of a pair of Langevin equations, see Fogedby (1994) [19] and Kleinhans and Friedrich (2007) [20]. For simulation of space-time fractional diffusion one so obtains a sequence of precise snapshots of a true particle trajectory, see for details Gorenflo *et al.* (2007) [18], and also Gorenflo and Mainardi (2011, 2012) [21,22].

However, we should note that already in the Sixties of the past century, Gnedenko and Kovalenko (1968) [23] obtained in disguised form the fractional Poisson process by properly rescaled infinite thinning (rarefaction) of a renewal process with power law waiting time. By "disguised" we mean that they found the Laplace transform of the Mittag-Leffler waiting time density, but being ignorant of the Mittag-Leffler function they only presented this Laplace transform. The same ignorance of the Mittag-Leffler function we again meet in a 1985 paper by Balakrishnan [24], who exhibited the Mittag-Leffler waiting time density in Laplace disguise as essential for approximating time-fractional diffusion for which he used the description in form of a fractional integro-differential equation. We have shown that the Mittag-Leffler waiting time density in a certain sense is asymptotically universal for power law renewal processes, see Gorenflo and Mainardi (2008) [25], Gorenflo (2010) [26].

The structure of our paper is as follows. In Section 2 we discuss the elements of the general renewal theory and the CTRW concept. In Section 3 we introduce the Poisson process and its fractional generalization then, in Section 4, the so-called Wright process related to the stable subordinator and its discretization. For both processes we consider the corresponding inverse processes, the Erlang processes. In Section 5 we briefly discuss the diffusion limit for all the above processes. Section 6 is devoted to conclusions. We have collected in Appendix A notations and terminology, in particular the basics on operators, integral transforms and special functions required for understanding our analysis. Finally, we provide in Appendix B an overview on the essential results.

For related aspects of subordination we refer the readers to our papers [18,21,22,25,27] and to papers by Bazhlekova [28], by Merschaert's team [10,29,30], and by Umarov [31].

2. Elements of Renewal Theory and CTRW

For the reader's convenience let us here present a brief introduction to renewal theory including the basics of continuous time random walk (CTRW).

2.1. The General Renewal Process

By a *renewal process* we mean an infinite sequence $0 = t_0 < t_1 < t_2 < \cdots$ of events separated by i.i.d. (independent and identically distributed) random waiting times $T_j = t_j - t_{j-1}$, whose probability density $\phi(t)$ is given as a function or generalized function in the sense of Gel'fand and Shilov [32] (interpretable as a measure) with support on the positive real axis $t \geq 0$, non-negative: $\phi(t) \geq 0$, and normalized: $\int_0^\infty \phi(t)\, dt = 1$, but not having a delta peak at the origin $t = 0$. The instant $t_0 = 0$ is not counted as an event. An important global characteristic of a renewal process is its mean waiting time $\langle T \rangle = \int_0^\infty t\, \phi(t)\, dt$. It may be finite or infinite. In any renewal process we can distinguish two processes, namely the *counting number process* and the process inverse to it, that we call the *Erlang process*. The instants $t_1, t_2, t_3, \cdots$ are often called *renewals*. In fact renewal theory is relevant in practice of maintenance or required exchange of failed parts, e.g., light bulbs.

2.2. The Counting Number Process and Its Inverse

We are interested in the *counting number process* $x = N = N(t)$

$$N(t) := \max\{\, n | t_n \leq t\} = n \text{for} t_n \leq t < t_{n+1} \,, n = 0, 1, 2, \cdots \tag{2.1}$$

where in particular $N(0) = 0$. We ask for the counting number probabilities in n, evolving in t,

$$p_n(t) := \mathcal{P}[N(t) = n] \,, \; n = 0, 1, 2, \cdots \tag{2.2}$$

We denote by $p(x, t)$ the sojourn density for the counting number having the value x. For this process the expectation is

$$m(t) := \langle N(t) \rangle = \sum_{n=0}^{\infty} n \, p_n(t) = \int_0^{\infty} x \, p(x, t) \, dx \tag{2.3}$$

[since $p(x, t) = \sum\limits_{n=0}^{\infty} p_n(t) \, \delta(x - n)$, see (2.12)] It provides the mean number of events in the half-open interval $(0, t]$, and is called the *renewal function*, see e.g., [33]. We also will look at the process $t = t(N)$, the inverse to the process $N = N(t)$, that we call the *Erlang process* in honour of the Danish telecommunication engineer A.K. Erlang (1878–1929), see Brockmeyer *et al.* (1948) [34]. It gives the value of time $t = t_N$ of the N-th renewal. We ask for the Erlang probability densities

$$q_n(t) = q(t, n) \,, \; n = 0, 1, 2, \ldots \tag{2.4}$$

For every n the function $q_n(t) = q(t, n)$ is a density in the variable of time having value t in the instant of the n-th event. Clearly, this event occurs after n (original) waiting times have passed, so that

$$q_n(t) = \phi^{*n}(t) \text{withLaplacetransform} \widetilde{q}_n(s) = \left(\widetilde{\phi}(s)^n \right) \tag{2.5}$$

In other words the function $q_n(t) = q(t, n)$ is a probability density in the variable $t \geq 0$ evolving in the variable $x = n = 0, 1, 2, ...$

2.3. The Continuous time Random Walk

A *continuous time random walk* (CTRW) is given by an infinite sequence of spatial positions $0 = x_0, x_1, x_2, \cdots$, separated by (i.i.d.) random jumps $X_j = x_j - x_{j-1}$, whose probability density function $w(x)$ is given as a non-negative function or generalized function (interpretable as a measure) with support on the real axis $-\infty < x < +\infty$ and normalized: $\int_0^{\infty} w(x) \, dx = 1$, this random walk being subordinated to a renewal process so that we have a random process $x = x(t)$ on the real axis with the property $x(t) = x_n$ for $t_n \leq t < t_{n+1}$, $n = 0, 1, 2, \cdots$.

We ask for the *sojourn probability density* $u(x, t)$ of a particle wandering according to the random process $x = x(t)$ being in point x at instant t.

Let us define the following cumulative probabilities related to the probability density function $\phi(t)$

$$\Phi(t) = \int_0^{t+} \phi(t') \, dt' \,, \Psi(t) = \int_{t+}^{\infty} \phi(t') \, dt' = 1 - \Phi(t) \tag{2.6}$$

For definiteness, we take $\Phi(t)$ as right-continuous, $\Psi(t)$ as left-continuous. When the non-negative random variable represents the lifetime of a technical system, it is common to call $\Phi(t) := \mathcal{P}(T \leq t)$ the *failure probability* and $\Psi(t) := \mathcal{P}(T > t)$ the *survival probability*, because $\Phi(t)$ and $\Psi(t)$ are the respective probabilities that the system does or does not fail in $(0, t]$. These terms, however, are commonly adopted for any renewal process.

In the Fourier-Laplace domain we have

$$\widetilde{\Psi}(s) = \frac{1 - \widetilde{\phi}(s)}{s} \tag{2.7}$$

and the famous Montroll-Weiss solution formula for a CTRW, see [35,36]

$$\widehat{\widetilde{u}}(\kappa, s) = \frac{1 - \widetilde{\phi}(s)}{s} \sum_{n=0}^{\infty} \left(\widetilde{\phi}(s) \, \widehat{w}(\kappa) \right)^n = \frac{1 - \widetilde{\phi}(s)}{s} \frac{1}{1 - \widetilde{\phi}(s) \, \widehat{w}(\kappa)} \tag{2.8}$$

In our special situation the jump density has support only on the positive semi-axis $x > 0$ and thus by replacing the space-Fourier transform $\widehat{w}(\kappa)$ with the space-Laplace transform $\widetilde{w}(\kappa)$, we obtain from (2.8) the Laplace-Laplace solution that we re-write as a new equation

$$\widetilde{\widetilde{u}}(\kappa, s) = \frac{1 - \widetilde{\phi}(s)}{s} \sum_{n=0}^{\infty} \left(\widetilde{\phi}(s) \, \widetilde{w}(\kappa) \right)^n = \frac{1 - \widetilde{\phi}(s)}{s} \frac{1}{1 - \widetilde{\phi}(s) \, \widetilde{w}(\kappa)} \tag{2.9}$$

Recalling from Appendix the definition of convolutions, in the physical domain we have for the solution $u(x, t)$ the *Cox-Weiss series*, see [36,37],

$$u(x, t) = \left(\Psi * \sum_{n=0}^{\infty} \phi^{*n} \, w^{*n} \right)(x, t) \tag{2.10}$$

This formula has an intuitive meaning: Up to and including instant t, there have occurred 0 jumps, or 1 jump, or 2 jumps, or $\cdots$, and if the last jump has occurred at instant $t' < t$, the wanderer is resting there for a duration $t - t'$.

From the rich literature on the concept of CTRW and its applications we recommend to study the surveys by Metzler and Klafter [38,39] and the original article by Chechkin, Hofmann and Sokolov [40].

2.4. *Renewal Process as a Special CTRW*

The essential trick of what follows consists in a rather non-conventional use of the CTRW concept. We treat renewal processes as continuous time random walks with waiting time density $\phi(t)$ and special jump density $w(x) = \delta(x - 1)$ corresponding to the fact that the counting number $N(t)$ increases by 1 at each positive event instant t_n. We then have $\widetilde{w}(\kappa) = \exp(-\kappa)$ and get for the counting number process $N(t)$ the sojourn density in the transform domain ($s \ge 0$, $\kappa \ge 0$),

$$\widetilde{\widetilde{p}}(\kappa, s) = \frac{1 - \widetilde{\phi}(s)}{s} \sum_{n=0}^{\infty} \left(\widetilde{\phi}(s) \right)^n \mathrm{e}^{-n\kappa} = \frac{1 - \widetilde{\phi}(s)}{s} \frac{1}{1 - \widetilde{\phi}(s) \, \mathrm{e}^{-\kappa}} \tag{2.11}$$

From this formula we can find formulas for the renewal function $m(t)$ and the probabilities $p_n(t) = P\{N(t) = n\}$. Because $N(t)$ assumes as values only the non-negative integers, the sojourn density $p(x, t)$ vanishes if x is not equal to one of these, but has a delta peak of height $p_n(t)$ for $x = n$ ($n = 0, 1, 2, 3, \cdots$). Hence

$$p(x, t) = \sum_{n=0}^{\infty} p_n(t) \, \delta(x - n) \tag{2.12}$$

Inverting (2.11) with respect to κ and s as

$$p(x, t) = \sum_{n=0}^{\infty} (\Psi * \phi^{*n})(t) \, \delta(x - n) \tag{2.13}$$

we identify

$$p_n(t) = (\Psi * \phi^{*n})(t) \tag{2.14}$$

According to the theory of Laplace transform we conclude from Equations (2.2) and (2.12)

$$m(t) = -\frac{\partial}{\partial \kappa} \widetilde{p}(\kappa, t)|_{\kappa=0} = \left(\sum_{n=0}^{\infty} n\, p_n(t)\, e^{-n\kappa} \right)\Bigg|_{\kappa=0} = \sum_{n=0}^{\infty} n\, p_n(t) \tag{2.15}$$

a result naturally expected, and

$$\widetilde{m}(s) = \sum_{n=0}^{\infty} n\, \widetilde{p}_n(s) = \widetilde{\Psi}(s) \sum_{n=0}^{\infty} n \left(\widetilde{\phi}(s)\right)^n = \frac{\widetilde{\phi}(s)}{s\left(1 - \widetilde{\phi}(s)\right)} \tag{2.16}$$

thereby using the identity

$$\sum_{n=0}^{\infty} n z^n = \frac{z}{(1-z)^2}, |z| < 1$$

Thus we have found in the Laplace domain the reciprocal pair of relationships

$$\widetilde{m}(s) = \frac{\widetilde{\phi}(s)}{s\left(1 - \widetilde{\phi}(s)\right)}, \widetilde{\phi}(s) = \frac{s\, \widetilde{m}(s)}{1 + s\, \widetilde{m}\,(s))} \tag{2.17}$$

saying that the waiting time density and the renewal function mutually determine each other uniquely. The first formula of Equation (2.17) can also be obtained as the value at $\kappa = 0$ of the negative derivative for $\kappa = 0$ of the last expression in Equation (2.11). Equation (2.17) implies the reciprocal pair of relationships in the physical domain

$$m(t) = \int_0^t [1 + m(t-t')]\, \phi(t')\, dt',\ m'(t) = \int_0^t [1 + m'(t-t')]\, \phi(t')\, dt' \tag{2.18}$$

The first of these equations usually is called the *renewal equation*.

Considering, formally, the counting number process $N = N(t)$ as CTRW (with jumps fixed to unit jumps 1), N running increasingly through the non-negative integers $x = 0, 1, 2, ...$, happening in natural time $t \in [0, \infty)$, we note that in the Erlang process $t = t(N)$, the roles of N and t are interchanged. The new "waiting time density" now is $w(x) = \delta(x-1)$, the new "jump density" is $\phi(t)$.

It is illuminating to consciously perceive the relationships for $t \geq 0$, $n = 0, 1, 2, \ldots$, between the counting number probabilities $p_n(t)$ and the Erlang densities $q_n(t)$. For Equation (2.5) we have $q_n(t) = \phi^{*n}(t)$, and then by (2.14)

$$p_n(t) = (\Psi * q_n)(t) = \int_0^t (q_n(t') - q_{n+1}(t))\, dt' \tag{2.19}$$

We can also express the q_n in another way by the p_n. Introducing the cumulative probabilities $Q_n(t) = \int_0^t q_n(t')\, dt'$, we have

$$Q_n(t) = \mathcal{P}\left(\sum_{k=1}^{n} T_k \leq t \right) = \mathcal{P}(N(t) \geq n) = \sum_{k=n}^{\infty} p_k(t) \tag{2.20}$$

finally

$$q_n(t) = \frac{d}{dt} Q_n(t) = \frac{d}{dt} \sum_{k=n}^{\infty} p_k(t) \tag{2.21}$$

All this is true for $n = 0$ as well, by the empty sum convention $\sum_{k=1}^{n} T_k = 0$ for $n = 0$.

3. The Poisson Process and Its Fractional Generalization

The most popular renewal process is the *Poisson process*. It is (uniquely) characterized by its *mean waiting time* $1/\lambda$ (equivalently by its *intensity* λ), which is a given positive number, and by its residual waiting time $\Psi(t) = \exp(-\lambda t)$ for $t \geq 0$, which corresponds to the *waiting time density* $\phi(t) = \lambda \exp(-\lambda t)$. With $\lambda = 1$ we have what we call the *standard Poisson process*. The general Poisson process arises from the standard one by rescaling the time variable t.

We generalize the standard Poisson process by replacing the exponential function by a function of Mittag-Leffler type. With $t \geq 0$ and a parameter $\beta \in (0, 1]$ we take

$$\begin{cases} \Psi(t) & = E_\beta(-t^\beta), \\ \phi(t) & = -\frac{d}{dt}E_\beta(-t^\beta) = \beta t^{\beta-1}E'_\beta(-t^\beta) = t^{\beta-1}\, E_{\beta,\beta}(-t^\beta) \end{cases} \tag{3.1}$$

The functions $\Psi(t)$ and $\phi(t)$ are plotted versus time for some values of β in Figure 1.

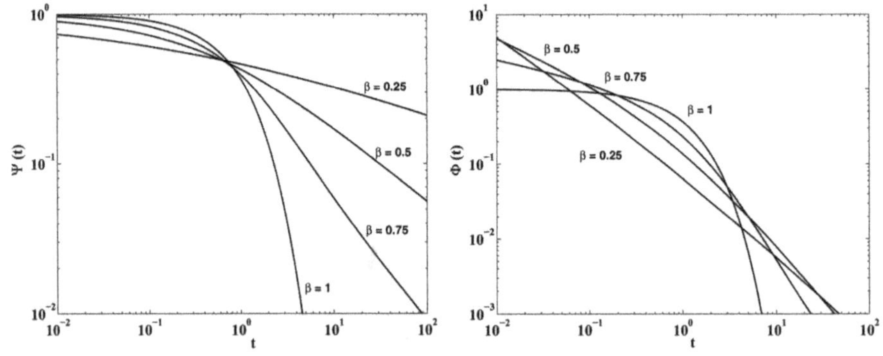

Figure 1. The functions $\Psi(t)$ (**left**) and $\phi(t)$ (**right**) *versus* t ($10^{-2} < t < 10^2$) for the renewal processes of Mittag-Leffler type with $\beta = 0.25, 0.50, 0.75, 1$.

We call this renewal process of Mittag-Leffler type the *fractional Poisson process*, see e.g., [1,4,6,8–11,27,41,42], and [7,43], or the *Mittag-Leffler renewal process* or the *Mittag-Leffler waiting time process*. To analyze it we go into the Laplace domain where we have

$$\widetilde{\Psi}(s) = \frac{s^{\beta-1}}{1+s^\beta}, \widetilde{\phi}(s) = \frac{1}{1+s^\beta} \tag{3.2}$$

If there is no danger of misunderstanding we will not decorate Ψ and ϕ with the index β. The special choice $\beta = 1$ gives us the standard Poisson process with $\Psi_1(t) = \phi_1(t) = \exp(-t)$.

Whereas the Poisson process has finite mean waiting time (that of its standard version is equal to 1), the *fractional Poisson process* ($0 < \beta < 1$) does not have this property. In fact,

$$\langle T \rangle = \int_0^\infty t\,\phi(t)\,dt = \beta \left.\frac{s^{\beta-1}}{(1+s^\beta)^2}\right|_{s=0} = \begin{cases} 1, & \beta = 1, \\ \infty, & 0 < \beta < 1. \end{cases} \tag{3.3}$$

Let us calculate the renewal function $m(t)$. Inserting $\widetilde{\phi}(s) = 1/(1+s^\beta)$ into Equation (2.11) and taking $w(x) = \delta(x-1)$ as in Section 2, we find for the sojourn density of the counting function $N(t)$ the expressions

$$\widetilde{\overline{p}}(\kappa, s) = \frac{s^{\beta-1}}{1+s^\beta - e^{-\kappa}} = \frac{s^{\beta-1}}{1+s^\beta} \sum_{n=0}^{\infty} \frac{e^{-n\kappa}}{(1+s^\beta)^n} \tag{3.4}$$

and

$$\widetilde{p}(\kappa,t) = E_\beta\left(-(1-e^{-\kappa})t^\beta\right) \tag{3.5}$$

and then

$$m(t) = -\frac{\partial}{\partial\kappa}\widetilde{p}(\kappa,t)|_{\kappa=0} = e^{-\kappa}t^\beta E'_\beta\left(-(1-e^{-\kappa})t^\beta\right)\Big|_{\kappa=0} \tag{3.6}$$

Using $E'_\beta(0) = 1/\Gamma(1+\beta)$ now yields

$$m(t) = \begin{cases} t, & \beta = 1 \\ \frac{t^\beta}{\Gamma(1+\beta)}, & 0<\beta<1 \end{cases} \tag{3.7}$$

This result can also be obtained by plugging $\widetilde{\phi}(s) = 1/(1+s^\beta)$ into the first equation in (2.17) which yields $\widetilde{m}(s) = 1/s^{\beta+1}$ and then by Laplace inversion Equation (3.7).

Using general Taylor expansion

$$E_\beta(z) = \sum_{n=0}^{\infty}\frac{E_\beta^{(n)}}{n!}(z-b)^n \tag{3.8}$$

in Equation (3.5) with $b = -t^\beta$ we get

$$\begin{aligned} \widetilde{p}(\kappa,t) &= \sum_{n=0}^{\infty}\frac{t^{n\beta}}{n!}E_\beta^{(n)}(-t^\beta)\,e^{-n\kappa} \\ p(x,t) &= \sum_{n=0}^{\infty}\frac{t^{n\beta}}{n!}E_\beta^{(n)}(-t^\beta)\,\delta(x-n) \end{aligned} \tag{3.9}$$

and, by comparison with Equation (2.12), the counting number probabilities

$$p_n(t) = \mathcal{P}\{N(t)=n\} = \frac{t^{n\beta}}{n!}E_\beta^{(n)}\left(-t^\beta\right) \tag{3.10}$$

Observing from Equation (3.4)

$$\widetilde{\widetilde{p}}(\kappa,s) = \frac{s^{\beta-1}}{1+s^\beta}\sum_{n=0}^{\infty}\frac{e^{-n\kappa}}{(1+s^\beta)^n} \tag{3.11}$$

and inverting with respect to κ,

$$\widetilde{p}(x,s) = \frac{s^{\beta-1}}{1+s^\beta}\sum_{n=0}^{\infty}\frac{\delta(x-n)}{(1+s^\beta)^n} \tag{3.12}$$

we finally identify

$$\widetilde{p}_n(s) = \frac{s^{\beta-1}}{(1+s^\beta)^{n+1}} \div \frac{t^{n\beta}}{n!}E_\beta^{(n)}\left(-t^\beta\right) = p_n(t) \tag{3.13}$$

En passant we have proved an often cited special case of an inversion formula by Podlubny (1999) [44], Equation (1.80).

For the Poisson process with intensity $\lambda > 0$ we have a well-known infinite system of ordinary differential equations (for $t \geq 0$), see e.g., Khintchine [45,46],

$$p_0(t) = e^{-\lambda t}, \frac{d}{dt}p_n(t) = \lambda(p_{n-1}(t) - p_n(t)), n \geq 1 \tag{3.14}$$

with initial conditions $p_n(0) = 0$, $n = 1, 2, \cdots$, which sometimes even is used to define the Poisson process. We have an analogous system of fractional differential equations for the fractional Poisson process. In fact, from Equation (3.13) we have

$$\left(1 + s^\beta\right) \widetilde{p}_n(s) = \frac{s^{\beta-1}}{(1+s^\beta)^n} = \widetilde{p}_{n-1}(s) \tag{3.15}$$

Hence

$$s^\beta \, \widetilde{p}_n(s) = \widetilde{p}_{n-1}(s) - \widetilde{p}_n(s) \tag{3.16}$$

so in the time domain

$$p_0(t) = E_\beta\left(-t^\beta\right), \; {}_*D_t^\beta \, p_n(t) = p_{n-1}(t) - p_n(t), n \geq 1 \tag{3.17}$$

with initial conditions $p_n(0) = 0$, $n = 1, 2, \cdots$, where ${}_*D_t^\beta$ denotes the time-fractional derivative of Caputo type of order β, see Appendix A. It is also possible to introduce and define the fractional Poisson process by this difference-differential system.

Let us note that by solving the system (3.17), Beghin and Orsingher in [8] introduce what they call the "first form of the fractional Poisson process", and in [10] Meerschaert *et al.* show that this process is a renewal process with Mittag-Leffler waiting time density as in (3.1), hence is identical with the fractional Poisson process.

Up to now we have investigated the fractional Poisson counting process $N = N(t)$ and found its probabilities $p_n(t)$ in Equation (3.10). To get the corresponding *Erlang probability densities* $q_n(t) = q(t, n)$, densities in t, evolving in $n = 0, 1, 2 \ldots$, we find by Equation (2.21) via telescope summation

$$q_n(t) = \beta \frac{t^{n\beta-1}}{(n-1)!} E_\beta^{(n)}\left(-t^\beta\right), 0 < \beta \leq 1 \tag{3.18}$$

We leave it as an exercise to the readers to show that in Equation (3.9) interchange of differentiation and summation is allowed.

Remark: With $\beta = 1$ we get the corresponding well-known results for the standard Poisson process. The counting number probabilities are

$$p_n(t) = \frac{t^n}{n!} \, \mathrm{e}^{-t}, n = 0, 1, 2, \ldots, \; t \geq 0 \tag{3.19}$$

and the Erlang densities

$$q_n(t) = \frac{t^{n-1}}{(n-1)!} \, \mathrm{e}^{-t}, n = 1, 2, 3, \ldots,, \; t \geq 0 \tag{3.20}$$

By rescalation of time we obtain

$$p_n(t) = \frac{(\lambda t)^n}{n!} \, \mathrm{e}^{-\lambda t}, n = 0, 1, 2, \ldots,, \; t \geq 0 \tag{3.21}$$

for the classical Poisson process with intensity λ and

$$q_n(t) = \lambda \frac{(\lambda t)^{n-1}}{(n-1)!} \, \mathrm{e}^{-\lambda t}, n = 1, 2, 3, \ldots,, \; t \geq 0 \tag{3.22}$$

for the corresponding Erlang process.

4. The Stable Subordinator and the Wright Process

Let us denote by $g_\beta(t)$ the *extremal Lévy stable density* of order $\beta \in (0,1]$ and support in $t \geq 0$ whose Laplace transform is $\widetilde{g}_\beta(s) = \exp(-s^\beta)$, that is

$$t \geq 0, g_\beta(t) \div \exp\left(-s^\beta\right), Re(s) \geq 0, 0 < \beta \leq 1 \tag{4.1}$$

The topic of Lévy stable distributions is treated in several books on probability and stochastic processes, see e.g., Feller (1971) [47], Sato (1999) [48]; an overview of the analytical and graphical aspects of the corresponding densities is found in Mainardi *et al.* (2001) [49], where an ad hoc notation is used.

From the Laplace transform correspondence (4.1) it is easy to derive the analytical expressions for $\beta = 1/2$ (the so called *Lévy Smirnov density*), $g_{1/2}(t) = \frac{1}{2\sqrt{\pi}} t^{-3/2} \exp(-1/(4t))$ and for the limiting case $\beta = 1$ (the time drift), $g_1(t) = \delta(t-1)$, where δ denotes the Dirac generalized function.

We note that the stable density (4.1) can be expressed in terms of a function of the Wright type. In fact, with the M-Wright function from Appendix A of this paper (see Appendix F of Mainardi's book [50] for more details), we have

$$g_\beta(t) = \frac{\beta}{t^{\beta+1}} M_\beta\left(t^{-\beta}\right) \tag{4.2}$$

The renewal process with waiting time density

$$\phi(t) = g_\beta(t) \tag{4.3}$$

was considered in detail by Mainardi *et al.* (2000,2005,2007) [14–16]. We call this process the *Wright renewal process* because the corresponding survival function $\Psi(t)$ and the waiting time density $\phi(t)$ are expressed in terms of certain Wright functions. So we distinguish it from the so called *Mittag-Leffler renewal process*, treated in the previous Section as *fractional Poisson process*. More precisely, recalling the Wright functions from the Appendix A, we have for $t \geq 0$,

$$\Psi(t) = \begin{cases} 1 - W_{-\beta,1}\left(-\frac{1}{t^\beta}\right), & 0 < \beta < 1, \\ \Theta(t) - \Theta(t-1), & \end{cases} \tag{4.4}$$

$$\phi(t) = \begin{cases} \frac{1}{t} W_{-\beta,0}\left(-\frac{1}{t^\beta}\right), & 0 < \beta < 1, \\ \delta(t-1), & \end{cases} \tag{4.5}$$

where Θ denotes the unit step Heaviside function. The functions $\Psi(t)$ and $\phi(t)$ are plotted versus time for some values of β in Figure 2.

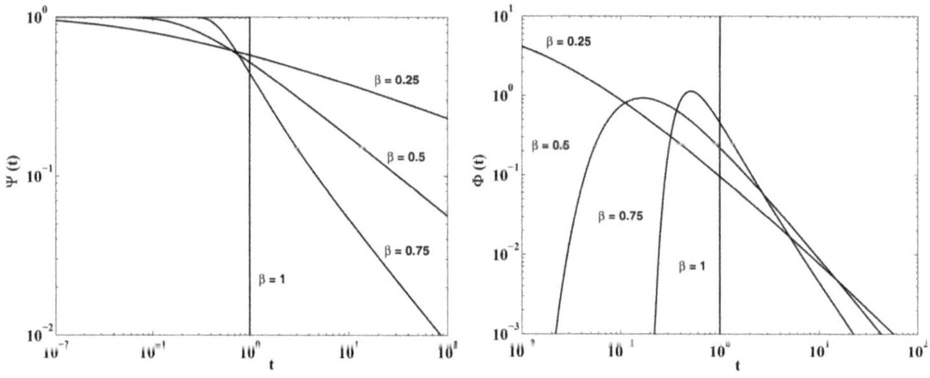

Figure 2. The functions $\Psi(t)$ **(left)** and $\phi(t)$ **(right)** *versus* t $(10^{-2} < t < 10^2)$ for the renewal processes of Wright type with $\beta = 0.25, 0.50, 0.75, 1$. For $\beta = 1$ the reader would recognize the Box function (extended up to $t = 1$) at left and the delta function (centred in $t = 1$) at right.

It is relevant to note the Laplace transform connecting the two transcendental functions M_β and E_β

$$M_\beta(t) \div E_\beta(-s), 0 < \beta \leq 1 \tag{4.6}$$

By the *stable subordinator* of order $\beta \in (0, 1]$ we mean the stochastic process $t = t(x)$ that has sojourn density in $t \geq 0$, evolving in $x \geq 0$ provided by the Laplace transform correspondence,

$$\widetilde{f}(s,x) = \mathrm{e}^{-xs^\beta} \div f(t,x) = x^{-1/\beta}\, g_\beta\left(x^{-1/\beta}\, t\right) = \frac{\beta}{t^{\beta+1}}\, x^{1+1/\beta}\, M_\beta\left(xt^{-\beta}\right) \tag{4.7}$$

This process is monotonically increasing: for this reason it is used in the context of time change and subordination in fractional diffusion processes.

We discretize the process $t = t(x)$ by restricting x to run through the integers $n = 0, 1, 2, \ldots$. The resulting discretized version is a renewal process happening in pseudo-time $x \geq 0$ with jumps in pseudo-space $t \geq 0$ having density $g_\beta(t)$. Inverting this discretized stable subordinator we obtain a counting number process $x = N = N(t)$ with waiting time density and jump density

$$\phi(t) = g_\beta(t), w(x) = \delta(x-1) \tag{4.8}$$

Because here the waiting time density is given by a function of Wright type we call this process the Wright renewal process, or simply the *Wright process*. Immediately we get its *Erlang densities* (in $t \geq 0$, evolving in $x = n = 0, 1, 2, \ldots$)

$$q_n(t) = \phi^{*n}(t) \div \mathrm{e}^{-ns^\beta} \tag{4.9}$$

so that, in view of (4.7) with $x = n$,

$$q_n(t) = f(t,n) = n^{-1/\beta}\, g_\beta\left(n^{-1/\beta}\, t\right) \tag{4.10}$$

In the special case $\beta = 1$ we have $q_n(t) = \delta(t-n)$.

We observe that this counting process gives us precise snapshots at $x = 0, 1, 2,$ of the stable subordinator $t = t(x)$.

Using (4.9) in (2.14) we find the *counting number probabilities* in time and Laplace domain

$$p_n(t) = (\Psi * \phi^{*n})(t) \div \widetilde{p}_n(s) = \frac{1-\mathrm{e}^{-s^\beta}}{s}\, \mathrm{e}^{-ns^\beta} = \frac{\mathrm{e}^{-ns^\beta} - \mathrm{e}^{-(n+1)s^\beta}}{s} \tag{4.11}$$

hence

$$p_n(t) = \int_0^t (q_n(t') - q_{n+1}(t'))dt' \tag{4.12}$$

according to (2.19).

With the probability distribution function

$$G_\beta(t) = \int_0^t g_\beta(t')\, dt' \tag{4.13}$$

we get

$$p_n(t) = G_\beta\left(n^{-1/\beta}t\right) - G_\beta\left((n+1)^{-1/\beta}t\right) \tag{4.14}$$

In the limiting case $\beta = 1$ we have

$$G_1(t) = \int_0^t \delta(t'-1)dt' = \begin{cases} 0 & \text{for } t < 1 \\ 1 & \text{for } t \geq 1 \end{cases} \tag{4.15}$$

as a function continuous from the right, and we calculate

$$p_n(t) = \begin{cases} 0 & \text{for } 0 < t < n, \text{ andfor } t \geq n+1 \\ 1 & \text{for } n \leq t < n+1 \end{cases} \tag{4.16}$$

For the renewal function we obtain its Laplace transform from (2.17)

$$\widetilde{m}(s) = \frac{e^{-s^\beta}}{s\left(1 - e^{-s^\beta}\right)} = \frac{1}{s}\sum_{n=1}^{\infty} e^{-ns^\beta} \tag{4.17}$$

so that

$$m(t) = \sum_{n=1}^{\infty} \int_0^t q_n(t')\, dt' = \sum_{n=1}^{\infty} G_\beta\left(n^{-1/\beta}t\right) \tag{4.18}$$

We do not know an explicit expression for this sum if $0 < \beta < 1$. However, in the limiting case $\beta = 1$ we obtain

$$m(t) = [t] = N(t) \tag{4.19}$$

Using (4.17) we investigate the asymptotic behaviour of $m(t)$ for $t \to \infty$. We have for $s \to 0$ $\widetilde{m}(s) \sim 1/s^{1+\beta}$ and thus, by Tauber theory, see e.g., Feller (1971) [47],

$$m(t) \sim \frac{t^\beta}{\Gamma(1+\beta)} \text{ for } t \to \infty \tag{4.20}$$

Remember, for the fractional Poisson process, we had found

$$m(t) = \frac{t^\beta}{\Gamma(1+\beta)} \text{ forall } t \geq 0 \,(4.21) \tag{4.21}$$

Remark: A rescaled version of the discretized stable subordinator can be used for producing closely spaced precise snapshots of a true particle trajectory of a space-time fractional diffusion process, see e.g., the recent chapter by Gorenflo and Mainardi (2011) [22] on parametric subordination.

5. The Diffusion Limits for the Fractional Poisson and the Wright Processes

In a CTRW we can, with positive scaling factor h and τ, replace the jumps X by jumps $X_h = h\,X$, the waiting times T by waiting times $T_\tau = \tau\,T$. This leads to the rescaled jump density $w_h(x) = w(x/h)/h$

and the rescaled waiting time density $\phi_\tau(t) = \phi(t/\tau)/\tau$ and correspondingly to the transforms $\hat{w}_h(\kappa) = \hat{w}(h\kappa)$, $\widetilde{\phi}_\tau(s) = \widetilde{\phi}(\tau s)$.

For the sojourn density $u_{h,\tau}(x,t)$, density in x evolving in t, we obtain from (2.9) in the transform domain

$$\hat{\widetilde{u}}_{h,\tau}(\kappa, s) = \frac{1-\widetilde{\phi}(\tau s)}{s} \frac{1}{1-\widetilde{\phi}(\tau s)\,\hat{w}(h\kappa)} \tag{5.1}$$

where, if $w(x)$ has support on $x \geq 0$ we can work with the Laplace transform instead of the Fourier transform (replace the $\hat{}$ by $\widetilde{}$). If there exists between h and τ a scaling relation $\mathcal{R}$ (to be introduced later) under which $u(x,t)$ tends for $h \to 0,\ \tau \to 0$ to a meaningful limit $v(x,t) = u_{0,0}(x,t)$, then we call the process $x = x(t)$ with this sojourn density a *diffusion limit*. We find it via

$$\hat{\widetilde{v}}(\kappa, s) = \lim_{h,\tau \to 0(\mathcal{R})} \hat{\widetilde{u}}_{h,\tau}(\kappa, s) \tag{5.2}$$

and Fourier-Laplace (or Laplace-Laplace) inversion.

Remark: This diffusion limit is a limit in the weak sense (convergence in distribution of the CTRW to the diffusion limit). The mathematical background consists in the application of the Fourier (or Laplace) continuity theorem of probability theory for fixed time t.

We will now find that the counting numbers of the fractional Poisson process and the Wright process have the same diffusion limit, namely the inverse stable subordinator. The two corresponding Erlang processes have the same diffusion limit, namely the stable subordinator. For $t \to \infty$ the renewal functions have the same asymptotic behaviour, namely $m(t) \sim t^\beta/\Gamma(1+\beta)$. Here, in the case of the fractional Poisson process, we can replace the sign $\sim$ of asymptotics by the sign = of equality for all $t \geq 0$.

To prove these statements we need the Laplace transform of the relevant functions $\phi(t)$ and $w(x)$. For the fractional Poisson process we have

$$\phi(t) = \frac{d}{dt} E_\beta\left(-t^\beta\right) \div \widetilde{\phi}(s) = \frac{1}{1+s^\beta}, w(x) = \delta(x-1) \div \widetilde{w}(\kappa) = \exp(-\kappa)$$

For the Wright process we have

$$\phi(t) = g_\beta(t) \div \widetilde{\phi}(s) = \exp\left(-s^\beta\right), w(x) = \delta(x-1) \div \widetilde{w}(\kappa) = \exp(-\kappa)$$

In all cases we have, for fixed s and κ

$$\widetilde{\phi}(\tau s) \sim 1 - (\tau s)^\beta \text{ as } \tau \to 0, \widetilde{w}(h\kappa) \sim 1 - (h\kappa) \text{ as } h \to 0,$$

and straightforwardly we obtain for the sojourn densities in both cases, by use of (5.1) with p in place of u and $\hat{}$ replaced by $\widetilde{}$

$$\widetilde{\widetilde{p}}_{h,\tau}(\kappa, s) \sim \frac{\tau^\beta\, s^{\beta-1}}{\tau^\beta\, s^\beta + h\,\kappa}, \text{for } \tau \to 0,\ h \to 0 \tag{5.3}$$

Using the scaling relation $\mathcal{R}$

$$h = \tau^\beta \tag{5.4}$$

we obtain

$$\widetilde{\widetilde{p}}_{0,0}(\kappa, s) = \frac{s^{\beta-1}}{s^\beta + \kappa} \tag{5.5}$$

By partial Laplace inversions we get two equivalent representations

$$p_{0,0}(x,t) = \mathcal{L}_\kappa^{-1}\left\{E_\beta\left(-\kappa t^\beta\right)\right\} = \mathcal{L}_s^{-1}\left\{s^{\beta-1}\exp\left(-x s^\beta\right)\right\} \tag{5.6}$$

leading to the density of the *inverse stable subordinator*

$$p_{0,0}(x,t) = t^{-\beta}\, M_\beta\left(x/t^\beta\right) = J_t^{1-\beta} f(t,x) \tag{5.7}$$

where M_β and $J_t^{1-\beta}$ denote respectively the M-Wright function and the Riemann-Liouville fractional integral introduced in Appendix A, and $f(t,x)$ the stable subordinator given by Equation (4.7).

Remark: In (4.7) and (5.7) the densities of the stable and the inverse stable subordinator are both represented via the M Wright function.

The Diffusion Limit for the Erlang Process

In the Erlang process the roles of space and time, likewise of jumps and waiting times, are interchanged. In other words we treat $x \geq 0$ as a pseudo-time variable and $t \geq 0$ as a pseudo-space variable. For the resulting sojourn density $q(t,x)$, we have from interchanging in (5.1) for $h \to 0$ and $\tau \to 0$,

$$\widetilde{\widetilde{q}}_{h,\tau}(s,\kappa) = \frac{1-\widetilde{w}(h\kappa)}{k}\,\frac{1}{1-\widetilde{w}(h\kappa)\,\widetilde{\phi}(\tau s)} \sim \frac{h}{h\kappa + (\tau s)^\beta} \tag{5.8}$$

Again using the scaling relation $\mathcal{R}$ in Equation (5.4) we find

$$\widetilde{\widetilde{q}}_{0,0}(s,\kappa) = \frac{1}{\kappa + s^\beta}\,, \tag{5.5'}$$

which is the Laplace-Laplace transform of the density of stable subordinator of Section 4. In fact, by partial Laplace inversion,

$$\widetilde{q}_{0,0}(s,x) = \exp\left(-xs^\beta\right) = \widetilde{f}(s,x)\,, \tag{5.9}$$

and it follows that

$$q_{0,0}(t,x) = f(t,x)\,, x \geq 0\,,\ t \geq 0\,. \tag{5.10}$$

See (4.7) for its explicit representation as a rescaled stable density expressed via a M-Wright function.

We get the same result by continualization of the discretized stable subordinator. Replace in Equations (4.9) and (4.10) the discrete variable n by the continuous variable x.

6. Conclusions

The fractional Poisson process and the Wright process (as discretization of the stable subordinator) along with their diffusion limits play eminent roles in theory and simulation of fractional diffusion processes. Here we have analyzed these two processes, concretely the corresponding counting number and Erlang processes, the latter being the processes inverse to the former. Furthermore we have obtained the diffusion limits of all these processes by well-scaled refinement of waiting times and jumps.

Acknowledgments: The authors are grateful to Professor Mathai for several invitations to visit the Centre for Mathematical Sciences in Pala-Kerala for conferences, teaching and research. They luckily enjoyed there the friendly and stimulating environment, scientifically and geographically. The first-named author appreciates the stimulating working conditions he enjoyed during several ERASMUS visits in the Department of Physics of Bologna University. The authors are grateful for helpful comments of the referees.

Author Contributions: Rudolf Gorenflo and Francesco Mainardi each contributed 50% to this publication.

Conflicts of Interest: The authors declare no conflict of interest.

Appendix : Operators, Transforms and Special Functions

For the reader's convenience here we present a brief introduction to the basic notions required for the presentation and analysis of the renewal processes to be treated, including essentials on fractional calculus and special functions of Mittag-Leffler and Wright type.

Thereby we follow our earlier papers concerning related topics, see [6,18,21,22,25–27,49,51–56], and our recent monograph on Mittag-Leffler Functions and Related Topics [57].

For more details on general aspects the interested reader may consult the treatises, listed in order of publication time, by Podlubny [44], Kilbas and Saigo [58], Kilbas, Srivastava and Trujillo [59], Mathai and Haubold [60], Mathai, Saxena and Haubold [61], Mainardi [50], Diethelm [62], Baleanu, Diethelm, Scalas and Trujillo [41], Uchaikin [43], Atanacković, Pilipovíc, Stanković and Zorica [63].

A.1. Fourier and Laplace Transforms

By $\mathbb{R}$ ($\mathbb{R}^+, \mathbb{R}_0^+$)) we mean the set of all (positive, non-negative) real numbers, and by $\mathbb{C}$ the set of complex numbers. It is known that the Fourier transform is applied to functions defined in $L_1(\mathbb{R})$ whereas the Laplace transform is applied to functions defined in $L_{loc}(\mathbb{R}^+)$. In our cases the arguments of the original function are the space–coordinate x ($x \in \mathbb{R}$ or $x \in \mathbb{R}_0^+$) and the time–coordinate t ($t \in \mathbb{R}_0^+$). We use the symbol $\div$ for the juxtaposition of a function with its Fourier or Laplace transform. A look at the superscript $\hat{}$ for the Fourier transform, $\tilde{}$ for the Laplace transform reveals their relevant juxtaposition. We use x as argument (associated to real κ) for functions Fourier transformed, and x or t as argument (associated to complex κ or s, respectively) for functions Laplace transformed.

$$f(x) \div \hat{f}(\kappa) := \int_{-\infty}^{+\infty} e^{i\kappa x} f(x)\, dx\,, \text{Fouriertransform.}$$

$$f(x) \div \tilde{f}(\kappa) := \int_{0}^{\infty} e^{-\kappa x} f(x)\, dx\,, \text{space-Laplacetransform.}$$

$$f(t) \div \tilde{f}(s) := \int_{0}^{\infty} e^{-st} f(t)\, dx\,, \text{time-Laplacetransform}$$

A.2. Convolutions

$$(u * v)(x) := \int_{-\infty}^{+\infty} u(x - x')\, v(x')\, dx'\,, \text{Fourierconvolution}$$

$$(u * v)(t) := \int_{0}^{t} u(t - t')\, v(t')\, dt'\,, \text{Laplaceconvolution}$$

The meaning of the connective * will be clear from the context. For convolution powers we have:

$$u^{*0}(x) = \delta(x)\,,\ u^{*1}(x) = u(x)\,,\ u^{*(n+1)}(x) = (u^{*n} * u)(x)$$

$$u^{*0}(t) = \delta(t)\,,\ u^{*1}(t) = u(t)\,,\ u^{*(n+1)}(t) = (u^{*n} * u)(t)$$

where δ denotes the Dirac generalized function.

A.3. Fractional Integral

The Riemann-Liouville fractional integral of order $\alpha > 0$, for a sufficiently well-behaved function $f(t)$ ($t \geq 0$), is defined as

$$J_t^{\alpha} f(t) = \frac{1}{\Gamma(\alpha)} \int_0^t (t - \tau)^{\alpha - 1} f(\tau)\, d\tau\,, \alpha > 0$$

by convention as $f(t)$ for $\alpha = 0$. Well known are the *semi-group property*

$$J_t^\alpha J_t^\beta = J_t^{\alpha+\beta} = J_t^\beta J_t^\alpha, \alpha, \beta \geq 0$$

and the Laplace transform pair

$$J_t^\alpha f(t) \div \frac{\widetilde{f}(s)}{s^\alpha}, \alpha \geq 0$$

A.4. Fractional Derivatives

The Riemann-Liouville fractional derivative operator of order $\alpha > 0$, D_t^α, is defined as the *left inverse operator* of the corresponding fractional integral J_t^α. Limiting ourselves to fractional derivatives of order $\alpha \in (0,1)$ we have, for a sufficiently well-behaved function $f(t)$ $(t \geq 0)$,

$$D_t^\alpha f(t) := D_t^1 J_t^{1-\alpha} f(t) = \frac{1}{\Gamma(1-\alpha)} \frac{d}{dt} \int_0^t \frac{f(\tau)}{(t-\tau)^\alpha} d\tau, \; 0 < \alpha < 1$$

while the corresponding Caputo derivative is

$$_*D_t^\alpha f(t) := J_t^{1-\alpha} D_t^1 f(t) = \frac{1}{\Gamma(1-\alpha)} \int_0^t \frac{f^{(1)}(\tau)}{(t-\tau)^\alpha} d\tau = D_t^\alpha f(t) - f(0^+) \frac{t^{-\alpha}}{\Gamma(1-\alpha)}$$

Both derivatives yield the ordinary first derivative as $\alpha \to 1^-$ but for $\alpha \to 0^+$ we have

$$D_t^0 f(t) = f(t), \; _*D_t^0 f(t) = f(t) - f(0^+)$$

We point out the major utility of the Caputo fractional derivative in treating initial-value problems with Laplace transform. We have

$$\mathcal{L}[\,_*D_t^\alpha f(t); s] = s^\alpha \widetilde{f}(s) - s^{\alpha-1} f(0^+), 0 < \alpha \leq 1$$

In contrast the Laplace transform of the Riemann-Liouville fractional derivative needs the limit at zero of a fractional integral of the function $f(t)$.

Note that both types of fractional derivative may exhibit singular behaviour at the origin $t = 0^+$.

A.5. Mittag-Leffler and Wright Functions

The Mittag-Leffler function of parameter α is defined as

$$E_\alpha(z) := \sum_{n=0}^{\infty} \frac{z^n}{\Gamma(\alpha n + 1)}, \alpha > 0, z \in C$$

It is entire of order $1/\alpha$. Let us note the trivial cases

$$\begin{cases} E_1(\pm z) = \exp(\pm z) \\ E_2(+z^2) = \cosh(z), \; E_2(-z^2) = \cos(z) \end{cases}$$

Without changing the order $1/\alpha$ the Mittag-Leffler function can be generalized by introducing an additional (arbitrary) parameter β.

The Mittag-Leffler function of parameters α, β is defined as

$$E_{\alpha,\beta}(z) := \sum_{n=0}^{\infty} \frac{z^n}{\Gamma(\alpha n + \beta)}, \alpha > 0, \beta, z \in C$$

Laplace transforms of Mittag-Leffler functions

For our purposes we need, with $0 < \nu \leq 1$ and $t \geq 0$, the Laplace transform pairs

$$\begin{cases} \Psi(t) = E_{\beta}(-t^{\nu}) \div \widetilde{\Psi}(s) = \frac{s^{\nu-1}}{1+s^{\nu}} \\ \phi(t) = -\frac{d}{dt}E_{\nu}(-t^{\nu}) = t^{\nu-1}\,E_{\nu,\nu}(-t^{\nu}) \div \widetilde{\phi}(s) = \frac{1}{1+s^{\nu}} \end{cases}$$

Of high relevance is the algebraic decay of $\Psi(t)$ and $\phi(t)$ as $t \to \infty$:

$$\begin{cases} \Psi(t) \sim \frac{\sin(\nu\pi)}{\pi}\,\frac{\Gamma(\nu)}{t^{\nu}}, \\ \phi(t) \sim \frac{\sin(\nu\pi)}{\pi}\,\frac{\Gamma(\nu+1)}{t^{\nu+1}}, \end{cases} \quad t \to +\infty$$

Furthermore $\Psi(t) = E_{\nu}(-t^{\nu})$ is the solution of the fractional relaxation equation with the Caputo derivative

$$_{*}D_t^{\nu}u(t) = -u(t), t \geq 0, u(0^{+}) = 1$$

whereas $\phi(t) = -\frac{d}{dt}E_{\nu}(-t^{\nu})$ is the solution of the fractional relaxation equation with the Riemann-Liouville derivative

$$D_t^{\nu}u(t) = -u(t), t \geq 0, \lim_{t\to 0^{+}} J_t^{1-\nu}u(t) = 1$$

We refer to the survey paper by Haubold, Mathai and Saxena [64], to our recent monograph [57] and to our papers [21,22,25–27,51] for the relevance of Mittag-Leffler functions in probability theory and stochastic processes with particular regard to theory of continuous time random walk and space-time fractional diffusion, and in power law asymptotics. Particularly worth to be mentioned is the pioneering paper by Hilfer and Anton [13]. They show that for transforming a general evolution equation for continuous time random walk into the time fractional version of the Kolmogorov-Feller equation a waiting time law expressible via a Mittag-Leffler type function is required.

The Wright function is defined as

$$W_{\lambda,\mu}(z) := \sum_{n=0}^{\infty} \frac{z^n}{n!\,\Gamma(\lambda n + \mu)}, \lambda > -1, \mu \in C$$

We distinguish the Wright functions of *first kind* ($\lambda \geq 0$) and *second kind* ($-1 < \lambda < 0$). The case $\lambda = 0$ is trivial since $W_{0,\mu}(z) = e^{z}/\Gamma(\mu)$. The Wright function is entire of order $1/(1+\lambda)$ hence of exponential type only if $\lambda \geq 0$.

Laplace transforms of the Wright functions

For the Wright function of the first kind, being entire of exponential type, the Laplace transform can be obtained by transforming the power series term by term:

$$W_{\lambda,\mu}(t) \div \frac{1}{s}\,E_{\lambda,\mu}\left(\frac{1}{s}\right), \lambda \geq 0$$

For the Wright function of the second kind, denoting $\nu = |\lambda| \in (0,1)$ we have with $\mu > 0$ for simplicity, we have

$$W_{-\nu,\mu}(-t) \div E_{\nu,\mu+\nu}(-s), 0 < \nu < 1$$

We note the minus sign in the argument in order to ensure the the existence of the Laplace transform thanks to the Wright asymptotic formula valid in a certain sector symmetric to and including the negative real axis.

Stretched Exponentials as Laplace transforms of Wright functions We outline the following Laplace transform pairs related to the stretched exponentials in the transform domain, useful for our purposes,

$$\frac{1}{t}\,W_{-\nu,0}\left(-\frac{1}{t^{\nu}}\right) \div e^{-s^{\nu}}$$

$$W_{-\nu,1-\nu}\left(-\frac{1}{t^{\nu}}\right) \div \frac{e^{-s^{\nu}}}{s^{1-\nu}}$$

$$W_{-\nu,1}\left(-\frac{1}{t^{\nu}}\right) \div \frac{e^{-s^{\nu}}}{s}$$

For $\nu = 1/2$ we have the three sister functions related to the diffusion equation available in most Laplace transform handbooks

$$\frac{1}{2\sqrt{\pi}} t^{-3/2} e^{-1/(4t)} \div e^{-s^{1/2}}$$

$$\frac{1}{\sqrt{\pi}} t^{-1/2} e^{-1/(4t)} \div \frac{e^{-s^{1/2}}}{s^{1/2}}$$

$$\operatorname{erfc}\left(\frac{1}{2t^{1/2}}\right) \div \frac{e^{-s^{1/2}}}{s}$$

Among the Wright functions of the second kind a fundamental role in fractional diffusion equations is played by the so called M-Wright function, see e.g., [49,50,54].

The M-Wright function is defined as

$$M_{\nu}(z) := W_{-\nu,1-\nu}(-z) = \sum_{n=0}^{\infty} \frac{(-z)^n}{n!\Gamma[-\nu n + (1-\nu)]} = \frac{1}{\pi} \sum_{n=1}^{\infty} \frac{(-z)^{n-1}}{(n-1)!} \Gamma(\nu n) \sin(\pi \nu n)$$

with $z \in \mathbb{C}$
and $0 < \nu < 1$. Special cases are

$$M_{1/2}(z) = \frac{1}{\sqrt{\pi}} \exp\left(-z^2/4\right), \; M_{1/3}(z) = 3^{2/3}\mathrm{Ai}\left(z/3^{1/3}\right)$$

where Ai denotes the Airy function, see e.g., [65].

The asymptotic representation of the M-Wright function

Choosing as a variable t/ν rather than t, the computation of the asymptotic representation as $t \to \infty$ by the saddle-point approximation yields:

$$M_{\nu}(t/\nu) \sim a(\nu)\, t^{(\nu-1/2)/(1-\nu)} \exp\left[-b(\nu)\, t^{1/(1-\nu)}\right]$$

where

$$a(\nu) = \frac{1}{\sqrt{2\pi(1-\nu)}} > 0, b(\nu) = \frac{1-\nu}{\nu} > 0$$

Mittag-Leffler function as Laplace transforms of M-Wright function

$$M_{\nu}(t) \div E_{\nu}(-s), 0 < \nu < 1, t \geq 0, s \geq 0$$

Stretched Exponentials as Laplace transforms of M-Wright functions

$$\frac{\nu}{t^{\nu+1}} M_{\nu}(1/t^{\nu}) \div e^{-s^{\nu}}, \; 0 < \nu < 1, t \geq 0, s \geq 0$$

$$\frac{1}{t^{\nu}} M_{\nu}(1/t^{\nu}) \div \frac{e^{-s^{\nu}}}{s^{1-\nu}}, \; 0 < \nu < 1, t \geq 0, s \geq 0$$

Note that $\exp(-s^{\nu})$ is the Laplace transform of the extremal (unilateral) stable density $L_{\nu}^{-\nu}(t)$, which vanishes for $t < 0$, so that, introducing the Riemann-Liouville fractional integral, we have

$$\frac{1}{t^{\nu}} M_{\nu}(1/t^{\nu}) = J_t^{1-\nu}\{L_{\nu}^{-\nu}(t)\} = J_t^{1-\nu}\left\{\frac{\nu}{t^{\nu+1}} M_{\nu}(1/t^{\nu})\right\}$$

Appendix : Collection of Results

B.1. General Renewal Process

Waiting time density: $\phi(t)$; Survival function: $\Psi(t) = \int_t^{\infty} \phi(t')\, dt'$

(a) The counting number process $x = N(t)$ has probability density function (density in $x \geq 0$ and evolving in $t \geq 0$):

$$p(x,t) = \sum_{n=0}^{\infty} p_n(t)\, \delta(x-n)\,,$$

and counting probabilities

$$p_n(t) = (\Psi \, * \, \phi^{*n})(t)\,.$$

(b) The Erlang process $t = t(n)$, inverse to the counting process has probability density function (density in t, evolving in $n = 0, 1, 2, \cdots$)

$$q_n(t) = q(t,n) = \phi^{*n}(t)\,,$$

with

$$q_n(t) = \frac{d}{dt} Q_n(t)\,, Q_n(t) = \sum_{k=n}^{\infty} p_k(t)\,,$$

where $q_n(t)$, $Q_n(t)$ are the Erlang densities and probability distribution functions, respectively. Note that $p_n(t) = (\Psi \, * \, q_n)(t)$.

B.2. Special Cases

(α) The fractional Poisson process

$$\phi(t) = -\frac{d}{dt} E_\beta\left(-t^\beta\right) \div \widetilde{\phi}(s) = \frac{1}{1+s^\beta}\,,$$

$$p_n(t) = \frac{t^{n\beta}}{n!}\, E_\beta^{(n)}\left(-t^\beta\right).$$

The Erlang densities are

$$q_n(t) = \beta \frac{t^{n\beta-1}}{(n-1)!}\, E_\beta^{(n)}\left(-t^\beta\right).$$

(β) The Wright process

$$\phi(t) = g_\beta(t) \div \widetilde{g}_\beta(s) = \exp\left(-s^\beta\right),$$

$$p_n(t) = G_\beta\left(n^{-1/\beta} t\right) - G_\beta\left((n+1)^{-1/\beta} t\right).$$

The Erlang densities are

$$q_n(t) = n^{-1/\beta}\, g_\beta\left(n^{-1/\beta}\, t\right).$$

References

1. Repin, O.N.; Saichev, A.I. Fractional Poisson law. *Radiophys. Quantum Electron.* **2000**, *43*, 738–741. [CrossRef]
2. Wang, X.; Wen, Z. Poisson fractional processes. *Chaos Solitons Fractals* **2003**, *18*, 169–177. [CrossRef]
3. Wang, X.; Wen, Z.; Zhang, S. Fractional Poisson process (II). *Chaos Solitons Fractals* **2006**, *28*, 143–147. [CrossRef]
4. Laskin, N. Fractional Poisson process. *Commun. Nonlinear Sci. Numer. Simul.* **2003**, *8*, 201–213. [CrossRef]
5. Laskin, N. Some applications of the fractional Poisson probability distribution. *J. Math. Phys.* **2009**, *50*, 113513:1–113513:12. [CrossRef]

6. Mainardi, F.; Gorenflo, R.; Scalas, E. A fractional generalization of the Poisson processes. *Vietnam J. Math.* **2004**, *32*, 53–64. Available online: http://arxiv.org/abs/math/0701454 (accessed on 3 August 2015).
7. Uchaikin, V.V.; Cahoy, D.O.; Sibatov, R.T. Fractional processes: From Poisson to branching one. *Int. J. Bifurcation Chaos* **2008**, *18*, 1–9. [CrossRef]
8. Beghin, L.; Orsingher, E. Fractional Poisson processes and related random motions. *Electron. Journ. Prob.* **2009**, *14*, 1790–1826. [CrossRef]
9. Cahoy, D.O.; Uchaikin, V.V.; Woyczynski, W.A. Parameter estimation for fractional Poisson processes. *J. Stat. Plan. Inference* **2010**, *140*, 3106–3120. [CrossRef]
10. Meerschaert, M.M.; Nane, E.; Vellaisamy, P. The fractional Poisson process and the inverse stable subordinator. *Electron. J. Prob.* **2011**, *16*, 1600–1620. Available online: http://arxiv.org/abs/1007.505 (accessed on 3 August 2015). [CrossRef]
11. Politi, M.; Kaizoji, T.; Scalas, E. Full characterization of the fractional Poisson process. *Eur. Phys. Lett.* **2011**, *96*, 20004:1–20004:6. [CrossRef]
12. Kochubei, A.N. General fractional calculus, evolution equations, and renewal processes. *Integral Equ. Oper. Theory* **2011**, *71*, 583–600. Available online: http://arxiv.org/abs/1105.1239 (accessed on 3 August 2015). [CrossRef]
13. Hilfer, H.; Anton, L. Fractional master equations and fractal time random walks. *Phys. Rev. E* **1995**, *51*, R848–R851. [CrossRef]
14. Mainardi, F.; Raberto, M.; Gorenflo, R.; Scalas, E. Fractional calculus and continuous-time finance II: The waiting-time distribution. *Phys. A* **2000**, *287*, 468–481. Available online: http://arxiv.org/abs/cond-mat/0006454 (accessed on 3 August 2015). [CrossRef]
15. Mainardi, F.; Gorenflo, R.; Vivoli, A. Renewal processes of Mittag-Leffler and Wright type. *Fract. Calc. Appl. Anal.* **2005**, *8*, 7–38. Available online: http://arxiv.org/abs/math/0701455 (accessed on 3 August 2015).
16. Mainardi, F.; Gorenflo, R.; Vivoli, A. Beyond the Poisson renewal process: A tutorial survey. *J. Comp. Appl. Math* **2007**, *205*, 725–735. [CrossRef]
17. Barkai, E. CTRW pathways to the fractional diffusion equation. *Chem. Phys.* **2002**, *284*, 13–27. [CrossRef]
18. Gorenflo, R.; Mainardi, F.; Vivoli, A. Continuous time random walk and parametric subordination in fractional diffusion. *Chaos Solitons Fractals* **2007**, *34*, 87–103. Available online: http://arxiv.org/abs/cond-mat/0701126 (accessed on 3 August 2015). [CrossRef]
19. Fogedby, H.C. Langevin equations for continuous time Lévy flights. *Phys. Rev. E* **1994**, *50*, 1657–1660. [CrossRef]
20. Kleinhans, D.; Friedrich, R. Continuous-time random walks: Simulations of continuous trajectories. *Phys. Rev E* **2007**, *76*, 061102:1–061102:6. [CrossRef] [PubMed]
21. Gorenflo, R.; Mainardi, F. Subordination pathways to fractional diffusion. *Eur. Phys. J. Spec. Top.* **2011**, *193*, 119–132. Available online: http://arxiv.org/abs/1104.4041 (accessed on 3 August 2015). [CrossRef]
22. Gorenflo, R.; Mainardi, F. Parametric Subordination in Fractional Diffusion Processes. In *Fractional Dynamics*; Klafter, J., Lim, S.C., Metzler, R., Eds.; World Scientific: Singapore, 2012; Chapter 10; pp. 229–263. Available online: http://arxiv.org/abs/1210.8414 (accessed on 3 August 2015).
23. Gnedenko, B.V.; Kovalenko, I.N. *Introduction to Queueing Theory*; Israel Program for Scientific Translations: Jerusalem, Israel, 1968.
24. Balakrishnan, V. Anomalous diffusion in one dimension. *Phys. A* **1985**, *132*, 569–580. [CrossRef]
25. Gorenflo, R.; Mainardi, F. Continuous time random walk, Mittag-Leffler waiting time and fractional diffusion: Mathematical aspects. In *Anomalous Transport: Foundations and Applications*; Klages, R., Radons, G., Sokolov, I.M., Eds.; Wiley-VCH: Weinheim, Germany, 2008; Chapter 4; pp. 93–127. Available online: http://arxiv.org/abs/0705.0797 (accessed on 3 August 2015).
26. Gorenflo, R. Mittag-Leffler waiting time, power laws, rarefaction, continuous time random walk, diffusion limit. In Proceedings of the National Workshop on Fractional Calculus and Statistical Distributions, CMS Pala Campus, India, 25–27 November 2010; Pai, S.S., Sebastian, N., Nair, S.S., Joseph, D.P., Kumar, D., Eds.; pp. 1–22. Available online: http://arxiv.org/abs/1004.4413 (accessed on 3 August 2015).
27. Gorenflo, R.; Mainardi, F. Laplace-Laplace analysis of the fractional Poisson process. In *Analytical Methods of Analysis and Differential Equations*; Rogosin, S., Ed.; Belarusan State University: Minsk, Belarus, 2012; Kilbas Memorial Volume, pp. 43–58. Available online: http://arxiv.org/abs/1305.5473 (accessed on 3 August 2015).

28. Bazhlekova, E. Subordination principle for a class of fractional order differential equations. *Mathematics* **2015**, *2*, 412–427. [CrossRef]
29. Meerschaert, M.M. Fractional Calculus, Anomalous Diffusion, and Probability. In *Fractional Dynamics*; Lim, S.C., Klafter, J., Metzler, R., Eds.; World Scientific: Singapore, 2012; Chapter 11; pp. 265–284.
30. Meerschaert, M.M.; Benson, D.A.; Scheffler, H.-P.; Baeumer, B. Stochastic solution of space-time fractional diffusion equations. *Phys. Rev. E* **2002**, *65*, 41103:1–41103:4. [CrossRef] [PubMed]
31. Umarov, S. Continuous time random walk models for fractional space-time diffusion equations. *Fract. Calc. Appl. Anal.* **2015**, *18*, 821–837. [CrossRef]
32. Gel'fand, I.M.; Shilov, G.E. *Generalized Functions*; Academic Press: New York, NY, USA, 1964; Volume 1.
33. Ross, S.M. *Stochastic Processes*, 2nd ed.; Wiley: New York, NY, USA, 1996.
34. Brockmeyer, E.; Halstrøm, H.L.; Jensen, A. *The Life and Works of A.K. Erlang, Transactions of the Danish Academy of Technical Sciences, No 2*; The Copenhagen Telephone Company: Copenhagen, Denmark, 1948.
35. Montroll, E.W.; Weiss, G.H. Random walks on lattices, II. *J. Math. Phys.* **1965**, *6*, 167–181. [CrossRef]
36. Weiss, G.H. *Aspects and Applications of Random Walks*; North-Holland: Amsterdam, The Netherlands, 1994.
37. Cox, D.R. *Renewal Theory*, 2nd ed.; Methuen: London, UK, 1967.
38. Metzler, R.; Klafter, J. The random walker's guide to anomalous diffusion: A fractional dynamics approach. *Phys. Rep.* **2000**, *339*, 1–77. [CrossRef]
39. Metzler, R.; Klafter, J. The restaurant at the end of the random walk: Recent developments in the description of anomalous transport by fractional dynamics. *J. Phys. A. Math. Gen.* **2004**, *37*, R161–R208. [CrossRef]
40. Chechkin, A.V.; Hofmann, M.; Sokolov, I.M. Continuous-time random walk with correlated waiting times. *Phys. Rev. E* **2009**, *80*. [CrossRef] [PubMed]
41. Baleanu, D.; Diethelm, K.; Scalas, E.; Trujillo, J.J. *Fractional Calculus: Models and Numerical Methods*; World Scientific: Singapore, 2012.
42. Scalas, E. A class of CTRWs: Compound fractional Poisson processes. In *Fractional Dynamics*; Klafter, J., Lim, S.C., Metzler, R., Eds.; World Scientific: Singapore, 2012; Chapter 15; pp. 353–374.
43. Uchaikin, V.V. *Fractional Derivatives for Physicists and Engineers, Vol. I, Background and Theory*; Springer: Heidelberg, Germany, 2013.
44. Podlubny, I. *Fractional Differential Equations*; Academic Press: San Diego, CA, USA, 1999.
45. Khintchine, A.Y. *Mathematical Methods in the Theory of Queuing*; Charles Griffin: London, UK, 1960.
46. Rogosin, S.V.; Mainardi, F. *The Legacy of A.Ya. Khintchine's Work in Probability Theory*; Cambridge Scientific Publ.: Cambridge, UK, 2011; Available online: http://www.cambridgescientificpublishers.com/ (accessed on 3 August 2015).
47. Feller, W. *An Introduction to Probability Theory and its Applications*; Wiley: New York, NY, USA, 1971; Volume II.
48. Sato, K.-I. *Lévy Processes and Infinitely Divisible Distributions*; Cambridge University Press: Cambridge, UK, 1999.
49. Mainardi, F.; Luchko, Y.; Pagnini, G. The fundamental solution of the space-time fractional diffusion equation. *Fract. Calc. Appl. Anal.* **2001**, *4*, 153–192.
50. Mainardi, F. *Fractional Calculus and Waves in Linear Viscoelasticity*; Imperial College Press: London, UK, 2010.
51. Gorenflo, R.; Abdel-Rehim, E. From power laws to fractional diffusion: The direct way. *Vietnam J. Math.* **2004**, *32*, 65–75.
52. Gorenflo, R.; Mainardi, F. Fractional calculus: Integral and differential equations of fractional order. In *Fractals and Fractional Calculus in Continuum Mechanics*; Carpinteri, A., Mainardi, F., Eds.; Springer Verlag: Wien, Austria, 1997; pp. 223–276. Available online: http://arxiv.org/abs/0805.3823 (accessed on 3 August 2015).
53. Gorenflo, R.; Mainardi, F.; Scalas, E.; Raberto, M. Fractional calculus and continuous-time finance III: The diffusion limit. In *Mathematical Finance*; Kohlmann, M., Tang, S., Eds.; Birkhäuser Verlag: Basel, Swizerland, 2001; pp. 171–180.
54. Mainardi, F.; Mura, A.; Pagnini, G. The *M*-Wright function in time-fractional diffusion processes: A tutorial survey. *Int. J. Diff. Equ.* **2010**, *2010*, 104505:1–104505:29. Available online: http://arxiv.org/abs/1004.2950 (accessed on 3 August 2015). [CrossRef]
55. Scalas, E.; Gorenflo, R.; Mainardi, F. Fractional calculus and continuous-time finance. *Phys. A* **2000**, *284*, 376–384. Available online: http://arxiv.org/abs/cond-mat/0001120 (accessed on 3 August 2015). [CrossRef]
56. Scalas, E.; Gorenflo, R.; Mainardi, F. Uncoupled continuous-time random walks: Solution and limiting behavior of the master equation. *Phys. Rev. E* **2004**, *69*, 011107:1–011107:8. [CrossRef] [PubMed]

57. Gorenflo, R.; Kilbas, A.A.; Mainardi, F.; Rogosin, S.V. *Mittag-Leffler Functions, Related Topics and Applications*; Springer: Heidelberg, Germany, 2014.
58. Kilbas, A.A.; Saigo, M. *H-Transform: Theory and Applications*; Chapman and Hall/CRC: New York, NY, USA, 2004.
59. Kilbas, A.A.; Srivastava, H.M.; Trujillo, J.J. *Theory and Applications of Fractional Differential Equations*; Elsevier: Amsterdam, The Netherlands, 2006.
60. Mathai, A.M.; Haubold, H.J. *Special Functions for Applied Scientists*; Springer: New York, NY, USA, 2008.
61. Mathai, A.M.; Saxena, R.K.; Haubold, H.J. *The H-function, Theory and Applications*; Springer Verlag: New York, NY, USA, 2010.
62. Diethelm, K. *The Analysis of Fractional Differential Equations. An Application Oriented Exposition Using Differential Operators of Caputo Type*; Springer: Berlin, Germany, 2010.
63. Atanacković, T.M.; Pilipović, S.; Stanković, B.; Zorica, D. *Fractional Calculus with Applications in Mechanics: Vibrations and Diffusion Processes*; ISTE Ltd, London and John Wiley: Hoboken, NJ, USA, 2014.
64. Haubold, H.J.; Mathai, A.M.; Saxena, R.K. Mittag-Leffler functions and their applications. *J. Appl. Math.* **2011**, *2011*, 298628:1–298628:51. [CrossRef]
65. Abramowitz, M.; Stegun, I.A. *Handbook of Mathematical Functions*; Dover: New York, NY, USA, 1965.

axioms

MDPI

Review

An Overview of the Pathway Idea and Its Applications in Statistical and Physical Sciences

Nicy Sebastian [1,*], Seema S. Nair [2] and Dhannya P. Joseph [3]

[1] St.Thomas' College, Thrissur P.O., Kerala 680001, India
[2] St. Gregorios College, Pulamon P.O., Kottarakara, Kerala 691531, India; E-Mail: seema.cms@gmail.com
[3] Kuriakose Elias College, Mannanam P.O., Kottayam, Kerala 686561, India; E-Mail: dhannyapj@gmail.com
* E-Mail: nicyseb@yahoo.com; Tel.: +91-949-525-4050.

Academic Editor: Hans J. Haubold
Received: 18 October 2015 / Accepted: 7 December 2015 / Published: 19 December 2015

Abstract: Pathway idea is a switching mechanism by which one can go from one functional form to another, and to yet another. It is shown that through a parameter α, called the pathway parameter, one can connect generalized type-1 beta family of densities, generalized type-2 beta family of densities, and generalized gamma family of densities, in the scalar as well as the matrix cases, also in the real and complex domains. It is shown that when the model is applied to physical situations then the current hot topics of Tsallis statistics and superstatistics in statistical mechanics become special cases of the pathway model, and the model is capable of capturing many stable situations as well as the unstable or chaotic neighborhoods of the stable situations and transitional stages. The pathway model is shown to be connected to generalized information measures or entropies, power law, likelihood ratio criterion or $\lambda-$criterion in multivariate statistical analysis, generalized Dirichlet densities, fractional calculus, Mittag-Leffler stochastic process, Krätzel integral in applied analysis, and many other topics in different disciplines. The pathway model enables one to extend the current results on quadratic and bilinear forms, when the samples come from Gaussian populations, to wider classes of populations.

Keywords: pathway model; entropy measure; superstatistics; Tsallis statistics; beta family; generalized gamma; Dirichlet densities; λ-criterion; H- function; quadratic forms

MSC: 85A99; 82B31; 60E05; 62C10; 33C60; 44A15

---

1. Introduction

The pathway idea was originally prepared by Mathai in the 1970's in connection with population models, and later rephrased and extended, Mathai [1], to cover scalar as well as matrix cases as made suitable for modelling data from statistical and physical situations. For practical purposes of analyzing data of physical experiments and in building up models in statistics, we frequently select a member from a parametric family of distributions. But it is often found that the model requires a distribution with a thicker or thinner tail than the ones available from the parametric family, or a situation of right tail cut-off. The experimental data reveal that the underlying distribution is in between two parametric families of distributions. In order to create a pathway from one functional form to another, a pathway parameter is introduced and a pathway model is created in Mathai [1]. The main idea behind the derivation of this model is the switching properties of going from one family of functions to another and yet another family of functions. The model enables one to proceed from a generalized type-1 beta model to a generalized type-2 beta model to a generalized gamma model when the variable is restricted to be positive. Thus the pathway parameter α takes one to three different functional forms. This is the distributional pathway. More families are available when the variable is allowed to vary over the real line. Mathai [1] deals mainly with rectangular matrix-variate distributions and the scalar case is a particular case there. For the real scalar case the pathway model is the following:

$$f_1(x) = c_1 x^{\gamma-1}[1 - a(1-\alpha)x^{\delta}]^{\frac{\eta}{1-\alpha}} \tag{1}$$

$a > 0,\ \delta > 0,\ 1 - a(1-\alpha)x^{\delta} > 0,\ \gamma > 0,\ \eta > 0$ where $c_1 = \frac{\delta(a(1-\alpha))^{\frac{\gamma}{\delta}}\Gamma(\frac{\eta}{1-\alpha}+1+\frac{\gamma}{\delta})}{\Gamma(\frac{\gamma}{\delta})\Gamma(\frac{\eta}{1-\alpha}+1)}$ for $\alpha < 1$, is the normalizing constant if a statistical density is needed and α is the pathway parameter. For $\alpha < 1$ the model remains as a generalized type-1 beta model in the real case. Other cases available are the regular type-1 beta density, Pareto density, power function, triangular and related models. Observe that Equation (1) is a model with the right tail cut off. When $\alpha > 1$ we may write $1 - \alpha = -(\alpha - 1),\ \alpha > 1$ so that $f(x)$ assumes the form,

$$f_2(x) = c_2 x^{\gamma-1}[1 + a(\alpha-1)x^{\delta}]^{-\frac{\eta}{\alpha-1}},\ x > 0 \tag{2}$$

which is a generalized type-2 beta model for real x and $c_2 = \frac{\delta(a(\alpha-1))^{\frac{\gamma}{\delta}}\Gamma(\frac{\eta}{\alpha-1})}{\Gamma(\frac{\gamma}{\delta})\Gamma(\frac{\eta}{\alpha-1}-\frac{\gamma}{\delta})}$ for $\alpha > 1$, is the normalizing constant, if a statistical density is required. Beck and Cohen's superstatistics belong to this case Equation (2) (for more details see, [2,3]). Again, dozens of published papers are available on the topic of superstatistics in statistical mechanics. For $\gamma = 1,\ a = 1, \delta = 1$ we have Tsallis statistics for $\alpha > 1$ from Equation (2) (for more details [4,5]). Other standard distributions coming from this model are the regular type-2 beta, the F-distribution, Lévi models and related models. When $\alpha \to 1$, the forms in Equations (1) and (2) reduce to

$$f(x) = cx^{\gamma-1}e^{-bx^{\delta}},\ x > 0,\ b = a\eta \tag{3}$$

where $c = \frac{\delta b^{\frac{\gamma}{\delta}}}{\Gamma(\frac{\gamma}{\delta})}$, is the normalizing constant. This includes generalized gamma, gamma, exponential, chisquare, Weibull, Maxwell-Boltzmann, Rayleigh, and related models (for more details see, [6,7]). If x is replaced by $|x|$ in Equation (1) then more families of distributions are covered in Equation (1). The behavior of the pathway model for various values of the pathway parameter α can be seen from the following figures.

From the Figure 1(Left) we can see that, as α moves away from 1 the function $f_2(x)$ moves away from the origin and it becomes thicker tailed and less peaked. From the path created by α we note that we obtain densities with thicker or thinner tail compared to generalized gamma density. From Figure 1 (Right) we can see that when α moves from $-\infty$ to 1, the curve becomes thicker tailed and less peaked, see ([8,9]). Note that α is the most important parameter here for enabling one to more than one family of functions. The other parameters are the usual parameters within each family of functions. The following is a list of some particular cases and the transformations are listed to go from the extended versions to the regular cases.

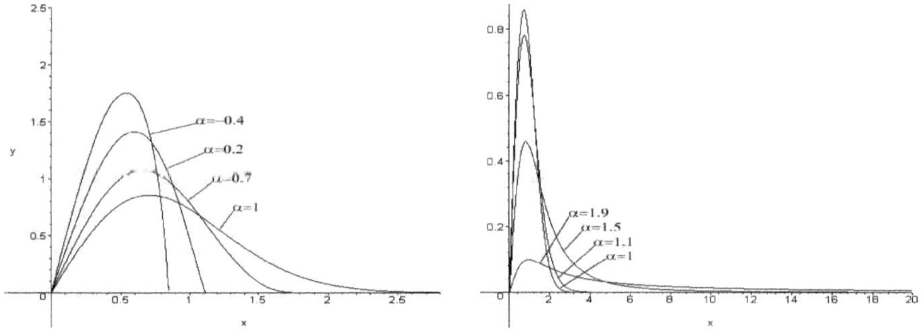

Figure 1. (**Left**) The graph of $f_1(x)$, for $\gamma = \eta = a = 1$, $\delta = 2$ and for various values of α. (**Right**) The graph of $f_2(x)$, for $\gamma = \eta = a = 1, \delta = 2$ and for various values of α.

$\alpha = 1,\ \gamma = 1,\ a = 1,\ \delta = 1$	Gaussian or normal density for $\infty < x < \infty$
$\alpha = 1, \gamma - 1 = \frac{3}{4},\ a = 1,\ \delta = 1$	Maxwell-Boltzmann density
$\alpha = 1,\ \gamma - 1 = \frac{1}{2},\ a = 1,\ \delta = 1$	Rayleigh density
$\alpha = 1,\ \gamma = \frac{\eta}{2},\ a = 1,\ \delta = 1$	Hermert density
$\alpha = 0,\ \gamma = 1,\ \eta = 1,\ \delta = 1$	U-shaped density
$\alpha = 2,\ \gamma = 1, \eta = \frac{v+1}{2}\ a = \frac{1}{v},\ \delta = 1$	Student-t for v degrees of freedom, $-\infty < x < \infty$
$\alpha = 2,\ \eta = 1,\ a = 1,\ \delta = 1$	Caushy density for $-\infty < x < \infty$
$\alpha < 1,\ a(1 - \alpha) = 1,\ \delta = 1$	Standard type-1 beta density

$\alpha > 1,\ a(1-\alpha) = 1,\ \delta = 1$	Standard type-2 beta density
$\gamma - 1 = \frac{1}{2},\ \eta = 1,\ a = 1,\ \delta = 1$	Tsallis statistics in Astrophysics, Power law, q-binomial density
$\alpha = 0,\ \gamma - 1 = \frac{1}{2},\ \eta = 1,\ \delta = 1$	Triangular density
$\alpha = 2,\ \gamma - \frac{1}{2} = \frac{m}{2}, a = \frac{m}{n}\ \eta = \frac{m+1}{2},\ \delta = 1$	F-density
$\alpha = 1,\ \gamma - 1 = \frac{1}{2},\ a = 1,\ \eta = \frac{mg}{KT},\ \delta = 1$	Helley's density in physics
$\alpha = 1,\ a = 1,\ \delta = 1$	Gamma density
$\alpha = 1,\ a = 1,\ \gamma - \frac{1}{2} = \frac{\nu}{2},\ \eta = \frac{1}{2},\ \delta = 1$	Chisquare density for ν degrees of freedom
$\alpha = 1,\ a = 1\ \gamma - 1 = \frac{1}{2},\ \delta = 1$	Exponential density (Laplace density with $x = \|z\|,\ -\infty < z < \infty$)
$\alpha = 1,\ a = 1$	Generalized gamma density
$\alpha = 1,\ a = 1,\ \gamma - 1 = \frac{1}{2}$	Weibull density
$\alpha = 2,\ a = 1,\ \gamma - 1 = \frac{1}{2},\ \eta = 2,\ \delta = 1,\ x = e^{y}$	Logistic density for $-\infty < y < \infty$
$\alpha = 2,\ a = 1,\ \gamma = 1,\ \eta = 1,\ \delta = 1,\ x = e^{\epsilon + \mu y}, \epsilon \neq 0,\ \mu > 0,$	Fermi-Dirac density, $0 \leq y < \infty$

1.1. Pathway Model from Mathai's Entropy Measure

In physical situations when an appropriate density is selected, one procedure is the maximization of entropy. Mathai and Rathie [10] consider various generalizations of Shannon entropy measure and describe various properties including additivity, characterization theorem *etc.* Mathai and Haubold ([11]) introduced a new generalized entropy measure which is a generalization of the Shannon entropy measure. For a multinomial population $P = (p_1, \ldots, p_k),\ p_i > 0,\ i = 1, \ldots, k,\ p_1 + p_2 + \cdots + p_k = 1$, the Mathai's entropy measure is given by the relation

$$M_{k,\alpha}(P) = \frac{\sum_{i=1}^{k} p_i^{2-\alpha} - 1}{\alpha - 1},\ \alpha \neq 1,\ -\infty < \alpha < 2. \quad \text{(discrete case)}$$

$$M_\alpha(f) = \frac{1}{\alpha - 1}\left[\int_{-\infty}^{\infty} [f(x)]^{2-\alpha} \mathrm{d}x - 1\right],\ \alpha \neq 1,\ \alpha < 2 \quad \text{(continuous case)}$$

By optimizing Mathai's entropy measure, one can arrive at pathway model of Mathai [1], which consists of many of the standard distributions in statistical literature as special cases. For fixed α, consider the optimization of $M_\alpha(f)$, which implies optimization of $\int_x [f(x)]^{2-\alpha}\mathrm{d}x$, subject to the following conditions:

(i) $f(x) \geq 0$, for all x
(i) $\int_x f(x)\mathrm{d}x < \infty$
(i) $\int_x x^{\rho(1-\alpha)} f(x)\mathrm{d}x =$ fixed for all f
(i) $\int_x x^{\rho(1-\alpha)+\delta} f(x)\mathrm{d}x =$ fixed for all f, where ρ and δ are fixed parameters

By using calculus of variation, one can obtain the Euler equation as

$$\frac{\partial}{\partial f}[f^{2-\alpha} - \lambda_1 x^{\rho(1-\alpha)} f + \lambda_2 x^{\rho(1-\alpha)+\delta} f] = 0$$

$$\Rightarrow (2-\alpha) f^{1-\alpha} = \lambda_1 x^{\rho(1-\alpha)}[1 - \frac{\lambda_2}{\lambda_1} x^{\delta}],\ \alpha \neq 1,2$$

$$\Rightarrow f_1 = c_1 x^{\rho}[1 - a(1-\alpha)x^{\delta}]^{\frac{1}{1-\alpha}} \tag{4}$$

for $\frac{\lambda_2}{\lambda_1} = a(1-\alpha)$ for some $a > 0$. For more details see ([12,13]).

When $\alpha \to 1$, the Mathai's entropy measure $M_\alpha(f)$ goes to the Shannon entropy measure and this is a variant of Havrda-Charvát entropy, and the variant form therein is Tsallis entropy. Then when α increases from 1, $M_\alpha(f)$ moves away from Shannon entropy. Thus α creates a pathway moving from one function to another, through the generalized entropy also. This is the entropic pathway. One can derive Tsallis statistics and superstatistics (for more details see, [2–5]) by using Mathai's entropy.

The pathway parameter α offers the differential pathway also. Let us consider

$$\begin{aligned} g(x) &= \frac{f_1(x)}{c_1} = x^{\beta}[1 - a(1-\alpha)x^{\delta}]^{\frac{1}{1-\alpha}},\ x > 0, \beta = \gamma - 1, \eta = 1 \\ \frac{d}{dx} g(x) &= \frac{\beta}{x} g(x) - a\delta x^{\delta - 1 + (1-\alpha)\gamma}[g(x)]^{\alpha} \\ &= -a[g(x)]^{\alpha} \text{ for } \beta = 0,\ \delta = 1 \end{aligned} \tag{5}$$

This is the power law. When $\eta = 1$ then the differential equation satisfied by $g_3 = \frac{f}{c}$ of Equation (3) is given by

$$\begin{aligned} \frac{d}{dx} g_3(x) &= \frac{\beta}{x} - a\delta x^{\delta-1} g_3(x) \\ &= -a[g_3(x)] \text{ for } \beta = \gamma - 1 = 0,\ \delta = 1 \end{aligned} \tag{6}$$

Thus when α moves to 1 the differential pathway is from the power law in Equation (5) to the maxwell-Boltzmann in Equation (6).

1.2. Laplacian Density and Stochastic Processes

The real scalar case of the pathway model in Equation (1), when x is replaced by $|x|$ and $\alpha \to 1$, takes the form

$$f_3(x) = c_3 |x|^{\gamma-1} e^{-a|x|^{\delta}}, -\infty < x < \infty, a > 0 \tag{7}$$

The density in Equation (7) for $\gamma = 1, \delta = 1$ is the simple Laplace density. For $\gamma = 1$ we have the symmetric Laplace density. A general Laplace density is associated with the concept of Laplacianness of quadratic and bilinear forms. For the concept of Laplacianness of bilinear forms, corresponding to the chisquaredness of quadratic forms, and for other details see [14,15]. Laplace density is also connected to input-output type models. Such models can describe many of the phenomena in nature. When two particles react with each other and energy is produced, part of it may be consumed or converted or lost and what is usually measured is the residual effect. The water storage in a dam is the residual effect of the water flowing into the dam minus the amount taken out of the dam. Grain storage in a sylo is the input minus the grain taken out. It is shown in Mathai ([16]) that when we have gamma type input and gamma type output the residual part $z = x - y$, $x =$ input variable, $y =$ output variable, then the special cases of the density of z is a Laplace density. In this case one can also obtain the asymmetric Laplace and generalized Laplace densities, which are currently used very frequently in stochastic processes, as special cases of the input-output model. Some aspects of the matrix version of the input-output model is also described in [16].

1.3. Mittag-Leffler Density and Processes

Recently there is renewed interest in Mittag-Leffler function as a model in many applied areas due to many reasons, one being that it gives a thicker tail compared to the exponential model. Fractional differential equations often lead to Mittag-Leffler functions and their generalizations as solutions, especially when dealing with fractional equations in reaction-diffusion problems. A large number of such situations are illustrated in [13,17,18]. The Mittag-Leffler density, associated with a 3-parameter Mittag-Leffler function is the following:

$$f(x) = \frac{x^{\alpha\beta-1}}{\delta^{\beta}}\sum_{k=0}^{\infty}\frac{(\beta)_k}{k!}\frac{(-x^{\alpha})^k}{\delta^k\Gamma(\alpha k+\alpha\beta)}, 0 \le x < \infty, \delta > 0, \beta > 0 \tag{8}$$

It has the Laplace transform

$$L_f(t) = [1+\delta t^{\alpha}]^{-\beta},\ 1+\delta t^{\alpha} > 0 \tag{9}$$

If δ is replaced by $\delta(q-1)$ and β by $\frac{\beta}{(q-1)}, q > 1$ and if we consider q approaching to 1 then we have

$$\lim_{q\to 1} L_f(t) = \lim_{q\to 1}[1+\delta(q-1)t^{\alpha}]^{-\frac{\beta}{q-1}} = e^{-\delta\beta t^{\alpha}} \tag{10}$$

But this is the Laplace transform of a constant multiple of a positive Lévy variable with parameter $\alpha, 0 < \alpha \le 1$, with the multiplicative constant being $(\delta\beta)^{\frac{1}{\alpha}}$, and thus the limiting form of a Mittag-Leffler distribution is a Lévy distribution. A connection of pathway model to Mittag-Leffler function is given in [19,20]. There is vast literature on Mittag-Leffler stochastic processes, see for example [21,22].

1.4. Laplace Transform of the Pathway Model

Let $L_{f_2}(t)$ be the Laplace transform of the pathway model $f_2(x)$ of Euqation (2). That is

$$L_{f_2}(t) = c_2\int_0^{\infty} e^{-tx}x^{\gamma-1}[1+a(\alpha-1)x^{\delta}]^{-\frac{\eta}{\alpha-1}}dx,\ a>0,\ b>0,\ \delta>0,\ \eta>0, \alpha>1 \tag{11}$$

Here the integrand can be taken as a product of positive integrable functions and then we can apply Mellin transform and inverse Mellin transform technique to evaluate the above integral. The integral in Equation (11) can be looked upon as the Mellin convolution of exponential density and superstatistics. Let us transform x_1 and x_2 to $u = \frac{x_1}{x_2}$ and $v = x_2$, then we can see that the marginal density of u obtained is actually the Laplace transform that we want to evaluate. Since the density is unique, in whatever way we evaluate the density we should get the same function or the functions will be identical. We can evaluate the density of u by the method of inverse Mellin transform, see [9]. Comparing the density obtained in these two different methods, we will get the Laplace transform of the pathway model given in Equation (2) as an H-function

$$L_{f_2}(t) = \frac{1}{\Gamma(\frac{\gamma}{\delta})\Gamma(\frac{\eta}{\alpha-1}-\frac{\gamma}{\delta})}H_{1,2}^{2,1}\left[\frac{t}{a^{\frac{1}{\delta}}(\alpha-1)^{\frac{1}{\delta}}}\Big|_{(0,1),(\frac{\eta}{\alpha-1}-\frac{1}{\delta}-\frac{1}{\delta},\frac{1}{\delta})}^{(1-\frac{\gamma}{\delta}\frac{1}{\delta})}\right], \tag{12}$$

for $\Re(\gamma) > 0$, $\Re(\frac{\eta}{\alpha-1}-\frac{\gamma}{\delta}) > 0$, $\alpha > 1$, where $H-$ function is defined as

$$H_{p,q}^{m,n}\left[z\Big|_{(b_1,\beta_1),\ldots,(b_q,\beta_q)}^{(a_1,\alpha_1),\ldots,(a_p,\alpha_p)}\right] = \frac{1}{2\pi i}\int_L \phi(s)\, z^{-s}ds \tag{13}$$

where

$$\phi(s) = \frac{\{\prod_{j=1}^{m}\Gamma(b_j+\beta_j s)\}\{\prod_{j=1}^{n}\Gamma(1-a_j-\alpha_j s)\}}{\{\prod_{j=m+1}^{q}\Gamma(1-b_j-\beta_j s)\}\{\prod_{j=n+1}^{p}\Gamma(a_j+\alpha_j s)\}}$$

where $\alpha_j,\ j=1,2,...,p$ and $\beta_j,\ j=1,2,...,q$ are real positive numbers, $a_j,\ j=1,2,...,p$ and b_j,j=1,2,...,q are complex numbers, L is a contour separating the poles of $\Gamma(b_j+\beta_j s),\ j=1,2,...,m$ from those of $\Gamma(1-a_j-\alpha_j s),\ j=1,2,...,n$. When $\alpha_1=1=\cdots=\alpha_p=\beta_1=1=\cdots=\beta_q$, then the $H-$function reduces to Meijer's $G-$ function, for more details see [18]. In a similar way we can evaluate the Laplace transform of the pathway model for $\alpha<1$ and is given by

$$\begin{aligned} L_{f_1}(t) &= c_1\int_0^{\left[\frac{1}{a(1-\alpha)}\right]^{\frac{1}{\delta}}} e^{-tx}x^{\gamma_1}[1-a(1-\alpha)x^{\delta}]^{\frac{\eta}{1-\alpha}}\mathrm{d}x \\ &= \frac{\Gamma(1+\frac{\eta}{1-\alpha}+\frac{\gamma}{\delta})}{\Gamma(\frac{\gamma}{\delta})}H_{1,2}^{1,1}\left[\frac{t}{a^{\frac{1}{\delta}}(1-\alpha)^{\frac{1}{\delta}}}\Big|_{(0,1),(-\frac{\eta}{1-\alpha}-\frac{\gamma}{\delta},\frac{1}{\delta})}^{(1-\frac{\gamma}{\delta},\frac{1}{\delta})}\right],\ \Re(\gamma)>0 \end{aligned} \tag{14}$$

THEOREM 1. *For $\Re(\gamma)>0,\ \Re(\frac{\eta}{\alpha-1}-\frac{\gamma}{\delta})>0,\ \delta>0,\ \eta>0,\ x>0,\ \alpha>1$, the Laplace transform (or the corresponding moment generating function) of the pathway model of the form $f_2(x)$, given in Euqation (12) goes to the Laplace transform (moment generating function) of the generalized gamma density given in Euqation (15) when $\alpha\to 1_+$.*

THEOREM 2. *For $\Re(\gamma)>0,\ \delta>0,\ \eta>0,\ x>0,\ \alpha<1$, the Laplace transform (or the corresponding moment generating function) of the pathway model of the form $f_2(x)$, given in Euqation (14) goes to the Laplace transform (moment generating function) of the generalized gamma density given in Euqation (15) when $\alpha\to 1_-$.*

The limiting case is given by

$$\begin{aligned} L_f(t) &= c\int_0^{\infty} e^{-tx}x^{\gamma}e^{-bx^{\delta}}\mathrm{d}x \\ &= \frac{1}{\Gamma(\frac{\gamma}{\delta})}H_{1,1}^{1,1}\left[\frac{t}{b^{\frac{1}{\delta}}}\Big|_{(0,1)}^{(1-\frac{\gamma}{\delta},\frac{1}{\delta})}\right] \end{aligned} \tag{15}$$

The Laplace transform of this density also provides the moment generating function of the extended gamma density thereby the moment generating functions of extended form of Weibull, chisquare, Reyleigh, Maxwell-Boltzmann, exponential and other densities in this general class. Usually we do not find the moment generating function or Laplace transform and the characteristic function of the generalized gamma density in the literature when $\delta\neq 1$. Here we obtained the Laplace transform of the generalized gamma density, besides giving an extension to this density.

1.5. Multivariate Generalizations

One generalization of the model in Euqation (1) for one scalar case is given by

$$\begin{aligned} f_1(x_1,x_2,\cdots,x_n) &= Kx_1^{\gamma_1-1}x_2^{\gamma_2-1}\cdots x_n^{\gamma_n-1} \\ &\quad\times[1-(1-\alpha)(a_1x_1^{\delta_1}+a_2x_2^{\delta_2}+\cdots+a_nx_n^{\delta_n})]^{\frac{\eta}{1-\alpha}} \end{aligned} \tag{16}$$

$\alpha<1,\ \eta>0,\ a_i>0,\ \delta_j>0,\ i=1,2,\cdots,n,\ 1-(1-\alpha)\sum_{i=1}^{n}a_ix_i^{\delta_i}>0$. We can see that Equation (16) is the Dirichlet family of densities. For $\alpha<1$, Equation (16) stays in the type-1 Dirichlet form and for $\alpha>1$ it stays as a type-2 Dirichlet form. This multivariate analogue can also produce multivariate extensions to Tsallis statistics and superstatistics. Here the variables are not independently distributed, but when $\alpha\to 1$ we have a surprising result that $x_1,\ x_2,\cdots,x_n$ will become independently distributed generalized gamma variables. Various generalizations of the pathway model are considered by Mathai and associates. The normalizing constants can be obtained by integrating out the variables one at a time, starting from x_n an going to x_1 (see [23]).

2. Connections to Astrophysics and Statistical Mechanics

2.1. Superstatistics Consideration and Pathway Model

Beck and Cohen [3] developed the concept of superstatistics in statistical mechanics. From a statistical point of view, the procedure is equivalent to starting with a conditional density for a random variable x at a given value of a parameter θ. Then assume the parameter θ has a prior density. Consider the conditional density of the form

$$f_{x|\theta}(x|\theta) = k_1 x^{\gamma} e^{-\theta x^{\delta}}, \ \gamma + 1 > 0, \ \theta > 0, \ \delta > 0, \ x > 0 \tag{17}$$

where $k_1 = \frac{\delta\theta^{\frac{\gamma+1}{\delta}}}{\Gamma(\frac{\gamma+1}{\delta})}$. Suppose that θ has an exponential density given by $f_\theta(\theta) = \lambda e^{-\lambda\theta}$, $\lambda > 0, \ \theta > 0$. Then the unconditional density of x is given by

$$\begin{aligned} f_x(x) &= \int_\theta f_{x|\theta}(x|\theta) f_\theta(\theta) d\theta \\ &= \frac{\delta\Gamma(\frac{\gamma+1}{\delta}+1)x^{\gamma}[1+\frac{x^\delta}{\lambda}]^{-(\frac{\gamma+1}{\delta}+1)}}{\lambda^{\frac{\gamma+1}{\delta}}\Gamma(\frac{\gamma+1}{\delta})} \end{aligned} \tag{18}$$

(see [2,3,11,18,24,25]). Equation (18) is the superstatistics of Beck and Cohen [3], in the sense of superimposing another distribution or the distribution of x with superimposed distribution of the parameter θ. In a physical problem the parameter θ may be something like temperature having its own distribution. Several physical interpretations of superstatistics are available from the papers of Beck and others. The factor $[1+\frac{x^\delta}{\lambda}]^{-(\frac{\gamma+1}{\delta}+1)}$ written as $[1+(\alpha-1)x^\delta]^{-\frac{1}{\alpha-1}}$, $\alpha > 1$ is the foundation for the current hot topic of Tsallis statistics in non-extensive statistical mechanics. Observe that only a form of the type $[1+\frac{x^\delta}{\lambda}]^{-(\frac{\gamma+1}{\delta}+1)}$ where $1+\frac{x^\delta}{\lambda} > 0, \ \lambda > 0, \ x^\delta > 0, \ \frac{\gamma+1}{\delta}+1 > 0$, that is, only a type-2 beta form can come from such a consideration. In other words a type-1 beta form cannot come because for the convergence of the integral in Euqation (18), $a + x^\delta$ must be positive with $\lambda > 0$ and $x^\delta > 0$. It is to be pointed out here that the superstatistics of Beck and Cohen [2] and Beck [3] are available from the procedure given above. It goes without saying that only type-2 beta form as given in Euqation (2) is available from superstatistics considerations, see [26]. The conditional density of the random variable θ, given x is the posterior probability density of θ and is given by

$$\begin{aligned} f_{\theta|x}(\theta|x) &= \frac{f_\theta(\theta) f_{x|\theta}(x|\theta)}{f_x(x)} \\ &= \frac{\lambda^{\frac{\gamma+1}{\delta}+1}[1+\frac{x^\delta}{\lambda}]^{\frac{\gamma+1}{\delta}+1}}{\Gamma(\frac{\gamma+1}{\delta}+1)} e^{-\theta(\lambda+x^\delta)}\theta^{\frac{\gamma+1}{\delta}}, \ \theta > 0 \end{aligned} \tag{19}$$

With the help of Euqation (19) we can obtain the Bayes' estimate of the parameter θ. To this extent, let

$$\begin{aligned} \Phi(x) &= E_{\theta|x}(\theta|x) = \int_0^\infty \theta f_{\theta|x}(\theta|x) d\theta \\ &= \frac{\frac{\gamma+1}{\delta}+1}{\lambda + x^\delta} \end{aligned} \tag{20}$$

Superstatistics and Tsallis statistics in statistical mechanics are given interpretations in terms of Bayesian statistical analysis. Subsequently superstatistics is extended by replacing each component of the conditional and marginal densities by Mathai's pathway model and further both components are replaced by Mathai's pathway model. This produces a wide class of mathematically and statistically interesting functions for prospective applications in statistical physics, see [19]. The same procedure can be used to look at the extended forms. Let the conditional density of x, given θ, be of the form

$$f_{\alpha_{x|\theta}}(x|\theta) = k_2 x^{\gamma}[1+\theta(\alpha-1)x^{\delta}]^{-\frac{1}{\alpha-1}},\ x>0,\ \theta>0,\ \alpha>1,\ \delta>0 \tag{21}$$

where $k_2 = \frac{\delta(\theta(\alpha-1))^{\frac{\gamma+1}{\delta}}\Gamma(\frac{1}{\alpha-1})}{\Gamma(\frac{\gamma+1}{\delta})\Gamma(\frac{1}{\alpha-1}-\frac{\gamma+1}{\delta})}$ and assume that the parameter θ has a prior density $f_{\theta}(\theta) = \lambda e^{-\lambda\theta},\ \lambda > 0,\ \theta > 0$. Then the unconditional density of x is given by

$$f_{\alpha_x}(x) = \frac{\lambda\delta x^{-(\delta+1)}}{(\alpha-1)\Gamma(\frac{\gamma+1}{\delta})\Gamma(\frac{1}{\alpha-1}-\frac{\gamma+1}{\delta})} G^{2,1}_{1,2}\left[\frac{\lambda}{x^{\delta}(\alpha-1)}\Bigg|^{-\frac{\gamma+1}{\delta}}_{0,\frac{1}{\alpha-1}-\frac{\gamma+1}{\delta}-1}\right],\ x>0 \tag{22}$$

where $G^{2,1}_{1,2}(\cdot)$ is a $G-$function. For the theory and application of the functions see Mathai [7]. The posterior probability density is given by

$$f_{\alpha_{\theta|x}}(\theta|x) = c_4^{-1}(\alpha-1)^{\frac{\gamma+1}{\delta}+1}x^{\gamma+\delta+1}\Gamma(\frac{1}{\alpha-1})\theta^{\frac{\gamma+1}{\delta}}e^{-\lambda\theta}[1+\theta(\alpha-1)x^{\delta}]^{-\frac{1}{\alpha-1}} \tag{23}$$

where $c_4 = G^{2,1}_{1,2}\left[\frac{\lambda}{x^{\delta}(\alpha-1)}\Bigg|^{-\frac{\gamma+1}{\delta}}_{0,\frac{1}{\alpha-1}-\frac{\gamma+1}{\delta}-1}\right]$. Then the Bayes' estimate of θ, at given x, defined by $\Phi_{\alpha}(x)$, is given by the following:

$$\Phi_{\alpha}(x) = E_{\alpha_{\theta|x}}(\theta|x) = \frac{c_4^{-1}}{(\alpha-1)x^{\delta}} G^{2,1}_{1,2}\left[\frac{\lambda}{x^{\delta}(\alpha-1)}\Bigg|^{-1-\frac{\gamma+1}{\delta}}_{0,\frac{1}{\alpha-1}-\frac{\gamma+1}{\delta}-2}\right] \tag{24}$$

LEMMA 1. *Let the conditional density of x given θ be $f_{x|\theta}(x|\theta)$ as in Equation (17) and assume that the parameter θ has a prior density $f_{\theta}(\theta) = \lambda e^{-\lambda\theta},\ \lambda > 0,\ \theta > 0$. Then the Bayes' estimate of θ is given by $\Phi(x)$ in Equation (20).*

THEOREM 3. *Let the conditional density of x given θ be $f_{\alpha_{x|\theta}}(x|\theta)$ as in Equation (21) and assume that the parameter θ has a prior density $f_{\theta}(\theta) = \lambda e^{-\lambda\theta},\ \lambda > 0,\ \theta > 0$. Then the Bayes' estimate of θ is given by $\Phi_{\alpha}(x)$ in Equation (24).*

Thus, the popular superstatistics in statistical mechanics can be considered as a special case of the pathway model in Equation (18) for $\eta = 1$ and $\delta = 1$.

2.2. *α-gamma Models Associated with Bessel Function*

Sebastian [27] deals with a new family of statistical distributions associated with Bessel function which gives an extension of the gamma density, which will connect the fractional calculus and statistical distribution theory through the special function. The idea is motivated by the fact that a non-central chi-square density is associated with a Bessel function. In order to make thicker or thinner tails in a gamma density we consider a density function of the following type:

$$f_{x|a}(x|a) = \frac{\rho\, a^{\frac{\gamma}{\rho}} e^{-\frac{\delta}{a}}}{\Gamma(\frac{\gamma}{\rho})} x^{\gamma-1}e^{-ax^{\rho}}{}_0F_1(\,;\frac{\gamma}{\rho};\delta x^{\rho});\ 0<x<\infty \tag{25}$$

where ${}_0F_1(\,;b;x) = \sum_{k=0}^{\infty}\frac{(x)^k}{(b)_k\, k!}$, $(b)_k$ is the Pochhammer symbol, $(b)_m = b(b+1)\cdots(b+m-1)$, $b \neq 0, (b)_0 = 1$. When $\delta = 0$ the Equation (25) reduces to generalized gamma density. Note that this is the generalization of some standard statistical densities such as gamma, Weibull, exponential, Maxwell-Boltzmann, Rayleigh and many more. When $\delta = 0, \rho = 2$, Equation (25) reduces to folded standard normal density. We can extend the generalized gamma model associated with Bessel function in Equation (25) by using the pathway model of Mathai [1], when $\alpha < 1$ we get the extended function as

$$g_\alpha(x) = k_1 x^{\gamma-1}[1-a(1-\alpha)x^\rho]^{\frac{1}{1-\alpha}} {}_0F_1(\,;\frac{\gamma}{\rho};\delta x^\rho);\ \alpha < 1,$$
$$a > 0, \rho > 0, 1-a(1-\alpha)x^\rho > 0, x > 0 \tag{26}$$

where k_1 is the normalizing constant. Note that $g_\alpha(x)$ is a generalized type-1 beta model associated with Bessel function. Observe that for $\alpha > 1$, writing $1-\alpha = -(\alpha-1)$ in Equation (25) produces extended type-2 beta form which is given by

$$f_\alpha(x) = k_2 x^{\gamma-1}[1+a(\alpha-1)x^\rho]^{-\frac{1}{\alpha-1}} {}_0F_1(\,;\frac{\gamma}{\rho};\delta x^\rho);\ \alpha > 1, a > 0, \rho > 0 \tag{27}$$

where k_2 is the normalizing constant. Note that in both the cases, when $\alpha \to 1$, we have Equation (25) and hence it can be considered to be an extended form of Equation (25). This model has wide potential applications in the discipline physical science especially in statistical mechanics, see [27,28]. For fixed values of $\gamma = 2, \rho = 1.2$ and $a = 1$, we can look at the graphs for $\delta = -0.5, q < 1$, $\delta = 0.5, q < 1$ as well as for $\delta = -0.5, q > 1$, $\delta = 0.5, q > 1$. From the Figures 2, we can see that when q moves from $-\infty$ to 1, the curve becomes less peaked. Similarly from Figure 3, we can see that when q moves from 1 to ∞, the curve becomes less peaked. It is also observed that when $\delta > 0$ the right tail of the density becomes thicker and thicker. Similarly when $\delta < 0$ the right tail gets thinner and thinner. Densities exhibiting thicker or thinner tail occur frequently in many different areas of science. For practical purposes of analyzing data from physical experiments and in building up models in statistics, we frequently select a member from a parametric family of distributions.

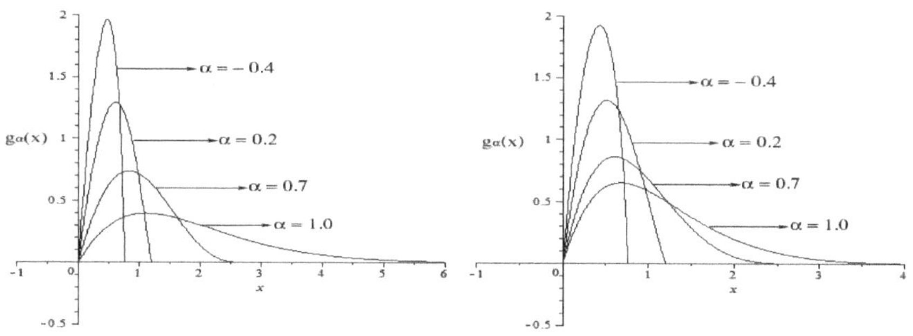

Figure 2. (**Left**) α gamma bessel model for $\delta = 0.5, \alpha < 1$; (**Right**) α gamma bessel model for $\delta = -0.5, \alpha < 1$.

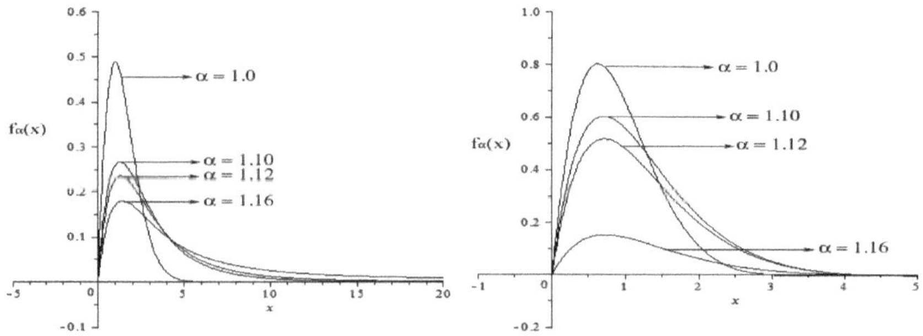

Figure 3. (**Left**) α gamma bessel model for $\delta = 0.5, \alpha > 1$; (**Right**) α gamma bessel model for $\delta = -0.5, \alpha > 1$.

2.3. *Tsallis Statistics*

Model Equation (1) for $\gamma = 1, \delta = 1, a = 1, \eta = 1$ is Tsallis statistics, which is the foundation for the newly created hot topic of non-extensive statistical mechanics. It is stated that over three thousand papers are written on Tsallis statistics so far. With α replaced in $-\alpha$ in Model Equation (1), with $\gamma = 0, a = 1, \eta = 1, \delta = 1$, one has an extension of the exponential function, known as q-exponential function [The parameter q is used instead of α and hence q-exponential]. The basis for the current hot topic of *q-calculus* is this q-exponential function. When $\gamma = 1, a = 1, \eta = 1, \delta = 1$ one has the following property if the resulting function of $f(x)$ is divided by the resulting normalizing factor c and obtain the function $g(x)$, that is, $g(x) = \frac{f(x)}{c}$. Then

$$\frac{\mathrm{d}g(x)}{\mathrm{d}x} = -[g(x)]^{\alpha}.$$

This is the power law in physics literature. Thus, power function law is a special case of the pathway model of Equations (1) and (2) for $\gamma = 1,\ \delta = 1,\ a = 1,\ \eta = 1$ and exactly the normalizing constants c_1 and c_2. Also Tsallis statistics can be looked upon as a special case of Equations (1) and (2) for $\gamma = 1,\ \delta = 1,\ a = 1,\ \eta = 1$.

2.4. *Extension of Thermonuclear Functions through Pathway Model*

Nuclear reactions govern major aspects of the chemical evolution of the universe. A proper understanding of the nuclear reactions that are going on in hot cosmic plasmas, and those in the laboratories as well, requires a sound theory of nuclear-reaction dynamics. The reaction probability integral is the probability per unit time that two particles, confined to a unit volume, will react with each other. Practically all applications of fusion plasmas are controlled in some way or other by the theory of thermonuclear reaction rates under specific circumstances. After several decades of effort, a systematic and complete theory of thermonuclear reaction rates has been developed [29–32].

The standard thermonuclear function in the Maxwell-Boltzmann case in the theory of nuclear reactions, is given by the following ([31,33]):
Nonresonant Case with Depleted Tail:

$$I_1 = \int_0^{\infty} x^{\gamma-1} e^{-ax^{\delta}-bx^{-\rho}} \mathrm{d}x,\ \ a > 0,\ b > 0,\ \delta > 0,\ \rho > 0 \tag{28}$$

Nonresonant case with depleted tail and high energy cut-off:

$$I_2 = \int_0^{d} x^{\gamma-1} e^{-ax^{\delta}-bx^{-\rho}} \mathrm{d}x,\ \ a > 0,\ b > 0,\ \delta > 0,\ \rho > 0,\ d < \infty \tag{29}$$

Note that if $\delta = 1$ is taken as the standard Maxwell-Boltzmannian behavior, then for $\delta > 1$ the right tail will deplete faster and if $\delta < 1$ then the depletion will be slower in Equation (29). We can extend the thermonuclear functions given in Equations (28) and (29) to more general classes by replacing $e^{-bx^{-\rho}}$ by a binomial factor $[1 - b(1-\alpha)x^{-\rho}]^{\frac{1}{1-\alpha}}$ ([13,34,35]). Thus we consider the general class of reaction rate integrals

$$I_{1\alpha} = \int_0^\infty x^{\gamma-1} e^{-ax^\delta}[1 + b(\alpha-1)x^{-\rho}]^{-\frac{1}{\alpha-1}} \mathrm{d}x,\ a > 0,\ b > 0,\ \delta > 0,\ \rho > 0,\ \alpha > 1 \tag{30}$$

$$I_{2\alpha} = \int_0^d x^{\gamma-1} e^{-ax^\delta}[1 - b(1-\alpha)x^{-\rho}]^{\frac{1}{1-\alpha}} \mathrm{d}x,\ a > 0,\ b > 0,\ \delta > 0,\ \rho > 0,\ \alpha < 1 \tag{31}$$

We can evaluate these extended integrals by using Mellin convolution property. Equation (30) can be looked upon as the Mellin convolution of two real positive scalar independently distributed random variables x_1 and x_2, where x_1 has a generalized gamma density and x_2 has an extended form of stretched exponential function. Make the transformation $u = x_1x_2$ and $v = x_1$, and then proceed as in the case of the evaluation of the Laplace transform of the pathway model, we can arrive at the value of the extended reaction rate integral as

$$\begin{aligned} I_{1\alpha} &= \int_0^\infty e^{-ax^\delta} x^{\gamma-1}[1 + b(\alpha-1)x^{-\rho}]^{-\frac{1}{\alpha-1}} \mathrm{d}x \\ &= \frac{1}{\rho\delta a^{\frac{\gamma}{\delta}}\Gamma(\frac{1}{\alpha-1})} H_{1,2}^{2,1}\left[a^{\frac{1}{\delta}}(b(\alpha-1))^{\frac{1}{\rho}}\Big|_{(0,\frac{1}{\rho}),(\frac{\gamma}{\delta},\frac{1}{\delta})}^{(1-\frac{1}{\alpha-1},\frac{1}{\rho})}\right],\ b = u^\rho \end{aligned} \tag{32}$$

Similarly we can evaluate Equation (31) by considering it as the Mellin convolution of exponential density and pathway model for $\alpha < 1$. When we take the limit as $\alpha \to 1$, in Equation (32), we will get the value of the reaction rate integral given in Equations (28) and (29), as an H- function.

2.4.1. Inverse Gaussian as a Particular Case of the Pathway Model

Note that one form of the inverse Gaussian probability density function is given by

$$g(x) = kx^{-\frac{3}{2}} e^{-\frac{\lambda}{2}(\frac{x}{\mu^2}+\frac{1}{x})},\ \mu \neq 0,\ x > 0,\ \lambda > 0$$

where k is the normalizing constant (see [7]). Put $\gamma = -\frac{5}{2}$, $\delta = 1$, $\rho = 1$, $a = \frac{\lambda}{2\mu^2}$, $b = \frac{\lambda}{2}$ in Equation (28), we can see that the inverse Gaussian density is the integrand in I_1. Hence I_1 in Equation (28) can be used to evaluate the moments of inverse Gaussian density. Hence the integrand in the extended integral $I_{1\alpha}$ can be considered as the extended form of the inverse Gaussian density.

2.4.2. An Interpretation of the Pathway Parameter α

Let us start with the pathway density in Equation (1) with $\eta = 1$. For this to remain a density we need the condition $1 - a(1-\alpha)x^\delta > 0$ or when x is positive then $0 < x < \frac{1}{[a(1-\alpha)]^{\frac{1}{\delta}}}$, $\alpha < 1$, or if we have the model in Equation (31) then

$$-\frac{1}{[a(1-\alpha)]^{\frac{1}{\delta}}} < x < \frac{1}{[a(1-\alpha)]^{\frac{1}{\delta}}},\ \alpha < 1$$

Outside this range, the density is zero. Thus for $x > 0$ the right tail of the density is cut-off at $\frac{1}{[a(1-\alpha)]^{\frac{1}{\delta}}}$. If d is this cut-off on the right then

$$d = \frac{1}{[a(1-\alpha)]^{\frac{1}{\delta}}} \Rightarrow \alpha = 1 - \frac{1}{ad^\delta},\ \text{for } \alpha < 1 \tag{33}$$

As α moves closer to 1 then the cut-off point d moves farther out and eventually when $\alpha \to 1$ then $d \to \infty$. In this case α is computed easily from the cut-off. Also the pathway parameter α can be estimated in terms of arbitrary moments, see for example [12].

3. Pathway Model and Fractional Calculus

When solving certain problems in reaction-diffusion, relaxation-oscillations, it is observed that the solution is obtained in terms of exponential function, see [33,36]. But if a fractional integration is under consideration then the residual reaction equation is given by

$$N(t) = N_0 - c^{\nu}{}_0D_t^{-\nu}N(t), \quad \nu > 0 \tag{34}$$

where ${}_0D_t^{-\nu}$ is the standard Riemann-Louville fractional integral operator, where $N(t)$ is the number density of the reacting particles and Equation (34) has the solution in terms of generalized Mittag-Leffler function. While fitting such a model, Mittag-leffler function will provide a better fit as compared to exponential function. If we consider the residual rate of change c, in Equation (34) is a random variable having a gamma type density

$$g(c) = \frac{w^{\mu}}{\Gamma(\mu)}c^{\mu-1}\mathrm{e}^{-wc}, \quad w > 0,\ 0 < c < \infty \tag{35}$$

where $\mu > 0,\ w,\ \mu$ are known and $\frac{\mu}{w}$ is the mean value of c. The residual rate of change may have small probabilities of it being too large or too small and the maximum probability may be for a medium range of values for the residual rate of change c. Equation (34) is the situation where the residual rate of change is such that the production rate dominates so that we have the form $-c^{\nu}, \nu > 0, c > 0$. If the destruction rate dominates then the constant will be of the form $c^{\nu},\ \nu > 0,\ c > 0$. By solving Equation (34), we have the unconditional number density of the following form:

$$N(t) = \frac{N_0}{\Gamma(\mu)}t^{\mu-1}(1+\frac{t^{\nu}}{w^{\nu}})^{-(\gamma+1)}, \quad 0 < t < \infty,\ w > 0 \tag{36}$$

If we make the substitution $\gamma + 1 = \frac{1}{\alpha-1},\ \alpha > 1 \Rightarrow \gamma = \frac{\alpha-2}{\alpha-1}$ and $w^{-\nu} = b(\alpha-1),\ b > 0$. Then we have

$$N(t) = \frac{N_0}{\Gamma(\mu)}t^{\mu-1}[1+b(\alpha-1)t^{\nu}]^{-\frac{1}{\alpha-1}} \tag{37}$$

for $\alpha > 1,\ t > 0,\ b > 0,\ \mu > 0$. For general values of μ and $\alpha > 1$ such that $\frac{1}{\alpha-1} - \frac{\mu}{\nu} > 0$, Euqation (37) corresponds to the pathway model of Mathai as well as the superstatistics of Beck and Cohen [3]. As an application of pathway model in fractional calculus, a general pathway fractional integral operator is introduced in [37], which generalizes the classical Riemann-Liouville fractional integration operator, see Section 4.5.

3.1. $\mathcal{P}$-Transform

Consider the generalized Krätzel function $D^{\nu,\alpha}_{\rho,\beta}(x)$, dealt with in [38], given by

$$D^{\nu,\alpha}_{\rho,\beta}(x) = \int_0^{\infty} y^{\nu-1}[1+a(\alpha-1)y^{\rho}]^{-\frac{1}{\alpha-1}}\mathrm{e}^{-xy^{-\beta}}\mathrm{d}y,\ x > 0 \tag{38}$$

with $\gamma \in \mathbb{C}, \beta > 0, \alpha > 1$. The generalized Krätzel function is obtained by using the pathway model introduced by Mathai [1]. The $\mathcal{P}$-transform or pathway transform introduced in [39] by using the pathway idea is defined as

$$(\mathcal{P}^{\rho,\beta,\alpha}_{\nu}f)(x) = \int_0^{\infty} D^{\nu,\alpha}_{\rho,\beta}(xt)f(t)\mathrm{d}t, x > 0 \tag{39}$$

where $D^{\nu,\alpha}_{\rho,\beta}(x)$ denotes the function given in Equation (38). The $\mathcal{P}$-transform is defined in the space $\mathbb{L}_{\nu,r}$ consisting of the lebesgue measurable complex valued functions f for which

$$\|f\|_{\nu,r} = \left\{ \int_0^\infty |t^\nu f(t)|^r \frac{\mathrm{d}t}{t} \right\}^{\frac{1}{r}} < \infty, \tag{40}$$

for $1 \leq r < \infty, \nu \in \mathbb{R}$. When $\beta = 1, a = 1$ and $\alpha \to 1$ the $\mathcal{P}$-transform reduces to the Krätzel tranform, introduced by Krätzel [40], which is given by

$$\mathcal{K}_\nu^{(\rho)} f(x) = \int_0^\infty Z_\rho^\nu(xt) f(t) \mathrm{d}t, \; x > 0 \tag{41}$$

where

$$Z_\rho^\nu(x) = \int_0^\infty y^{\nu-1} \mathrm{e}^{-y^\rho - xy^{-1}} \mathrm{d}y \tag{42}$$

The $\mathcal{P}$-transform reduces to the Meijer transform when $\rho = 1$ and $\alpha \to 1$. When $a = 1, \beta = 1$ and $\alpha \to 1$, then the generalized Krätzel function defined in Equation (38) reduces to the modified Bessel function of the third kind or McDonald function ([38,41,42]) have considered Equation (38) for $\beta = 1$ and established its composition with fractional operators and represented it in terms of various generalized special functions. The Krätzel function defined in Equation (42) for any real ρ, was studied by [43] and established its representations in terms of H-function and extended the function from positive $x > 0$ to complex z. Here we establishe connection between generalized Krätzel function and $\mathcal{P}$-transform with generalized special functions.

The particular case of the kernel of the $\mathcal{P}$-transform given in Equation (38) is the extended non-resonant thermonuclear function used in Astrophysics which is already discussed. As $\alpha \to 1$ we get the standard reaction rate probability integral in the Maxwell-Boltzmann case. The behavior of the generalized Krätzel function Equation (38) can be studied from the following graphs. We take $a = 1, \beta = 1, \nu = 2, \rho = 3$ and $a = 1, \beta = 1, \nu = 2, \rho = 5$ for example and investigate the behavior of $D^{2,\alpha}_{3,1}(x)$ and $D^{2,\alpha}_{5,1}(x)$ for various values of $\alpha > 1$ and $\alpha \to 1$. The graphs of these functions $D^{2,\alpha}_{3,1}(x)$ and $D^{2,\alpha}_{5,1}(x)$ at $\alpha = 1, \alpha = 1.25, \; \alpha = 1.5, \; \alpha = 1.8$ are given in Figure 4a,b, respectively. From these graphs we observe that if the value of the parameter α increases then the curves move away from the limiting situation for $\alpha \to 1$.

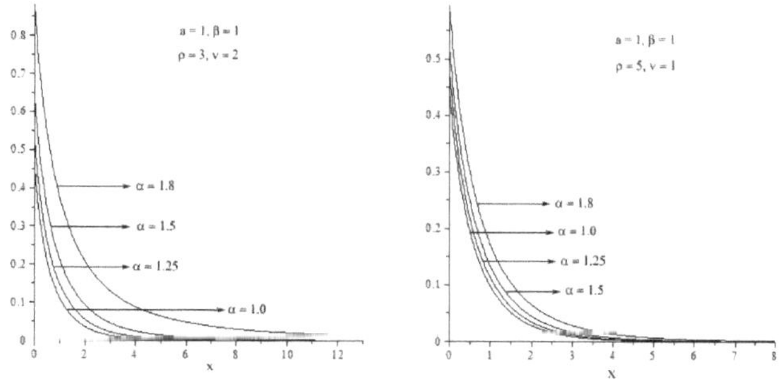

Figure 4. (**Left**) Behaviour of $D^{2,\alpha}_{3,1}(x)$ for various values of $\alpha > 1$ and $\alpha \to 1$; (**Right**) Behaviour of $D^{2,\alpha}_{5,1}(x)$ for various values of $\alpha > 1$ and $\alpha \to 1$.

4. A Matrix Variate Pathway Model

Let $X = (x_{ij})$, $i = 1, \ldots, p$, $j = 1, \ldots, q$, $q \geq p$, of rank p and of real scalar variables $x_{ij}'s$ for all i and j, subject to the condition that the rank of X is p, having the density $f(X)$, where $f(X)$ is a scalar function of X is given by

$$f(X) = C|A^{\frac{1}{2}}(X-M)B(X-M)'A^{\frac{1}{2}}|^{\gamma}|I - a(1-\alpha)A^{\frac{1}{2}}(X-M)B(X-M)'A^{\frac{1}{2}}|^{\frac{\eta}{1-\alpha}} \quad (43)$$

where C is the normalizing constant, $A = A' > 0, B = B' > 0, A$ is $p \times p, B$ is $q \times q$, A and B are constant positive definite matrices, and M is a $p \times q$ constant matrix. $A^{\frac{1}{2}}$ denotes the positive definite square root of A, $a > 0$ and α is the pathway parameter. For keeping non-negativity of the determinant in Equation (43) we need the condition

$$I - a(1-\alpha)A^{\frac{1}{2}}(X-M)B(X-M)'A^{\frac{1}{2}} > 0$$

where, for example, the notation $A < Y < B \Rightarrow A = A' > 0, B = B' > 0, Y = Y' > 0, Y - A > 0, B - Y > 0$. In Equation (43) the constant matrix M can act as a relocation matrix or as the mean value matrix so that $E(X) = M$ where E denotes the expected value.

4.1. The Normalizing Constants

This requires certain transformations and the knowledge of the corresponding Jacobians. Here we will use a few multi-linear and some nonlinear transformations and the corresponding Jacobians. The details of the derivations of these Jacobians are available from [44]. Put

$$Y = A^{\frac{1}{2}}(X-M)B^{\frac{1}{2}} \Rightarrow \mathrm{d}Y = |A|^{\frac{q}{2}}|B|^{\frac{p}{2}}\mathrm{d}X$$

where $|(\cdot)|$ denotes the determinant of $(\cdot)$, $\mathrm{d}X$ is defined as the following wedge product of differentials:

$$\mathrm{d}X = \mathrm{d}x_{11} \wedge \mathrm{d}x_{12} \wedge \cdots \wedge \mathrm{d}x_{1q} \wedge \mathrm{d}x_{22} \wedge \cdots \wedge \mathrm{d}x_{2q} \wedge \cdots \wedge \mathrm{d}x_{pq}$$

see [44] for the Jacobian. Since $f(X)$ is assumed to be a density,

$$1 = \int_X f(X)\mathrm{d}X = C|A|^{-\frac{q}{2}}|B|^{-\frac{p}{2}} \int_Y |YY'|^{\gamma}|I - a(1-\alpha)YY'|^{\frac{\eta}{1-\alpha}}\mathrm{d}Y$$

Now, make suitable transformations and integrating out over the Stiefel manifold (for details, see [1]) we have

$$\begin{aligned} 1 &= \int_X f(X)\mathrm{d}X \\ &= C|A|^{-\frac{q}{2}}|B|^{-\frac{p}{2}} \frac{\pi^{\frac{pq}{2}}}{\Gamma_p(\frac{q}{2})} \int_Z |Z|^{\gamma+\frac{q}{2}-\frac{p+1}{2}}|I - a(1-\alpha)Z|^{\frac{\eta}{1-\alpha}}\mathrm{d}Z \end{aligned} \quad (44)$$

where, for example,

$$\Gamma_p(\beta) = \pi^{\frac{p(p-1)}{4}}\Gamma(\beta)\Gamma(\beta - \frac{1}{2})...\Gamma(\beta - \frac{p-1}{2}), \Re(\beta) > \frac{p-1}{2} \quad (45)$$

is the real matrix-variate gamma and $\Re(\cdot)$ denotes the real part of $(\cdot)$.

Further evaluation of the integral in Equation (44) depends on the value of α. If $\alpha < 1$ then Equation (44) belongs to the real matrix-variate type-1 beta. Then integrating out with the help of a matrix variate type-1 beta integral we get, for $q \geq p, \alpha < 1$,

$$C = |A|^{\frac{q}{2}}|B|^{\frac{p}{2}}[a(1-\alpha)]^{p(\gamma+\frac{q}{2})}\frac{\Gamma_p(\frac{q}{2})\Gamma_p(\gamma+\frac{q}{2}+\frac{\eta}{1-\alpha}+\frac{p+1}{2})}{\pi^{\frac{pq}{2}}\Gamma_p(\gamma+\frac{q}{2})\Gamma_p(\frac{\eta}{1-\alpha}+\frac{p+1}{2})} \tag{46}$$

for $\alpha < 1, \eta > 0, a > 0, \Re(\gamma+\frac{q}{2}) > \frac{p-1}{2}, q \geq p, p, q = 1, 2, ..., A = A' > 0, B = B' > 0$. For $\alpha > 1$, write $1-\alpha = -(\alpha-1), \alpha > 1$ then the integral in Equation (44) goes into a type-2 beta form. Then integrating out with the help of a real matrix-variate type-2 beta integral we have for $\alpha > 1$,

$$C = \frac{|A|^{\frac{q}{2}}|B|^{\frac{p}{2}}[a(\alpha-1)]^{p(\gamma+\frac{q}{2})}\Gamma_p(\frac{q}{2})\Gamma_p(\frac{\eta}{\alpha-1})}{\pi^{\frac{pq}{2}}\Gamma_p(\gamma+\frac{q}{2})\Gamma_p(\frac{\eta}{\alpha-1}-\gamma-\frac{q}{2})} \tag{47}$$

for $\alpha > 1, \Re(\gamma+\frac{q}{2}) > \frac{p-1}{2}, \Re(\frac{\eta}{\alpha-1}-\gamma-\frac{q}{2}) > \frac{p-1}{2}, a > 0, \eta > 0, A = A' > 0, B = B' > 0$. When $\alpha \to 1$ we have

$$\lim_{\alpha\to 1}|I-a(1-\alpha)Z|^{\frac{\eta}{1-\alpha}} = e^{-a\eta\,\mathrm{tr}(Z)}$$

and hence, evaluating with the help of a real matrix-variate gamma integral, we will get C for $\alpha \to 1$,

$$C = \frac{|A|^{\frac{q}{2}}|B|^{\frac{p}{2}}(a\eta)^{p(\gamma+\frac{q}{2})}\Gamma_p(\frac{q}{2})}{\pi^{\frac{pq}{2}}\Gamma_p(\gamma+\frac{q}{2})} \tag{48}$$

for $\Re(\gamma+\frac{q}{2}) > \frac{p-1}{2}, a > 0, \eta > 0$.

A model corresponding to Equation (43) in the complex domain, along with its properties and connections to various other fields, is studied in [23].

4.2. *Density of the Volume Content*

The squared volume of the p-parallelotope in the q-space,

$$v^2 = |A^{\frac{1}{2}}(X-M)B(X-M)'A^{\frac{1}{2}}| \tag{49}$$

The density of $u = v^2$ in Equation (49) can be evaluated by looking at the h-th moment of v^2 for an arbitrary h. That is, the expected value of $(v^2)^h$, is given by

$$E(v^2)^h = \int_X (v^2)^h f_5(X)\mathrm{d}X$$

This is available from the normalizing constants in Equations (46)–(48) by observing that the only change is that the parameter γ is changed to $\gamma + h$. Hence from Equation (46) for $\alpha < 1$,

$$E(v^2)^h = [a(1-\alpha)]^{-ph}\frac{\Gamma_p(\gamma+\frac{q}{2}+h)}{\Gamma_p(\gamma+\frac{q}{2})}\frac{\Gamma_p(\gamma+\frac{q}{2}+\frac{\eta}{1-\alpha}+\frac{p+1}{2})}{\Gamma_p(\gamma+\frac{q}{2}+\frac{\eta}{1-\alpha}+\frac{p+1}{2}+h)} \tag{50}$$

for $\alpha < 1, \Re(\gamma+\frac{q}{2}) > \frac{p-1}{2}, \Re(\gamma+\frac{q}{2}+h) > \frac{p-1}{2}, \eta > 0, a > 0$. Let $u_1 = [a(1-\alpha)]^p v^2$. Then

$$\begin{aligned} E(u_1^h) &= \prod_{j=1}^{p}\frac{\Gamma(\gamma+\frac{q}{2}-\frac{j-1}{2}+h)}{\Gamma(\gamma+\frac{q}{2}-\frac{j-1}{2})}\frac{\Gamma(\gamma+\frac{q}{2}+\frac{\eta}{1-\alpha}+\frac{p+1}{2}-\frac{j-1}{2})}{\Gamma(\gamma+\frac{q}{2}+\frac{\eta}{1-\alpha}+\frac{p+1}{2}-\frac{j-1}{2}+h)} \\ &= E(v_1^h)E(v_2^h)...E(v_p^h) \end{aligned} \tag{51}$$

where v_j is a real scalar type-1 beta random variable with the parameters $(\gamma+\frac{q}{2}-\frac{j-1}{2}, \frac{\eta}{1-\alpha}+\frac{p+1}{2})$, $j = 1, ..., p$, and further, $v_1, ..., v_p$ are statistically independently distributed. Hence structurally

$$u_1 = v_1 v_2 ... v_p \tag{52}$$

and the density of u_1 is that of $v_1 ... v_p$. This is available by treating Equation (51) as coming from the Mellin transform of the density $g_1(u_1)$ of u_1. Even though the density $g_1(u_1)$ is unknown, its Mellin transform is available from Equation (51) for $h = s - 1$. Hence from the unique inverse Mellin transform

$$g_1(u_1) = u_1^{-1} \frac{1}{2\pi i} \int_L [E(u_1^h)] u_1^{-h} \mathrm{d}h$$

where $i = \sqrt{-1}$, L is a suitable contour and $E(u_1^h)$ is given in Equation (51). But the structure in Equation (51) is that of a Mellin transform of a G-function of the type $G^{p,0}_{p,p}(\cdot)$. Hence

$$g_1(u_1) = u_1^{-1} \left[\prod_{j=1}^{p} \frac{\Gamma(\gamma + \frac{q}{2} + \frac{\eta}{1-\alpha} + \frac{p+1}{2} - \frac{j-1}{2}}{\Gamma(\gamma + \frac{q}{2} - \frac{j-1}{2})} \right]$$
$$\times G^{p,0}_{p,p} \left[\Big|^{\frac{\eta}{1-\alpha} + \frac{p+1}{2}, j=1,\ldots,p}_{\gamma + \frac{q}{2} - \frac{j-1}{2}, j=1,\ldots,p} \right], 0 < u_1 < 1 \tag{53}$$

For $\alpha > 1$, from Equation (47), $u_2 = [a(\alpha - 1)]^p v^2$ is a product of p statistically independently distributed real scalar type-2 beta random variables and hence the density of u_2 can be written in terms of a G-function of the type $G^{p,p}_{p,p}(\cdot)$.

Similarly for $\alpha \to 1$, $u_3 = (a\eta)^p v^2$ is structurally a product of p statistically independently distributed real gamma random variables and the density of u_3 can be written in terms of a G-function of the type $G^{p,0}_{0,p}(\cdot)$.

A form such as the one in Equation (52) is connected to the $\lambda-$criterion in the likelihood ratio principle of testing statistical hypothesis on the parameters of one or more multivariate Gaussian populations. The pathway model for $\alpha < 1$ can be structurally identified with a constant multiple of the one-to-one function of the $\lambda-$criterion in many situations in multivariate statistical analogues.

4.3. Connection to Likelihood Ratio Criteria

Consider the case $\alpha < 1$.

From Equations (51) and (52) one can see a structural representation in the form of a product of p statistically independently distributed type-1 beta random variables. Such a structure can also arise from a determinant of the type

$$\lambda = \frac{|G_1|}{|G_1 + G_2|} \tag{54}$$

where G_1 and G_2 are independently distributed real matrix-variate gamma variables with the same scale parameter matrix. Such matrix representations are recently examined in [45]. A particular case of a real matrix-variate gamma matrix is a Wishart matrix. When G_1 and G_2 are independently distributed Wishart matrices λ in Equation (54) corresponds to the likelihood ratio test criterion or a one-to-one function of it in multivariate statistical analysis. A large number of test criteria based on the principle of maximum likelihood have the structure in Equation (54), with a representation as in Equation (52) in the null case or when the statistical hypothesis is true. Distribution of the test statistic, when the null hypothesis is true, is known as the null distribution. Thus, a large number of null distributions, associated with the tests of hypotheses on the parameters of one or more multivariate normal populations, have the structure in Equation (54) and thus a large number of cases are covered by our discussion above. Hence a direct link can be established between the volume of a random p-parallelotope and the λ-criterion in Equation (54).

Various types of generalizations of the Dirichlet distribution are recently studied by [46–48]. In these papers there are several theorems characterizing or uniquely determining these generalized Dirichlet models through a product of independently distributed real scalar type-1 beta random variables. In all these cases one can establish a direct link from Equation (52) to Equation (54) providing explicit representations of the densities in terms of generalized Dirichlet integrals. As byproducts, these can also produce several results connecting G-functions and Dirichlet integrals. One such characterization theorem is recently explored by [49].

4.4. Quadratic Forms

For $p = 1$ and $q > 1$, Equation (49) gives

$$Q - (X - M)B(X - M)' \tag{55}$$

the quadratic form.

Current theory of quadratic form in random variables is based on the assumption of a multivariate Gaussian population. But from Equation (43) we have a generalized quadratic form in X, where X has the density in Equation (43), and the density of Q is available from the step in Equation (41) to Equation (53). The theory of quadratic form in random variables can be extended to the samples from the density in Equation (43), rather than confining the study to Gaussian population. For the study of quadratic and bilinear forms see [15,23]. When $p = 1$ the quadratic form in Equation (55) can be given many interpretations. Let $B = V^{-1}$ where V is the covariance matrix in X, that is, $V = \text{cov}(X) = E[(X - \mu)'(X - \mu)]$ where $X - \mu$ is $1 \times q, q > 1$. Since $Y = V^{-\frac{1}{2}}(X - \mu)'$ is the standardized X, $YY' = y_1^2 + ... + y_q^2 = (X - \mu)V^{-1}(X - \mu)'$ is the square of the generalized distance between X and μ. Also $(X - \mu)V^{-1}(X - \mu)' = c > 0$ is the ellipsoid of concentration in X or a scalar measure of the extent of dispersion in $X - \mu$. The components in X can be given physical interpretations. Consider a growing and moving rain droplet in a tropical cloud. Then x_1 could be the surface area, x_2 could be the energy content, x_3 the velocity in a certain direction and so on. The components in this case have joint variations. When $p = 1$ one can derive the pathway density in Equation (43) through an optimization of a certain measure of entropy also.

4.5. Pathway Fractional Integral

Recently, an extension of classical fractional integral operators of scalar functions of scalar variables to the matrix-variate cases has been given by [50]. Real-valued scalar functions of matrix argument, where the argument matrix is real and positive definite, are used in the extensions. In this regard, a matrix-variate pathway fractional integral operator is introduced, see [51] which may be regarded as a generalization of matrix-variate Riemann-Liouville fractional integral operator. Moreover, from this operator one can figure out all the matrix-variate fractional integrals and almost all the extended densities for the pathway parameter $\alpha < 1$ and $\alpha \to 1$. Through this new fractional integral operator, one can go to matrix-variate gamma to matrix-variate Gaussian or normal density with appropriate parametric values. In the present paper we bring out the idea of matrix-variate pathway to the corresponding fractional integral transform. Consequently a scalar version of pathway fractional integral operator can also be deduced, which generalizes the classical Reimann-Liouville fractional integration operator. The pathway fractional integral operator has found applications in reaction-diffusion problems, non-extensive statistical mechanics, non-linear waves, fractional differential equations, non-stable neighborhoods of physical system *etc.*

The following definition and notation is given for the matrix-variate pathway fractional integral:

$$(P_{O+}^{(\eta,\alpha)} f)(X) = |X|^{\eta - \frac{p+1}{2}} \int_T |I - a(1-\alpha)X^{-\frac{1}{2}} T X^{-\frac{1}{2}}|^{\frac{\eta}{(1-\alpha)} - \frac{p+1}{2}} f(T) \mathrm{d}T \tag{56}$$

where O stands for a null matrix, and $T = T' > 0$ and O are $p \times p$ matrices and $f(T)$ is a real-valued integrable scalar function of T and $\mathrm{d}T$ is the wedge product of differentials. Also a, α scalars, $\eta \in C$ $a > 0$, $I - a(1-\alpha)X^{-\frac{1}{2}}TX^{-\frac{1}{2}} > 0$ (positive definite) and α is the pathway parameter, $\alpha < 1$. It is hoped that the matrix-variate extensions of the operator will enable researchers working in physical, chemical and engineering sciences to extend their theories to the corresponding matrix-variate situations.

When $\alpha \to 1_-$, $|I - a(1-\alpha)X^{-\frac{1}{2}}TX^{-\frac{1}{2}}|^{\frac{\eta}{(1-\alpha)}-\frac{P+1}{2}} \to e^{-a\eta\,\mathrm{tr}[X^{-\frac{1}{2}}TX^{-\frac{1}{2}}]}$. This follows from the following facts: Since $X^{-\frac{1}{2}}TX^{-\frac{1}{2}}$ is real symmetric there exists an orthogonal matrix Q such that $Q'X^{-\frac{1}{2}}TX^{-\frac{1}{2}}Q$ will be a diagonal matrix with eigenvalues being the diagonal elements. Then when the limit is applied, each factor goes to the exponent and sum up to equate to the trace. Thus the operator will become

$$(P_{O+}^{(\eta,1)}f)(X) = |X|^{\eta-\frac{P+1}{2}} \int_{T=T'>0} e^{-\mathrm{tr}[a\eta X^{-1}T]} f(T)\,\mathrm{d}T = |X|^{\eta-\frac{P+1}{2}} L_f(a\eta X^{-1})$$

When $\alpha = 0$, $a = 1$ in Equation (56), the integral will become,

$$\begin{aligned}(P_{O+}^{(\eta,0)}f)(X) &= |X|^{\eta-\frac{P+1}{2}} \int_{O<T<X} |I - X^{-\frac{1}{2}}TX^{-\frac{1}{2}}|^{\eta-\frac{P+1}{2}} f(T)\mathrm{d}T,\ \Re(\eta) > \frac{p-1}{2} \\ &= \int_{O<T<X} |X-T|^{\eta-\frac{P+1}{2}} f(T)\mathrm{d}T = \Gamma_p(\eta)\ {}_OD_X^{-\eta}f \qquad (57)\end{aligned}$$

where ${}_OD_X^{-\eta}$ is the matrix-variate extension of the standard left-sided Reimann-Liouville fractional integral and is defined as

$${}_OD_X^{-\eta}f = \frac{1}{\Gamma_p(\eta)} \int_{O<T<X} |X-T|^{\eta-\frac{P+1}{2}} f(T)\mathrm{d}T,\ \ \Re(\eta) > \frac{p-1}{2} \qquad (58)$$

When $p = 1$ in Equation (56), one can obtain the scalar version of the pathway fractional integral operator, see [37] and is defined by the following:

Let $f(x) \in L(a,b)$, $\eta \in C$, $\Re(\eta) > 0$, $a > 0$ and $\alpha < 1$, then

$$(P_{0+}^{(\eta,\alpha)}f)(x) = x^{\eta-1} \int_0^{[\frac{x}{a(1-\alpha)}]} [1 - \frac{a(1-\alpha)t}{x}]^{\frac{\eta}{(1-\alpha)}-1} f(t)\mathrm{d}t \qquad (59)$$

where α is the pathway parameter and $f(t)$ is an arbitrary function. In [37], it is shown that as $\alpha \to 1_-$, Equation (59) takes the form of Laplace transform of the arbitrary function $f(t)$. Again when $\alpha = 0$, $a = 1$, Equation (59) reduces to the left-sided Reimann-Liouville fractional integral operator. By the way an idea of thicker or thinner tail model associated with Mittag-Leffler function is obtained. As a result, generalized gamma Mittag-Leffler density can be obtained as a limiting case of pathway operator. It is also shown that under the conditions $\alpha = 0$, $a = 1$ and as $f(t)$ changes to ${}_2F_1(\eta+\beta, -\gamma; \eta; 1-\frac{t}{x})f(t)$, Equation (59) yields the Saigo fractional integral operator. Thus we can obtain all the generalizations, like [41,52,53], of left-sided fractional integrals by suitable substitutions, so that we call it the pathway fractional operator, a path through α, leading to the above known fractional operators.

The importance of the operator is that a connection is established to wide classes of statistical distributions, to several types of situations in physics, chemistry, to the input-output situations in social sciences, in reaction-diffusion problems etc. The pathway parameter α establishes a path of going from one family of distributions to another family and to different classes of distributions. Thus, the pathway fractional operator, will enable us to derive a number of results covering wide range of distributions. The "fractional integration" nature of the operator will then extend the corresponding results to wider ranges, where when the pathway parameter α goes to 1 the corresponding results on generalized gamma type functions are obtained.

5. Open Problem

The main idea to introduce the pathway model is the switching property of the binomial function to the corresponding exponential function. That is

$$\lim_{\alpha\to1} {}_1F_0\left(\frac{1}{1-\alpha};\ ; -a(1-\alpha)x^\delta\right) = \lim_{\alpha\to1}\left(1 + a(1-\alpha)x^\delta\right)^{-\frac{1}{1-\alpha}} = e^{-ax^\delta},\ a > 0 \qquad (60)$$

Similar to Equation (60) we can develop a pathway connecting Bessel to exponential, as given below,

$$\lim_{\alpha\to 1} {}_0F_1\left(;\frac{1}{1-\alpha};-\frac{ax^{\delta}}{(1-\alpha)}\right) = {}_0F_0\left(;;-ax^{\delta}\right) = \mathrm{e}^{-ax^{\delta}},\ a>0$$

Thus α can provide a pathway between Bessel functions and exponential functions. Thus, in all types of applications where Bessel functions are used, one can extend the scope of applications into an exponential form. In a physical system, if the exponential form gives the stable situation, then the parameter α will provide a pathway between stable and chaotic situations. So far, this area which offers a wide scope of possibilities, has yet to be explored.

Acknowledgments: Authors acknowledge gratefully the encouragement given by Professor A. M. Mathai, Department of Mathematics and Statistics, McGill University, Montreal, Canada H3A 2K6 and Professor H. J. Haubold, Office of Outer Space Affairs, United Nations, Vienna International Centre, Austria.

Conflicts of Interest: The authors declare that there is no conflict of interests regarding the publication of this article.

References

1. Mathai, A.M. A Pathway to matrix-variate gamma and normal densities. *Linear Algebra Appl.* **2005**, *396*, 317–328.
2. Beck, C. Stretched exponentials from superstatistics. *Phys. A* **2006**, *365*, 96–101.
3. Beck, C.; Cohen, E.G.D. Superstatistics. *Phys. A* **2003**, *322*, 267–275.
4. Tsallis, C. Possible generalizations of Boltzmann-Gibbs statistics. *J. Stat. Phys.* **1988**, *52*, 479–487.
5. Tsallis, C. What should a statistical mechanics satisfy to reflect nature? *Phys. D* **2004**, *193*, 3–34.
6. Honerkamp, J. *Stochastic Dynamical Systems: Concepts, Numerical Methods, Data Analysis*; VCH Publishers: New York, NY, USA, 1994.
7. Mathai, A.M. *A Handbook of Generalized Special Functions for Statistical and Physical Sciences*; Clarendon Press: Oxford, UK, 1993.
8. Haubold, H.J.; Kumar, D.; Nair, S.S.; Joseph, D.P. Special functions and pathways for problems in astrophysics: An essay in honor of A.M. Mathai. *Fract. Calc. Appl. Anal.* **2010**, *13*, 133–158.
9. Joseph, D. P. Gamma distribution and extensions by using pathway idea. *Stat. Pap.* **2009**, doi:10.1007/s00362-009-0231-y.
10. Mathai, A.M.; Rathie, P.N. *Basic Concepts in Information Theory and Statistics: Axiomatic Foundations and Applications*; Wiley Halsted: New York, NY, USA, 1975.
11. Mathai, A.M.; Haubold, H.J. On generalized entropy measures and pathways. *Phys. A Stat. Mech. Appl.* **2007**, 385, 493–500.
12. Mathai, A.M.; Haubold, H.J. Pathway model Pathway model, superstatistics Tsallis statistics and a generalized measure of entropy. *Phys. A* **2007**, *375*, 110–122.
13. Mathai, A.M.; Haubold, H.J. Pathway parameter and thermonuclear functions. *Phys. A Stat. Mech. Appl.* **2008**, *387*, 2462–2470.
14. Mathai, A.M. On non-central generalized Laplacianness of quadratic forms in normal variables. *J. Multivar. Anal.* **1993**, *45*, 239–246.
15. Mathai, A.M.; Provost, S.B.; Hayakawa, T. *Bilinear Forms and Zonal Polynomials, Lecture Notes*; Springer-Verlag: New York, NY, USA, 1995.
16. Mathai, A.M. The residual effect of a growth-decay mechanism and the distributions of covariance structures. *Can. J. Stat.* **1993**, *21*, 277–283.
17. Mathai, A. M. Some properties of Mittag- Leffler functions and matrix variate analogues: A statistical perspective. *Fract. Cal. Appl. Anal.* **2010**, *30*, 113–132.
18. Mathai, A.M.; Saxena, R.K.; Haubold, H.J. *The H-Function Theory and Applications*; Springer: New York, NY, USA, 2010.
19. Mathai, A.M.; Haubold, H.J. A pathway from bayesian statistical analysis to superstatistics. **2010**, arXiv:1011.5658v1.

20. Mathai, A.M.; Moschopoulos, P. A Pathway Idea for Model Building. *J. Stat. Appl. Proc.* **2012**, *1*, 15–20.
21. Pillai, R.N. On Mittag-Leffler functions and related distributions. *Ann. Inst. Statist. Math.* **1990**, *42*, 157–161.
22. Shanoja Naik, R. *Pathway Distributions, Autoregressive Processes and Their Applications.* PhD Thesis., Mahatma Gandhi University, Kottayam, Kerala, India, 2008.
23. Mathai, A.M.; Provost, S.B. *Quadratic Forms in Random Variables: Theory and Applications*; Marcel Dekker: New York, NY, USA, 2005.
24. Han, J.H. Gamma function to Beck-Cohen superstatistics. *Phys. A* **2013**, *392*, 4288–4298.
25. Sebastian, N. Limiting Approach to Generalized Gamma Bessel Model via Fractional Calculus and its Applications in Various Disciplines. **2013**, arXiv:1307.7949.
26. Seema Nair, S.; Kattuveettil, A. Some remarks on the paper "On the q-type Distributions". *Proc. Astrophys. Space Sci.* **2010**, 11–15.
27. Sebastian, N. A Generalized Multivariate Gamma Model Associated with Bessel Function. *Integral Transforms and Special Functions* **2011**, *22*, 601–615.
28. Sebastian, N. A generalized gamma model associated with Bessel function and its applications in statistical mechanics. In Proceedings of AMADE-09, Institute of Mathematics, Belarusian State Mathematics, Minsk, Belarus, 2009.
29. Anderson, W.J.; Haubold, H.J.; Mathai, A.M. Astrophysical thermonuclear functions. *Astrophys. Space Sci.* **1994**, *214*, 49–70.
30. Haubold, H.J.; John, R.W. On the evaluation of an integral connected with the thermonuclear reaction rate in the closed form. *Astron. Nachr.* **1978**, *299*, 225–232.
31. Mathai, A.M.; Haubold, H.J. *Modern Problems in Nuclear and Neutrino Astrophysics*; Akademie-Verlag: Berlin, Germany, 1988.
32. Mathai, A.M.; Haubold, H.J. Review of Mathematical techniques applicable in astrophysical reaction rate theory. *Astrophys. Space Sci.* **2002**, *282*, 265–280.
33. Haubold, H.J.; Mathai, A.M. The fractional kinetic equation and thermonuclear functions. *Astrophys. Space Sci.* **2000**, *273*, 53–63.
34. Haubold, H.J.; Kumar, D. Extension of thermonuclear functions through the pathway model including Maxwell-Boltzmann and Tsallis distributions. *Astropart. Phys.* **2008**, *29*, 70–76.
35. Joseph, D. P.; Haubold, H.J. Extended reaction rate integral as solutions of some general differential equations. *Proc. Astrophys. Space Sci.* **2009**, 41–51.
36. Haubold, H.J.; Mathai, A.M. A heuristic remark on the periodic variation in the number of solar neutrinos detected on Earth. *Astrophys. Space Sci.* **1995**, *228*, 113–134.
37. Seema Nair, S. Pathway fractional integration operator. *Fract. Calc. Appl. Anal.* **2009**, *12*, 237–252.
38. Kilbas, A.A.; Kumar, D. On generalized Krätzel functions. *Integral Transform. Spec. Funct.* **2009**, *20*, 836–845.
39. Kumar, D. Type-2 P-transforms. In Proceedings of AMADE-09, Belarusian State University, Minsk, Belarus, 2009.
40. Krätzel, E. Integral transformations of Bessel type. *Generalized Functions Operational Calculus* **1979**, 148–155.
41. Erdélyi, A.; Magnus, W.; Oberhettinger, F.; Tricomi, F.G. *Higher Transcendental Functions*; McGraw-Hill: New York, NY, USA, 1953; Volume 1.
42. Kilbas, A.A.; Shlapakov, S.A. On Bessel-type integral transformation and its compositions with integral and differential operators. *Dokl. Akad. Nauk Belarusi* **1993**, *37*, 10–14.
43. Kilbas, A.A.; Saxena, R.K.; Trujillo, J.J. Krätzel function as a function of hypergeometric type. *Fract. Calc. Appl. Anal.* **2006**, *2*, 109–130.
44. Mathai, A.M. *Jacobians of Matrix Transformations and Functions of Matrix Argument*; World Scientific Publishing: New York, NY, USA, 1997.
45. Mathai, A.M. Random volumes under a general matrix-variate model. *Linear Algebra Appl.* **2007**, *425*, 162–170.
46. Jacob, J.; Jose, K.K.; Mathai, A.M. Some properties of real matrix-variate inverted generalized Dirichlet integral. *J. Indian Acad. Math.* **2004**, *26*, 175–189.
47. Jacob, J.; Sebastian George, S.; Mathai, A.M. Some properties of complex matrix-variate generalized Dirichlet integrals. *Proc. Indian Acad. Sci.* **2005**, *15*, 1–9.
48. Kurian, K.M.; Benny, K.; Mathai, A.M. A matrix-variate extension of inverted Dirichlet integral. *Proc. Natl. Acad. Sci. (India)* **2004**, *74*, 1–10.

49. Thomas, S.; Thannippara, A.; Mathai, A.M. On a matrix-variate generalized type-2 Dirichlet model. *Adv. Appl. Stat.* **2008**, *8*, 37–56.
50. Mathai, A.M. Fractional integrals in the matrix-variate cases and connection to statistical distributions. *Integral Transforms Spec. Funct.* **2009**, *20*, 871–882.
51. Seema Nair, S. Pathway fractional integral operator and matrix-variate functions. *Integr. Transform. Spec. Funct.* **2011**, *22*, 233–244.
52. Kober, H. On fractional integrals and derivatives. *Q. J. Math.* **1940**, *11*, 193–211.
53. Saigo, M. A remark on integral operators involving the Gauss hypergeometric functions. *Math. Rep. Kyushu Univ.* **1978**, *11*, 135–143.

axioms

MDPI

Article

Entropy Production Rate of a One-Dimensional Alpha-Fractional Diffusion Process

Yuri Luchko

Department of Mathematics, Physics, and Chemistry, Beuth Technical University of Applied Sciences, Luxemburger Str. 10, 13353 Berlin, Germany; luchko@beuth-hochschule.de

Academic Editor: Hans J. Haubold
Received: 30 December 2015; Accepted: 2 February 2016; Published: 5 February 2016

Abstract. In this paper, the one-dimensional α-fractional diffusion equation is revisited. This equation is a particular case of the time- and space-fractional diffusion equation with the quotient of the orders of the time- and space-fractional derivatives equal to one-half. First, some integral representations of its fundamental solution including the Mellin-Barnes integral representation are derived. Then a series representation and asymptotics of the fundamental solution are discussed. The fundamental solution is interpreted as a probability density function and its entropy in the Shannon sense is calculated. The entropy production rate of the stochastic process governed by the α-fractional diffusion equation is shown to be equal to one of the conventional diffusion equation.

Keywords: Caputo fractional derivative; Riesz fractional derivative; α-fractional diffusion equation; Mellin transform; fundamental solution; Mellin-Barnes integrals; entropy; entropy production rate

MSC: 26A33; 35C05; 35E05; 35L05; 45K05; 60E99

1. Introduction

In the literature, many different kinds of the time-, space, and time- and space-fractional diffusion equations have been already introduced and analyzed. The fractional derivatives contained in the fractional diffusion equations are defined in the Riemann-Liouville, Weyl, Caputo, Riez, or Riesz-Feller sense to mention only some of the most used types of the fractional derivatives. From the mathematical viewpoint, all of these fractional diffusion equations can be seen as generalizations of the conventional diffusion equation and thus are worth to be investigated. On the other hand, it is not clear at all, what kinds of the fractional diffusion equations could/should be employed as mathematical models, say, for describing the phenomena of the anomalous diffusion (see, e.g., the recent survey paper [1] for about three hundred references to the relevant works). Usually, the anomalous diffusion processes are defined as those that do not longer follow the Gaussian statistics on the long time intervals. Especially, the linear time dependence of the mean squared displacement of the diffusing particles does not hold any more and has to be either replaced with a different (mainly power-law) dependence or the mean squared displacement does not exist at all. As a rule, the stochastic processes governed by different kinds of the fractional diffusion equations are non Gaussian and thus can be seen as potential candidates for the role of mathematical models for anomalous diffusion processes.

One more important characteristic of the diffusion processes is their entropy and the entropy production rate. The concept of entropy was first introduced in the macroscopic thermodynamics and then extended for description of some phenomena in statistical mechanics, information theory, ergodic theory of dynamical systems, *etc.* Historically, many definitions of entropy were proposed and applied in different knowledge areas. In this paper, we employ the statistical concept of entropy that goes back to Shannon and was introduced by him in the theory of communication and transmission of information (see [2]). The entropy of the processes governed by the time- and space-fractional diffusion equations has been discussed in [3–6], respectively. It is worth mentioning that according to [3,5] the entropy production rates for the time- and the space-fractional diffusion equations depend on the derivative order α of the time- or space-fractional derivative, respectively, and increase with increasing of α from 1 (diffusion) to 2 (wave propagation) that results in the so called entropy production paradox. In [7], entropy behavior of solutions to the one-dimensional neutral-fractional equation that contains fractional derivatives of the same order α, $1 \le \alpha \le 2$ both in space and in time, has been considered. It was shown in [7] that the entropy production rate of solutions to the neutral-fractional equation does not depend on the equation order α and is twice as much as the entropy production rate of solutions to the conventional diffusion equation. In this paper, we show that the entropy production rate of the fundamental solution to the α-fractional diffusion equation is exactly the same as in the case of the conventional diffusion equation. Thus the α-fractional diffusion equation combines the properties of the anomalous diffusion (the mean squared displacement of the diffusing particles does not exists) and of the conventional diffusion (the same entropy production rate) and could be considered to be a kind of a "natural fractionalization" of the diffusion equation. In this paper we restrict ourselves to the classical Shannon entropy, other kinds of generalized entropies (see e.g., [8–10] and the references therein) will be considered elsewhere.

From the mathematical viewpoint, the α-fractional diffusion equation is a particular case of the one-dimensional space-time fractional diffusion equation that has been considered in [11] in detail. The equation studied in [11] contains the Riesz-Feller derivative of order $\alpha \in (0,2]$ and skewness θ and the Caputo fractional derivative of order $\beta \in (0,2]$. In particular, it was shown in [11] that the fundamental solution to the space-time fractional diffusion equation can be interpreted as a spatial probability density function evolving in time if $\{0 < \alpha \le 2\} \cap \{0 < \beta \le 1\}$ or if $\{1 < \beta \le \alpha \le 2\}$. The α-fractional diffusion equation we deal with in this paper corresponds to the case $\alpha = 2\beta$, $0 \le \beta \le 1$, $\theta = 0$ in the space-time fractional diffusion equation considered in [11] and possesses some remarkable properties that do not hold true for solutions of the general equation.

The rest of the paper is organized as follows. In the 2nd section, the basic definitions, problem formulation, and some analytical results for the initial-value problems for the one-dimensional α-fractional equation are presented. Among other things, the Mellin-Barnes integral representation of the fundamental solution as well as its series representation and asymptotics are given. The last section is devoted to a probabilistic interpretation of the fundamental solution to the one-dimensional α-fractional diffusion equation. In particular, the Shannon entropy and the entropy production rate are calculated. The entropy production rate of the stochastic process governed by the α-fractional diffusion equation is shown to be independent on the equation order α and is exactly the same as the entropy production rate of the conventional diffusion process.

2. Alpha-Fractional Diffusion Equation

2.1. Problem Formulation

In this paper, we deal with the one-dimensional α-fractional diffusion equation in the form

$$D_t^{\alpha} u(x,t) \;=\; -\,(-\Delta)^{\alpha} u(x,t), \;\; x \in \mathbb{R}, \;\; t \in \mathbb{R}_+, \; 0 < \alpha \le 1. \tag{1}$$

In the Equation (1), D_t^{α} is the Caputo time-fractional derivative of order α defined by

$$(D^{\alpha} f)(t) = (I^{n-\alpha} f^{(n)})(t), \; n-1 < \alpha \le n, \; n \in \mathbb{N}, \tag{2}$$

I^{α}, $\alpha \ge 0$ being the Riemann-Liouville fractional integral

$$(I^{\alpha} f)(t) = \begin{cases} \frac{1}{\Gamma(\alpha)} \int_0^t (t-\tau)^{\alpha-1} f(\tau)\, d\tau, \; \alpha > 0, \\ f(t), \; \alpha = 0, \end{cases}$$

and Γ the Euler gamma function. For $\alpha = n,\ n \in \mathbf{N}$, the Caputo fractional derivative coincides by definition with the derivative of order n.

For a sufficiently well-behaved function f, the Riesz fractional derivative $-(-\Delta)^{\alpha}$ is defined as a pseudo-differential operator with the symbol $-|\kappa|^{2\alpha}$ (see e.g., [11–14]):

$$(\mathcal{F} - (-\Delta)^{\alpha} f)(\kappa) = -|\kappa|^{2\alpha} (\mathcal{F} f)(\kappa), \tag{3}$$

$\mathcal{F}$ being the Fourier transform of a function f defined by the formula

$$(\mathcal{F} f)(\kappa) = \hat{f}(\kappa) = \int_{-\infty}^{+\infty} f(x) e^{ix\kappa}\, dx.$$

The Riesz fractional derivative (3) can be represented as a hypersingular integral under the condition $0 < \alpha < 1$ (see [14] for the case $\alpha \neq \frac{1}{2}$ and [12] for the general case)

$$-(-\Delta)^{\alpha} f(x) = \frac{1}{\pi}\Gamma(1+2\alpha)\sin(\alpha\pi) \int_0^{\infty} \frac{f(x+\xi) - 2f(x) + f(x-\xi)}{\xi^{2\alpha+1}} d\xi. \tag{4}$$

For $\alpha = \frac{1}{2}$, the relation (4) can be interpreted in terms of the Hilbert transform

$$-(-\Delta)^{\frac{1}{2}} f(x) = -\frac{1}{\pi}\frac{d}{dx} \int_{-\infty}^{+\infty} \frac{f(\xi)}{x-\xi} d\xi,$$

where the integral is understood in the sense of the Cauchy principal value as first noted in [15] and then revisited and stated more precisely in [12,16].

Let us note that the Riesz fractional derivative is a symmetric operator with respect to the space variable x. Because of the relation $-|\kappa|^{2\alpha} = -(\kappa^2)^{\alpha}$ it can be formally interpreted as

$$\frac{d^{2\alpha}}{d|x|^{2\alpha}} = -\left(-\frac{d^2}{dx^2}\right)^{\alpha},$$

i.e., as a power of the self-adjoint and positive definite operator $-\frac{d^2}{dx^2}$.

For $\alpha = 1$, Equation (1) is reduced to the one-dimensional diffusion equation. In what follows, we focus on the case $0 \leq \alpha < 1$ because the case $\alpha = 1$ (diffusion equation) is well studied in the literature.

In the rest of the paper, we consider the initial-value problem

$$u(x, 0) = \varphi(x),\ \ x \in \mathbf{R} \tag{5}$$

for the Equation (1). In doing so, we are mostly interested in behavior and properties of the fundamental solution (Green function) $G_{\alpha} = G_{\alpha}(x,t)$ of the Equation (1), *i.e.*, in its solution with the initial condition $\varphi(x) = \delta(x)$, δ being the Dirac delta function.

2.2. Fundamental Solution of the Alpha-Fractional Diffusion Equation

In this subsection, we follow the derivations presented in [11] for the more general case of the one-dimensional time-space fractional diffusion equation with some minor modifications. To determine the fundamental solution G_{α} let us apply the Fourier transform to the Equation (1) and to the initial conditions (5) with $\varphi(x) = \delta(x)$. Using definition of the Riesz fractional derivative, for the Fourier transform $\hat{G}_{\alpha}$ we get then the initial-value problem

$$\hat{G}(\kappa, 0) = 1 \tag{6}$$

for the fractional differential equation

$$(D^{\alpha}\hat{G}_{\alpha})(t) + |\kappa|^{2\alpha}\hat{G}_{\alpha}(\kappa, t) = 0. \tag{7}$$

The unique solution of (6), (7) is given by the expression (see e.g., [17,18])

$$\hat{G}_{\alpha}(\kappa, t) = E_{\alpha}(-|\kappa|^{2\alpha}t^{\alpha}) \tag{8}$$

in terms of the Mittag-Leffler function E_α that is defined as an convergent power series

$$E_{\alpha}(z) = \sum_{k=0}^{\infty} \frac{z^k}{\Gamma(1+\alpha k)}, \ \alpha > 0. \tag{9}$$

As follows from the well-known asymptotic formula

$$E_{\alpha}(-x) = -\sum_{k=1}^{m} \frac{(-x)^{-k}}{\Gamma(1-\alpha k)} + O(x^{-1-m}), \ m \in \mathbf{N}, \ x \to +\infty, \ 0 < \alpha < 2,$$

the Fourier transform $\hat{G}_\alpha$ belongs to $L_1(\mathbf{R})$ with respect to κ under the condition $\alpha > \frac{1}{2}$. In the further discussions we suppose that this conditions holds true. Then we can apply the inverse Fourier transform and get the representation

$$G_{\alpha}(x, t) = \frac{1}{2\pi} \int_{-\infty}^{+\infty} e^{-i\kappa x} E_{\alpha}(-|\kappa|^{2\alpha}t^{\alpha})\, d\kappa, \ x \in \mathbf{R}, \ t > 0 \tag{10}$$

that can be rewritten as the cos-Fourier transform:

$$G_{\alpha}(x, t) = \frac{1}{\pi} \int_{0}^{\infty} \cos(\kappa x)\, E_{\alpha}(-\kappa^{2\alpha}t^{\alpha})\, d\kappa, \ x \in \mathbf{R}, \ t > 0. \tag{11}$$

Now we are going to apply the technique of the Mellin integral transform to deduce the Mellin-Barnes representation of the fundamental solution. For the reader's convenience, some basic elements of the Mellin integral transform theory are presented below.

The Mellin integral transform of a function f is defined by the formula

$$f^*(s) = (\mathcal{M}f)(s) = \int_0^{\infty} f(t)\, t^{s-1}\, dt, \ \ \gamma_1 < \Re(s) < \gamma_2, \tag{12}$$

and the inverse Mellin integral transform by the formula

$$f(t) = (\mathcal{M}^{-1}f^*)(t) = \frac{1}{2\pi i} \int_{\gamma - i\infty}^{\gamma + i\infty} f^*(s)\, t^{-s} ds, \ t > 0, \ \Re(s) = \gamma, \ \gamma_1 < \gamma < \gamma_2. \tag{13}$$

The Mellin integral transform and its inversion exist under the following sufficient conditions (see e.g., [19]): Let $f \in L^c(\epsilon, E)$, $0 < \epsilon < E < \infty$ be a function continuous on the intervals $(0, \epsilon]$ and $[E, \infty)$, and let $|f(r)| \leq M r^{-\gamma_1}$ for $0 < r < \epsilon$, $|f(r)| \leq M r^{-\gamma_2}$ for $r > E$, where M is a constant. If $\gamma_1 < \gamma_2$, then the Mellin transform (12) of the function f exists and is analytic in the vertical strip $\gamma_1 < \gamma = \Re(s) < \gamma_2$.

If f is piecewise differentiable and $f(r)\, r^{\gamma-1} \in L^c(0, \infty)$, then the Formula (13) holds true in all points where f is continuous. The integral in (13) must be understood in the sense of the Cauchy principal value.

Let us denote by $\overset{\mathcal{M}}{\leftrightarrow}$ the juxtaposition of a function f with its Mellin transform f^*. With this notation the convolution theorem for the Mellin convolution reads as follows:

$$\int_0^\infty g(\kappa)f(y/\kappa)\frac{d\kappa}{\kappa} \overset{\mathcal{M}}{\leftrightarrow} g^*(s)f^*(s). \tag{14}$$

Now let us return back to the integral representation (11) of the fundamental solution G_α and consider the cases a) $x = 0$ and b) $x \neq 0$.

(a) For $x = 0$, the integral at the right-hand side of (11) can be interpreted as the Mellin integral transform of the Mittag-Leffler function at the point $s = \frac{1}{2\alpha}$. It converges under the condition $\alpha > \frac{1}{2}$ and its value is given by the formula (see e.g., [19])

$$\frac{1}{\pi}\int_0^\infty E_\alpha(-\kappa^{2\alpha}t^\alpha)\,d\kappa = \frac{1}{2\pi\alpha\sqrt{t}}\int_0^\infty E_\alpha(-u)u^{\frac{1}{2\alpha}-1}\,du$$

$$= \frac{1}{2\pi\alpha\sqrt{t}}\frac{\Gamma\left(\frac{1}{2\alpha}\right)\Gamma\left(1-\frac{1}{2\alpha}\right)}{\Gamma\left(1-\alpha\frac{1}{2\alpha}\right)} = \frac{1}{\alpha\sqrt{4\pi t}\sin\left(\frac{\pi}{2\alpha}\right)},\ t > 0.$$

For $\alpha = 1$, the Green function G_α is a time-dependent Gaussian distribution $G_1(x,t) = \frac{1}{\sqrt{4\pi t}}\exp\left(-\frac{x^2}{4t}\right)$ and thus $G_1(0,t) = \frac{1}{\sqrt{4\pi t}}$ that is in accordance with the above formula for the value of $G_\alpha(0,t)$.

(b) In the case $x \neq 0$, we recognize that the integral at the right-hand side of (11) can be interpreted as the Mellin convolution of the functions

$$g(\kappa) = E_\alpha(-\kappa^{2\alpha}t^\alpha) \text{ and } f(\kappa) = \frac{1}{\pi|x|\kappa}\cos\left(\frac{1}{\kappa}\right)$$

at the point $y = \frac{1}{|x|}$.

Using the known Mellin integral transforms of the cos-function and the Mittag-Leffler function as well as some elementary properties of the Mellin integral transform (see [19,20]) we get the formulas:

$$g^*(s) = \frac{1}{2\alpha t^{\frac{s}{2}}}\frac{\Gamma\left(\frac{s}{2\alpha}\right)\Gamma\left(1-\frac{s}{2\alpha}\right)}{\Gamma\left(1-\frac{s}{2}\right)},\ 0 < \Re(s) < 2\alpha,$$

$$f^*(s) = \frac{1}{\sqrt{\pi}|x|2^s}\frac{\Gamma\left(\frac{1}{2}-\frac{s}{2}\right)}{\Gamma\left(\frac{s}{2}\right)},\ 0 < \Re(s) < 1.$$

These formulas together with the convolution theorem and the inverse Mellin integral transform lead to the following Mellin-Barnes representation of the fundamental solution G_α:

$$G_\alpha(x,t) = \frac{1}{2\alpha\sqrt{\pi}|x|}\frac{1}{2\pi i}\int_{\gamma-i\infty}^{\gamma+i\infty}\frac{\Gamma(\frac{1}{2}-\frac{s}{2})\Gamma(\frac{s}{2\alpha})\Gamma(1-\frac{s}{2\alpha})}{\Gamma(\frac{s}{2})\Gamma(1-\frac{s}{2})}\left(\frac{2\sqrt{t}}{|x|}\right)^{-s} ds,\ 0 < \gamma < \min\{1, 2\alpha\}. \tag{15}$$

The linear variables substitution $s \to 2s$ in the integral at the right-hand side of (15) leads to the representation

$$G_\alpha(x,t) = \frac{1}{\alpha\sqrt{\pi}|x|}\frac{1}{2\pi i}\int_{\gamma-i\infty}^{\gamma+i\infty}\frac{\Gamma\left(\frac{1}{2}-s\right)\Gamma\left(\frac{s}{\alpha}\right)\Gamma\left(1-\frac{s}{\alpha}\right)}{\Gamma(s)\Gamma(1-s)}\left(\frac{4t}{x^2}\right)^{-s} ds,\ 0 < \gamma < \min\{1/2, \alpha\} \tag{16}$$

and then to the formula

$$G_\alpha(x,t) = \frac{1}{\alpha\sqrt{\pi}|x|}\frac{1}{2\pi i}\int_{\gamma-i\infty}^{\gamma+i\infty}\Gamma\left(\frac{1}{2}-s\right)\frac{\sin(\pi s)}{\sin(\pi s/\alpha)}\left(\frac{4t}{x^2}\right)^{-s} ds,\ 0 < \gamma < \min\{1/2, \alpha\} \tag{17}$$

by using the reflection formulas for the gamma function.

In the next section, we need one more Mellin-Barnes integral representation of the fundamental solution G_α that is obtained from the Formula (16) using the linear variables substitution $s \to -s$:

$$G_\alpha(x,t) = \frac{1}{\alpha\sqrt{\pi}|x|}\frac{1}{2\pi i}\int_{\gamma-i\infty}^{\gamma+i\infty}\frac{\Gamma\left(\frac{1}{2}+s\right)\Gamma\left(-\frac{s}{\alpha}\right)\Gamma\left(1+\frac{s}{\alpha}\right)}{\Gamma(-s)\Gamma(1+s)}\left(\frac{x^2}{4t}\right)^{-s} ds, \; -\min\{1/2,\alpha\} < \gamma < 0. \tag{18}$$

From Equation (18), a useful representation

$$G_\alpha(x,t) = \frac{1}{|x|}L_\alpha\left(\frac{x^2}{4t}\right), \; x \neq 0, \; t > 0 \tag{19}$$

of the fundamental solution G_α in terms of an auxiliary function L_α defined by its Mellin-Barnes representation

$$L_\alpha(\tau) = \frac{1}{\alpha\sqrt{\pi}}\frac{1}{2\pi i}\int_{\gamma-i\infty}^{\gamma+i\infty}\frac{\Gamma\left(\frac{1}{2}+s\right)\Gamma\left(-\frac{s}{\alpha}\right)\Gamma\left(1+\frac{s}{\alpha}\right)}{\Gamma(-s)\Gamma(1+s)}\tau^{-s} ds, \; -\min\{1/2,\alpha\} < \gamma < 0 \tag{20}$$

can be obtained.

Because the auxiliary function L_α is defined in Equation (20) as an inverse Mellin transform, its Mellin transform is given by the formula

$$L_\alpha^*(s) = \int_0^\infty L_\alpha(\tau)\,\tau^{s-1}\,d\tau = \frac{1}{\alpha\sqrt{\pi}}\frac{\Gamma\left(\frac{1}{2}+s\right)\Gamma\left(-\frac{s}{\alpha}\right)\Gamma\left(1+\frac{s}{\alpha}\right)}{\Gamma(-s)\Gamma(1+s)}, \; -\min\{1/2,\alpha\} < \Re(s) < 0. \tag{21}$$

Now we derive a series representation of the fundamental solution G_α by employing the Formula (16) and the general theory of the Mellin-Barnes integrals ([19]). To arrive at a series representation, the contour of integration in the integral at the right-hand side of (16) has to be transformed to the infinite loop $\mathcal{L}_{+\infty}$ starting and ending at $+\infty$ and encircling all poles $s_k = 1/2 + k, \; k = 0,1,2,\ldots$ of the gamma function $\Gamma\left(\frac{1}{2}-s\right)$ and all poles $s_k = \alpha + \alpha k, \; k = 0,1,2,\ldots$ of the gamma function $\Gamma\left(1-\frac{s}{\alpha}\right)$.

For the sake of simplicity let us restrict ourselves to the case of the simple poles, *i.e.*, we suppose that the conditions $1/2 + k \neq \alpha + \alpha n$ are fulfilled for all $k, n \in \mathbb{N}$, *i.e.*, that the derivative order α cannot be represented in the form $\alpha = \frac{1/2+k}{1+n}, \; k,n \in \mathbb{N}$. In particular, evidently it is the case if α is not a rational number or a rational number in the form $\alpha = \frac{p}{2q+1}, \; p, q \in \mathbb{N}$.

Taking into account the known formula

$$\mathrm{res}_{s=-k}\Gamma(s) = \frac{(-1)^k}{k!}, \; k = 0,1,2,\ldots,$$

the Jordan lemma and the Cauchy residue theorem provide us with the desired series representation of G_α:

$$G_\alpha(x,t) = \frac{1}{\alpha\sqrt{\pi}|x|}\left(\Sigma_1(y) + \Sigma_2(y)\right), \; y = \frac{4t}{x^2}, \tag{22}$$

where

$$\Sigma_1(y) = \sum_{k=0}^{\infty}\frac{(-1)^k}{k!}\frac{\Gamma\left(\frac{\frac{1}{2}+k}{\alpha}\right)\Gamma\left(1-\frac{\frac{1}{2}+k}{\alpha}\right)}{\Gamma\left(\frac{1}{2}+k\right)\Gamma\left(1-\left(\frac{1}{2}+k\right)\right)}y^{-k-\frac{1}{2}},$$

$$\Sigma_2(y) = \alpha\sum_{k=0}^{\infty}\frac{(-1)^k}{k!}\frac{\Gamma\left(\frac{1}{2}-\alpha(k+1)\right)\Gamma(k+1)}{\Gamma(\alpha(k+1))\Gamma(1-\alpha(k+1))}y^{-\alpha(k+1)}.$$

Using the reflection formula for the gamma function and some elementary trigonometric formulas, the series for Σ_1 and Σ_2 can be represented in the following shorter form:

$$\Sigma_1(y) = y^{-\frac{1}{2}} \sum_{k=0}^{\infty} \frac{\sin(\pi(1/2+k))}{\sin(\pi(1/2+k)/\alpha))} \frac{(-y)^{-k}}{k!} = y^{-\frac{1}{2}} \sum_{k=0}^{\infty} \frac{1}{\sin(\pi(1/2+k)/\alpha))} \frac{y^{-k}}{k!}$$

$$\Sigma_2(y) = \alpha y^{-\alpha} \sum_{k=0}^{\infty} \frac{\tan(\pi\alpha(k+1))}{\Gamma(1/2+\alpha(k+1))} (-y^{\alpha})^{-k}.$$

In the case $\alpha = 1$, we get

$$\Sigma_1(y) = y^{-\frac{1}{2}} \sum_{k=0}^{\infty} \frac{(y)^{-k}}{(-1)^k k!} = y^{-\frac{1}{2}} \exp(-1/y), \; \Sigma_2(y) \equiv 0$$

and thus the well-known formula

$$G_1(x,t) = \frac{1}{\sqrt{\pi}|x|} \left(\Sigma_1(y) + \Sigma_2(y)\right) = \frac{1}{\sqrt{\pi}|x|} y^{-\frac{1}{2}} \exp(-1/y) = \frac{1}{\sqrt{4\pi t}} \exp\left(-\frac{x^2}{4t}\right)$$

for the fundamental solution of the one-dimensional diffusion equation.

Let us now introduce another auxiliary variable, namely, $z = \frac{1}{y} = \frac{x^2}{4t}$. Then the formulas from above can be rewritten in the following form:

$$G_\alpha(x,t) = \frac{1}{\alpha\sqrt{\pi}|x|} \left(\Sigma_3(z) + \Sigma_4(z)\right), \; z = \frac{x^2}{4t}, \tag{23}$$

where

$$\Sigma_3(z) = z^{\frac{1}{2}} \sum_{k=0}^{\infty} \frac{1}{\sin(\pi(1/2+k)/\alpha))} \frac{z^k}{k!},$$

$$\Sigma_4(z) = \alpha z^{\alpha} \sum_{k=0}^{\infty} \frac{\tan(\pi\alpha(k+1))}{\Gamma(1/2+\alpha(k+1))} (-z^{\alpha})^k.$$

It follows from the last formula that the asymptotic behavior of G_α as $z \to 0$ (e.g., as $x \to 0$ with a fixed $t > 0$ or $t \to +\infty$ with a fixed $x \neq 0$) is of a power law type:

$$G_\alpha(x,t) \approx \frac{1}{\alpha\sqrt{4\pi}\sin\left(\frac{\pi}{2\alpha}\right)} t^{-\frac{1}{2}} + \frac{\tan(\pi\alpha)}{4^{\alpha}\sqrt{\pi}\Gamma(1/2+\alpha)} t^{-\alpha} |x|^{2\alpha-1}. \tag{24}$$

We remember the readers that all derivations above are valid only under the condition $\alpha > 1/2$ that we assumed to hold true.

To get the asymptotic behavior of the fundamental solution G_α as $z \to +\infty$ (e.g., as $|x| \to +\infty$ with a fixed $t > 0$ or $t \to 0$ with a fixed $x \neq 0$) we again employ the Mellin-Barnes representation (16). This time, the contour of integration in the integral at the right-hand side of (16) has to be transformed to the infinite loop $\mathcal{L}_{-\infty}$ starting and ending at $-\infty$ and encircling all poles $s_k = -\alpha k$, $k = 0, 1, 2, \ldots$ of the gamma function $\Gamma\left(\frac{s}{\alpha}\right)$.

In doing so we first get an asymptotic series

$$G_\alpha(x,t) \approx -\frac{1}{\pi^{\frac{3}{2}}|x|} \sum_{k=1}^{\infty} \Gamma\left(\frac{1}{2} + \alpha k\right) \sin(\pi\alpha k) \left(-\left(\frac{1}{z}\right)^{\alpha}\right)^k, \; z \to +\infty$$

and then the asymptotic formula

$$G_\alpha(x,t) \approx \frac{\Gamma\left(\frac{1}{2}+\alpha\right)\sin(\pi\alpha)}{4^{-\alpha}\pi^{\frac{3}{2}}}\, t^\alpha\, |x|^{-2\alpha-1}, \quad \frac{x^2}{4t} \to \infty. \tag{25}$$

3. Entropy Production Rate of the Alpha-Fractional Diffusion Process

For $\alpha = 1$, the fundamental solution G_α is a Gaussian probability density function evolving in time:

$$G_1(x,t) = \frac{1}{\sqrt{4\pi t}} \exp\left(-\frac{x^2}{4t}\right).$$

As has been shown in [11], the fundamental solution G_α to the α-fractional diffusion Equation (1) can be interpreted as a probability density function evolving in time for all values of α between 0 and 1, too. Let us mention that it follows from the asymptotic Formula (25) that the second spatial moment of the probability density function G_α does not exist and the mean squared displacement of the diffusing particles in the framework of the diffusion process that is governed by the α-fractional diffusion Equation (1) is not finite. Thus the α-fractional diffusion Equation (1) describes a kind of an anomalous diffusion. Still, we show in this section that the entropy production rate of a diffusion process that is governed by the α-fractional diffusion Equation (1) is exactly the same as the one of the conventional diffusion process.

Let us start with some definitions and examples. In the case of a one-dimensional continuous random variable with the probability density function $p(x)$, $x \in \mathcal{X} \subseteq \mathbf{R}$, we adopt the Shannon definition of the entropy:

$$S(p) = -k \int_{-\infty}^{\infty} p(x)\, \ln(p(x))\, dx, \tag{26}$$

where the constant k can be set to be equal to one without loss of the generality. The Shannon entropy (26) is a special case of the more general definitions by Mathai, Tsallis or Rényi and these entropies will be considered elsewhere.

Let us mention that the Shannon entropy of a Gaussian random variable defined by the probability density function

$$\mathcal{N}(\mu;\sigma^2) = \frac{1}{\sqrt{2\pi\sigma^2}} \exp\left(-\frac{(x-\mu)^2}{2\sigma^2}\right)$$

has the form

$$S(\mathcal{N}(\mu;\sigma^2)) = \frac{1}{2}(1+\ln(2\pi\sigma^2)). \tag{27}$$

Thus the entropy increases with the width σ^2 of the probability density function $\mathcal{N}(\mu;\sigma^2)$, *i.e.*, the broader the distribution (uncertainty of the event), the larger the entropy, so that the Shannon entropy can be interpreted as a measure of uncertainty of an event that is governed by a probability density function $p(x)$.

When a probability density function is time-dependent, the entropy (26) depends on time, too:

$$S(p,t) = -\int_{-\infty}^{\infty} p(x,t)\, \ln(p(x,t))\, dx. \tag{28}$$

For such time-dependent random processes, the entropy production rate R defined by

$$R(p,t) = \frac{d}{dt} S(p,t)$$

is a very important characteristic that can be interpreted as a natural measure of the irreversibility of a process. Say, in the case of a diffusion process that is described by the one-dimensional diffusion equation with the diffusion coefficient taken to be equal to one and that is therefore governed by the Gaussian distribution $\mathcal{N}(0; 2t)$, the Formula (27) leads to the following result:

$$R(\mathcal{N}(0;2t)) = \frac{d}{dt} S(\mathcal{N}(0;2t)) = \frac{1}{2t}, \tag{29}$$

i.e., the entropy production rate is strictly positive for $t > 0$ and the diffusion process can be classified as an irreversible process.

Otherwise, a wave propagation described by the wave equation is a reversible process with the entropy production rate equal to zero for $t > 0$.

It is worth mentioning that the entropy production rates for the time- and the space-fractional diffusion equations that were calculated in [3] and [5], respectively, depend on the derivative order α and increase with increasing of α from 1 (diffusion) to 2 (wave propagation) that results in the so called entropy production paradox (see [3,5] for attempts of resolving this paradox).

To calculate the entropy of the process governed by the one-dimensional α-fractional diffusion Equation (1), let us employ the representation (19) of its fundamental solution in terms of the auxiliary function L_α. Substituting (19) into (28) and after some elementary transformations, we get the following chain of equalities:

$$S(\alpha,t) = -\int_{-\infty}^{\infty} \frac{1}{|x|} L_\alpha\left(\frac{x^2}{4t}\right) \ln\left(\frac{1}{|x|} L_\alpha\left(\frac{x^2}{4t}\right)\right) dx = \tag{30}$$

$$-\int_0^\infty \frac{2}{x} L_\alpha\left(\frac{x^2}{4t}\right) \ln\left(\frac{1}{x} L_\alpha\left(\frac{x^2}{4t}\right)\right) dx = -\int_0^\infty \frac{1}{\tau} L_\alpha(\tau) \ln\left(\frac{1}{2} t^{-\frac{1}{2}} \tau^{-\frac{1}{2}} L_\alpha(\tau)\right) d\tau =$$

$$-\int_0^\infty \frac{1}{\tau} L_\alpha(\tau) \left(-\frac{1}{2}\ln(t) + \ln\left(\frac{1}{2}\tau^{-\frac{1}{2}} L_\alpha(\tau)\right)\right) d\tau = A_\alpha \ln(t) + B_\alpha,$$

where

$$A_\alpha = \frac{1}{2}\int_0^\infty L_\alpha(\tau)\,\tau^{-1}\,d\tau, \quad B_\alpha = -\int_0^\infty \frac{L_\alpha(\tau)}{\tau} \ln\left(\frac{1}{2}\tau^{-\frac{1}{2}} L_\alpha(\tau)\right) d\tau. \tag{31}$$

To determine the entropy production rate of G_α, the constant A_α is determined in explicit form. The integral that defines A_α can be interpreted as the Mellin transform of the auxiliary function L_α at the point $s = 0$. Because the Formula (21) for the Mellin transform of L_α was derived under the condition $-\min\{1/2, \alpha\} < \Re(s) < 0$, the Mellin transform of L_α at the point $s = 0$ will be calculated as the limit of the right-hand side of (21) as $s \to 0$. Thus we get the following chain of equalities:

$$A_\alpha = \frac{1}{2}\int_0^\infty L_\alpha(\tau)\,\tau^{-1}\,d\tau = \frac{1}{2}\lim_{s\to 0} \frac{1}{\alpha\sqrt{\pi}} \frac{\Gamma\left(\frac{1}{2}+s\right)\Gamma\left(-\frac{s}{\alpha}\right)\Gamma\left(1+\frac{s}{\alpha}\right)}{\Gamma(-s)\,\Gamma(1+s)} =$$

$$\frac{1}{2\alpha\sqrt{\pi}} \lim_{s\to 0} \Gamma\left(\frac{1}{2}+s\right) \frac{\sin(\pi s)}{\sin(\pi s/\alpha)} = \frac{\Gamma\left(\frac{1}{2}\right)}{2\alpha\sqrt{\pi}} \lim_{s\to 0} \frac{\pi s}{\pi s/\alpha} = \frac{1}{2}.$$

This formula along with the Formula (30) leads to the following expression for the entropy production rate $R(t)$ of the α-fractional diffusion process described by the Equation (1):

$$R(t) = \frac{d}{dt} S(\alpha,t) = \frac{1}{2t}. \tag{32}$$

This formula shows that $R(t)$ does not depend on the equation order α and is exactly the same as the entropy production rate of the conventional one-dimensional diffusion equation.

4. Conclusions

In this paper, a special case of the time- and space-fractional diffusion equation has been considered, namely, the case when the quotient of the orders of the time- and space-fractional derivatives is equal to one half as it is the case in the conventional diffusion equation. This special choice of the derivative orders leads to several important consequences. On the one hand, the fundamental solution to the α-fractional equation can be expressed via an auxiliary function of the argument $\frac{x^2}{4t}$ like in the case of the conventional diffusion equation. Another important property of the α-fractional diffusion equation is that its entropy production rate is exactly the same as the one of the diffusion process. Thus the α-fractional diffusion equation can be considered to be a "natural fractionalization" of the diffusion equation. On the other hand, the mean squared displacement of the diffusing particles governed by the α-fractional diffusion equation is infinite and thus this equation describes an anomalous diffusion process.

Conflicts of Interest: The author declares no conflict of interest.

References

1. Metzler, R.; Jeon, J.-H.; Cherstvy, A.G.; Barkai, E. Anomalous diffusion models and their properties: Non-stationarity, non-ergodicity, and ageing at the centenary of single particle tracking. *Phys. Chem. Chem. Phys.* **2014**, *16*, 24128–24164.
2. Shannon, C.E. A mathematical theory of communication. *Bell Syst. Tech. J.* **1948**, *27*, 379–423.
3. Hoffmann, K.H.; Essex, C.; Schulzky, C. Fractional diffusion and entropy production. *J. Non-Equilib. Thermodyn.* **1998**, *23*, 166–175.
4. Li, X.; Essex, C.; Davison, M.; Hoffmann, K.H.; Schulzky, C. Fractional diffusion, irreversibility and entropy. *J. Non-Equilib. Thermodyn.* **2003**, *28*, 279–291.
5. Prehl, J.; Essex, C.; Hoffmann, K.H. The superdiffusion entropy production paradox in the space-fractional case for extended entropies. *Phy. A* **2010**, *389*, 214–224.
6. Prehl, J.; Essex, C.; Hoffmann, K.H. Tsallis relative entropy and anomalous diffusion. *Entropy* **2012**, *14*, 701–716.
7. Luchko, Y. Wave-diffusion dualism of the neutral-fractional processes. *J. Comput. Phys.* **2015**, *293*, 40–52.
8. Mathai, A.M.; Rathie, P.N. *Basic Concepts in Information Theory and Statistics: Axiomatic Foundations and Applications*; Wiley Eastern: New Delhi, India, 1975.
9. Mathai, A.M.; Haubold, H.J. Pathway model, superstatistics, Tsallis statistics and a generalized measure of entropy. *Phys. A* **2007**, *375*, 110–122.
10. Mathai, A.M.; Haubold, H.J. On generalized entropy measures and pathways. *Phys. A* **2007**, *385*, 493–500.
11. Mainardi, F.; Luchko, Yu.; Pagnini, G. The fundamental solution of the space-time fractional diffusion equation. *Fract. Calc. Appl. Anal.* **2001**, *4*, 153–192.
12. Gorenflo, R.; Mainardi, F. Random walk models approximating symmetric space-fractional diffusion processes. In *Problems in Mathematical Physics*; Elschner, J., Gohberg, I., Silbermann, B., Eds.; Birkhäuser Verlag: Basel, Switzerland, 2001; pp. 120–145.
13. Saichev, A.; Zaslavsky, G. Fractional kinetic equations: Solutions and applications. *Chaos* **1997**, *7*, 753–764.
14. Samko, S.G.; Kilbas, A.A.; Marichev, O.I. *Fractional Integrals and Derivatives: Theory and Applications*; Gordon and Breach: New York, NY, USA, 1993.
15. Feller, W. On a generalization of Marcel Riesz' potentials and the semi-groups generated by them. *Comm. Sém. Mathém. Univ. Lund.* **1952**, 73–81.
16. Gorenflo, R.; Mainardi, F. Random walk models for space-fractional diffusion processes. *Frac. Calc. Appl. Anal.* **1998**, *1*, 167–191.
17. Luchko, Y. Operational method in fractional calculus. *Fract. Calc. Appl. Anal.* **1999**, *2*, 463–489.
18. Luchko, Y.; Gorenflo, R. An operational method for solving fractional differential equations with the Caputo derivatives. *Acta Math. Vietnam.* **1999**, *24*, 207–233.

19. Marichev, O.I. *Handbook of Integral Transforms of Higher Transcendental Functions, Theory and Algorithmic Tables*; Ellis Horwood: Chichester, UK, 1983.
20. Luchko, Yu.; Kiryakova, V. The Mellin integral transform in fractional calculus. *Fract. Calc. Appl. Anal.* **2013**, *16*, 405–430.

axioms

MDPI

Article

On the q-Laplace Transform and Related Special Functions

Shanoja R. Naik [1,*] **and Hans J. Haubold** [2]

[1] Department of Mathematics and Statistics, University of Regina, Regina, SK S4S 0A2, Canada
[2] Office for Outer Space Affairs, United Nations, Vienna International Centre, Vienna 1400, Austria; hans.haubold@gmail.com
* Correspondence: shanoja.naik@yahoo.ca; Tel.: +1-519-883-1478

Academic Editor: Javier Fernandez
Received: 12 July 2016; Accepted: 30 August 2016; Published: 6 September 2016

Abstract: Motivated by statistical mechanics contexts, we study the properties of the q-Laplace transform, which is an extension of the well-known Laplace transform. In many circumstances, the kernel function to evaluate certain integral forms has been studied. In this article, we establish relationships between q-exponential and other well-known functional forms, such as Mittag–Leffler functions, hypergeometric and H-function, by means of the kernel function of the integral. Traditionally, we have been applying the Laplace transform method to solve differential equations and boundary value problems. Here, we propose an alternative, the q-Laplace transform method, to solve differential equations, such as as the fractional space-time diffusion equation, the generalized kinetic equation and the time fractional heat equation.

Keywords: convolution property; G-transform; Gauss hypergeometric function; generalized kinetic equation; Laplace transform; Mittag–Leffler function; versatile integral

1. Introduction

The classical Laplace, Fourier and Mellin transforms have been widely used in mathematical physics and applied mathematics. The theory of the Laplace transform is well-known [1], and its generalization was considered by many authors [2–6]. Various existence conditions and detailed study about the range and invertibility were studied by Rooney [7]. The Laplace transform and Mellin transform are widely used together to solve the fractional kinetic equations and thermonuclear equations [8,9]. Different types of integral transforms, like the Hankel transform, Erdély–Kober type fractional integration operators, the Gauss hypergeometric function as a kernel, the Bessel-type integral transform, etc. [10], are introduced in the literature to solve the boundary value problems for models of ordinary and partial differential equations. In some situations, the solutions of the differential equation cannot be tractable using the classical integral transforms, but may be characterized by many integral transforms with various special functions as kernels. Many of the integral transforms can be interpreted in terms of the G-transform and H-transform [11–16].

In physical situations when an appropriate density is selected, the best practice is to maximize the entropy. Mathai and Rathie [17] considered various generalizations of the Shannon entropy measure and describe various properties, including additivity, the characterization theorem, etc. Mathai and Haubold [18] introduced a new generalized entropy measure, which is a generalization of the Shannon entropy measure. For a multinomial population $P = (p_1, \ldots, p_k), p_i \geq 0$, $i = 1, \ldots, k, p_1 + p_2 + \cdots + p_k = 1$, the Mathai's entropy measure (discrete case) is given by the relation:

$$M_{k,\alpha}(P) = \frac{\sum_{i=1}^{k} p_i^{2-\alpha} - 1}{\alpha - 1}, \ \alpha \neq 1, \ -\infty < \alpha < 2.$$

When $\alpha \rightarrow 1$, the above measure goes to the Shannon entropy measure, and this is a variant of Havrda–Charvat entropy and Tsallis entropy. One can derive Tsallis statistics and superstatistics [19–22] by using Mathai's entropy. By optimizing Mathai's entropy measure, a new pathway model, which consists of many of the standard distributions in the statistical literature as special cases (see [23]), is derived. The main idea behind the derivation of this model is the switching properties of the special functions, like ${}_1F_1$ and ${}_1F_0$, which means the binomial to exponential function. Thus, the pathway between the exponential function e^{-cx} and the binomial function $[1 - c(1-\alpha)x]^{\frac{1}{1-\alpha}}$ can be created with the parameter α named as the pathway parameter. For the real scalar case, the pathway density can be written in the form:

$$f_1(x) \quad = \quad c|x|^{\gamma}[1 - a(1-\alpha)|x|^{\delta}]^{\frac{\eta}{1-\alpha}}, a > 0, 1 - a(1-\alpha)|x|^{\delta} \geq 0, \eta > 0, \alpha < 1$$

where c is the normalizing constants. One can assume the Type 2 model by replacing $(1-\alpha)$ by $-(\alpha - 1)$. These distributions include Type 1 beta, Type 2 beta, gamma, Weibull, Gaussian, Cauchy, exponential, Rayleigh, Student t, Fermi–Dirac, chi-square, logistic, etc. The corresponding asymmetric generalization was introduced and studied in the paper [24]. By representing the entropy function in terms of a density function $f(\cdot)$ for the continuous case and giving the suitable constraints therein, the generalized entropy is maximized. There are restrictions, such as the $[(\gamma - 1)(1-\alpha)]$-th moment, and the $[(\gamma - 1)(1-\alpha) + \delta]$-th moments are constants for fixed $\gamma > 0$ and $\delta > 0$. Maximizing Mathai's entropy by using the calculus of variations, we get the basic function of the model, and when the range of x is restricted over the positive real line and by evaluating the normalizing constant, we get the pathway model introduced by Mathai [23]. As $q \rightarrow 1$, $f_1(x)$ tend to $f_2(x)$, which is the generalized gamma distribution, where $f_2(x)$ is given by:

$$f_2(x) = \frac{\delta(a\beta)^{\frac{\alpha}{\delta}}}{2\Gamma\left(\frac{\alpha}{\delta}\right)} \mid x \mid^{\alpha-1} \ \exp(-a\beta \mid x \mid^{\delta}); \quad -\infty < x < \infty; a, \alpha, \beta, \delta > 0. \tag{1}$$

For different values of parameters in the pathway model, we get different distributions like Weibull, gamma, beta Type 1, beta Type 2, etc. By taking $\delta = \alpha$, $\beta = 1$, $a = \lambda^{\alpha}$ in $f_1(x)$, the pathway model reduces to the q-Weibull distribution, which facilitates a transition to the Weibull distribution [25]. The connection of pathway models and Tsallis statistics with the q-extended versions of various functions is also considered. To this extent, we generalize the Laplace transform using the switching property of ${}_0F_0$ to ${}_0F_1$. Here, the q-exponential function is the kernel, and we call the extension as the q-Laplace transform; as q approaches to unity, we get the Laplace transform of the original function.

The article is organized as follows. In Section 2, we introduce the q-Laplace transform and the obtained various properties of the transform. Section 3 deals with the q-Laplace transform of some basic functions, which includes special functions, like the hypergeometric function, the Mittag–Leffler function and the H-function. In Section 4, this transform is connected to other known integral transforms, like the Mellin transform, the G-transform and the Henkel transforms. In Section 5, we obtain the solution of the fractional space-time diffusion equation, the generalized kinetic equation and the time fractional heat equation through the q-Laplace transform in terms of the Mittag–Leffler function.

2. The q-Laplace Transform and Basic Properties

The Laplace transform L of a function $f(\cdot)$ is given by:

$$L[f(x)](s) \equiv \int_0^\infty f(x)\, \mathrm{e}^{-sx}\, \mathrm{d}x$$

where $f(x)$ is defined over the positive real line and $s \in \mathbb{C}, \Re(s) > 0$, $\Re(\cdot)$ denotes the real part of $(\cdot)$. This Laplace transform plays a major role in pure and applied analysis, especially in solving differential equations. Now, we define the extended Laplace transform concept, namely the q-Laplace transform of a function, which will play a similar role in mathematical analysis, as well as mathematical physics. Instead of the exponential function, here, we consider the e_q^{-sx} the q-exponential defined as:

$$\mathrm{e}_q^{-x} \equiv c \begin{cases} [1-(1-q)x]^{\frac{1}{1-q}} \text{ for } 0 < x < \frac{1}{1-q}, q < 1 \\ [1+(q-1)x]^{-\frac{1}{q-1}} \text{ for } x \geq 0, q > 1 \end{cases} \tag{2}$$

with $\mathrm{e}_1^x \equiv \mathrm{e}^x$ and c is the normalizing constant. More precisely, for given function $f(\cdot)$ and for $s \in \mathbb{C}$ with support over $(0,\infty)$, we define its q-Laplace transform as:

$$L_q[f(x)](s) = \int_0^\infty [\mathrm{e}_q^{-sx}] f(x) \mathrm{d}x \text{ for }, \Re(s) > 0 \tag{3}$$

where e_q^{-x} is defined as in Equation (2). This Laplace transform can be written in the form,

$$L_q[f(x)](s) = \begin{cases} \int_0^{\frac{1}{(1-q)s}} [1-(1-q)sx]^{\frac{1}{1-q}} f(x)\mathrm{d}x & \text{for } \Re(1-(1-q)sx) > 0, \Re(s) > 0 \\ \int_0^\infty [1+(q-1)xs]^{-\frac{1}{q-1}} f(x)\mathrm{d}x & \text{for } \Re(s) > 0. \end{cases}$$

The q-Laplace transform of a function $f(\cdot)$ is valid at every point at which $f(\cdot)$ is continuous provided that the function is defined in $(0,\infty)$, is piecewise continuous and of bounded variation in every finite subinterval in $(0,\infty)$, and the integral is finite. Some basic properties of the q-Laplace transform are given below.

1. Scaling: For a real constant k, $L_q[kf(x)](s) = kL_q[f(x)](s)$.
2. Linearity : $L_q[mf(x) + ng(x)](s) = mL_q[f(x)](s) + nL_q[g(x)](s)$, where $m, n \in \Re$.
3. Transform of derivatives: For $\Re(s) > 0$, $L_q[\frac{\mathrm{d}}{\mathrm{d}x} f(x)](s) = sL_q(f)(sq)$ for all $q \in \Re/\{1\}$.

Proof. Let $g(x) = \frac{\mathrm{d}}{\mathrm{d}x} f(x)$. Then:

$$L_q\,[g(x)]\,(s) = \int_0^\infty [1+(q-1)xs]^{-\frac{1}{q-1}} \frac{\mathrm{d}}{\mathrm{d}x}[f(x)]\mathrm{d}x.$$

By applying integration by parts, we get:

$$\int_0^\infty [1+(q-1)xs]^{-\frac{1}{q-1}} \frac{\mathrm{d}}{\mathrm{d}x}[f(x)]\mathrm{d}x = f(x)[1+(q-1)xs]^{-\frac{1}{q-1}}\Big|_0^\infty - \int_0^\infty f(x)(-s)[1+(q-1)xs]^{-\frac{1}{q-1}-1}\mathrm{d}x$$

which implies:

$$L_q[g(x)] = -f(0) + sL_{\frac{2q-1}{q}}(f)(sq).$$

As a consequence, we get:

$$D^n\{L_q\left[f(x)\right](s)\} = s^n L_q\left[f(x)\right](sA_n(q)) - \sum_{j=1}^{n} s^{n-j} D^j f(0) \tag{4}$$

where $A_n(q) = \prod_{j=1}^{n}(j + (j-1)q)$, for $q > 1$, $D = \frac{d}{dx}f(x)$. □

4. Derivatives of transforms: The nth derivative of the q-Laplace transform is given by $D^n\left[L_q(f)(s)\right] = A_{n-1}(q)\{L_q\left[(-x)^n\ {}_nA^{-1}(q)f(x\ {}_nA^{-1})\right](s)\}$ where ${}_nA^{-1}$ is the reciprocal of the nth term of $A_n(q)$.

Proof. For $q > 1$:

$$\begin{aligned}
D^2(\mathrm{e}_q^{-sx}) &= qx^2[\mathrm{e}_q^{-sx}]^{1-2q} \\
D^3(\mathrm{e}_q^{-sx}) &= q(1-2q)(-x^3)[\mathrm{e}_q^{-sx}]^{2-3q} \\
&\vdots \\
D^n(\mathrm{e}_q^{-sx}) &= \prod_{j=1}^{n-1} A_j \int_0^\infty (-x)^n \mathrm{e}_q^{-s\,{}_nA^{-1}(q)x} f({}_nA^{-1}(q)x)\mathrm{d}x \\
&= A_{n-1}(q)\{L_q\left[(-x)^n\ {}_nA^{-1}(q)f(x\ {}_nA^{-1})\right]\}(s).
\end{aligned}$$

□

5. Transforms of integrals: For $\Re(s) > 0$, $L_q\left[\int_0^x f(t)\mathrm{d}t\right](s) = \frac{1}{s}L_q(f)(s)$.

Proof. For $q > 1$, we have:

$$\begin{aligned}
L_q\left[\int_0^x f(t)\mathrm{d}t\right](s) &= \int_0^\infty [\mathrm{e}_q^{-sx}]\{\int_0^x f(t)\mathrm{d}t\}\mathrm{d}x \\
&= -\frac{1}{s}\int_0^\infty \{\int_0^x f(t)\mathrm{d}t\}\frac{\mathrm{d}}{\mathrm{d}x}[\mathrm{e}_q^{-s(2-q)x}]\mathrm{d}x \\
&= -\frac{1}{s}\left\{\int_0^x f(t)\mathrm{d}t[\mathrm{e}_q^{-s(2-q)x}]\Big|_0^\infty - \int_0^\infty f(x)\mathrm{e}_q^{-sx}\mathrm{d}x\right\} \\
&= \frac{1}{s}L_q(f)(s) \text{ for } \Re(s) > 0.
\end{aligned}$$

□

6. Convolution property: Let $f_1(x)$ and $f_2(x)$ be two positive real scalar functions of x, and let $g_1(t)$ and $g_2(t)$ be their q-Laplace transform. Then,

$$L_q[f_1(x) * f_2(x)](s) = g_1(x)g_2(x)$$

where $f_1(x) * f_2(x) = \int_0^x f_1(t)f_2(x-t)\mathrm{d}t$.

Proof.

$$\begin{aligned}
L_q[f_1(x) * f_2(x)](s) &= \int_0^\infty \mathrm{e}_q^{-sx}\left\{\int_0^x f_1(t)f_2(x-t)\mathrm{d}t\right\}\mathrm{d}x \\
&= \int_0^\infty \int_0^x [1+(q-1)xs]^{-\frac{1}{q-1}} f_1(t)f_2(x-t)\mathrm{d}t\mathrm{d}x \\
&= \int_{t=0}^\infty f_1(t)\left\{\int_{x=t}^\infty [1+(q-1)xs]^{-\frac{1}{q-1}} f_2(x-t)\mathrm{d}x\right\}\mathrm{d}t.
\end{aligned}$$

Now, let us consider the integral $I = \int_{x=t}^{\infty}[1+(q-1)xs]^{-\frac{1}{q-1}} f_2(x-t)\mathrm{d}t$. Substitute $x - t = u$, and manipulate the integral; we get:

$$I = [1+(q-1)ts]^{-\frac{1}{q-1}} \quad \int_{x=0}^{\infty} \quad [1+(q-1)us]^{-\frac{1}{q-1}}[1+(q-1)ts]^{\frac{1}{q-1}}[1+(q-1)us]^{\frac{1}{q-1}} \\ \times \quad [1+(q-1)(t+u)s]^{-\frac{1}{q-1}} f_2(u)\mathrm{d}u.$$

Let $[1+(q-1)ts]^{\frac{1}{q-1}}[1+(q-1)us]^{\frac{1}{q-1}}[1+(q-1)(t+u)s]^{-\frac{1}{q-1}} f_2(u) = f_2^*(u)$, then $f_2(x-t) = [1+(q-1)ts]^{-\frac{1}{q-1}}[1+(q-1)(x-t)s]^{-\frac{1}{q-1}}[1+(q-1)xs]^{\frac{1}{q-1}} f_2^*(x-t)$. Then:

$$L_q[f_1(x) * f_2(x)](s) = \int_{t=0}^{\infty}[1+(q-1)ts]^{-\frac{1}{q-1}} f_1(t) \\ \left\{\int_{x=t}^{\infty}[1+(q-1)(x-t)s]^{-\frac{1}{q-1}} f_2^*(x-t)\mathrm{d}x\right\}\mathrm{d}t.$$

On substituting $x - t = u$, the integral can be separated, and hence, we have:

$$L_q[f_1(x) * f_2(x)](s) = L_q[f_1(x)]L_q[f_2^*(x)].$$

□

3. The q-Laplace Transform of Some Basic Functions

Let us introduce a new notation, $\Gamma_{(q)}(\alpha)$, such that:

$$\Gamma_{(q)}(\alpha) = \int_0^{\frac{1}{1-q}} x^{\alpha-1}[1-(1-q)x]^{\frac{1}{1-q}}\mathrm{d}x \text{ for } \Re(\alpha) > 0, q < 1.$$

If we replace $(1-q)$ by $-(q-1)$, then the function assumes the form:

$$\Gamma_{(q)}(\alpha) = \int_0^{\infty} x^{\alpha-1}[1+(q-1)x]^{-\frac{1}{q-1}}\mathrm{d}x \text{ for } \Re(\alpha) > 0, q > 1$$

and for $q = 1$ in the sense $q \to 1$, the q-gamma function is the usual classical gamma function defined as

$$\Gamma(\alpha) = \int_0^{\infty} x^{\alpha-1}\mathrm{e}^{-x}\mathrm{d}x.$$

Now, the q-gamma function can be explicitly written as:

$$\Gamma_{(q)}(\alpha) = \begin{cases} \frac{1}{(1-q)^{\alpha}} \frac{\Gamma(\alpha)\,\Gamma(\frac{1}{1-q}+1)}{\Gamma(\frac{1}{1-q}+\alpha+1)} & \text{for } q < 1 \\ \Gamma(\alpha) & \text{for } q = 1 \\ \frac{1}{(q-1)^{\alpha}} \frac{\Gamma(\alpha)\,\Gamma(\frac{1}{q-1}-\alpha)}{\Gamma(\frac{1}{q-1})} & \text{for } q > 1, \frac{1}{q-1} - \alpha > 0 \end{cases} \tag{5}$$

for $\Re(\alpha) > 0$. Here, $q = 1$ in the sense $q \to 1$ the q-gamma function $\Gamma_{(q)}(\cdot) \to \Gamma(\cdot)$, which can be easily proven using the asymptotic expansion of the gamma function:

$$\Gamma(z+a) \approx \sqrt{2\pi} z^{z+a-\frac{1}{2}}\mathrm{e}^{-z}.$$

Mathai [26] introduced a general class of integrals, known as the versatile integrals, which are connected to the reaction rate in kinetic theory. The integral is in the form:

$$I = \int_0^{\infty} x^{\gamma-1}[1+z_1^{\delta}(\alpha-1)x^{\delta}]^{-\frac{1}{\alpha-1}}[1+z_2^{\delta}(\beta-1)x^{-\rho}]^{-\frac{1}{\beta-1}} \tag{6}$$

for $\alpha, \beta > 1, z_1, z_2 \geq 0, \delta, \rho > 0, \Re(\gamma+1) > 0, \Re(\frac{1}{\alpha-1} - \frac{\gamma+1}{\delta}) > 0, \Re(\frac{1}{\beta-1} - \frac{1}{\rho}) > 0$, and the solution is obtained in terms of the H-function as follows:

$$I = c\, H_{2,2}^{2,2}\left[z_1 z_2 (\alpha-1)^{\frac{1}{\delta}} (\beta-1)^{\frac{1}{\rho}} \Big|_{(\frac{\gamma}{\delta},\frac{1}{\delta}),(0,\frac{1}{\rho})}^{(1-\frac{1}{\alpha-1}+\frac{\gamma}{\delta},\frac{1}{\delta}),(1-\frac{1}{\beta-1},\frac{1}{\rho})} \right]$$

where $c = \delta\rho z_1^{\gamma}(\alpha-1)^{\frac{\gamma}{\delta}}$ and $H_{p,r}^{m,n}$ is a H-function. Here, we provide the definition of H-function as follows:

$$H_{p,q}^{m,n}\left[z \Big|_{(b_1,\beta_1),(b_2,\beta_2),\cdots,(b_q,\beta_q)}^{(a_1,\alpha_1),(a_2,\alpha_2),\cdots,(a_k,\alpha_k)} \right] = \frac{1}{2\pi i}\int_L h(s) z^{-s} ds$$

where:

$$h(s) = \frac{\left\{\prod_{j=1}^{m}\Gamma(b_j+\beta_j s)\right\}\left\{\prod_{j=1}^{n}\Gamma(1-a_j-\alpha_j s)\right\}}{\left\{\prod_{j=m+1}^{q}\Gamma(1-b_j-\beta_j s)\right\}\left\{\prod_{j=n+1}^{p}\Gamma(a_j+\alpha_j s)\right\}}$$

and L is a suitable path. An empty product is interpreted as unity, and it is assumed that the poles of $\Gamma(b_j+\beta_j s), j = 1,2,\ldots,m$ are separated from the poles of $\Gamma(1-a_j-\alpha_j s), j = 1,2,\ldots,n$. Here, $a_1, a_2, \ldots, a_p; b_1, b_2, \ldots, b_q$ are complex numbers and $\alpha_1, \alpha_2, \ldots, \alpha_p, \beta_1, \beta_2, \ldots, \beta_q$ are positive real numbers. The poles of $\Gamma(b_j+\beta_j s),\ j = 1,2,\ldots,m$ are at the points $s = -\frac{b_j+\nu}{\beta_j}$ where $j = 1,2,\ldots,m$, $\nu = 0,1,\ldots$, and the poles of $\Gamma(1-a_j-\alpha_j s), j = 1,2,\ldots,n$ are at $s = \frac{1-a_k+\lambda}{\alpha_k}$ where $k = 1,2,\ldots,n$, $\lambda = 0,1,\ldots$. For more details about the theory and applications, refer to [27]. This integral includes the q-Laplace transform of gamma function and q-gamma function as special cases. Now, as $q \to 1$ in any of the functions $[1+z_1^{\delta}(\alpha-1)x^{\delta}]^{-\frac{1}{\alpha-1}}$ or $[1+z_2^{\delta}(\beta-1)x^{-\rho}]^{-\frac{1}{\beta-1}}$, we get the q-Laplace transform of some basic functions. The following table gives the q-Laplace transform of some basic functions with $q > 1$, which are special cases of the above integral. The results are obtained in terms of hypergeometric function. The Gaussian hyper geometric function is defined as:

$${}_mF_n\left[\begin{matrix} a_1,a_2,\cdots,a_m \\ b_1,b_2,\cdots,b_n \end{matrix} \Big| x \right] = \sum_{k=0}^{\infty} \frac{(a_1)_k (a_2)_k \cdots (a_m)_k}{(b_1)_k (b_2)_k \cdots (b_n)_k} \frac{x^k}{k!}$$

where $(a)_m$ denotes the Pochhammer symbol expressed in the form:

$$(a)_m = a(a+1)...(a+m-1),\ a \neq 0,\ m = 1,2... \tag{7}$$

Lemma 1. *For $\alpha, s \in C$, $\Re(s) > 0$ and for $q \neq 1$, the q- Laplace transform of $x^{\alpha-1}$ is given by $L_q[x^{\alpha-1}](s) = \frac{\Gamma_q(\alpha)}{s^{\alpha}}$ for $q \neq 1$.*

Proof. For $q > 1$,

$$\begin{aligned} L_q[x^{\alpha-1}](s) &= \int_0^{\infty} x^{\alpha-1} e_q^{-sx} dx \\ &= \int_0^{\infty} x^{\alpha-1}[1+(q-1)sx]^{-\frac{1}{q-1}} dx. \end{aligned}$$

Now, substitute $(q-1)sx = t$, and $dx = \frac{1}{s(q-1)} du$. Then:

$$L_q[x^{\alpha-1}](s) = \frac{\Gamma_q(\alpha)}{s^{\alpha}},\ \alpha, s \in C, \Re(s) > 0.$$

□

Lemma 2. *For $s \in C$, $\Re(s) > 0$, there holds the formula:*

$$L_q[e^{-ax}](s) = \frac{1}{(2-q)s}\,{}_1F_1\left[1; \frac{2q-3}{q-1}; \frac{a}{s(q-1)}\right]$$

for $a > 0$, $\frac{3}{2} < q < 2$.

Proof. For $q > 1$,

$$\begin{aligned} L_q[e^{-ax}](s) &= \int_0^\infty [1+(q-1)sx]^{-\frac{1}{q-1}} e^{-ax} dx \\ &= \sum_{k=0}^{\infty} \frac{(-a)^k}{k!} \int_0^\infty x^k [1+(q-1)sx]^{-\frac{1}{q-1}} dx \\ &= \frac{1}{s} \sum_{k=0}^{\infty} \frac{(-\frac{a}{s})^k}{k!} \frac{\Gamma_{(q)}(k+1)}{s^{k+1}}, \Re(\frac{1}{q-1} - k - 1) > 0 \\ &= \frac{1}{s(2-q)}\,{}_1F_1\left[1; \frac{2q-3}{q-1}; \frac{a}{s(q-1)}\right] \text{ for } \frac{3}{2} < q < 2, \Re(s) > 0, a > 0. \end{aligned}$$

□

Lemma 3. *For $a \in \Re$, $\Re(s) > 0$, the q-Laplace transform of the function e_q^{-ax} is given by $L_q[e_q^{-ax}](s) = \frac{1}{(s+a)(2-q)}\,{}_2F_1\left[1, \frac{1}{2}; \frac{2q-3}{q-1}; -\frac{4as}{(a+s)^2}\right]$ for $\frac{3}{2} < q < 2$, $\left|\frac{4as}{(a+s)^2}\right| < 1$.*

Proof. For $q > 1$, the q-Laplace transform of the q-exponential function is given by:

$$\begin{aligned} L_q[e_q^{-ax}](s) &= \int_0^\infty [1+(q-1)sx]^{-\frac{1}{q-1}} [1+(q-1)ax]^{-\frac{1}{q-1}} dx \\ &= \int_0^\infty \{[1+(q-1)sx][1+(q-1)ax]\}^{-\frac{1}{q-1}} dx \\ &= \int_0^\infty \left[1+(q-1)sx+(q-1)ax+(q-1)^2asx^2\right]^{-\frac{1}{q-1}} dx \\ &= \frac{1}{(s+a)(2-q)}\,{}_2F_1\left[1, \frac{1}{2}; \frac{2q-3}{q-1}; -\frac{4as}{(a+s)^2}\right] \\ &\quad \text{provided } \frac{3}{2} < q < 2, \left|\frac{4as}{(a+s)^2}\right| < 1. \end{aligned}$$

□

Lemma 4. *For $\Re(s) > 0$ and for $\frac{3}{2} < q < 2$, $a \in \Re$, there holds the formula, $L_q[\cos(ax)](s) = \frac{1}{s(2-q)}\,{}_1F_2\left(1; \frac{2q-3}{2(q-1)}, \frac{3q-4}{2(q-1)}; -\frac{a^2}{[2s(q-1)]^2}\right)$.*

Proof. For $q > 1$, $\Re(s) > 0$, $a \in \Re$, the q-Laplace transform of the trigonometric function $\cos(ax)$ is given by:

$$\begin{aligned} L_q[\cos(ax)](s) &= \int_0^\infty [1+(q-1)sx]^{-\frac{1}{q-1}} \cos(ax) dx \\ &= \sum_{k=0}^{\infty} \frac{(-1)^k a^{2k}}{(2k)!} \int_0^\infty x^{2k} [1+(q-1)sx]^{-\frac{1}{q-1}} dx \end{aligned}$$

$$= \frac{1}{s(2-q)} \sum_{k=0}^{\infty} \frac{(a^2)^k}{[4(q-1)^2 s^2]^k} \frac{(1)_k}{k! \left(\frac{2q-3}{2(q-1)}\right)_k \left(\frac{3q-4}{2(q-1)}\right)_k} \text{ for } q > \frac{3}{2}.$$

By applying the properties of the beta function and integral evaluations, we get:

$$L_q[\cos(ax)](s) = \frac{1}{s(2-q)} {}_1F_2\left(1; \frac{2q-3}{2(q-1)}, \frac{3q-4}{2(q-1)}; -\frac{a^2}{[2s(q-1)]^2}\right),$$
$$\text{for } \frac{3}{2} < q < 2, q > 1, \Re(s) > 0, a \in \Re.$$

□

One can easily check that as $q \to 1$, the above function gives a direct connection to the Laplace transforms of the original function simply by applying Sterling's approximation for the gamma function involved in the hypergeometric function involved in the equation.

Lemma 5. *The q-Laplace transform of the Gauss hypergeometric function is given by:*

$$L_q[{}_mF_n](s) = \frac{1}{s(2-q)} \; {}_{m+1}F_{n+1}\left[\begin{matrix} a_1, a_2, \cdots, a_m, 1 \\ b_1, b_2, \cdots, b_n, \frac{2q-3}{q-1} \end{matrix} \middle| \frac{1}{(q-1)s}\right]$$

for $\Re(s) > 0, \frac{3}{2} < q < 2$.

Proof. For $q > 1$, the q-Laplace transform of the Gauss hyper geometric function is given by:

$$\begin{aligned} L_q[{}_mF_n](s) &= \sum_{k=0}^{\infty} \frac{(a_1)_k (a_2)_k \cdots (a_m)_k}{(b_1)_k (b_2)_k \cdots (b_n)_k} \frac{1}{k!} \int_0^{\infty} x^k [1 + (q-1)sx]^{-\frac{1}{q-1}} \\ &= \frac{1}{(2-q)s} \sum_{k=0}^{\infty} \frac{(a_1)_k (a_2)_k \cdots (a_m)_k, (1)_k}{(b_1)_k (b_2)_k \cdots (b_n)_k, (\frac{2q-3}{q-1})_k} \frac{\left(\frac{1}{(q-1)s}\right)^k}{k!} \\ &= \frac{1}{s(2-q)} \; {}_{m+1}F_{n+1}\left[\begin{matrix} a_1, a_2, \cdots, a_m, 1 \\ b_1, b_2, \cdots, b_n, \frac{2q-3}{q-1} \end{matrix} \middle| \frac{1}{(q-1)s}\right] \text{ for } \Re(s) > 0, \frac{3}{2} < q < 2. \end{aligned}$$

□

Corollary: When $m = n = 0$, we get the exponential function, and the q-Laplace transform is the confluent hypergeometric function ${}_1F_1$.

3.1. The q-Laplace Transform of the Mittag–Leffler Function

The single parameter Mittag–Leffler function is defined as follows:

$$E_\alpha(z) = \sum_{k=0}^{\infty} \frac{z^k}{\Gamma(1+\alpha k)}, \text{for } \alpha \in C, \Re(\alpha) > 0.$$

Lemma 6. *For* $q > 1, \Re(s) > 0$, *the q-Laplace transform of* $E_\alpha(x^\alpha)$ *is given by:*

$$L_q[E_\alpha(x^\alpha)](s) = \frac{1}{s(q-1)\Gamma(\frac{1}{q-1})} \; H^{2\,1}_{1\,2}\left[\frac{1}{s^\alpha (q-1)^\alpha} \middle|\begin{matrix} (0,1) \\ (0,1), (\frac{2-q}{q-1}, \alpha) \end{matrix}\right]$$

with suitable restrictions for the existence of Mittag–Leffler function.

Proof. For $q > 1, \Re(s) > 0$, the q-Laplace transform of the Mittag–Leffler function is given by:

$$\begin{aligned} L_q[E_\alpha(x^\alpha)](s) &= \sum_{k=o}^{\infty} \frac{1}{\Gamma(1+\alpha k)} \int_0^\infty x^{\alpha k}[1+(q-1)sx]^{-\frac{1}{q-1}} dx \\ &= \sum_{k=o}^{\infty} \frac{\Gamma\left(\frac{2-q}{q-1} - \alpha k\right)}{\Gamma(\frac{1}{q-1})} \left(\frac{1}{s(q-1)}\right)^{\alpha k+1} \\ &= \frac{1}{s(q-1)\Gamma(\frac{1}{q-1})} H^{2\,1}_{1\,2}\left[\frac{1}{s^\alpha(q-1)^\alpha}\Bigg|^{(0,1)}_{(0,1),(\frac{2-q}{q-1},\alpha)}\right]. \end{aligned}$$

□

The generalized Mittag–Leffler function introduced by Prabhakar is defined as follows:

$$E^\delta_{\beta,\gamma}(z) = \sum_{k=0}^{\infty} \frac{(\delta)_n z^k}{\Gamma(\beta k+\gamma)}, \text{for } \beta,\gamma,\delta \in C, \Re(\gamma) > 0, \Re(\delta) > 0.$$

Lemma 7. *Let $\beta,\gamma,\delta \in C, \Re(\beta) > 0, \Re(\gamma) > 0, \Re(\delta) > 0, \Re(\frac{1}{q-1} - \gamma) > 0$, and for $1 < q < 2$, there holds the formula:*

$$L_q[E^\delta_{\beta,\gamma}(ax^\beta)](s) = \frac{1}{s^\gamma(q-1)^\gamma\Gamma(\delta)\Gamma(\frac{1}{q-1})} H^{2\,1}_{1\,2}\left[{}^{(1-\delta,1)}_{(0,1),(\frac{1}{q-1}-\gamma,\beta)}\Bigg|\frac{1}{s^\beta(q-1)^\beta}\right]$$

for $1 < q < 2, \Re(\beta) > 0, \Re(\gamma) > 0, \Re(\delta) > 0, \Re(\frac{1}{q-1} - \gamma) > 0.$

The proof is similar to Lemma (7).

The details of the existence conditions, various properties and applications of H-functions are available in [27].

3.2. The q-Laplace Transform of the Fox H Function

Lemma 8. *For $q < 1$, consider the following restrictions. Let $a^* = \sum_{i=1}^{n} a_i - \sum_{i=n+1}^{p} \alpha_i + \sum_{j=1}^{m} \beta_j - \sum_{j=m+1}^{r} \beta_j$, $\Delta = \sum_{j=1}^{r} \beta_j - \sum_{i=1}^{p} \alpha_i$ and $\mu = \sum_{j=1}^{r} b_j - \sum_{i=1}^{p} a_i + \frac{p-r}{2}$ from the basic definition of the H-function.*

If either $a^* > 0, a^* = 0, \Re(\mu) < -1, \min\limits_{1\le j\le m} \dfrac{\Re(b_j)}{\beta_j} > -1$

when $a^* > 0, a^* = 0, \Delta \ge 0, \min\limits_{1\le j\le m} \left[\dfrac{\Re(b_j)}{\beta_j}, \dfrac{\Re(\mu)+\frac{1}{2}}{\Delta}\right] > -1$

when $a^* = 0, \Delta < 0$, then for $1 < q < 2$, the q-Laplace transform of the H-function exists, and the formula:

$$L_q[H^{m,n}_{p,r}](s) = \frac{1}{s(q-1)\Gamma(\frac{1}{q-1})} H^{m+1,n+1}_{p+1,r+1}\left[\frac{1}{s(q-1)}\Bigg|^{(a_1,\alpha_1),(a_2,\alpha_2),\cdots,(a_k,\alpha_k),(0,1)}_{(b_1,\beta_1),(b_2,\beta_2),\cdots,(b_r,\beta_r),(\frac{2-q}{q-1},1)}\right]$$

holds for $s \in \mathbb{C}, \Re(s) > 0$.

Proof. For $q > 1$,

$$\begin{aligned} L_q[H^{m,n}_{p,r}](s) &= \frac{1}{2\pi i}\int_L h(t) \int_0^\infty x^{-t}[1+(q-1)sx]^{-\frac{1}{q-1}} dx \wedge dt \\ &= \frac{1}{s(q-1)\Gamma(\frac{1}{q-1})} \frac{1}{2\pi i}\int_L h(t)\, \Gamma(1-t)\, \Gamma\left(\frac{2-q}{q-1}+t\right) dt \end{aligned}$$

$$= \frac{1}{s(q-1)\Gamma(\frac{1}{q-1})} H^{m+1,n+1}_{p+1,r+1}\left[\frac{1}{s(q-1)}\Big|^{(a_1,\alpha_1),(a_2,\alpha_2),\cdots,(a_k,\alpha_k),(0,1)}_{(b_1,\beta_1),(b_2,\beta_2),\cdots,(b_r,\beta_r),(\frac{2-q}{q-1},1)}\right]$$

with suitable existing conditions. □

4. Connection to Other Integral Transforms

In this section, we consider connections of the q-Laplace transform of a function $f(\cdot)$ to other integral transforms. The following theorem gives a relation between the Mellin transform of the q-Laplace transform of a function, where the Mellin transform of the function $f(x)$ for $x > 0$ is defined by $(Mf)(t) = \int_0^\infty x^{t-1} f(x) dx,\ t \in C$.

Theorem 1. *For $t \in C$, $\Re(t) < \frac{1}{q-1}, q > 1$, the Mellin transform $L_q[x^{\gamma-1}f(x)](s)$ is given by:*

$$\mathfrak{M}(L_q(x^{\gamma-1}f);t) = \frac{\Gamma(t)\Gamma(\frac{1}{q-1}-t)}{(q-1)^t\Gamma(\frac{1}{q-1})}\mathfrak{M}(f;\gamma-t).$$

Proof. For $q > 1$:

$$\begin{aligned}
\mathfrak{M}L_q[x^{\gamma-1}f(x)](s) &= \int_0^\infty s^{t-1}\int_0^\infty x^{\gamma-1} e_q^{-sx} f(x) dx \\
&= \int_0^\infty x^{\gamma-1} f(x) \frac{1}{(q-1)^t x^t}\frac{\Gamma(t)\Gamma(\frac{1}{q-1}-t)}{\Gamma(\frac{1}{q-1})};\ \Re(\frac{1}{q-1}-t) > 0 \\
&= \frac{\Gamma(t)\Gamma(\frac{1}{q-1}-t)}{\Gamma(\frac{1}{q-1})}\mathfrak{M}(f;\gamma-t);\ \Re(t) < \frac{1}{q-1}
\end{aligned}$$

hence the result. □

Remark 1. *For $\gamma = 1$ and $t \in C$, it directly implies that the Mellin transform of the q-Laplace transform is given by:*

$$\mathfrak{M}(L_q(f);t) = \frac{\Gamma(t)\Gamma(\frac{1}{q-1}-t)}{(q-1)^t\Gamma(\frac{1}{q-1})}\mathfrak{M}(f;1-t) \text{ for } q > 1, \Re(t) < \frac{1}{q-1}.$$

The G-transform of the function $f(x)$ is given in the form:

$$(Gf)(t) = \int_0^\infty G^{m,n}_{p,r}\left[xt\Big|^{(a_i)_{1,p}}_{(b_i)_{1,r}}\right] f(x) dx$$

where the Meijers G-function is considered as the kernel, with suitable existence conditions. The following theorem helps to evaluate the G-transform of $L_q(f(x))$.

Theorem 2. *The G-transform of $L_q(f(x))$ is given by the following relation:*

$$\mathbb{G}^{m,n}_{p,r}\{L_q[f(x)]\}(t) = \frac{1}{(q-1)\Gamma(\frac{1}{q-1})}\mathbb{G}^{m+1,n+1}_{p+1,r+1}[f(x)](t)$$

with suitable existing conditions.

Proof.

$$\mathbb{G}^{m,n}_{p,r}\{L_q[f(x)]\}(t) = \int_0^\infty G^{m,n}_{p,r}\left[st\Big|^{(a_i)_{1,p}}_{(b_j)_{1,r}}\right] f(s) ds$$

$$
\begin{aligned}
&= \frac{1}{2\pi i}\int_L \int_0^\infty h(\omega)(st)^{-\omega} L_q[f(x)](s)\mathrm{d}s\mathrm{d}\omega \\
&= \frac{1}{2\pi i}\int_L \int_0^\infty \int_0^\infty h(\omega)(st)^{-\omega}[1+(q-1)sx]^{-\frac{1}{q-1}} f(x)\mathrm{d}x\mathrm{d}s\mathrm{d}\omega \\
&= \frac{1}{2\pi i}\int_L \int_0^\infty h(\omega)x^{\omega-1}\frac{\Gamma(1-\omega)\Gamma(\frac{1}{q-1}-1+\omega)}{[x(q-1)]^{1-\omega}\Gamma(\frac{1}{q-1})} f(x)\mathrm{d}x\mathrm{d}\omega \\
&= \frac{1}{(q-1)\Gamma(\frac{1}{q-1})}\int_0^\infty G^{m+1,n+1}_{p+1,r+1}\left[st\left|\begin{matrix}(a_i)_{1,p},\frac{1}{q-1}\\ (b_j)_{1,r},1\end{matrix}\right.\right] f(s)\mathrm{d}s \\
&= \frac{1}{(q-1)\Gamma(\frac{1}{q-1})}\mathbb{G}^{m+1,n+1}_{p+1,r+1}[f(x)](t).
\end{aligned}
$$

□

Remark 2. *The q-Laplace transform can be converted in terms of the G-transform in the sense that the q-exponential can be converted as* $\frac{1}{\Gamma(\frac{1}{q-1})}G^{1,0}_{0,1}[-(q-1)sx|^{1-\frac{1}{q-1}}]$ *for* $|(q-1)sx| \le 1$. *That is:*

$$
L_q[f(x)](s) = \int_0^\infty G^{1,0}_{0,1}[-(q-1)sx|_0^{1-\frac{1}{q-1}}] f(x)\mathrm{d}x \text{ for } s > 0, |(q-1)sx| \le 1. \tag{8}
$$

Now, the integral transforms is of the form:

$$
(Hf)(t) = \int_0^\infty H^{m,n}_{p,r}\left[xt\left|\begin{matrix}(a_i,\alpha_i)_{1,p}\\ (b_i,\beta_i)_{1,r}\end{matrix}\right.\right] f(x)\mathrm{d}x
$$

which is known as the H-transform with suitable existence conditions.

The Hankel transform of a function $f(x)$ for $x > 0$ is defined by:

$$
(H_n f)(t) = \int_0^\infty (xt)^{\frac{1}{2}} J_n(xt) f(x)\mathrm{d}x
$$

where $J_n(z)$ is the Bessel function of the first kind of order $\eta \in \mathbb{C}$, such that $\Re(\eta) > -1$, which is given by:

$$
J_n(z) = \sum_{k=0}^\infty \frac{(-1)^k}{\Gamma(\eta+k+1)k!}\left(\frac{z}{2}\right)^{2k+\eta}.
$$

Theorem 3. *The Hankel transform of the q-Laplace transform* $(H_n L_q(f))(t)$ *can be expressed in terms of the H-transform.*

Proof. The integral transform with the Hankel kernel, which is operated on the q-Laplace transform, is given by:

$$
\begin{aligned}
H_n L_q(f)(t) &= \int_0^\infty (st)^{\frac{1}{2}} J_n(st) \int_0^\infty e_q^{-sx} f(x)\mathrm{d}x\mathrm{d}s \\
&= \int_0^\infty t^{\frac{1}{2}} \sum_{k=0}^\infty \frac{(-1)^k}{\Gamma(\eta+k+1)k!}\left(\frac{t}{2}\right)^{2k+\eta} \frac{\Gamma(2k+\eta+\frac{3}{2})\Gamma(\frac{1}{q-1}-2k-\eta-\frac{3}{2})}{[(q-1)x]^{2k+\eta+\frac{3}{2}}\Gamma(\frac{1}{q-1})} f(x)\mathrm{d}x \\
&= \frac{t^{\eta+\frac{1}{2}}}{2^\eta (q-1)^{\eta+\frac{3}{2}}\Gamma(\frac{1}{q-1})}\int_0^\infty H^{2\,1}_{1\,3}\left[\left(\frac{t}{2(q-1)x}\right)^2\left|\begin{matrix}(0,1),(\frac{1}{q-1}-\eta-\frac{3}{2}),(\eta,1)\\ (-\eta-\frac{1}{2},2)\end{matrix}\right.\right] x^{-\eta-\frac{3}{2}} f(x)\mathrm{d}x
\end{aligned}
$$

which is the H-transform of $x^{-\eta-\frac{3}{2}}f(x)$. □

Remark 3. *The q-Laplace transform of $f(\cdot)$ for $q < 1$ can be considered as a general case of the Riemann–Liouville integral operator, since for $q = 0$ and for $x = \frac{u}{t}$, we get the general form of the Riemann–Liouville operator.*

Remark 4. *We can extend the q-Laplace transform to its generalized version by considering the function $f(\cdot)$ with support over $(0,\infty)$ with:*

$$L_q[f(x)](s) = \int_0^\infty (xs)^{-\alpha}[e_q^{-sx}]f(x)dx \text{ for } \Re(s) > 0, \Re(\alpha) > 0 \tag{9}$$

where e_q^{-x} is defined as in 2. Now, as $q \to 1$, we get the generalized Laplace transform of the function f, with support over the positive real line defined as:

$$(Lf)(t) = \int_0^\infty (xt)^{-\alpha}e^{-(tx)^k}f(x)dx$$

that has interesting application in various fields.

5. Differential Equations by Means of the q-Laplace Transform

In this section, we apply the properties of the q-Laplace transform to solve the fractional space-time diffusion equation, the kinetic equation and the time-fractional heat equation.

5.1. Fractional Space-Time Diffusion: Laplace Transform and H-Function

We consider the following diffusion model with fractional-order spatial and temporal derivatives:

$${}_0D_t^\beta N(x,t) = \eta \, {}_xD_\theta^\alpha N(x,t), \tag{10}$$

with the initial conditions ${}_0D_t^{\beta-1}N(x,0) = \sigma(x), 0 \leq \beta \leq 1, \lim_{x\to\pm\infty} N(x,t) = 0$, where η is a diffusion constant; $\eta, t > 0, x \in R; \alpha, \theta, \beta$ are real parameters with the constraints:

$$0 < \alpha \leq 2, |\theta| \leq min(\alpha, 2-\alpha),$$

and $\delta(x)$ is the Dirac-delta function. Then, for the fundamental solution of (1) with initial conditions, there holds the formula:

$$N(x,t) = \frac{t^{\beta-1}}{\alpha|x|}H_{3,3}^{2,1}\left[\frac{|x|}{(\eta t^\beta)^{1/\alpha}}\middle|\begin{matrix}(1,1/\alpha),(\beta,\beta/\alpha),(1,\rho)\\(1,1/\alpha),(1,1),(1,\rho)\end{matrix}\right], \alpha > 0 \tag{11}$$

where $\rho = \frac{\alpha-\theta}{2\alpha}$. The following special cases of (1) are of special interest for fractional diffusion models:

(i) For $\alpha = \beta$, the corresponding solution of (1), denoted by N_α^θ, can be expressed in terms of the H-function as given below and can be defined for $x > 0$:

Non-diffusion: $0 < \alpha = \beta < 2; \theta \leq \min\{\alpha, 2-\alpha\}$,

$$N_\alpha^\theta(x) = \frac{t^{\alpha-1}}{\alpha|x|}H_{3,3}^{2,1}\left[\frac{|x|}{t\eta^{1/\alpha}}\middle|\begin{matrix}(1,1/\alpha),(\alpha,1),(1,\rho)\\(1,1/\alpha),(1,1),(1,\rho)\end{matrix}\right], \rho = \frac{\alpha-\theta}{2\alpha}. \tag{12}$$

(ii) When $\beta = 1, 0 < \alpha \leq 2; \theta \leq \min\{\alpha, 2-\alpha\}$, then (1) reduces to the space-fractional diffusion equation, which is the fundamental solution of the following space-time fractional diffusion model:

$$\frac{\partial N(x,t)}{\partial t} = \eta \; {}_xD_\theta^\alpha N(x,t), \eta > 0, x \in R, \tag{13}$$

with the initial conditions $N(x, t = 0) = \sigma(x), \lim_{x \to \pm\infty} N(x,t) = 0$, where η is a diffusion constant and $\sigma(x)$ is the Dirac-delta function. Hence, for the solution of (1), there holds the formula:

$$L_\alpha^\theta(x) = \frac{1}{\alpha(\eta t)^{1/\alpha}} H_{2,2}^{1,1}\left[\frac{(\eta t)^{1/\alpha}}{|x|}\middle|_{(\frac{1}{\alpha},\frac{1}{\alpha}),(\rho,\rho)}^{(1,1),(\rho,\rho)}\right], \; 0 < \alpha < 1, |\theta| \leq \alpha, \tag{14}$$

where $\rho = \frac{\alpha-\theta}{2\alpha}$. The density represented by the above expression is known as α-stable Lévy density. Another form of this density is given by:

$$L_\alpha^\theta(x) = \frac{1}{\alpha(\eta t)^{1/\alpha}} H_{2,2}^{1,1}\left[\frac{|x|}{(\eta t)^{1/\alpha}}\middle|_{(0,1),(1-\rho,\rho)}^{(1-\frac{1}{\alpha},\frac{1}{\alpha}),(1-\rho,\rho)}\right], 1 < \alpha < 2, |\theta| \leq 2-\alpha. \tag{15}$$

(iii) Next, if we take $\alpha = 2, 0 < \beta < 2; \theta = 0$, then we obtain the time-fractional diffusion, which is governed by the following time-fractional diffusion model:

$$\frac{\partial^\beta N(x,t)}{\partial t^\beta} = \eta \frac{\partial^2}{\partial x^2} N(x,t), \eta > 0, x \in R, \; 0 < \beta \leq 2, \tag{16}$$

with the initial conditions ${}_0D_t^{\beta-1} N(x,0) = \sigma(x), {}_0D_t^{\beta-2} N(x,0) = 0$, for $x \in r$, $\lim_{x \to \pm\infty} N(x,t) = 0$, where η is a diffusion constant and $\sigma(x)$ is the Dirac-delta function, whose fundamental solution is given by the equation:

$$N(x,t) = \frac{t^{\beta-1}}{2|x|} H_{1,1}^{1,0}\left[\frac{|x|}{(\eta t^\beta)^{1/2}}\middle|_{(1,1)}^{(\beta,\beta/2)}\right]. \tag{17}$$

(iv) If we set $\alpha = 2, \beta = 1$ and $\theta \to 0$, then for the fundamental solution of the standard diffusion equation:

$$\frac{\partial}{\partial t} N(x,t) = \eta \frac{\partial^2}{\partial x^2} N(x,t), \tag{18}$$

with initial condition:

$$N(x, t = 0) = \sigma(x), \lim_{x \to \pm\infty} N(x,t) = 0, \tag{19}$$

there holds the formula:

$$N(x,t) = \frac{1}{2|x|} H_{1,1}^{1,0}\left[\frac{|x|}{\eta^{1/2} t^{1/2}}\middle|_{(1,1)}^{(1,1/2)}\right] = (4\pi\eta t)^{-1/2} \exp[-\frac{|x|^2}{4\eta t}], \tag{20}$$

which is the classical Gaussian density.

5.2. Solution of the Generalized Kinetic Equation

Consider the generalized kinetic equation derived by Haubold and Mathai [8],

$$N(t) - N_0 = -c_0 \; {}_0D_t^{-\alpha} N(t) \text{ for } \alpha > 0, \tag{21}$$

where ${}_0D_t^{-\alpha} N(t)$ is the Riemann–Liouville integral operator, in the form:

$${}_0D_t^{-\alpha} N(t) = \frac{1}{\Gamma(\alpha)} \int_0^t (t-u)^{\alpha-1} f(u) du$$

with the assumption that ${}_aD_t^0 g(t) = g(t)$.

Lemma 9. *The solution of the kinetic Equation (21) is given by:*

$$N(t) = \frac{N(0)}{2-q} E_\alpha \left(\frac{c_0 \Gamma_{(q)}(\alpha) t^\alpha}{\Gamma(\alpha)(2-q)} \right)$$

where $E_\alpha(\cdot)$ *represents the two parameter Mittag–Leffler function.*

Proof. The q-Laplace transform of the Riemann–Liouville integral operator is given by $L_q[f(s)] = \frac{\Gamma_{(q)}(\alpha)\tilde{f}(u)}{s^\alpha(2-q)\Gamma(\alpha)}$ using the convolution property of the q-Laplace transform, and $\tilde{f}(u)$ is the q-Laplace transform of $f(u)$. Now, by applying the q-Laplace transform on both sides of (21), we get:

$$\tilde{N}(t) - \frac{N(0)}{s(2-q)} = -c_0 \frac{\Gamma_{(q)}(\alpha)}{s^\alpha(2-q)\Gamma(\alpha)} \tilde{N}(t)$$

where $\tilde{N}(t) = L_q[N(t)]$, the q-Laplace transform of $N(t)$. Simplifying the equation we get

$$\tilde{N}(t) = \frac{N(0)}{s(2-q)} \left\{ 1 + \frac{c_0 \Gamma_{(q)}(\alpha)}{s^\alpha(2-q)\Gamma(\alpha)} \right\}^{-1}$$

This can be expanded as an infinite sum, and on finding the inverse q-Laplace transform, we get:

$$\begin{aligned} N(t) &= \frac{N(0)}{(2-q)} \sum_{k=0}^{\infty} \left[\frac{c_0 \Gamma_{(q)}(\alpha)}{\Gamma(\alpha)(2-q)} \right]^k \frac{t^{\alpha k}}{\Gamma(\alpha k + 1)} \\ &= \frac{N(0)}{2-q} E_\alpha \left(\frac{c_0 \Gamma_{(q)}(\alpha) t^\alpha}{\Gamma(\alpha)(2-q)} \right) \end{aligned}$$

for $\left| \frac{c_0 \Gamma_{(q)}(\alpha)}{\Gamma(\alpha)(2-q)} \right| < 1$ where $E_\alpha(\cdot)$ represents the two-parameter Mittag–Leffler function. □

5.3. Solution of the Time-Fractional Heat Equation

The standard heat equation is:

$$\frac{\partial u(x,t)}{\partial t} = \frac{\partial^2 u(x,t)}{\partial x^2}$$

where $u(x,t)$ represents the temperature, which is a function of time t and space x. Let us write the equation in terms of the derivative operator D, such as:

$$D_t(u) = D_x^2(u) \tag{22}$$

where $u = u(x,t)$. Then, for $t \geq 0$, the boundary conditions are that $u(t,0) = u(t,L) = 0$ where L represents the length of a heating rod and an initial condition:

$$u(0,x) = -\frac{4a}{L^2}x^2 + \frac{4a}{L}x$$

where $a = u(0, \frac{L}{2})$. The general solution for Equation (22) assumed to be in the form $u(t, x) = w(t)v(x)$ yields:

$$D(w(t))v(x) = w(t)D^2(v(x)) \Rightarrow \frac{D_t(w(t))}{w(t)} = \frac{D_x^2(v(x))}{w(x)} = K(say)$$

obtained from the general Equation (22). Let θ be the temperature decaying rate, and let $K = -\theta^2$ for $\theta \in \Re$; then, the ordinary differential equations $D(w(t)) = -\theta^2 w(t)$ and $D^2(v(x)) = -\theta^2 v(x)$ provide the general solution of Equation (22) of the form:

$$u(t, x) = K_1 \cos(\theta x) e^{-\theta^2 t} + K_2 \sin(\theta x) e^{-\theta^2 t}.$$

Now, let us consider the time fractional heat equation of the form:

$$D_t^{\alpha}(u) = D_x^2(u) \; 0 \leq \alpha < 2. \tag{23}$$

By considering similar steps as in the general solution and using the Laplace transform method to solve the differential equation $D_t^{\alpha}(w(t)) = -\theta^2 w(t)$, this yields the Mittag–Leffler function (similar steps as in Section 5.) as in the form:

$$w(t) = \sum_{k=0}^{\infty} \frac{(-\theta^2 t^{\alpha})^k}{\Gamma(\alpha k + 1)}. \tag{24}$$

Now, motivated from the same, we apply the q-Laplace transform for Equation (23) to obtain the solution for $D_t^{\alpha}(w(t)) = -\theta^2 w(t)$. The solution turns out to be:

$$w(t) = \frac{1}{(2-q)} \sum_{k=0}^{\infty} \left[\frac{-\theta^2 \Gamma_{(q)}(\alpha)}{\Gamma(\alpha)(2-q)} \right]^k \frac{t^{\alpha k}}{\Gamma(\alpha k + 1)} = \frac{1}{2-q} E_{\alpha} \left(\frac{-\theta^2 \Gamma_{(q)}(\alpha) t^{\alpha}}{\Gamma(\alpha)(2-q)} \right) \tag{25}$$

and hence, the general solution can be derived accordingly. Throughout the derivation, we consider the Laplace transformation for $q > 1$. Similar derivation exists, when $q < 1$.

6. Conclusions

In this article, we have proposed the q-Laplace transform as a suitable extension of the well-known Laplace transform. Despite the fact that it is difficult to evaluate some of the H-function numerically due to the constraints, the proposed method is an improvement over the regular practice of evaluating the Laplace transform within boundary values. The numerical illustration is not incorporated in this article; however, the methodology proposed here would be to generalize the result obtained in the regular sense of the Laplace transform. Another enhancement in this theory is that we applied the method of q-Laplace transforms in the generalized functional forms, such as Mittag–Leffler, hyper geometric, etc., so that applicability for particular functions, such as exponential, gamma, etc., can be easily deductible. The natural extension of the existing methodology explained in this article would further be considered for its generalized form, and it is an avenue for further research that could flow from this work.

Acknowledgments: The authors would like to thank Arakaparampil Mathai Mathai, Department of Mathematics and Statistics, McGill University, Canada, Jose Kanichukattu, Department of Mathematics and Statistics, Central University of Rajasthan, India and unknown referees for their valuable suggestions and modifications. Furthermore, the first author would like to thank the Department of Mathematics and Statistics, University of Regina, for all facilities provided during the period of research.

Author Contributions: Shanoja R. Naik and Hans J. Haubold contributed materials; Shanoja R. Naik wrote the paper.

Conflicts of Interest: The authors declare no conflict of interest.

References

1. Sneddon, I.N. *Fourier Transforms*; McGraw-Hill: New York, NY, USA, 1951.
2. Zemanian, A.H. *Generalized Integral Transforms*; Interscience Publ.: New York, NY, USA, 1968.
3. Ernst, T. The History of q-Calculus and a New Method. Available online: http://www.math.uu.se/thomas/Lics.pdf (accessed on 12 July 2016).
4. Rao, G.L.N. The generalized Laplace transform of generalized functions. *Ranchi Univ. Math. J.* **1974**, *5*, 76–88.
5. Saxena, R.K. Some theorems on generalized Laplace transform, I. *Proc. Natl. Inst. Sci. India Part A* **1960**, *26*, 400–413.
6. Saxena, R.K. Some theorems on generalized Laplace transform, II. *Riv. Mat. Univ. Parma* **1961**, *2*, 287–299.
7. Rooney, P.G. On integral transformation with G-function kernels. *Proc. R. Soc. Edinb. Sect. A* **1982**, *93*, 265–297.
8. Haubold, H.J.; Mathai, A.M. The fractional kinetic equation and thermonuclear functions. *Astrophys. Space Sci.* **2000**, *327*, 53–63.
9. Mathai, A.M.; Haubold, H.J. The fractional kinetic equation and thermonuclear functions. *Astrophys. Space Sci.* **2007**, *273*, 53–63.
10. Samko, S.G.; Kilbas A.A.; Marichev, O.I. *Fractional Integrals and Derivatives: Theory and Applications*; Gordon and Breach Science Publ.: Yverdon, Switzerland, 1993.
11. Abdi, W.H. On certain q-difference equations and q-Laplace transform. *Proc. Natl. Inst. Sci. India A* **1962**, *28*, 1–15.
12. Borges, E.P. A possible deformed algebra and calculus inspired in nonextensive thermo-statistics. *Phys. A Stat. Mech. Appl.* **2004**, *340*, 95–101.
13. Borges, E.P. A q-generalization of circular and hyperbolic functions. *Phys. A Math. Gen.* **1998**, *31*, 5281–5288.
14. Ferhan, M.A.; Eloe, W.P. Fractional q-Calculus on a time scale. *J. Nonlinear Math. Phys.* **2007**, *14*, 341–352.
15. Kilbas, A.A.; Saigo, M.; Shlapakov, S.A. Integral transforms with Fox's H-function in space of summable functions. *Integral Transform. Spec. Funct.* **1993**, *1*, 87–103.
16. Kilbas, A.A.; Saigo, M.; Shlapakov, S.A. Integral transforms with Fox's H-function in $\mathcal{L}_{v,r}$ -spaces II. *Fukuoka Univ. Sci. Rep.* **1994**, *24*, 13–38.
17. Mathai, A.M.; Rathie, P.N. *Basic Concepts in Information Theory and Statistics: Axiomatic Foundations and Applications*; Wiley Publications: New York, NY, USA, 1975.
18. Mathai, A.M.; Haubold, H.J. Pathway model, superstatistics, Tsallis statistics and a generalized measure of entropy. *Phys. A* **2007**, *375*, 110–122.
19. Beck, C. Stretched exponential from superstatistics. *Phys. A Stat. Mech. Appl.* **2006**, *365*, 96–101.
20. Beck, C.; Cohen, E.G.D. Superstatistics. *Phys. A Stat. Mech.* **2003**, *322*, 267–275.
21. Tsallis, C. Possible generalizations of Boltzmann-Gibbs statistics. *J. Stat. Phys.* **1988**, *52*, 479-487.
22. Wilk, G.; Wlodarczyk, Z. Non-exponential decays and nonextensivity. *Phys. Lett. A* **2001**, *290*, 55–58.
23. Mathai, A.M. A pathway to matrix-variate gamma and normal densities. *Linear Algebra Appl.* **2005**, *396*, 317–328.
24. Jose, K.K.; Naik, S.R. A class of asymmetric pathway distributions and an entropy interpretation. *Phys. A* **2008**, *387*, 6943–6951.
25. Jose, K.K.; Naik, S.R. On the q-Weibull distribution and its applications. *Commun. Stat. Theory Methods* **2009**, *38*, 912–926.
26. Mathai, A.M. *A Versatile Integral*; Centre for Mathamatical and Statistical Sciences: Kottayam, India, 2007.
27. Mathai, A.M. *A Handbook of Generalized Special Functions for Statistical and Physical Sciences*; Oxford University Press: Oxford, UK, 1993.

axioms

MDPI

Article

Operational Solution of Non-Integer Ordinary and Evolution-Type Partial Differential Equations

Konstantin V. Zhukovsky [1,*] and Hari M. Srivastava [2,3]

1 Faculty of Physics, Moscow State University, Leninskie Gory, Moscow 119991, Russia
2 Department of Mathematics and Statistics, University of Victoria, Victoria, BC V8W 3R4, Canada; harimsri@math.uvic.ca
3 Department of Medical Research, China Medical University Hospital, China Medical University, Taichung 40402, Taiwan
* Correspondence: zhukovsk@physics.msu.ru; Tel.: +7-495-939-3177; Fax: +7-495-939-2991

Academic Editor: Hans J. Haubold
Received: 27 October 2016; Accepted: 6 December 2016; Published: 13 December 2016

Abstract: A method for the solution of linear differential equations (DE) of non-integer order and of partial differential equations (PDE) by means of inverse differential operators is proposed. The solutions of non-integer order ordinary differential equations are obtained with recourse to the integral transforms and the exponent operators. The generalized forms of Laguerre and Hermite orthogonal polynomials as members of more general Appel polynomial family are used to find the solutions. Operational definitions of these polynomials are used in the context of the operational approach. Special functions are employed to write solutions of DE in convolution form. Some linear partial differential equations (PDE) are also explored by the operational method. The Schrödinger and the Black–Scholes-like evolution equations and solved with the help of the operational technique. Examples of the solution of DE of non-integer order and of PDE are considered with various initial functions, such as polynomial, exponential, and their combinations.

Keywords: inverse operator; derivative; differential equation; special functions; Hermite and Laguerre polynomials

PACS: 02.30 Gp; Hq; Jr; Mv; Nw; Tb; Uu; Vv; Zz; 41.85.Ja; 03.65.Db; 05.60.Cd

1. Introduction

Differential equations (DE) play an important role in pure mathematics and physics. They describe a broad range of physical processes and finding their solutions is of great importance. Only a few types of DE allow explicit analytical solutions. A vast literature is dedicated to the topic, and the contribution of scientists such as A.M. Mathai can hardly be overestimated (see, for example, [1,2]). Fractional calculus has rapidly drawn increasing attention from researchers in the last decade. They study the solutions of fractional reaction-diffusion, statistical, and other equations (see, for example, [3–6]. In many cases, expansion in series of orthogonal polynomials and their generalized forms with many indexes and variables as well as the usage of integral transforms are the most common tools to analytically solve DE.

The method of operational solution of DE demonstrated in [7–10] is applicable to a wide spectrum of physical problems, described by linear partial differential equations (PDE), such as propagation and radiation from charged particles [11–19], heat diffusion [20–22], including processes not described by Fourier law, and others [23–25]. In the context of the operational approach, the operational definitions for the polynomials through the operational exponent are very useful [26].The operational exponent is also applied when describing the fundamentals of structures in nature, including elementary particles and quarks [27–29]; such modern mathematical instruments are also used for the theoretical study of neutrino mixing [30–32] and for analysis of relevant experimental data [33–35]. The obtained solutions were formulated in terms of series of generalized forms of orthogonal polynomials of Hermite, Laguerre, more general Appèl, and some other polynomials [36,37], special functions of hyperbolic, elliptic Weierstrass and Jacobi-type, cylindrical Bessel-type, and generalized Airy-type functions.

While the role of various parameters in the solutions of DE and their physical meaning is most clear in the analytical form of the solutions, this last is not always available. Modern computer methods help to solve DE. The numerical approach is widely applied nowadays due to the revolutionary breakthrough in computational technique and technical support. Advanced numerical methods for the solution of fractional differential equations, formulated, for example, in [38–41], can be effectively executed with modern computers. In this context we note also semi-analytical models and numerical simulations of relaxation of hot electrons and holes [42], the diffusion of charge carriers, and the energy relaxation and transfer with respect to the electron excited states in crystals [43,44].

Different from these numerical computations, analytical solutions, when available, give clearer insight into the underlying physical processes. In the following we will apply the operational method to obtain exact solutions for some linear ordinary DE with non-integer derivatives and for evolution-type PDE, giving examples of solutions of Schrödinger-type and Black–Scholes-type equations, and their generalized forms with the Laguerre derivative operator.

The structure of the manuscript is as follows. In the first section we will explore generalized Hermite and Laguerre polynomials, the inverse derivative operator, the Laguerre derivative, and the relations between them; we will also touch on the Appèl polynomials. In the second section we will apply the orthogonal polynomials and inverse differential operators to find the solution of some non-integer order DE. In the third section we will construct convolution forms of solutions for DE with the help of special functions and integral transforms. In the fourth section we will consider the operational solutions for some PDE; in particular, we will consider the evolution partial differential equations of Schrödinger and Black–Scholes types. In every section we will consider examples of solutions with various initial functions, such as the functions $f(x) = x^n$, $f(x) = \sum_n c_n x^n$, $f(x) = e^{-x^2}$, $f(x) = e^{-\gamma x}$, $f(x) = \sum_k x^k e^{\gamma x}$, $f(x) = W_0(-x^2, 2)$. Eventually, we will provide the results and the conclusions.

2. Operational Approach and Orthogonal Polynomials

First of all, we note that an inverse function is one that undoes another function: For $f(x) = y$ the inverse is $g(y) = x$, $g(f(x)) = x$. The differential operators can be treated similarly. For the DE $\psi(D)F(x) = f(x)$, where $\psi(D)$ is a differential operator, the inverse differential operator $(\psi(D))^{-1}$ is defined, which undoes $\psi(D)$, $\psi(D)(\psi(D))^{-1}f(x) = f(x)$, so that $F(x) = (\psi(D))^{-1}f(x)$. Consider a common differential operator d_x: $F'(x) = f(x)$. Its inverse is $d_x{}^{-1}f(x) = F(x)$, which is an integral operator $\int f(x) = F(x) + C$, where C is the integration constant. The inverse derivative operator of the n-th order acts according to its definition:

$$D_x^{-n} f(x) = \frac{1}{(n-1)!} \int_0^x (x - \xi)^{n-1} f(\xi) d\xi, \ (n \in N = \{1, 2, 3, \ldots\}), \tag{1}$$

which is complemented by the definition for its zeroth order action:

$$D_x^0 f(x) = f(x) \tag{2}$$

and its action on the unity gives

$$D_x^{-n} \mathbf{1} = \frac{x^n}{n!}, \ (n \in N_0 = N \cup \{0\}). \tag{3}$$

It is elementary to demonstrate that, for example, the DE $\psi(D)F(x) = e^{\alpha x}$ has the following particular integral $F(x) = (\psi(D))^{-1}e^{\alpha x} = e^{\alpha x}(\psi(\alpha))^{-1}$, and to prove the following identity:

$$(\psi(D))^{-1}e^{\alpha x}f(x) = e^{\alpha x}(\psi(D+\alpha))^{-1}f(x). \tag{4}$$

With the help of the above identity the action of the shifted inverse differential operator $(\psi(D+\alpha))^{-1}$ on $f(x)$ can be expressed via the inverse differential operator $(\psi(D))^{-1}$, as follows:

$$F(x) = (\psi(D+\alpha))^{-1}f(x) = e^{-\alpha x}(\psi(D))^{-1}e^{\alpha x}f(x). \tag{5}$$

Equation (5) might seem trivial, but it is particularly useful for the solution of a broad class of DE with shifted differential operators.

Traditionally, polynomial families are defined by their expansion in series. However, they can be defined operationally through the relationship with the exponential differential operators. We recall that, in general, an exponential of an operator can be viewed as the series expansion $e^{\hat{A}} = \sum_{n=0}^{\infty} \hat{A}^n/n!$. The Hermite polynomials of two variables [45], if considered in the context of the operational approach [37], can be explicitly defined by the following operational rule [36] in addition to their series expansion [46]:

$$H_n^{(m)}(x,y) = e^{y\frac{\partial^m}{\partial x^m}}[x^n],\ H_n^{(m)}(x,y) = n!\sum_{r=0}^{[n/m]}\frac{x^{n-mr}y^r}{(n-mr)!\,r!}. \tag{6}$$

For the first-order polynomial we obtain simply

$$H_n{}^{(1)}(x,y) = (x+y)^n, \tag{7}$$

and for the second-order polynomial we have the two-variable Hermite polynomials $H_n(x,y)$:

$$H_n{}^{(2)}(x,y) = H_n(x,y) = e^{y\frac{\partial^2}{\partial x^2}}[x^n],\ H_n(x,y) = n!\sum_{r=0}^{[n/2]}\frac{x^{n-2r}y^r}{(n-2r)!\,r!}. \tag{8}$$

Thus, the Hermite polynomials of two variables are defined through the action of the heat operator $\hat{S}$:

$$\hat{S} = e^{t\partial_x^2} \tag{9}$$

on the monomial x^n. The heat operator (9) was thoroughly studied, for example, in [47]. The Hermite polynomials of two variables have the following generating function:

$$e^{xt+yt^2} = \sum_{n=0}^{\infty}\frac{t^n}{n!}H_n(x,y) \tag{10}$$

and they actually represent another form of the common Hermite polynomials of one variable:

$$H_n(x,y) = (-i)^n y^{n/2} H_n\left(\frac{ix}{2\sqrt{y}}\right) = i^n(2y)^{n/2} He_n\left(\frac{x}{i\sqrt{2y}}\right). \tag{11}$$

Direct application of the operational definition (8) to the Hermite polynomials yields the following identity:

$$e^{t\frac{\partial^2}{\partial x^2}}H_n(x,y) = H_n(x,y+t), \tag{12}$$

which consists of a shift in the y variable.

With the help of the following relation for Hermite polynomials:

$$z^n H_n(x,y) = H_n(xz,yz^2) \tag{13}$$

and with the operational identity:

$$e^{y\frac{\partial^m}{\partial x^m}} f(x) = f\left(x + my\frac{\partial^{m-1}}{\partial x^{m-1}}\right)\{1\}, \tag{14}$$

applied together with the operational rule (5), we obtain for the action of the heat diffusion operator $\hat{S}$ on the polynomial-exponential function the following result:

$$e^{y\partial_x^2} x^k e^{\alpha x} = e^{(\alpha x+\alpha^2 y)} H_k(x+2\alpha y, y). \tag{15}$$

The Hermite, Laguerre, and some other polynomials belong to a more general family of Appèl polynomials [48], if viewed in the framework of the operational approach. For example, the two-variable Hermite polynomials belong to the family of Appèl polynomials $a_n(x)$, which can be defined through the following generating function [47]:

$$\sum_{n=0}^{\infty} \frac{t^n}{n!} a_n(x) = A(t)e^{xt}, \tag{16}$$

where it is assumed that a finite region of t exists, in which $A(t)$ is expandable in Taylor series and this expansion converges. Then, with the help of the obvious identity: $te^{xt} = \hat{D}_x e^{xt}$, $\hat{D}_x = d/dx$, we can rewrite Equation (16) in the following operational form:

$$\sum_{n=0}^{\infty} \frac{t^n}{n!} a_n(x) = A(\hat{D}_x)e^{xt}. \tag{17}$$

Now, expanding the exponential in Equation (17) in series and equating the terms on the right- and left-hand sides of (17), we obtain the following definition for $a_n(x)$:

$$a_n(x) = A(\hat{D}_x)\, x^n, \tag{18}$$

where $A(\hat{D}_x)$ is the Appèl operator. In the case of the two-variable Hermite polynomials, the identity (18) becomes the operational definition (8) and Appèl operator for Hermite polynomials is realized by the exponential

$$A(\hat{D}_x)\big|_{a_n(x)=H_n(x,y)} = e^{y\hat{D}_x^2}. \tag{19}$$

Let us assume that the inverse of the Appèl operator $\left[A(\hat{D}_x)\right]^{-1}$ can be defined as $\left[A(\hat{D}_x)\right]^{-1} A(\hat{D}_x) = \hat{1}$. The main properties of the Appèl polynomials arise from the operational definition (18). For example, if the operators $A(\hat{D}_x)$ and $\hat{D}_x$ commute: $\left[A(\hat{D}_x), \hat{D}_x\right] = 0$, then, by acting with $\hat{D}_x$ on both sides of (18), we obtain the following relation for $a_n(x)$ and $a_{n-1}(x)$:

$$\hat{D}_x a_n(x) = n a_{n-1}(x). \tag{20}$$

Moreover, it follows from (18) that $a_{n+1}(x)$ and $a_n(x)$ are related to each other as follows:

$$a_{n+1}(x) = \left[A(\hat{D}_x)x\right]x^n. \tag{21}$$

This allows us to introduce the multiplicative operator $\hat{M}$ for Appèl polynomials:

$$a_{n+1} = \hat{M} a_n(x), \tag{22}$$

where $\hat{M}$ is given by the Appèl operator as follows:

$$\hat{M} = A(\hat{D}_x)xA(\hat{D}_x)^{-1}, \tag{23}$$

and on account of $\left[f(\hat{D}_x), x\right] = f'(\hat{D}_x)$, where f' is the derivative of f, we write:

$$\hat{M} = x + \left[A(\hat{D}_x)\right]^{-1} A'(\hat{D}_x). \tag{24}$$

For the Appèl polynomials $a_n(x)$ the operators $\hat{M}$ and $\hat{D}_x$ stand for the multiplicative and derivative operators and this set of operators: $\hat{M}, \hat{D}, \hat{1}$, realizes the Weyl–Heisenberg algebra. From the following relation for Appèl polynomials:

$$\hat{M}\hat{D}_x a_n(x) = n a_n(x) \tag{25}$$

it is easy to derive the following differential equation for Appèl polynomials:

$$x\hat{D}_x a_n(x) + \frac{A'(\hat{D}_x)}{A(\hat{D}_x)}\hat{D}_x a_n(x) = n a_n(x), \tag{26}$$

where A' is the derivative of A. This equation is valid for all of the polynomials belonging to the Appèl family. Moreover, it is easy to recognize that Appèl polynomials satisfy the following recurrence:

$$a_{n+1}(x) = \left(x + \frac{A'(\hat{D}_x)}{A(\hat{D}_x)}\right) a_n(x). \tag{27}$$

In the context of the Appèl polynomial family we obtain for the Hermite polynomials the following multiplicative operator $\hat{M}$:

$$\hat{M} = x + 2y\hat{D}_x. \tag{28}$$

The differential equation for the Hermite polynomials as part of the Appèl polynomial family reads as follows:

$$2y\hat{D}_x^2 H_n(x,y) + x\hat{D}_x H_n(x,y) = nH_n(x,y). \tag{29}$$

The Laguerre polynomials of two variables can be defined in the operational way or as a finite sum:

$$L_n(x,y) = e^{-y\frac{\partial}{\partial x}x\frac{\partial}{\partial x}}\left[\frac{(-x)^n}{n!}\right] = n!\sum_{r=0}^{n}\frac{(-1)^r y^{n-r} x^r}{(n-r)!(r!)^2}. \tag{30}$$

The Laguerre polynomials of two variables, as well as the Hermite polynomials of two variables, are just another way for writing proper polynomials of one variable [47]:

$$L_n(x,y) = y^n\, L_n\left(\frac{x}{y}\right),\; L_n(x) = y^{-n}\, L_n(xy,y) = L_n(x,1). \tag{31}$$

However, there is more than just another notation behind the introduction of this form with two variables in Hermite and Laguerre polynomials. It allows us to consider proper polynomials as solutions of partial differential equations (PDE) with proper initial conditions:

$$\partial_y L_n(x,y) = -(\partial_x x \partial_x)\, L_n(x,y) \text{ with } L_n(x,0) = \frac{(-x)^n}{n!} \tag{32}$$

for Laguerre polynomials $L_n(x,y)$ and

$$\partial_y H_n(x,y) = \partial_x^2 H_n(x,y) \text{ with } H_n(x,0) = x^n \tag{33}$$

for Hermite polynomials $H_n(x,y)$. We introduce the Laguerre derivative ${}_LD_x$ and then the two variable Laguerre polynomials can be operationally defined as follows:

$$L_n(x,y) = e^{y\,{}_L\hat{D}_x}\left[\frac{(-x)^n}{n!}\right], {}_L\hat{D}_x = -\hat{D}_x x \hat{D}_x. \tag{34}$$

This operational definition is equivalent to the summation definition (30), which can be easily proved by direct execution of the action of ${}_L\hat{D}_x$ on $(-x)^n/n!$:

$${}_L\hat{D}_x\left[\frac{(-x)^n}{n!}\right] = n\left[\frac{(-x)^{n-1}}{(n-1)!}\right]. \tag{35}$$

The differential and multiplicative operators are formed by the operators

$${}_LD_x = \partial_x x \partial_x = -\hat{P} \text{ and } \hat{M} = y - D_x^{-1}, \tag{36}$$

which do not commute:

$$\left[{}_LD_x, D_x^{-1}\right] = -1. \tag{37}$$

Moreover, in the framework of the inverse derivative (see (1)) the following operational relationship exists between them 10:

$${}_LD_x = \frac{\partial}{\partial D_x^{-1}}, \tag{38}$$

which immediately raises associations with the relationship between the momentum and the coordinate in quantum mechanics. This relationship allows us to solve operationally the differential equations with the Laguerre derivative operator $\partial_x x \partial_x$, as we will demonstrate in what follows. Directly from (36) and (38) we conclude that the Laguerre polynomials $L_n(x,y)$, (30) and (34), can be expressed in terms of the inverse derivative operator (1) as follows:

$$L_n(x,y) = n!\sum_{k=0}^{n}\frac{(-x)^k y^{n-k}}{(n-k)!(k!)^2} = (y - D_x^{-1})^n\{1\}. \tag{39}$$

This relation is also particularly useful for solution of some types of DE, involving the Laguerre derivative. Moreover, the operational definition (34) and the relations (38) and (39) yield the following operational rule for the Laguerre polynomials:

$$\exp\left(\alpha\frac{\partial}{\partial D_x^{-1}}\right)L_n(x,y) = L_n(x,y-\alpha). \tag{40}$$

Framing classical polynomials in the Appèl family should be done with some caution. Strictly speaking, the two-variable Laguerre polynomials can be considered members of the Appèl family with respect to the y variable only. Indeed, they are not Appèl polynomials with respect to x. However, the Laguerre polynomial family can be introduced in the context of the Appèl family in a way similar to (16) and (18) by the following substitution:

$$x^n \to \frac{(-x)^n}{n!}, \quad \hat{D}_x \to {}_L\hat{D}_x. \tag{41}$$

In this way we obtain the following formula:

$$l_n(x) = A\left({}_L\hat{D}_x\right)\left[\frac{(-x)^n}{n!}\right]. \tag{42}$$

With respect to the y variable, the Laguerre polynomials, as defined in (30), certainly belong the Appèl family as they can be given by the following operational rule:

$$L_n(x,y) = C_0(x\hat{D}_y)[y^n], C_n(x) = \sum_{r=0}^{\infty} \frac{(-1)^r x^r}{r!(r+n)!} = x^{-\frac{n}{2}} J_n(2\sqrt{x}). \tag{43}$$

Moreover, a hybrid family of polynomials exists, defined by the Appèl operator $C_0(x\hat{D}_y^2)$ or, alternatively, defined by the following sum:

$$P_n(x,y) = n! \sum_{r=0}^{[n/2]} \frac{(-1)^r x^r y^{n-2r}}{(r!)^2 (n-2r)!}. \tag{44}$$

Further study of their properties is beyond the scope of the present paper, but they are quite interesting, being in between those of Laguerre and Hermite polynomials. Moreover, for $x = 1 - y^2/4$ these polynomials reduce to the Legendre family. Studies of these and relevant polynomials were recently performed in [49–51].

Eventually, let us note that umbral calculus can provide a common framework for known and new identities for orthogonal polynomials. Let us recall the identity [52]

$$\frac{1}{n+1} = \sum_{m=0}^{n} (-1)^m \binom{n}{m} \frac{1}{m+1}, \tag{45}$$

which, after defining the umbral variable $\hat{a}$ 53, reads in its terms as follows:

$$\hat{a}^m 1 = \frac{1}{m+1}, \quad \hat{a}^0 = 1. \tag{46}$$

Therefore, as a consequence of the binomial theorem and of definition (46), we can write Equation (45) in the following useful form:

$$\frac{1}{n+1} = (1-\hat{a})^n 1. \tag{47}$$

Now with the help of identity (47), it is easy to generate new identities, such as the obvious consequence of Equation (47):

$$\frac{1}{2n+1} = (1-\hat{a})^{2n} 1 = (1-\hat{a})^n (1-\hat{a})^n 1 == \sum_{r=0}^{n} (-1)^r \binom{n}{r} \hat{a}^r \sum_{s=0}^{n} (-1)^s \binom{n}{s} \hat{a}^s 1, \tag{48}$$

which, together with

$$\hat{a}^s \hat{a}^r 1 = \hat{a}^{s+r} 1 = \frac{1}{s+r+1}, \tag{49}$$

yields the relation

$$\frac{1}{2n+1} = \sum_{r=0}^{n} (-1)^r \binom{n}{r} \sum_{s=0}^{n} (-1)^s \frac{1}{s+r+1}. \tag{50}$$

Moreover, from (50) and (47) more identities follow:

$$\frac{1}{m+n+1} = \sum_{p=0}^{m} (-1)^p \binom{m}{p} \sum_{s=0}^{n} (-1)^s \binom{n}{s} \frac{1}{p+s+1}, \tag{51}$$

$$\frac{1}{m\,n+1} = \sum_{s_m=0}^{n} (-1)^{s_m} \binom{n}{s_m} \ldots \sum_{s_1=0}^{n} (-1)^{s_1} \binom{n}{s_1} \frac{1}{1+\sum_{r=1}^{m} s_r}, \quad m \in Z \cup m > 0. \tag{52}$$

We can define the operator of umbral derivative $\hat{\Delta}_a$ [53] by the following rule:

$$\hat{\Delta}_a \hat{a}^n = n\, \hat{a}^{n-1}, \tag{53}$$

which, together with the multiplication condition:

$$\hat{a} \cdot \hat{a}^n = \hat{a}^{n+1} \tag{54}$$

yields the following result for the commutator of the two operators [53]:

$$\left[\hat{\Delta}_a, \hat{a}\right] = \hat{\Delta}_a \hat{a} - \hat{a}\, \hat{\Delta}_a = 1, \tag{55}$$

Equation (55) allows us to use Weyl–Heisenberg algebra when needed. We can define the associated Hermite polynomials of two variables, with the operator $\hat{a}$ as one of the variables, and thus we come to the following sum:

$$H_n(\hat{a}, y)1 = n! \sum_{r=0}^{\left[\frac{n}{2}\right]} \frac{\hat{a}^{n-2r} y^r}{(n-2r)!r!} 1 = n! \sum_{r=0}^{\left[\frac{n}{2}\right]} \frac{y^r}{(n-2r)!(n-2r+1)r!}. \tag{56}$$

The multiplication condition (54) does not define any new polynomial family and such Hermite polynomials $H_n(\hat{a}, y)$ satisfy the following relation

$$H_n(\hat{a}, y)\, 1 = \frac{1}{n+1} H_{n+1}(1, y) - y^{\left[\frac{n+1}{2}\right]}, \tag{57}$$

and the following recurrences:

$$\left(\hat{\Delta}_a H_n(\hat{a}, y)\right) 1 = n H_{n-1}(\hat{a}, y)\, 1, \tag{58}$$

$$\left(\partial_y H_n(\hat{a}, y)\right) 1 = \hat{\Delta}_a^2 H_n(\hat{a}, y)\, 1, \tag{59}$$

which are direct generalizations of the relevant terms for the two-variable Hermite polynomials $H_n(x, y)$ equations. Indeed, Equation (59) is the umbral heat equation—the direct generalization of the heat Equation (33). It can be used to define the associated polynomials (56) in terms of the following operational equation:

$$H_n(\hat{a}, y)\, 1 = e^{y \hat{\Delta}^2{}_a}\, \hat{a}^n\, 1, \tag{60}$$

which is the generalization of definition (8). Further study of this topic represents stand-alone research; it will be addressed elsewhere.

3. Operational Solution of Some Non-Integer Ordinary DE

Let us consider a differential equation where ν is not necessarily an integer, shifted by the constant α derivative d_x:

$$\left(\beta^2 - \widetilde{D}^2\right)^{\nu} F(x) = f(x),\ \widetilde{D} \equiv d_x + \alpha,\ \alpha,\ \beta = \text{constant}. \tag{61}$$

Its particular integral formally reads

$$F(x) = \left(\beta^2 - \widetilde{D}^2\right)^{-\nu} f(x), \tag{62}$$

and it can be found in the form of the integral if the well-known operational identity [26,47] is applied:

$$\hat{q}^{-\nu} = \frac{1}{\Gamma(\nu)} \int_0^{\infty} e^{-\hat{q}t} t^{\nu-1} dt,\ \min\{\mathrm{Re}(q), \mathrm{Re}(\nu)\} > 0. \tag{63}$$

For $\hat{q} = \beta^2 - \widetilde{D}^2$ we obtain the following particular solution, involving the integrated weighted action of the operator $e^{t\widetilde{D}^2}$ on the initial function $f(x)$:

$$F(x) = \left(\beta^2 - \widetilde{D}^2\right)^{-\nu} f(x) = \frac{1}{\Gamma(\nu)} \int_0^\infty e^{-\beta^2 t} t^{\nu-1} e^{t\widetilde{D}^2} f(x) dt. \tag{64}$$

Example 1. *It is inviting to choose the initial function for Equation (61) in the form of the monomial $f(x) = x^n$. The action of the heat diffusion operator $\hat{S}$ on the monomial gives the Hermite polynomials according to their operational definition (30); the action of $\hat{S}$ on the polynomial-exponential function is given in (15). With account for the generating function (10), we directly write the particular integral (62) for $f(x) = x^n$ as follows:*

$$F(x)|_{f(x)=x^n} = \left(\beta^2 - (d_x + \alpha)^2\right)^{-\nu} x^n = \frac{1}{\Gamma(\nu)} \int_0^\infty e^{-t((\beta^2-\alpha^2)} t^{\nu-1} H_n(x + 2\alpha t, t) dt. \tag{65}$$

The resulting function (65) with the Hermite polynomial of two variables is characterized by the shift of the argument $x \to x + 2\alpha t$. Evidently, for $\alpha = 0$ we have Equation (66):

$$\left(\beta^2 - d_x^2\right)^\nu Q(x) = f(x), \tag{66}$$

whose solution in the integral form is nothing but a particular case of (64) with $\widetilde{D} \to D$:

$$Q(x) = \frac{1}{\Gamma(\nu)} \int_0^\infty e^{-\beta^2 t} t^{\nu-1} \hat{S} f(x) dt \tag{67}$$

and involves the action of the heat diffusion operator $\hat{S} = e^{t\partial_x^2}$ (9).

Example 2. *Let us choose the Gaussian initial function $f(x) = e^{-x^2}$. Then, by means of the operational rule 47,*

$$\hat{S} f(x) = e^{y\partial_x^2} e^{-x^2} = e^{-\frac{x^2}{1+4y}} \Big/ \sqrt{1+4y}, \tag{68}$$

we immediately get the desired solution:

$$Q(x)|_{f(x)=e^{-x^2}} = \frac{1}{\Gamma(\nu)} \int_0^\infty \frac{dt}{\sqrt{1+4t}} e^{-\beta^2 t} t^{\nu-1} e^{-\frac{x^2}{1+4t}}. \tag{69}$$

Now, let us consider the following equation with the Laguerre derivative ${}_L D_x$, where ν is not necessarily an integer:

$$(\beta - {}_L D_x)^\nu Y(x) = f(x), \ \nu \in \text{Reals}. \tag{70}$$

Its operational solution is:

$$Y(x) = (\beta - {}_L D_x)^{-\nu} f(x) = \frac{1}{\Gamma(\nu)} \int_0^\infty e^{-\beta t} t^{\nu-1} e^{t {}_L D_x} f(x) dt. \tag{71}$$

The common change of variables $t \to e^t$ in such cases transforms the solution of the fractional Laguerre Equation (70) into

$$Y(x) = (\beta - {}_LD_x)^{-\nu} f(x) = \frac{1}{\Gamma(\nu)} \int_{-\infty}^{\infty} e^{t\nu} e^{-\beta e^t} e^{e^t {}_LD_x} f(x) dt, \tag{72}$$

and in the particular case of $\beta = 1$, $\nu = 1$ we obtain the Laplace transforms $1/(1 - \hat{a}) = \int_0^\infty e^{-s(1-\hat{a})} ds$ for the Laguerre derivative operator ${}_LD_x$, where the substitution $a \to {}_LD_x$ has been performed.

Example 3. *For Equation (71) with the initial monomial $f(x) = x^n$ the following particular integral arises:*

$$Y(x)|_{f(x)=x^n} = (\beta - {}_LD_x)^{-\nu} x^n = \frac{(-1)^n n!}{\Gamma(\nu)} \int_0^\infty e^{-\beta t} t^{n+\nu-1} L_n(x/t) dt. \tag{73}$$

Suppose the initial function $f(x)$ is expandable in series of the Laguerre polynomials $L_n(x)$:

$$f(x) = \sum_{n=0}^{\infty} c_n L_n(x). \tag{74}$$

Then with the help of (40), we readily write the solution (71) of Equation (70) in the integral form:

$$Y(x)\Big|_{f(x)=\sum\limits_{n=0}^{\infty} c_n L_n(x)} = \frac{1}{\Gamma(\nu)} \int_0^\infty e^{-\beta t} t^{\nu-1} \sum_{n=0}^{\infty} c_n L_n(x, 1-t) dt. \tag{75}$$

Example 4. *Let us consider the initial exponential function $f(x) = e^{-\gamma x}$. Then, the usage of the generalized Gleisher operational rule 10,*

$$e^{-t_L D_x} e^{-\gamma x} = e^{-\frac{\gamma x}{1-\gamma t}} \Big/ (1 - \gamma t), \tag{76}$$

gives the solution:

$$Y(x) = (\beta - {}_LD_x)^{-\nu} e^{-\gamma x} = \frac{1}{\Gamma(\nu)} \int_0^\infty e^{-\beta t} t^{\nu-1} e^{-\frac{\gamma x}{1+\gamma t}} \frac{dt}{1+\gamma t}. \tag{77}$$

Now, let us consider an ordinary DE like (61), with shifted Laguerre derivative ${}_LD_x$ instead of the common derivative d_x:

$$\left(\beta^2 - ({}_LD_x + \alpha)^2\right)^\nu Z(x) = f(x). \tag{78}$$

Let us choose the initial function for (78) in the form of the particular case of the Bessel–Wright function:

$$f(x) = W_0(-x^2, 2), \tag{79}$$

where

$$W_n(x, m) = \sum_s \frac{x^s}{s!(ms+n)!}, \tag{80}$$

is the particular case of the Bessel–Wright function [47]. In complete analogy with (63) we readily write the operational integral solution:

$$Z(x) = \left(\beta^2 - {}_LD_x{}^2\right)^{-\nu} W_0(-x^2, 2) = \frac{1}{\Gamma(\nu)} \int_0^\infty e^{-\beta^2 t} t^{\nu-1} e^{t_L D_x^2} f(x) dt. \tag{81}$$

Now we should compute the action of the heat operator with Laguerre derivative ${}_L D_x$ on the initial function $f(x) = W_0(-x^2, 2)$. With the help of the operational definition of Laguerre polynomials (30) and of the Gleisher operational rule [10],

$$e^{{}_L D_x^2} W_0(-x^2, 2) = W_0(-1/(1+4t), 2) \Big/ \sqrt{1+4t}, \tag{82}$$

we obtain the particular integral as follows:

$$Z(x)|_{f(x)=W_0(-x^2,2)} = \frac{1}{\Gamma(\nu)} \int_0^\infty e^{-\beta^2 t} t^{\nu-1} W_0(-1/(1+4t), 2) \frac{dt}{\sqrt{1+4t}}. \tag{83}$$

The operational definitions of the polynomials and relevant operational rules allow writing solutions with ease for other types of equations too. For example, consider the following fractional order DE:

$$\left(x d_x^2 + (\alpha+1) d_x\right)^\nu V(x) = f(x) \tag{84}$$

Usage of the operational rule (63) immediately yields the integral solution for (84):

$$V(x) = \overline{D}_x^{-\nu} f(x) = \frac{1}{\Gamma(\nu)} \int_0^\infty e^{-\beta t} t^{\nu-1} e^{t \overline{D}_x} f(x) dt, \tag{85}$$

which involves the operational exponent action: $e^{t\overline{D}_x} f(x)$, where we denoted the differential operator

$$\overline{D}_x = x {d^2}_x + (\alpha+1) d_x. \tag{86}$$

Consider the initial condition $f(x) = x^n$. Direct application of the operational definition of the generalized Laguerre polynomials $L_n^{(\alpha)}(x, y)$,

$$L_n^{(\alpha)}(x, y) = \exp\left[-y \overline{D}_x\right] \left\{ \frac{(-x)^n}{n!} \right\}, \tag{87}$$

immediately gives results in the particular integral of the generalized Laguerre polynomials with the exponential power weight:

$$V(x)|_{f(x)=x^n} = \overline{D}_x^{-\nu} x^n = \frac{(-1)^n n!}{\Gamma(\nu)} \int_0^\infty e^{-\beta t} t^{\nu-1} L_n^{(\alpha)}(x, -t) dt. \tag{88}$$

Now consider the other initial condition function: $f(x) = e^{-\gamma x}$. To obtain the solution we exploit the generalized form of the Gleisher operational rule 47, which yields the solution

$$V(x)|_{f(x)=e^{-\gamma x}} = \overline{D}_x^{-\nu} e^{-\gamma x} = \frac{1}{\Gamma(\nu)} \int_0^\infty \frac{dt}{(1+\gamma t)^{\alpha+1}} e^{-\beta t} t^{\nu-1} e^{-\frac{\gamma x}{1+\gamma t}}. \tag{89}$$

We have demonstrated that the usage of the inverse derivative, combined with the operational formalism, provides a straightforward and easy way of solving some classes of linear DE. In what follows we will demonstrate how this technique allows solutions of partial differential equations (PDE).

4. Convolution Forms for Solution of DE

In what follows, we will apply the inverse differential operators in order to obtain the convolution forms of solution for Equation (61). The operational approach to the solution of Equation (61) involves the exponential operator technique, the inverse derivative formalism, and integral transforms. In general, for solution of equations with $D+\alpha$ operational rule (5) can be applied, where

$$\psi^{-1}(D)=\left(\beta^2-d_x^2\right)^{-\nu}. \tag{90}$$

We continue with account for (66), (67) and (9), and make use of the action of the heat diffusion operator $\hat{S}$ (9) on $e^{\alpha x}g(x)$ with the help of the following chain rule:

$$e^{y\partial_x^2}e^{\alpha x}g(x)=e^{\alpha x}e^{\alpha^2 y}e^{2\alpha y\partial_x}e^{y\partial_x^2}g(x), \tag{91}$$

where y and α are the parameters. This results in the following particular solution for Equation (61), expressed as the integral:

$$F(x)=\frac{1}{\Gamma(\nu)}\int_0^{\infty}t^{\nu-1}e^{-(\beta^2-\alpha^2)t}\hat{\Theta}\hat{S}f(x)dt, \tag{92}$$

where $\hat{S}$ is the heat operator (9) and $\hat{\Theta}$ is the well-known operator of translation:

$$\hat{\Theta}=e^{2\alpha t\partial_x},\quad \hat{\Theta}f(x)=f(x+2\alpha t). \tag{93}$$

The action of the operator $\hat{S}=e^{t\partial_x^2}$ can be written in the form of the Gaussian integral transform:

$$\Omega(x,t)\equiv\hat{S}f(x)=\frac{1}{2\sqrt{\pi t}}\int_{-\infty}^{\infty}\exp\left\{-\frac{(x-\xi)^2}{4t}\right\}f(\xi)d\xi. \tag{94}$$

Therefore, apart from the phase factor, the solution (92) of Equation (61) consists of the integrated action of the heat operator $\hat{S}$ and in the consequent translation by $\hat{\Theta}$ of the initial function $f(x)$:

$$F(x)=\frac{1}{\Gamma(\nu)}\int_0^{\infty}t^{\nu-1}e^{-t(\beta^2-\alpha^2)}U(x,t)dt, \tag{95}$$

where the integrand function $U(x,t)$ is (94), shifted by $\hat{\Theta}$:

$$U(x,t)\equiv\hat{\Theta}\hat{S}f(x)=\Omega(x+2\alpha t,t)=\frac{1}{2\sqrt{\pi t}}\int_{-\infty}^{\infty}\exp\left\{-\frac{(x+2\alpha t-\xi)^2}{4t}\right\}f(\xi)d\xi. \tag{96}$$

Example 5. *The example of the Gaussian initial condition $f(x)=\exp(-x^2)$ can be the illustration of the operational solution, described above. Accounting for (68), we directly write the solution of Equation (61), which is in turn a Gaussian:*

$$F(x)|_{f(x)=\exp(-x^2)}=\frac{1}{\Gamma(\nu)}\int_0^{\infty}\frac{t^{\nu-1}e^{-(\beta^2-\alpha^2)t}}{\sqrt{1+4t}}e^{-\frac{(x+2\alpha t)^2}{1+4t}}dt. \tag{97}$$

Note that from the general form of the solution (92), using the operational definition of the Hermite polynomials (8), we can directly obtain the solution (65) of the DE (61).

For the solution (64) of DE (61) with given initial function $f(x)$ we have to calculate the action of the exponential differential operator $e^{t\tilde{D}^2}$ in the exponential. This can be performed in a number of different ways. One of them consists in direct application of operational definitions, as we did in the case of the initial monomial x^n. However, this is a rare case. The exponential operator of the second-order derivative can be reduced to the exponential of the first-order derivative if we apply the integral presentation for the exponential of a square of an operator $\hat{q}$ [54]:

$$e^{\hat{q}^2} = \int_{-\infty}^{\infty} e^{-\xi^2+2\xi\hat{q}} d\xi \Big/ \sqrt{\pi}, \tag{98}$$

in our case $\hat{q} = \sqrt{t}\tilde{D}$. The above formula then reads as follows:

$$e^{t\tilde{D}^2} f(x) = \int_{-\infty}^{\infty} e^{-\xi^2+2\xi\sqrt{t}\tilde{D}} f(x) d\xi \Big/ \sqrt{\pi}. \tag{99}$$

Accounting for the action of the translation operator $e^{\eta(\partial_x+\alpha)} f(x) = e^{\eta\alpha} f(x+\eta)$, we obtain the following particular integral (62) for the DE on non-integer order (61):

$$F(x) = \frac{1}{\sqrt{\pi}\Gamma(\nu)} \int_0^{\infty} t^{\nu-1} \exp\left((\alpha^2-\beta^2)t\right) \int_{-\infty}^{\infty} \exp\left(-(\xi-\sqrt{t}\alpha)^2\right) f(x+2\xi\sqrt{t}) d\xi dt \tag{100}$$

Now, upon subject to the change of variables

$$\eta = x + 2\xi\sqrt{t} \text{ and } t = \tau^2, \tag{101}$$

we end up with the following form of the particular solution for Equation (61):

$$F(x) = \frac{1}{\sqrt{\pi}\Gamma(\nu)} \int_0^{\infty} \tau^{2(\nu-1)} \exp\left(-(\beta\tau)^2\right) \int_{-\infty}^{\infty} \exp\left(-\left(\frac{\eta-x}{2\tau}\right)^2 + \alpha(\eta-x)\right) f(\eta) d\eta d\tau \tag{102}$$

Several convolution forms are possible for the solution of (61). Indeed, for an arbitrary function $f(x)$ in the r.h.s. of (61) and the real values of α and $\nu > 0$ we can involve the generating function for Hermite polynomials (10) to disentangle two integrals in (102):

$$F(x) = \frac{1}{\sqrt{\pi}\Gamma(\nu)} \sum_{n=0}^{\infty} \int_0^{\infty} \tau^{2(\nu-1)} \exp(-\beta^2\tau^2) H_n\left(\alpha, -\frac{1}{4\tau^2}\right) d\tau \frac{1}{n!} \int_{-\infty}^{\infty} (\eta-x)^n f(\eta) d\eta \tag{103}$$

It follows from Equation (103) that the solution of DE (61) can be written in the form of series

$$F(x) = \sum_{n=0}^{\infty} C_n \phi(x), \tag{104}$$

involving the convolution $\int_{-\infty}^{\infty} \Phi(x-\eta) f(\eta) d\eta \equiv \Phi(x) * f(x)$, $\Phi(x-\eta) = (\eta-x)^n$ with the power kernel:

$$\phi(x) = \Phi(x) * f(\eta), \ \Phi(x) = x^n. \tag{105}$$

The respective coefficients in the series depend on the order of the equation, which can be a non-integer, and on the constants α, β as follows:

$$C_n = \frac{(-1)^n}{n!\sqrt{\pi}\Gamma(\nu)} \int_0^{\infty} \tau^{2(\nu-1)} e^{-\beta^2\tau^2} H_n\left(\alpha, -\frac{1}{4\tau^2}\right) d\tau. \tag{106}$$

Other convolution forms with different kernels are possible. Among them the Gaussian frequency kernel form is, perhaps, the most compact. Indeed, the integral form (102) of the solution of DE (61) can be viewed as the integral of the following convolution:

$$F(x) = \frac{1}{\sqrt{\pi}\Gamma(\nu)}\int_0^\infty \tau^{2(\nu-1)}e^{-(\beta\tau)^2}\varphi(x,\tau)d\tau - \frac{1}{\sqrt{\pi}\Gamma(\nu)}\int_{-\infty}^\infty e^{2\nu\tau-\beta^2e^{2\tau}}\varphi(x,e^\tau)d\tau, \tag{107}$$

where $\varphi = \int_{-\infty}^\infty G(x-\eta)f(\eta)d\eta$ has the kernel

$$\varphi(x,\tau) = G(x,\tau) * f(x),\ G(x,\tau) = e^{-(x/2\tau)^2-\alpha x}. \tag{108}$$

The above expression involves the convolution with the Gauss frequency function kernel. Furthermore, the remaining integral can be taken, and it gives the Bessel function of the second kind $K_\varkappa(x)$:

$$\int_0^\infty \tau^{2(\nu-1)}e^{-(\beta\tau)^2-\frac{(x-\eta)^2}{4\tau^2}}d\tau = \left(\frac{|x-\eta|}{2\beta}\right)^{\nu-1/2}K_{\nu-\frac{1}{2}}(\beta|x-\eta|). \tag{109}$$

Note that for the integer order of the equation, $\nu \in Z$, we have semi-integer index of the Bessel function of the second kind, $K_{n-1/2}(x)$, the latter easily expressed in elementary functions, for example: $K_{1/2}(x) = \sqrt{\frac{\pi}{2x}}e^{-x}$, etc.

Thus, we have obtained the particular solution $F(x) = \left(\beta^2 - (D+\alpha)^2\right)^{-\nu}f(x)$ for DE (61) in the form of the integral, which appears in the form of the convolution with the initial function $f(x)$:

$$\begin{aligned} F(x) &= \frac{1}{\sqrt{\pi}\Gamma(\nu)}\int_{-\infty}^\infty \left(\frac{|x-\eta|}{2\beta}\right)^{\nu-1/2}e^{-\alpha(x-\eta)}K_{\nu-\frac{1}{2}}(\beta|x-\eta|)f(\eta)d\eta \\ &= \frac{1}{\sqrt{\pi}\Gamma(\nu)}\int_{-\infty}^\infty \chi(x-\eta)f(\eta)d\eta \end{aligned} \tag{110}$$

with the kernel, containing the Bessel function of the second kind $K_{\nu-1/2}$, the exponential, and the n power of x:

$$\chi(x-\eta) = \left(\frac{|x-\eta|}{2\beta}\right)^{\nu-1/2}e^{-\alpha(x-\eta)}K_{\nu-\frac{1}{2}}(\beta|x-\eta|) \tag{111}$$

Finally can we write the compact convolution form of the solution of DE (61) as follows:

$$\begin{aligned} F(x) &= \frac{1}{\sqrt{\pi}\Gamma(\nu)}\chi * f, \\ \chi &= \left(\frac{|x|}{2\beta}\right)^{\nu-1/2}e^{-\alpha x}K_{\nu-\frac{1}{2}}(\beta|x|). \end{aligned} \tag{112}$$

So far we have demonstrated that the usage of the inverse derivative and of the inverse differential operators constitutes a straightforward and easy way to solve some classes of linear DE. In what follows we will apply this concept to solve more complicated problems, formulated in terms of PDE.

5. Operational Solution for Evolution-Type Partial Differential Equations

The technique of the inverse differential and exponential operators is useful for finding solutions to a broad range of mathematical and physical problems. In what follows we shall demonstrate the solution of the evolution-type DE by the operational approach. Let us consider the Schrödinger equation for an electric charge in a constant electric field in imaginary time. It effectively corresponds to the case when the charge diffuses under a potential barrier in the electric field, so that the charge energy is lower than the height of the barrier. This process is governed by the Schrödinger equation upon the $t \to i\tau$, $\beta \to -\beta$ change:

$$\partial_t F(x,t) = \alpha \partial_x^2 F(x,t) + \beta\, x\, F(x,t),\ F(x,0) = f(x), \tag{113}$$

which is the common heat equation $\partial_t F(x,t) = \partial_x^2 F(x,t)$ with the linear term βx in the r.h.s. The solution of DE (113) can be obtained operationally:

$$\begin{aligned} F(x,t) &= e^{\Phi(x,t;\beta)}\, \hat{\Theta}\hat{S} f(x) = e^{\Phi(x,t;\beta)} f(x+\beta t^2, t), \\ \Phi(x,t;\beta) &= \tfrac{1}{3}\alpha t(\beta t)^2 + \beta t x, \end{aligned} \tag{114}$$

and consists in the transform of the initial function $F(x,0) = f(x)$ by the operators $\hat{S} = e^{\alpha t \partial_x^2}$ and $\hat{\Theta} = e^{\alpha\beta t^2 \partial_x}$. Note that, although the solution for the Schrödinger equation in the electric field in real and in imaginary time, i.e., over and under the barrier, has the same structure (114), there is a fundamental difference between them. Indeed, the $F(x,t)$ function for a particle in quantum mechanics is the amplitude of the probability of finding it at point x at moment t: $F(x,t) = \Psi(x,t)$. For the charge over the barrier, the solution $F(x,t) \to \Psi(x,\tau)$ of the Schrödinger equation is complex due to the complex phase $\Phi(x,\tau;\beta)$; this does not trouble the probability $|\Psi(x,\tau)|^2$ over the barrier for $t \to \infty$, which regularly converges.

Example 6. *Let us consider the initial polynomial $f(x) = \sum_n c_n x^n$ in the context of the Fourier heat conduction of DE (113). The operational definition of the Hermite polynomials (8) gives $e^{a\partial_x} x^n = H_n(x,a)$, and the operator $\hat{\Theta} = e^{b\partial_x}$ gives the shift: $F(x) \propto H_n(x+b,a)$. The solution immediately appears in terms of the sum of the Hermite polynomials:*

$$F(x,t) = e^{\Phi} \sum_n c_n H_n\left(x + \alpha\beta t^2, \alpha t\right). \tag{115}$$

Example 7. *Now let us choose the initial condition $f(x) = \sum_k x^k e^{\gamma x}$. This function for $\gamma < 0$ represents a pulse, the shape of which depends on the values of k and γ, and varies from a sudden surge to a flat, smooth spatial wave. This choice of the initial function allows for modeling heat pulses for experimental tests (see, for example, [55]). Now applying the operational rule (15), where, in our case $y \to \alpha t$, and the shift by the translation operator $\hat{\Theta} = e^{\alpha\beta t^2 \partial_x}$, we obtain the solution in the form of the Hermite polynomials*

$$F(x,t) = e^{\Phi+\Delta_1} \sum_k H_k\left(x + 2t\alpha\gamma + t^2\alpha\beta, \alpha t\right) \tag{116}$$

with the common phase Φ written in the solution (114) and $\Delta_1 = \gamma\ (x + \gamma\alpha\ t + \alpha\beta t^2)$. For $\gamma = 0$ it immediately returns the result (115). For pure Fourier heat conduction $\beta = 0$ and the solution further simplifies:

$$F(x,t)|_{\beta=0, f(x)=x^k e^{\gamma x}} = e^{\gamma x + \gamma^2 \alpha t} \sum_k H_k(x + 2t\alpha\gamma, \alpha t). \tag{117}$$

It is easy to follow its evolution in time: for $t >> x/\alpha\gamma$ the coordinate dependence fades out: $F(x,t)|_{t>>x/\alpha\gamma} \cong H_k(2t\alpha\gamma, \alpha t)\exp\{\gamma^2 \alpha t\}$, and the time dependence prevails. For relatively short times of the evolution of the initial heat pulse $f(x) = \sum_k x^k e^{\gamma x}$, such that $t << x/\alpha\gamma$, the solution is approximated by $F(x,t)|_{t<<x/\alpha\gamma} \cong e^{x\gamma}(1 + t\alpha\gamma^2) H_k(x,\alpha t)$ and for very short times $\alpha t \to 0$ the Hermite polynomials tend to $H_k(x,0) = x^k$, which is in perfect agreement with our initial condition $f(x) = x^k e^{\gamma x}$.

Deeper consideration of the above topic is beyond the scope of the present paper. In forthcoming publications we will apply the operational method to explore and solve relativistic heat equations and other non-local extensions of the heat conduction.

Let us consider the following modification of the common Black–Scholes differential equation with the Laguerre derivative (see (30) and (36)) and the initial function $g(x) = A(x,0)$:

$$\frac{1}{\rho}\frac{\partial}{\partial t}A(x,t) = (\partial_x x \partial_x)^2 A(x,t) + \lambda(\partial_x x \partial_x)A(x,t) - \mu A(x,t), \; A(x,0) = g(x), \tag{118}$$

where ρ, λ, and μ are constants. Equation (118) is in fact the general form of the equation, which unifies the Laguerre heat equation and the matter diffusion equation with Laguerre derivative, as previously explored in [10,37]. In order to solve Equation (118), we employ the operational method. As usually in Black–Scholes DE, we distinguish the perfect square of the derivative, in this case of ${}_L D_x = \partial_x x \partial_x$. Then the solution takes the form of the exponential

$$A(x,t) = \exp\left\{\rho t\left(({}_L D_x + \lambda/2)^2 - \varepsilon\right)\right\} g(x), \; \varepsilon = \mu + (\lambda/2)^2. \tag{119}$$

With the help of the operational identity (98), we reduce $e^{(a_L D_x)^2}$ to the first-order Laguerre derivative in the exponential and thus the following solution for $A(x,t)$ arises:

$$\begin{aligned} A(x,t) &= \tfrac{\exp(-\varepsilon\alpha^2)}{\sqrt{\pi}} \int_{-\infty}^{\infty} \exp(-\sigma^2 - \sigma\alpha\lambda - 2\sigma\alpha_L D_x) g(x) d\sigma, \\ \alpha &= \alpha(t) = \sqrt{\rho t}. \end{aligned} \tag{120}$$

The above integral form of the solution, provided the integral converges, contains the exponential Laguerre derivative, which acts on the initial function: $e^{-a_L D_x} g(x)$.

Example 8. *Let us consider the example of the polynomial initial function* $A(x,0) = g(x) = \sum_n c_n x^n$. *Following the operational definition of the Laguerre polynomials (30), we directly write the solution for DE (118):*

$$A(x,t)|_{g(x)=\sum_n c_n x^n} = \sum_n (-1)^n \frac{n!}{\sqrt{\pi}} e^{-\varepsilon\alpha^2/4} \int_{-\infty}^{\infty} e^{-\sigma^2 - \sigma\alpha\lambda} L_n(x, 2\sigma\alpha) d\sigma. \tag{121}$$

Consequent integration results in the finite sum, involving gamma function Γ and hypergeometric function ${}_1F_1$:

$$\begin{aligned} A(x,t)|_{g(x)=\sum_n c_n x^n} &= \tfrac{e^{-\alpha^2\mu}}{\sqrt{\pi}} \sum_n (-1)^n (n!)^2 \sum_{r=0}^{n} \tfrac{(-x)^r (2\alpha)^{n-r}}{(n-r)!(r!)^2} \times \\ &\left(\tfrac{\alpha\lambda}{2}\left(e^{i(n-r)\pi} - 1\right) I + \tfrac{1}{2}\left(e^{i(n-r)\pi} + 1\right) J\right), \\ I &= \Gamma(1 + \tfrac{n-r}{2})\,{}_1F_1\left(\tfrac{1-(n-r)}{2}, \tfrac{3}{2}, -\left(\tfrac{\alpha\lambda}{2}\right)^2\right), \\ J &= \Gamma\left(\tfrac{1+n-r}{2}\right){}_1F_1\left(-\tfrac{n-r}{2}, \tfrac{1}{2}, -\left(\tfrac{\alpha\lambda}{2}\right)^2\right). \end{aligned} \tag{122}$$

Now suppose the initial function $A(x,0)$ can be expanded in series of the Laguerre polynomials: $g(x) = \sum_n a_n L_n(x)$. The operational relationships (40) and (31) in this case immediately propose the solution of DE (118) in the following form:

$$A(x,t)|_{g(x)=\sum_n a_n L_n(x)} = \frac{1}{\sqrt{\pi}} e^{-\varepsilon\alpha^2/4} \sum_n a_n \int_{-\infty}^{\infty} e^{(-\sigma^2 - \sigma\alpha\lambda)} L_n(x, 2\sigma\alpha + 1) d\sigma. \tag{123}$$

Now let us consider the general case of the initial function $A(x,0) = g(x)$. Then the solution of DE (118), $A(x,t)$ can be obtained in the following steps. With the help of the operational definitions

(36) and with the inverse derivative Formula (1), we write the solution in terms of the operator of the inverse derivative D_x^{-1} and of the function φ:

$$A(x,t) = \frac{1}{\sqrt{\pi}} e^{-\varepsilon\alpha^2} \int_{-\infty}^{\infty} e^{-\sigma^2 - \sigma\alpha\lambda} e^{-2\sigma\alpha \frac{\partial}{\partial D_x^{-1}}} \varphi(D_x^{-1}) \mathbf{1} d\sigma, \ \varphi(D_x^{-1})\,\mathbf{1} = g(x), \tag{124}$$

where φ is the image function, determined by the integral. $\varphi(x) = \int_0^\infty \exp(-\kappa) g(x\kappa) d\kappa$. The exponential of the Laguerre derivative, acting on the initial function, yields the solution of the Laguerre diffusion equation 10:

$$\partial_t f(x,t) = -{}_L D_x f(x,t), \ f(x,0) = g(x), \ f(x,t) = e^{-t_L D_x} g(x). \tag{125}$$

Hence, by applying the exponential differential operator to the function $g(x) = \varphi(D_x^{-1})\mathbf{1}$ in $e^{-t\frac{\partial}{\partial D_x^{-1}}} \varphi(D_x^{-1})\mathbf{1}$, we obtain the solution of the Laguerre diffusion equation:

$$f(x,t) = e^{-t\frac{\partial}{\partial D_x^{-1}}} g(x) = \varphi\left(D_x^{-1} - t\right)\mathbf{1}. \tag{126}$$

With account for the above relation (126) the solution of DE (118) becomes:

$$A(x,t) = \frac{1}{\sqrt{\pi}} e^{-\varepsilon\alpha^2} \int_{-\infty}^{\infty} e^{-\sigma^2 - \sigma\alpha\lambda} g(x,t) d\sigma, \tag{127}$$

$$g(x,t) = \varphi(D_x^{-1} - 2\sigma\alpha)\,\mathbf{1} = e^{-2\sigma\alpha \frac{\partial}{\partial D_x^{-1}}} \varphi\left(D_x^{-1}\right)\mathbf{1}. \tag{128}$$

Example 9. *Let us consider the particular case of the Bessel–Wright function [47]* $W_n(x,m) = \sum_{r=0}^{\infty} \frac{x^r}{r!(mr+n)!}$, $m \in, n \in_0$, *for* $m = 2$, $n = 0$ *as the initial function:* $g(x) = W_0(-x^2, 2) = \sum_{r=0}^{\infty} \frac{(-1)^r x^{2r}}{r!(2r)!}$. *Its image is* $\varphi(x) = e^{-x^2}$. *The operational identity (98) and the function (128) together yield, in accordance with the previously computed in [10], the result:*

$$g(x,t) = \frac{1}{\sqrt{\pi}} \int_{-\infty}^{\infty} e^{-\xi^2 + 4i\sigma\alpha\xi} C_0(2i\xi x) d\xi \tag{129}$$

where C_0 *is the particular case of the Bessel–Tricomi function* [56]: $C_n(x) = \sum_{r=0}^{\infty} \frac{(-x)^r}{r!(r+n)!}$, $n \in_0$. *A relationship with the Bessel functions exists:* $C_n(x) = x^{-n/2} J_n(2\sqrt{x})$. *Finally, we have obtained the solution of DE (118) with the initial condition* $g(x) = W_0(-x^2, 2)$ *in the form of the integral (127) of the exponentially weighted function (129).*

6. Results

We have obtained solutions for some ordinary DE of non-integer order with shifted derivatives. In particular, we derived the integral form of the particular solution $F(x) = \left(\beta^2 - (\partial_x + \alpha)^2\right)^{-\nu} f(x)$ for real values of ν. The integrand involves the operators of heat propagation $\hat{S}$ and translation $\hat{\Theta}$, which act on the function $f(x)$. Moreover, the convolution form of these solutions $\phi(x,\tau) = G(x,\tau) * f(\eta)$ and the integrals of other convolutions with several kernels different from each other are obtained. The comprehensive solution with the kernel, involving the Bessel function of the second kind with power-exponential weight, is obtained. Other integral forms of the solution with the convolutions with the Gaussian frequency kernel and with the monomial kernel are also obtained. We considered the examples of the Gaussian distribution $f(x) = e^{-x^2}$ and of the monomial $f(x) = x^k$ and found explicit solutions for them in terms of integrals and series of Hermite polynomials. We operationally solved the DE with Laguerre derivatives: $(x\partial^2{}_x + (\alpha+1)\partial_x)^{\nu} F(x) = f(x)$ and demonstrated the examples of solutions for the functions $f(x) = \exp(-\gamma x)$, $f(x) = x^k$ and for the Bessel–Wright function $f(x) = W_0(-x^2, 2)$. The obtained operational solutions are expressed in terms of the integrals of generalized Laguerre polynomials and Bessel functions.

The linear evolution-type PDEs were solved by the operational technique. In particular, the Black–Scholes equation with the Laguerre derivative ${}_LD_x = \partial_x x \partial_x$ was solved operationally. The example of the initial polynomial was considered. By using the operational definitions for Hermite polynomials we obtained explicit solutions in the form of the polynomials of x with the coefficients, given by Γ and ${}_1F_1$ functions. The solution of the Black–Scholes type equation with Laguerre derivative ${}_LD_x$ for the Bessel–Wright function $f(x) = W_0(-x^2, 2)$ is obtained in the integral form, involving Bessel–Tricomi function $C_n(x)$. We have obtained the operational solution of a Fourier-type heat equation with an additional term, describing the heat exchange with the environment, for the initial distribution $f(x) = \sum_k x^k e^{\gamma x}$, which describes a heat pulse for $\gamma < 0$. We also obtained the solution of the Schrödinger equation for a charge in electric field in real and in imaginary time, i.e., over and under the potential barrier, and demonstrated that in real time, i.e. under the barrier, the solution is purely real, contrary to that over the barrier. Thus $|F|^2$ diverges for $t \to \infty$ in the case of the real solution and $\beta \neq 0$, but converges otherwise, as a square of the amplitude $|\Psi(x,t)|^2$ of the probability function should behave (see also [57]).

7. Conclusions

In the present work we advocate the operational approach for solution of linear DE and the use of inverse differential operators, which allow direct and straightforward finding of solutions. The latter include the action of the operator of heat conduction and the operator of shift and dilatation. The operational approach involves operational definitions for Hermite and Laguerre orthogonal polynomials. In this way, we avoid cumbersome calculations and directly obtain the results of the action of proper exponential differential operators on the initial functions. If the DE contains the Laguerre derivatives, the commutation relationship between the inverse derivative operator and the Laguerre derivative operator helps in solving DE. Complemented by the usage of the integral transforms where needed, the operational technique yielded solutions of relatively complicated DE, such as Black–Scholes-type DE with Laguerre derivative, etc. Thus, our research demonstrates that the operational approach for solution of linear DEs is advantageous for its ease. The solutions are derived directly based on the operational definitions and on commutation relationships. Operational study of more complicated equations, describing heat propagation accounting for wave and ballistic heat transfer and, for equations, modeling other physical processes, is possible. It will be performed in forthcoming publications.

Author Contributions: Konstantin V. Zhukovsky performed the calculations and wrote the article; Hari M. Srivastava provided the scientific advising.

Conflicts of Interest: The authors declare no conflict of interest.

References

1. Mathai, M.; Saxena, R.K.; Haubold, H.J. *The H-Function: Theory and Applications*; Springer Science & Business Media: Berlin, Germany, 2009.
2. Mathai, A.M.; Haubold, H.J. *Special Functions for Applied Scientists*; Springer Science & Business Media: Berlin, Germany, 2008.
3. Mathai, M. M-convolutions of products and ratios, statistical distributions and fractional calculus. *Analysis* **2016**, *36*, 15–26. [CrossRef]

4. Saxena, R.K.; Mathai, A.M.; Haubold, H.J. Computational solutions of unified fractional reaction-diffusion equations with composite fractional time derivative. *Commun. Nonlinear Sci. Numer. Simul.* **2015**, *27*, 1–11. [CrossRef]
5. Mathai, M. Fractional differential operators in the complex matrix-variate case. *Linear Algebra Appl.* **2015**, *478*, 200–217. [CrossRef]
6. Saxena, R.K.; Mathai, A.M.; Haubold, H.J. Solutions of certain fractional kinetic equations and a fractional diffusion equation. *J. Math. Phys.* **2010**, *51*, 103506. [CrossRef]
7. Zhukovsky, K. Solution of Some Types of Differential Equations: Operational Calculus and Inverse Differential Operators. *Sci. World J.* **2014**, *2014*, 454865. [CrossRef] [PubMed]
8. Zhukovsky, K.V. A method of inverse differential operators using ortogonal polynomials and special functions for solving some types of differential equations and physical problems. *Moscow Univ. Phys. Bull.* **2015**, *70*, 93–100. [CrossRef]
9. Zhukovsky, K. Operational solution for some types of second order differential equations and for relevant physical problems. *J. Math. Anal. Appl.* **2017**, *446*, 628–647. [CrossRef]
10. Dattoli, G.; Srivastava, H.M.; Zhukovsky, K.V. Operational Methods and Differential Equations with Applications to Initial-Value problems. *Appl. Math. Comput.* **2007**, *184*, 979–1001. [CrossRef]
11. Denisov, V.I.; Denisova, I.P.; Pimenov, A.B.; Sokolov, V.A. Rapidly rotating pulsar radiation in vacuum nonlinear electrodynamics. *Eur. Phys. J. C* **2016**, *76*, 612. [CrossRef]
12. Denisov, V.I.; Shvilkin, B.N.; Sokolov, V.A.; Vasiliev, M.I. Pulsar radiation in post-Maxwellian vacuum nonlinear electrodynamics. *Phys. Rev. D* **2016**, *94*, 045021. [CrossRef]
13. Zhukovsky, K.V.; Dattoli, G. Evolution of non-spreading Airy wavepackets in time dependent linear potentials. *Appl. Math. Comp.* **2011**, *217*, 7966–7974. [CrossRef]
14. Zhukovsky, K. Analytical account for a planar undulator performance in a constant magnetic field. *J. Electromagn. Waves Appl.* **2014**, *28*, 1869–1887. [CrossRef]
15. Zhukovsky, K.V. Harmonic Radiation in a Double-Frequency Undulator with Account for Broadening. *Moscow Univ. Phys. Bull.* **2015**, *70*, 232–239. [CrossRef]
16. Zhukovsky, K. High harmonic generation in undulators for FEL. *Nuclear Instrum. Methods B* **2015**, *369*, 9–14. [CrossRef]
17. Zhukovsky, K.V. Harmonic generation by ultrarelativistic electrons in a planar undulator and the emission-line broadening. *J. Electromagn. Waves Appl.* **2015**, *29*, 132–142. [CrossRef]
18. Zhukovsky, K. High harmonic generation in the undulators for free electron lasers. *Opt. Commun.* **2015**, *353*, 35–41. [CrossRef]
19. Zhukovsky, K. Emission and tuning of harmonics in a planar two-frequency undulator with account for broadening. *Laser Part Beams* **2016**, *34*, 447–456. [CrossRef]
20. Zhukovsky, K.V.; Srivastava, H.M. Analytical solutions for heat diffusion beyond Fourier law. *Appl. Math. Comput.* **2017**, *293*, 423–437. [CrossRef]
21. Zhukovsky, K.V. Operational solution of differential equations with derivatives of non-integer order, Black–Scholes type and heat conduction. *Moscow Univ. Phys. Bull.* **2016**, *71*, 237–244. [CrossRef]
22. Zhukovsky, K.V. Violation of the maximum principle and negative solutions with pulse propagation in Guyer-Krumhansl model. *Int. J. Heat Mass Transf.* **2016**, *98*, 523–529. [CrossRef]
23. Haimo, D.T.; Markett, C. A representation theory for solutions of a higher-order heat equation. I. *J. Math. Anal. Appl.* **1992**, *168*, 89–107. [CrossRef]
24. Haimo, D.T.; Markett, C. A representation theory for solutions of a higher-order heat equation. II. *J. Math. Anal. Appl.* **1992**, *168*, 289–305. [CrossRef]
25. Zhukovsky, K.V. Exact solution of Guyer-Krumhansl type heat equation by operational method. *Int. J. Heat Mass Transf.* **2016**, *96*, 132–144. [CrossRef]
26. Erdélyi, A.; Magnus, W.; Oberhettinger, F.; Tricomi, F.G. *Higher Transcendental Functions*; McGraw-Hill Book Company: New York, NY, USA; Volume II.
27. Dattoli, G.; Zhukovsky, K. Quark Flavour Mixing and the Exponential Form of the Kobayashi-Maskawa Matrix. *Eur. Phys. J. C* **2007**, *50*, 817–821. [CrossRef]
28. Dattoli, G.; Zhukovsky, K. Quark Mixing in the Standard Model and the Space Rotations. *Eur. Phys. J. C* **2007**, *52*, 591–595. [CrossRef]

29. Zhukovsky, K.V.; Dattoli, G. Quark Mixing and Exponential Form of the Cabibbo-Kobayashi-Maskawa Matrix. *Phys. At. Nucl.* **2008**, *71*, 1807–1812. [CrossRef]
30. Dattoli, G.; Zhukovsky, K.V. Neutrino Mixing and the exponential form of the Pontecorvo-Maki-Nakagawa-Sakata matrix. *Eur. Phys. J. C* **2008**, *55*, 547–552. [CrossRef]
31. Zhukovsky, K.; Borisov, A. Exponential parameterization of the neutrino mixing matrix: Comparative analysis with different data sets and CP violation. *Eur. Phys. J. C* **2016**, *76*, 637. [CrossRef]
32. Zhukovsky, K.; Melazzini, F. Exponential parameterization of neutrino mixing matrix with account of CP-violation data. *Eur. Phys. J. C* **2016**, *76*, 462. [CrossRef]
33. Haubold, H.J.; Mathai, A.M.; Saxena, R.K. Analysis of solar neutrino data from Super-Kamiokande I and II. *Entropy* **2014**, *16*, 1414–1425. [CrossRef]
34. Mathai, A.M.; Haubold, H.J. On a generalized entropy measure leading to the pathway model with a preliminary application to solar neutrino data. *Entropy* **2013**, *15*, 4011–4025. [CrossRef]
35. Mathai, A.M.; Saxena, R.K.; Haubold, H.J. Back to the Solar Neutrino Problem. *Space Res. Today* **2012**, *185*, 112–123.
36. Dattoli, G. Generalized polynomials, operational identities and their applications. *J. Comput. Appl. Math.* **2000**, *118*, 111. [CrossRef]
37. Dattoli, G.; Srivastava, H.M.; Zhukovsky, K. Orthogonality properties of the Hermite and related polynomials. *J. Comput. Appl. Math.* **2005**, *182*, 165–172. [CrossRef]
38. Abd-Elhameed, W.M.; Youssri, Y.H. A Novel Operational Matrix of Caputo Fractional Derivatives of Fibonacci Polynomials: Spectral Solutions of Fractional Differential Equations. *Entropy* **2016**, *18*, 345. [CrossRef]
39. Abd-Elhameed, W.M.; Youssri, Y.H. Spectral solutions for fractional differential equations via a novel Lucas operational matrix of fractional derivatives. *Romanian J. Phys.* **2016**, *61*, 795–813.
40. Abd-Elhameed, W.M.; Youssri, Y.H. New spectral solutions of multi-term fractional order initial value problems with error analysis. *Comp. Model. Eng. Sci.* **2015**, *105*, 375–398.
41. Abd-Elhameed, W.M.; Youssri, Y.H. New ultraspherical wavelets spectral solutions for fractional Riccati differential equations. *Abstr. Appl. Anal.* **2014**, *2014*, 626275. [CrossRef]
42. Spassky, D.; Omelkov, S.; Mägi, H.; Mikhailin, V.; Vasil'ev, A.; Krutyak, N.; Tupitsyna, I.; Dubovik, A.; Yakubovskaya, A.; Belsky, A. Energy transfer in solid solutions $Zn_xMg_{1-x}WO_4$. *Opt. Mater.* **2014**, *36*, 1660–1664. [CrossRef]
43. Gridin, S.; Belsky, A.; Dujardin, C.; Gektin, A.; Shiran, N.; Vasil'ev, A. Kinetic Model of Energy Relaxation in Csl:A (A = Tl and In) Scintillators. *J. Phys. Chem. C* **2015**, *119*, 20578–20590. [CrossRef]
44. Krutyak, N.R.; Mikhailin, V.V.; Vasil'ev, A.N.; Spassky, D.A.; Tupitsyna, I.A.; Dubovik, A.M.; Galashov, E.N.; Shlegel, V.N.; Belsky, A.N. The features of energy transfer to the emission centers in $ZnWO_4$ and $ZnWO_4$:Mo. *J. Lumin.* **2013**, *144*, 105–111. [CrossRef]
45. Hermite, C. *Sur un Nouveau Développement en Série de Fonctions*; Gauthier Villar: Paris, France, 1864.
46. Gould, H.W.; Hopper, A.T. Operational formulas connected with two generalizations of Hermite polynomials. *Duke Math. J.* **1962**, *29*, 51–63. [CrossRef]
47. Srivastava, H.M.; Manocha, H.L. *A Treatise on Generating Functions*; Halsted Press (Ellis Horwood Limited): Chichester, UK, 1984.
48. Appell, P.; Kampé de Fériet, J. *Fonctions Hypergéométriques et Hypersphériques, Polynômes d'Hermite*; Gauthier-Villars: Paris, France, 1926.
49. Khan, S.; Yasmin, G.; Khan, R.; Hassan, N.A.M. Hermite-based Appell polynomials: Properties and applications. *J. Math. Anal. Appl.* **2009**, *351*, 756–764. [CrossRef]
50. Khan, S.; Al-Saad, M.W.; Khan, R. Laguerre-based Appell polynomials: Properties and applications. *Math. Comput. Model.* **2010**, *52*, 247–259. [CrossRef]
51. Khan, S.; Raza, N. Hermite-Laguerre matrix polynomials and generating relations. *Rep. Math. Phys.* **2014**, *73*, 137–164. [CrossRef]
52. Comtet, L. *Advanced Combinatorics: The Art of Finite and Infinite Expansions*, Revised English ed.; Reidel: Dordrecht, The Netherlands, 1974.
53. Roman, S. *The Umbral Calculus*; Academic Press: New York, NY, USA, 1984.
54. Wolf, K.B. *Integral Transforms in Science and Engineering*; Plenum Press: New York, NY, USA, 1979.

55. Ván, P. Theories and heat pulse experiments of non-Fourier heat conduction. *Commun. Appl. Ind. Math.* **2016**, *7*, 150–166. [CrossRef]
56. Watson, G.N. *A Treatise on the Theory of Bessel Functions*, 2nd ed.; Cambridge University Press: Cambridge, UK, 1944.
57. Kilbas, A.A.; Srivastava, H.M.; Trujillo, J.J. *Theory and Applications of Fractional Differential Equations*; North-Holland Mathematical Studies; Elsevier Science Publishers: Amsterdam, The Netherland; London, UK; New York, NY, USA, 2006; Volume 204.

axioms

MDPI

Article

Operational Approach and Solutions of Hyperbolic Heat Conduction Equations

Konstantin Zhukovsky

Faculty of Physics, Moscow State University, Leninskie Gory, Moscow 119991, Russia; zhukovsk@physics.msu.ru; Tel.: +7-495-939-3177; Fax: +7-495-939-2991

Academic Editor: Hans J. Haubold
Received: 27 October 2016; Accepted: 6 December 2016; Published: 12 December 2016

Abstract: We studied physical problems related to heat transport and the corresponding differential equations, which describe a wider range of physical processes. The operational method was employed to construct particular solutions for them. Inverse differential operators and operational exponent as well as operational definitions and operational rules for generalized orthogonal polynomials were used together with integral transforms and special functions. Examples of an electric charge in a constant electric field passing under a potential barrier and of heat diffusion were compared and explored in two dimensions. Non-Fourier heat propagation models were studied and compared with each other and with Fourier heat transfer. Exact analytical solutions for the hyperbolic heat equation and for its extensions were explored. The exact analytical solution for the Guyer-Krumhansl type heat equation was derived. Using the latter, the heat surge propagation and relaxation was studied for the Guyer-Krumhansl heat transport model, for the Cattaneo and for the Fourier models. The comparison between them was drawn. Space-time propagation of a power–exponential function and of a periodic signal, obeying the Fourier law, the hyperbolic heat equation and its extended Guyer-Krumhansl form were studied by the operational technique. The role of various terms in the equations was explored and their influence on the solutions demonstrated. The accordance of the solutions with maximum principle is discussed. The application of our theoretical study for heat propagation in thin films is considered. The examples of the relaxation of the initial laser flash, the wide heat spot, and the harmonic function are considered and solved analytically.

Keywords: Guyer-Krumhansl equation; Cattaneo equation; heat propagation; analytical solution; inverse differential operator; Hermite polynomials

---

1. Introduction

The differential equations (DE) are of paramount importance both in pure mathematics and in physics since they describe a very broad range of physical processes. Rapid development of computer methods and machine calculations in the 21st century facilitated equation solving. A good description of the major numerical methods is given, for example, in [1–6]. Here they allow numerical modelling of complicated physical processes [7–19], including multidimensional heat transfer in rectangles and cylinders [20–23]. However, proper understanding of the solutions and of the obtained results can be best done when they are obtained in analytical form. Analytical studies generally are more suitable for the analysis of the undergoing physical processes rather than numerical models, the latter giving precise description of the performance of the devices and of the specific studied cases. Analytical solutions are highly appreciated, but only a few types of DE allow explicit, if any, exact analytical solutions. Recently some fractional ordinary DE and partial differential equations (PDE) were analyzed and analytically solved in [24–37]. They benefit from the use of special functions [38–40]. The mathematical instruments, used to solve DE, generally range from a variety of integral transforms [41,42] to expansion in a series of generalized orthogonal polynomials [43] with many variables and indices [44–46], which arise naturally in studies of physical problems, such as the radiation and dynamics of beams of charges [47–54], heat and mass transfer [55–59], etc. Moreover, exponential operators and matrices are currently used also for description of such nature fundamentals as neutrino and quarks in theoretical [60–65] and in experimental [66–68] frameworks. The method of inverse differential and exponential operators has multiple applications for treating the above mentioned problems and related processes; some examples of DE solution by the inverse derivative method with regard to the heat equation, the diffusion equation, and their extensions, involving the Laguerre derivative, were given in [46,69–73]. Orthogonal polynomials can be defined in forms through operational relations [74], although we will also use their series presentations.

In what follows we treat the problem of heat conduction by the operational method in the framework of classical thermodynamics. We obtain, compare, and explore exact solutions for relevant DE in the framework of heat conduction models for Fourier [75], Cattaneo [76], and Guyer-Krumhansl [77] heat laws.

2. Fourier Heat Equation and Its Operational Solution

Fourier's law of heat propagation imposes a linear relation between the temperature gradient and the heat flux. This is one of the most popular laws in continuum physics and it is in excellent agreement with everyday life and with more than 90% experiments. Recently the Fourier heat equation with a linear term

$$\partial_t F(x,t) = \left(\alpha\,\partial_x^2 + \beta\, x\right) F(x,t) \tag{1}$$

was studied operationally as a special case of the Schrödinger equation in imaginary time [70,73]. The solution of such DE with the initial condition $F(x,0) = f(x)$ is given by the Gauss transform:

$$\begin{aligned} F(x,t) &= e^{\Phi(x,t;\alpha,\beta)}\,\hat{\Theta}\hat{S} f(x) = e^{\Phi(x,t;\alpha,\beta)} f(x+\alpha\beta\, t^2, t), \\ f(x+\alpha\beta\, t^2, t) &= \frac{1}{2\sqrt{\pi\alpha t}} \int\limits_{-\infty}^{\infty} e^{-\frac{(x+\alpha\beta t^2-\xi)^2}{4t\alpha}} f(\xi) d\xi, \end{aligned} \tag{2}$$

where $\Phi(x,t;\alpha,\beta) = \frac{1}{3}\alpha\beta^2 t^3 + \beta\, t\, x$, $\hat{\Theta} = e^{\alpha\beta t^2 \partial_x}$, $\hat{S} = e^{\alpha t\,\partial_x^2}$. The heat diffusion operator

$$\hat{S} = e^{t\partial_x^2} \tag{3}$$

was thoroughly explored by Srivastava in [78]. The exponential differential operator $e^{\partial_x^2}$ (3) reduces to the exponential differential operator e^{∂_x} upon the application of the following integral presentation [41]:

$$e^{\hat{p}^2} = \frac{1}{\sqrt{\pi}} \int\limits_{-\infty}^{\infty} e^{-\xi^2 + 2\xi\hat{p}} d\xi, \tag{4}$$

where $\hat{p} = \sqrt{t}\widetilde{D}$ in our case. Thus, the above formula reads as follows:

$$e^{t\widetilde{D}^2} f(x) = \frac{1}{\sqrt{\pi}} \int\limits_{-\infty}^{\infty} e^{-\xi^2 + 2\xi\sqrt{t}\widetilde{D}} f(x) d\xi. \tag{5}$$

The action of the operator of translation $\exp(\eta\widetilde{D})$ for $\widetilde{D} = D + \alpha$ produces a shift

$$e^{\eta(D+\alpha)} f(x) = e^{\eta\alpha} f(x+\eta), \tag{6}$$

The solution (2) consists of the action of the evolution operator on the initial condition $F(x,0) = f(x)$, which is transformed by $\hat{\hat{S}}$ and $\hat{\hat{\Theta}}$. Interestingly, for the Airy initial condition

$f(x) = Ai(\frac{x}{C}) = \frac{1}{\pi}\int\limits_0^\infty \cos\left(\frac{1}{3}\zeta^3 + \frac{x}{C}\zeta\right)d\zeta$ the operational method readily yields the fading oscillating solution without any spread [71], while the solution for the Gauss initial function demonstrates the spread of the packet.

In direct analogy with (1) the following two dimensional DE

$$\partial_t F(x,y,t) = \left\{\left(\alpha\partial_x^2 + \gamma\partial_y^2\right) + bx + cy\right\}F(x,y,t),\ \min(\alpha,\beta,\gamma) > 0, \tag{7}$$

with the initial condition $F(x,y,0) = f(x,y)$ has the solution, which reads:

$$F(x,y,t) = e^{\Psi}\,\hat{\Theta}_x\hat{\Theta}_y\hat{S}_x\hat{S}_y f(x,y) \propto f\left(x + t^2\alpha b, y + t^2\gamma c, t\right), \tag{8}$$

and it includes the phase $\Psi = (\alpha b^2 + \gamma c^2)t^3/3 + t(bx + cy)$. The solution consists of the two-dimensional Gauss transform due to the action of the heat diffusion operators (3) $\hat{S}_x\hat{S}_y$, and of the consequent shift along both coordinates, executed by the translation operators $\hat{\Theta}_x = e^{t^2(\alpha b/2)\partial_x}$, $\hat{\Theta}_y = e^{t^2(\gamma c/2)\partial_y}$.

In the context of the operational method the orthogonal polynomials are useful and their operational definitions are necessary. Indeed, the action of the heat diffusion operator on the monomial x^n yields $e^{a\partial_x^2}x^n = H_n(x,a)$ according to the operational definition of the Hermite polynomials of two variables $H_n(x,y)$ [74]

$$H_n(x,y) = e^{y\frac{\partial^2}{\partial x^2}}x^n = n!\sum_{r=0}^{[n/2]}\frac{x^{n-2r}y^r}{(n-2r)!r!},\quad e^{xt+yt^2} = \sum_{n=0}^{\infty}\frac{t^n}{n!}H_n(x,y), \tag{9}$$

and we obtain the solution of DE (1) $F(x) \propto H_n(x + ab, a)$, where $a = \alpha t$, $b = \beta t$ as follows:

$$F(x,t) = e^{\Phi}H_n\left(x + \alpha\beta t^2, \alpha t\right). \tag{10}$$

It is easy to demonstrate that for the initial condition $f(x) = x^k e^{\delta x}$ the following operational rule applies:

$$e^{y\partial_x^2}x^k e^{\alpha x} = e^{(\alpha x + \alpha^2 y)}H_k(x + 2\alpha y, y), \tag{11}$$

so that we obtain $\hat{S}f(x) = e^{\delta(x+\delta a)}H_k(x + 2\delta a, a) = f(x,t)$, $a = \alpha t$. The consequent action of the translation operator $\hat{\Theta}$ yields the shift along the x argument and results in

$$F(x,t) = e^{\Phi + \Delta_1}H_k\left(x + 2t\alpha\delta + t^2\alpha\beta, \alpha t\right), \tag{12}$$

where $\Delta_1 = \delta\left(x + \delta\alpha t + \alpha\beta t^2\right)$.

Now we can easily solve the following extended heat equation:

$$\partial_t F(x,t) = \left(\alpha\partial_x^2 + 2\delta\partial_x + \beta x + \gamma\right)F(x,t),\ F(x,0) = f(x), \tag{13}$$

which can be considered as the generalization of Equation (1) upon the substitution of the derivative $\partial_x \to \partial_x + \delta/\alpha$. To address Equation (13) we note that if $G(x,t)$ is the solution of Equation $\psi(\partial_x)G(x,t) = \hat{D}(t)G(x,t)$ with the initial function $g(x) = G(x,0) = e^{\lambda x}F(x,0) = e^{\lambda x}f(x)$, then $F(x,t) = e^{-\lambda x}G(x,t)$ is the solution of the equation $\psi(\partial_x + \lambda)F(x,t) = \hat{D}(t)F(x,t)$ with the initial function $f(x) = F(x,0)$. Distinguishing the perfect square of the operator $\partial_x + \delta/\alpha$, we come to the solution of Equation (13) in the following form:

$$F(x,t) = e^{t(\gamma - \frac{\delta^2}{\alpha}) - \frac{x\delta}{\alpha}}\,G(x,t), \tag{14}$$

where $G(x,t)$ satisfies Equation (1) for G with the initial condition $g(x) = G(x,0) = e^{\delta x/\alpha} f(x)$. Let us set $\alpha = 1$ without substantial loss of generality and choose the monomial $f(x) = x^k$ initial condition for (13). Then $g(x) = x^k e^{\delta x}$ is the initial condition for the Equation $\partial_t G = (\partial_x^2 + \beta x)G$. Its solution is given by (12) upon the substitution of $F \to G$ and we end up with the following solution of the extended heat Equation (13)

$$F(x,t) = e^{\Phi+\Delta_2} H_k\left(x + 2t\delta + t^2\beta, t\right), \tag{15}$$

where $\Delta_2 = t\gamma + t^2\delta\beta$ is the additional phase. This solution reduces to (10) in the proper limiting case.

3. Propagation of a Heat Surge with Fourier Heat Diffusion Type Equation

In heat conduction experiments the relaxation of the instant point-like heat surge is the standard technique, approved for thermal diffusivity measurements [79]. In practical terms it is executed with the flash method—the standard engineering procedure for measuring thermal diffusivity with intense and ultra-short laser heat pulses. The latter can be modelled by the initial δ-function space distribution. Let us consider this example now: $f(x,y) = \delta(x,y)$. With the help of Equation (5) we immediately obtain the action of the $\hat{S}$ operator on δ-function as follows:

$$\hat{S}\delta(x) \equiv e^{t\partial_x^2}\delta(x) = \frac{1}{2\sqrt{\pi t}} e^{-\frac{x^2}{4t}}. \tag{16}$$

In two dimensions Equation (16) becomes $f(x,y,t) = \hat{S}_x\hat{S}_y\delta(x,y) = e^{-(x^2/\alpha + y^2/\gamma)/4t}/4\pi t\alpha\gamma$. Then the operators $\hat{\Theta}_{x,y}$ induce the space shift and yield the particular solution (8) for the initial function $f(x,y) = \delta(x,y)$, which evolves in the Gaussian as follows:

$$F(x,y,t) = e^{\Psi}\,\hat{\Theta}_x\hat{\Theta}_y\hat{S}_x\hat{S}_y\delta(x,y) = \frac{e^{\Psi}}{4\pi t\alpha\gamma} e^{-\frac{\frac{(x+t^2\alpha b)^2}{\alpha} + \frac{(y+t^2\gamma c)^2}{\gamma}}{4t}}. \tag{17}$$

One-dimensional solution for the initial flash condition $f(x) = \delta(x)$ simply reads $F(x,t) = e^{\Psi}\frac{e^{-\frac{(x+abt^2)^2}{4t\alpha}}}{2\sqrt{\pi t\alpha}}$, and unsurprisingly resembles the solution $F(x,t) = \frac{e^{\Phi(x,t)}}{\sqrt{1+4t}} e^{-\frac{(x+\beta t^2)^2}{1+4t}}$ [71] for the initial Gaussian function $f(x) = e^{-x^2}$.

Consider the Equation (13) for $\beta = \delta = 0$, $\gamma \neq 0$. This case represents the Fourier heat propagation, including heat exchange with the environment. Consider the flash initial condition $f(x) = \delta(x)$ for it. The solution is shown in Figure 1 and it evidences significant spread of the initial function already for the times $t \in [10^{-3}, 10^{-2}]$.

The plot in Figure 1 is compiled for $\alpha = 1$, $\gamma = 1$, which represents the Fourier heat solution with some heat exchange with the environment. At the moment $t = 0.001$ we see the Gaussian as the result of the evolution of the δ-function; at the moment $t = 0.01$ this Gaussian has faded and the contribution of the heat exchange term γ is noticeable. The relaxed solution slowly grows due to non-zero heat exchange with $\gamma \neq 0$.

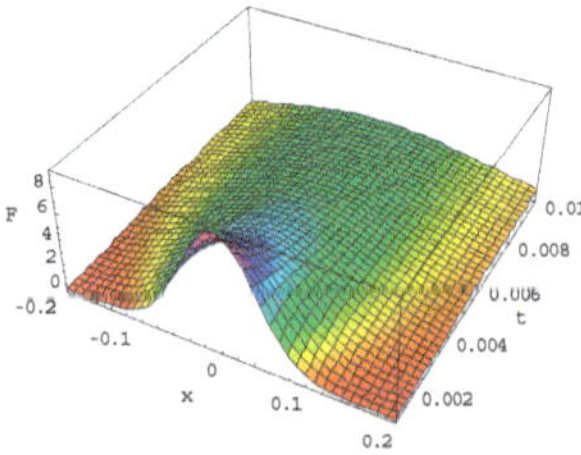

Figure 1. Evolution of the initial $\delta(x)$ function as the solution of the Fourier Equation for $\alpha = \gamma = 1$, $\beta = \delta = 0$, for the interval of time $t \in [10^{-3}, 10^{-2}]$.

4. Operational Solution of the Hyperbolic Heat Conduction Equation

The Fourier heat law [75] has some shortcomings, noted by L. Onsager in 1931, who said [80] that it "contradicts the principle of microscopic reversibility, but this contradiction is removed if we recognize that it is only an approximate description of the process of conduction, neglecting the time needed for acceleration of the heat flow". The Fourier law does not properly describe heat conduction at low temperature <25 K in dielectric crystals and in systems with reduced dimensions. Moreover, it has some unphysical properties, such as lack of inertia. In other words, if an instant temperature perturbation is applied at a point in the solid, it will be felt everywhere instantaneously. This contradicts the phenomenon of the second sound, when the temperature perturbation propagates like a wave with damping. To overcome these problems Cattaneo [76] proposed a time-dependent relaxational model, which yielded the following equation: $(\tau\partial_t^2 + \partial_t)T = D_T\nabla^2 T$, where D_T is the heat conductivity and the relaxation time τ in heat conduction is extremely small ($\tau \approx 10^{-13}$ s) at room temperature. This equation models the phenomenon of the second sound, first observed in liquid helium [81]. With the development of the second sound theoretical background [77], it was then later detected also in solid crystals [82–85]. In the relevant tests heat flash technology [79] was used for the sensitive measurement of the thermal diffusivity.

The application of the operational method for solution of the second order PDE was given in [72]. The following PDE with initial conditions:

$$\left(\frac{\partial^2}{\partial t^2} + \hat{\varepsilon}(x)\frac{\partial}{\partial t}\right)F(x,t) = \hat{D}(x)F(x,t),\ F(x,0) = f(x),\ F(x,\infty) = 0, \tag{18}$$

where $\hat{D}(x)$ and $\hat{\varepsilon}(x)$ are differential operators, acting over the coordinate, dependently on the specific form of the operators $\hat{\varepsilon}(x)$ and $\hat{D}(x)$, describes a variety of physical processes in thermodynamics, electrodynamics etc. The particular fading with time solution of Equation (18) formally reads as follows [72]:

$$F(x,t) = f(x)e^{-\frac{t\hat{\varepsilon}(x)}{2}}e^{-\frac{t}{2}\sqrt{\hat{\varepsilon}^2(x)+4\hat{D}(x)}}, \tag{19}$$

The other solution of Equation (18) has a positive sign in the exponential: $F_2 \propto e^t$, and does not satisfy the requirement $F(x,\infty) = 0$, which otherwise can be formulated as $F(x,\infty) < \infty$, i.e., the solution converges at infinite time. Both of these assumptions are reasonable for physical applications. We perform the Laplace transforms [41] in (19):

$$e^{-t\sqrt{V}} = \frac{t}{2\sqrt{\pi}}\int_0^\infty \frac{d\xi}{\xi^{3/2}}e^{-\frac{t^2}{4\xi}-\xi V},\ t > 0, \tag{20}$$

and, provided the integral converges, we obtain for Equation (18) the following vanishing at infinite time solution:

$$F(x,t) = e^{-\frac{t}{2}\hat{\varepsilon}(x)} \frac{t}{4\sqrt{\pi}} \int_0^\infty \frac{d\xi}{\xi\sqrt{\xi}} e^{-\frac{t^2}{16\xi}} e^{-\xi\hat{\varepsilon}^2(x)} e^{-4\xi\hat{D}(x)} f(x) \tag{21}$$

whose particular form depends on the differential operators $\hat{D}(x)$, $\hat{\varepsilon}(x)$ and on the initial condition $f(x)$. In what follows, we consider some examples.

Let us first of all consider $\hat{\varepsilon}(x) = \varepsilon = \text{const}$ and $\hat{D}(x) = \alpha\partial_x^2 + \kappa$, keeping the linear term κ in the r.h.s. of Equation (18). Then we end with the telegraph equation, also known as a hyperbolic heat conduction equation (HHE):

$$\left(\partial_t^2 + \varepsilon\partial_t\right) F(x,t) = \left(\alpha\partial_x^2 + \kappa\right) F(x,t), \; F(x,0) = f(x), \; F(x,\infty) = 0. \tag{22}$$

Its fading at $t \to \infty$ solution reads as follows:

$$F(x,t) = e^{-\frac{\varepsilon t}{2}} \frac{t}{4\sqrt{\pi}} \int_0^\infty \frac{d\xi}{\xi^{3/2}} e^{-\frac{t^2}{16\xi} - \xi(\varepsilon^2 + 4\kappa)} \hat{S} f(x), \; \hat{S} f(x) = e^{-4\alpha\xi\partial_x^2} f(x). \tag{23}$$

The action of the heat diffusion operator $\hat{S}$ can be accomplished with the help of the identity (4), resulting in

$$F(x,t) = e^{-\frac{\varepsilon t}{2}} \frac{t}{4\sqrt{\pi}} \int_0^\infty \frac{du}{u^{3/2}} e^{-\frac{t^2}{16u} - u(\varepsilon^2 + 4\kappa)} \frac{1}{\sqrt{\pi}} \int_{-\infty}^\infty e^{-v^2} f\left(x + 2iv\sqrt{\zeta}\right) dv, \tag{24}$$

where $\zeta = 4u\alpha$. Consider an example of the initial function $f(x) = e^{\gamma x} x^n$, which, is useful for the description of heat pulses of a variety of shapes, custom modelled by the sum $\sum_{n,\gamma} e^{\gamma x} x^n$, $\gamma < 0$. With the help of the operational identity (11) we obtain the exact form of the solution:

$$F(x,t) = \frac{t e^{-\frac{t\varepsilon}{2} + \gamma x}}{4\sqrt{\pi}} \int_0^\infty \frac{du}{u^{3/2}} e^{-\frac{t^2}{16u} - u\delta} H_n(x - 2\gamma\zeta, -\zeta), \quad \delta = \varepsilon^2 + 4\left(\kappa + \alpha\gamma^2\right). \tag{25}$$

We omit the bulky result of the above integration for arbitrary $n \in$ Integers, $\gamma \in$ Reals, which can be computed with account for (9). In the particular case of given n and γ, for the example for $n = -\gamma = 1$, we obtain

$$F(x,t)|_{f(x)=xe^{-x}} = e^{-x - \frac{t\varepsilon}{2} - \frac{t}{2}\sqrt{4\alpha+\varepsilon^2}} \left(x + \frac{2t\alpha}{\sqrt{4\alpha + \varepsilon^2}}\right). \tag{26}$$

For $n = 2$, $\gamma = -1$, $\kappa = 0$ we have the following solution:

$$F(x,t)|_{f(x)=x^2e^{-x}} = e^{-x - \frac{t\varepsilon}{2} - \frac{t}{2}\sqrt{4\alpha+\varepsilon^2}} \left(\frac{4t^2\alpha^2}{4\alpha + \varepsilon^2} + x^2 - \frac{2t\alpha\varepsilon^2}{(4\alpha + \varepsilon^2)^{3/2}} + \frac{4t\alpha x}{\sqrt{4\alpha + \varepsilon^2}}\right). \tag{27}$$

The solution for the initial function $f(x) = |x|^3 e^{-|x|}$ is too bulky to be presented in its analytical form; it describes the damped wave propagation as shown in Figure 2.

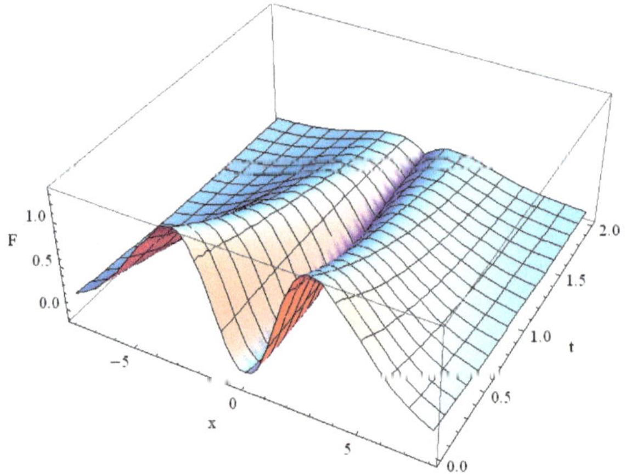

Figure 2. Solution of the Cattaneo heat equation for $\mu = 0$, $\alpha = 1$, $\varepsilon = 1$ for the initial function $F(x,0) = e^{-|x|}|x|^3$.

By choosing $\alpha = \varepsilon = 5$ in (22), i.e., $D_T = \alpha/\varepsilon = 1$, $\tau = 1/\varepsilon = 0.2$, the influence of the second time derivative ∂_t^2 in the Cattaneo equation is reduced, but the heat conductivity $D_T = \alpha/\varepsilon = 1$ remains unchanged, as compared with the case $\alpha = \varepsilon = 1$. The relaxation of the solution occurs much earlier for $\alpha = \varepsilon = 5$, than for $\alpha = \varepsilon = 1$, as follows from the comparison of the plot in Figure 2 with that in Figure 3; the reason is that the diffusive heat transfer in this case prevails over the wave heat transfer.

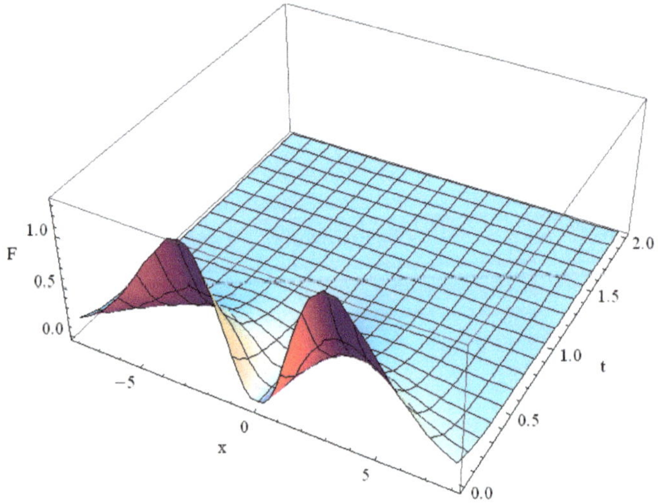

Figure 3. Solution of the Cattaneo heat equation for $\mu = 0$, $\alpha = 5$, $\varepsilon = 5$ for the initial function $F(x,0) = e^{-|x|}|x|^3$.

5. Propagation of a Heat Surge with the Hyperbolic Heat Conduction Equation

We now consider the example of initial δ-function $f(x) = \delta(x)$ for Equation (22). The action of the operator $\hat{S} = e^{-4\alpha\xi\partial_x^2}$ in our solution (23) can be accomplished with the help of Equation (16), where

$\nu = -4\alpha\xi$, as follows: $\hat{S}\delta(x) = \frac{e^{\frac{x^2}{16\alpha\xi}}}{4\sqrt{\pi|\alpha\xi|}}$. Now the integration in (23) can be completed for the times $t > x/\sqrt{\alpha}$, where the solution exists:

$$F(x,t) = \frac{te^{-\frac{\varepsilon t}{2}}}{16\pi\alpha^{1/2}} \int_0^\infty \frac{d\xi}{\xi^2} e^{-\xi(\varepsilon^2+4\kappa) - \frac{t^2 - x^2/\alpha}{16\xi}} = \frac{te^{-\frac{\varepsilon t}{2}}}{2\pi\alpha^{1/2}} \sqrt{\frac{\omega}{v}} K_1\left(\frac{\sqrt{\omega v}}{2}\right), \quad \omega = \varepsilon^2 + 4\kappa > 0, \; v = t^2 - x^2/\alpha > 0. \tag{28}$$

Physical meaning of the condition $v = t^2 - x^2/\alpha > 0$, i.e., $t > x/\sqrt{\alpha}$ is that the time, needed for the initial $\delta(x)$ function to reach the point x in space, is exactly $t_0 = x/\sqrt{\alpha}$ and before that the heat surge is just not felt in the point x at all. The value $\sqrt{\alpha}$ is the velocity of the surge propagation, finite in the Cattaneo model, contrary to the infinite signal propagation speed in the Fourier law. The solution of Equation (22) for $\varepsilon = 10$, $\kappa = 1$, $\alpha = 100$ for the initial function $f(x) = \delta(x)$ is shown in Figure 4 for the moments of time $t = 0.01$ (light blue line), $t = 0.1$ (lilac line), $t = 0.3$ (pink line), $t = 0.6$ (blue line) and $t = 1.1$ (yellow–green line). Differently from the Fourier law, the initial flash reaches the point x at the moment $t_0 = x/\sqrt{\alpha}$ and there is still no non-trivial solution in the point x before this moment of time. The initial δ-function is not shown in Figure 4 because of its infinite amplitude.

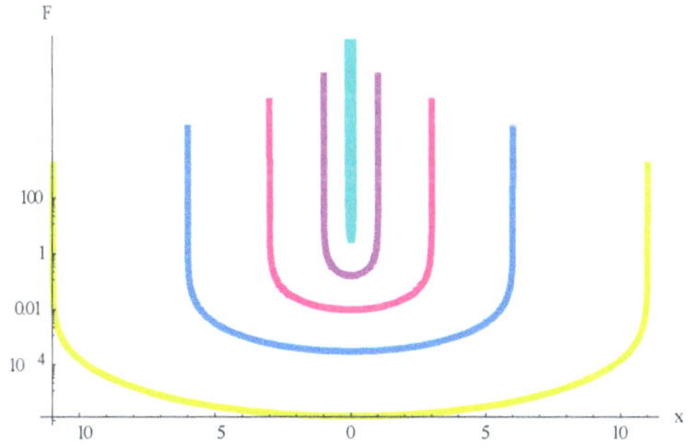

Figure 4. Solution $F(x, t)$ of hyperbolic heat equation with initial $\delta(x)$ function for $\alpha = 100$, $\varepsilon = 10$, $\kappa = 1$ in the moments of time $t = 0.01$ (light blue line), $t = 0.1$ (lilac line), $t = 0.3$ (pink line), $t = 0.6$ (blue line) and $t = 1.1$ (yellow–green line).

Higher values of α and ε reduce the contribution of the second order time-derivative ∂_t^2 in the Cattaneo Equation respectively to the role of other terms in (22). The damping of the solution in this case occurs sooner. We omit the proper plot for conciseness. Note that in the case of negative κ, i.e., when positive κ is in the l.h.s. of the hyperbolic Equation (22), the integral in (28) converges for $\varepsilon^2 > 4\kappa$, $\kappa < 0$. Interestingly, despite the fact that the solution (28) was obtained for $\omega > 0$ and $v > 0$, it holds true for any $\omega \neq 0$ and $v \neq 0$. For $\omega < 0$ and $v < 0$ we obtain positive values for $F(x,t)$, while for $\omega < 0$ and $v > 0$ as well as for $\omega > 0$ and $v < 0$ we obtain complex values for the resulting function $F(x,t)$. For $\omega \to 0$, $v \neq 0$ we obtain $F(x,t)|_{\omega \to 0} = e^{-\frac{\varepsilon t}{2}} \frac{t}{2\pi\alpha^{1/2}} \frac{2}{|v|}$. For $\omega \neq 0$, $v \to 0$ the solution (28) diverges.

Heat propagation in three spatial dimensions in the Cattaneo model is governed by the direct three-dimensional generalization of (22) for $\kappa = 0$: $\nabla^2 T = \frac{1}{D_T}\frac{\partial T}{\partial t} + \frac{\tau}{D_T}\frac{\partial^2 T}{\partial t^2}$, where D_T is the heat conductivity. The ratio $C = \sqrt{D_T/\tau}$ has the dimension of velocity and it stands for the speed of the heat wave propagation in the medium, τ is the material parameter, describing the time, needed for the initiation of a heat flow after a temperature gradient was imposed at the boundary of the domain. Thus, it determines the time lag for the appearance or disappearance of the heat flow after the temperature gradient is imposed or removed. This relaxation time is associated in the framework of the Cattaneo model with the linkage time of phonon-phonon collision and it measures the thermal inertia of the medium through the time, needed for the heat flow to fade in or out. The role of the constant term in the HHE becomes clearer in the context of the signal propagation in long telephone lines. We will consider it in what follows. At the moment we will keep for generality the constant term in the three-dimensional telegraph equation, thus writing it as follows:

$$\left(\tau\partial_t^2 + \partial_t\right)F(x,t) = \left(D_T\nabla^2 + \mu\right)F(x,t),\ F(x,y,z,0) = f(x,y,z),\ F(x,y,z,\infty) = 0, \tag{29}$$

where $D_T = \alpha/\varepsilon = k/c_p\rho$ is the thermal diffusivity, k is the thermal conductivity, ρ is the mass density, c_p is the specific heat capacity, $\tau = 1/\varepsilon$ is the relaxation time, often related to the speed of the second sound C in media: $\tau = D_T/C^2$. The operational solution of (29) includes the action of the heat diffusion operators for each coordinate on the initial function:

$$F(x,y,z,t) = e^{-\frac{\varepsilon t}{2}}\frac{t}{4\sqrt{\pi}}\int_0^\infty \frac{d\xi}{\xi\sqrt{\xi}}e^{-\frac{t^2}{16\xi}-\xi(\varepsilon^2+4\kappa)}\hat{S}_x\hat{S}_y\hat{S}_z f(x,y,z),\ f(x,y,z) = F(x,y,z,0) \tag{30}$$

The action of $\hat{S}_x\hat{S}_y\hat{S}_z f(x,y,z)$, where $\hat{S}_i = e^{-4\alpha\xi\partial_i^2}f(i)$, $i = x,y,z$ is easy to obtain: the result for each coordinate heat operator action is given by the inner integral in (24) and then the integration over $d\xi$ can be performed if the integral converges.

For experimental measurement of the thermal conductivity the flash technique is commonly approved. For this reason we choose the initial function $f(x,y,z) = \delta(x,y,z)$, modelling an intense instant point-like volume heating. The application of the heat operators $\hat{S}_i = e^{-4\alpha\xi\partial_{x_i}^2}$ gives $\hat{S}_x\hat{S}_y\hat{S}_z\delta(x,y,z) = e^{-\frac{x^2+y^2+z^2}{16\alpha\xi}}/16(\pi|\alpha\xi|)^{3/2}$ yields the following solution for HHE (29) with $f(x,y,z) = \delta(x,y,z)$:

$$F(x,y,z,t) = e^{-\frac{\varepsilon t}{2}}\frac{t}{64\pi^2\alpha^{3/2}}\int_0^\infty \frac{d\xi}{\xi^3}e^{-\xi(\varepsilon^2+4\kappa)-\frac{t^2-(x^2+y^2+z^2)/\alpha}{16\xi}}. \tag{31}$$

The above integral converges for $t > \sqrt{(x^2+y^2+z^2)/\alpha}$ and we obtain the following simple analytical expression, involving modified Bessel functions K_2:

$$\begin{gathered} F(x,y,z,t) = e^{-\frac{\varepsilon t}{2}}\frac{t}{32\pi^2\alpha^{3/2}}\frac{\omega}{v}K_2(2\sqrt{\omega v}), \\ \omega = \varepsilon^2 + 4\kappa > 0,\ v = \frac{t^2-(x^2+y^2+z^2)/\alpha}{16} > 0. \end{gathered} \tag{32}$$

Note, that the above solution also holds for $\omega \neq 0$ and $v \neq 0$. As well as in the one-dimensional case (28) for $\omega < 0$, $v < 0$, that is for the times, inferior $t_0 = \sqrt{(x^2+y^2+z^2)/\alpha}$, when the heat wave reaches point x, we obtain positive values for the solution $F(x,t)$ for negative κ, such that $4\kappa < -\varepsilon^2$. For $\omega < 0$, $v > 0$ as well as for $\omega > 0, v < 0$ we obtain complex solution $F(x,t)$. In the limiting case of $\omega \to 0$, $v \neq 0$ we obtain the finite limit for the solution: $F(x,y,z,t)|_{\omega\to 0} = e^{-\frac{\varepsilon t}{2}}\frac{4t}{\pi^2\alpha^{3/2}}\frac{1}{v^2}$, but for $\omega \neq 0$, $v \to 0$ the solution diverges.

Despite that the Cattaneo heat Equation (22) for $\kappa = 0$ is reversible on the time scale of the thermal relaxation time τ, and despite that it predicts a finite value for the heat wave velocity $C = \sqrt{D_T/\tau}$, its quantitative value disagrees with the experimental data at high frequencies and low temperatures. However, the HHE is widely used in radio engineering.

6. Propagation of a Harmonic Signal with the Telegraph Equation

In the context of signal propagation in cable lines without radiation loss the HHE (22) is perfectly usable for the description of the voltage and current 86. For this reason we choose the initial harmonic function $f(x) = e^{inx}$. The action of the operator $\hat{S}$ with account for (5) yields $\hat{S}e^{inx} = e^{v\partial_x^2}e^{inx} = e^{inx-n^2v}$,

$\nu = -4\alpha\xi$. Integration in (24) in turn yields the following solution for the telegraph Equation (22) for $F(x,0) = f(x) = e^{inx}$:

$$F(x,t) = e^{inx - \frac{t}{2}(\varepsilon + \sqrt{V})}, \quad V = \varepsilon^2 + 4\left(\kappa - \alpha n^2\right). \tag{33}$$

The harmonic solution (33) is not spreading in space, but only fading in time. Its physical meaning is clarified if we attribute the constants $\alpha, \varepsilon, \kappa$ in HHE (22) the values according to the schematic diagram of the electric circuit in Figure 5, where the electric line with finite resistance, inductivity, capacitance, and leakage is shown.

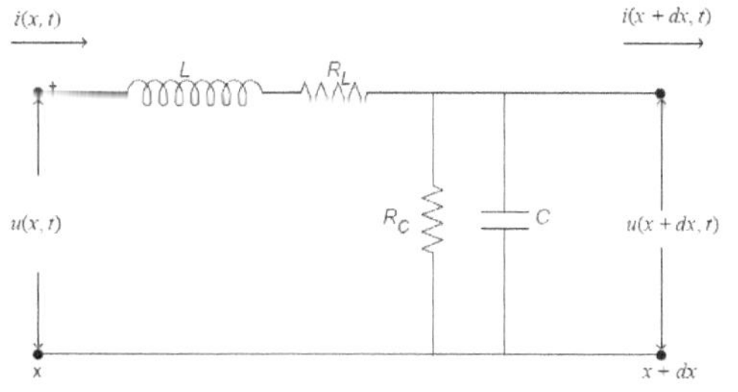

Figure 5. Schematic diagram of an electric cable line with leakage.

In terms of the resistance R_L, inductivity L, capacitance C and leakage resistance R_C the voltage $u(x,t)$ along the transmission line, shown in Figure 5, is described by the one-dimensional telegraph equation [86]:

$$\begin{array}{c} (\partial_t^2 + (a+b)\partial_t)u(x,t) = (c^2\partial_x^2 - ab)u(x,t), \\ a = \frac{R_C}{C} > 0, \quad b = \frac{R_L}{L} > 0, \quad c^2 = \frac{1}{LC} > 0. \end{array} \tag{34}$$

In our notations $\alpha = c^2 = 1/LC$, $\varepsilon = (a+b) = R_C/C + R_L/L$, $\kappa = -ab = -R_C R_L/LC < 0$, $V = (a-b)^2 - (2cn)^2$ and $\varepsilon^2 + 4\kappa = a^2 + b^2 - 2ab = (a-b)^2 = (R_C/C - R_L/L)^2 \geq 0$. The solutions (28) and (32) for the initial δ-surge are valid and converge also for $R_c/C = R_L/L$ provided $t \neq t_0 = \sqrt{(x^2 + y^2 + z^2)/\alpha}$.

For $(R_C/C - R_L/L)^2 \geq 4n^2/LC$ the periodic solution $u(x,t) = e^{inx - \frac{t}{2}(\varepsilon + \sqrt{V})}$, $V \geq 0$ (33) is realized; it fades without the space shift (see examples in Figure 6). In the case of $(R_C/C - R_L/L)^2 < 4n^2/LC$ the voltage behavior in the circuit has space shift and its time fading depends exclusively on ε (see examples in Figure 7):

$$u(x,t) = e^{i(nx - \frac{t}{2}\sqrt{|V|}) - \frac{t}{2}\varepsilon}, \; V < 0. \tag{35}$$

The essential difference between the plots in Figures 6 and 7 is in the time dependent spatial phase shift, which is seen in Figure 7 for $n = 4$, 5 and is absent in Figure 6 for $n = 2$, 3.

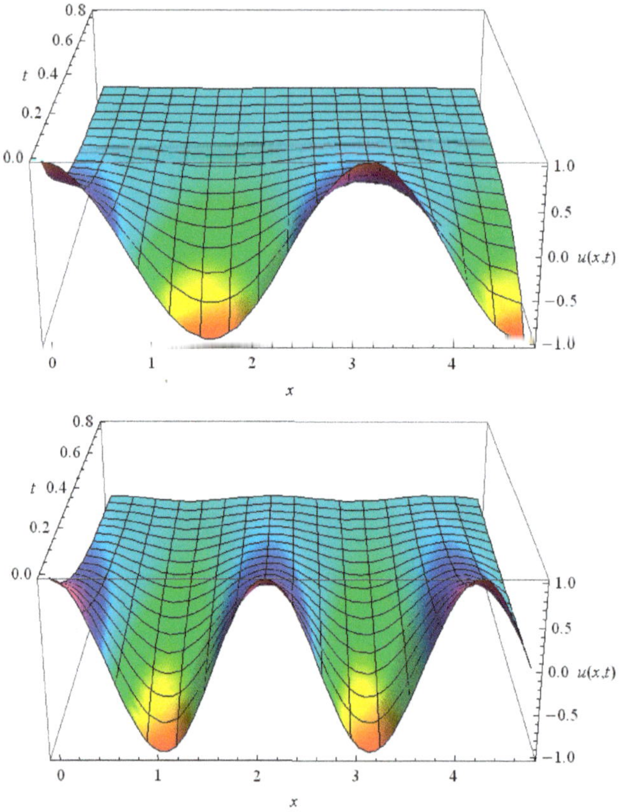

Figure 6. Space-time distribution of $\mathrm{Re}[u(x,t)]$ for $n = 2, 3$ harmonic $f(x) = e^{nix}$ for $R_C = 7$, $C = L = R_L = 1$.

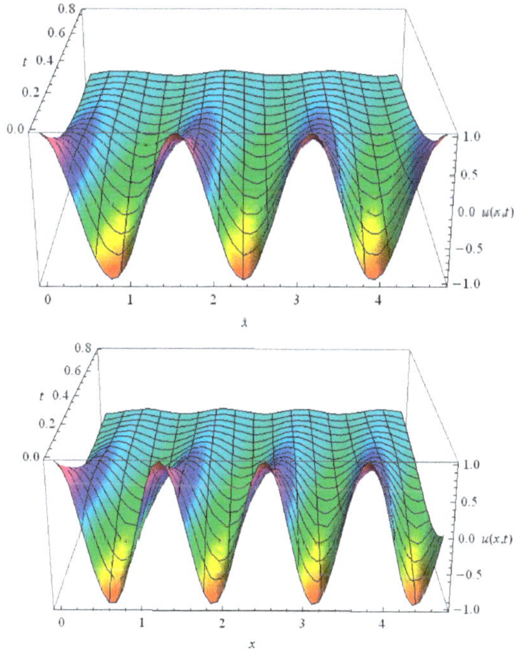

Figure 7. Space—time distribution of $\mathrm{Re}[u(x,t)]$ for $n = 4, 5$ harmonic $f(x) = e^{nix}$ for $R_C = 7$, $C = L = R_L = 1$.

Thus the first three harmonics for $R_C = 7$, $C = L = R_L = 1$ develop in time without the space shift with decent fading (see examples for $n = 2, 3$ in Figure 6); for them the solution (33) is realized. The harmonics with $n \geq 4$ are governed by the solution (35), they have a time-dependent space shift (see examples for $n = 4, 5$ in Figure 7). With increase of n the relaxation time of the harmonics also increases, but from a certain harmonic, for which the space shift appears, the relaxation time stabilizes and remains equal for all higher harmonics. In the context of heat conduction it means that the heat conduction for higher harmonics is lower than the heat conduction for low harmonics. It may also occur that, dependent on the values of the parameters in the HHE all harmonics have time-dependent space shift, i.e., the solution (35) is realized for $n \geq 1$. This frequency dependent heat conductivity is typical for HHE and it is absent in the Fourier law.

Concluding the study of the hyperbolic heat Equation (22) we note that it can be easily modified by adding non-commuting with ∂_x terms and mixed derivatives over time and coordinate. Such a modified hyperbolic equation can be solved with the help of the above developed operational technique.

7. Operational Solution of Guyer-Krumhansl Type Heat Equation and Heat Conduction in Thin Films

The Cattaneo hyperbolic equation gives a qualitatively correct description of the second sound through the heat wave propagation at finite velocity, but numerical results do not match the experiment. The predicted values for the speed of the heat wave in matter disagree with the data on ultrasonic wave propagation in dilute gases. Neither the heat pulse propagation at very low temperature is described correctly in non-metallic very pure crystals of Bi and Na F. In response to this disagreement, further generalizations of the Fourier law were developed. Among them the Guyer-Krumhansl (GK) equation [87] is distinguished for simplicity and coherence with observed data. In addition to the heat waves the GK model takes into consideration the so-called ballistic transport, which can be observed when the mean free path of a particle significantly exceeds the dimension of the medium in which it travels. The mean free path increases at low temperatures For example, for electrons in a medium with negligible electrical resistance, their motion is altered by collisions with the walls. Ballistic conduction applies also to phonons. It is typically observed in low dimensional structures, such as very thin films, silicon nanowires, carbon nanotubes, graphene etc. (see, for example, [88–92]). When the bulk phonon mean free path l is comparable to the structure size L, neither the Casimir phonon theory [93] nor Fourier diffusion [75] describe the heat transfer well, which is affected by boundary as well as internal scattering. The common rule that determines which type of transport dominates is the following: when $l << L$, the heat diffusion prevails, but for $l >> L$, or when the temperature gradient becomes large, the ballistic transport cannot be ignored. Recently the conditions for the transition region between the ballistic heat transport and the diffusion were described in [92]. Whether the heat transfer is ballistic or diffusive is also important for experimentalists, because the account for the ballistic phonon transport requires the knowledge of the boundary quality, while the wave and diffusive transport constants are intrinsic material properties [94–96]. The three dimensional Guyer-Krumhansl (GK) heat law was derived from the solution of the linearized Boltzmann equation for a phonon field in dielectric crystals at low temperature. It is good for the description of phonon gases in low temperature samples and even for some heat propagation processes on the microscopic [97] and macroscopic [98–102] level at room temperature [103–105]. The relaxation of the laser flash was shown to follow the GK law [106], which was reconsidered in the framework of weakly non-local thermodynamics in [107]. Due to the wide interest in the GK equation and its relatively good agreement with an array of experiments over a broad temperature range, conducted in various materials, we will consider the GK equation solution for propagation of spatial heat waves and flash pulses. To this end we will use the operational technique.

Let us choose the operators $\hat{\varepsilon} = \varepsilon - \delta\partial_x^2$ and $\hat{D} = \alpha\partial_x^2 + \kappa$ in (18), which results in the following GK type equation with mixed derivative:

$$\begin{gathered}(\partial_t^2 + \varepsilon\partial_t - \delta\partial_t\partial_x^2)F(x,t) = (\alpha\partial_x^2 + \kappa)F(x,t)\,, \\ \text{with} \\ F(x,0) = f(x),\ F(x,\infty) = 0,\ \alpha, \varepsilon, \delta, \kappa = \text{const}.\end{gathered} \tag{36}$$

The one-dimensional GK equation is essentially Equation (36) with $\kappa = 0$; we will keep $\kappa \neq 0$ for the sake of generality. Its role is not indifferent and it will be discussed in what follows. Introducing the operators $\hat{\Delta}_1 = \partial_t - (\alpha/\varepsilon)\partial_x^2$ and $\hat{\Delta}_2 = \partial_t - \delta\partial_x^2$, we can rewrite (36) with positive coefficients $\alpha, \varepsilon, \delta > 0$ in the following form:

$$(\partial_t\hat{\Delta}_2 + \varepsilon\hat{\Delta}_1)F(x,t) = \kappa F(x,t). \tag{37}$$

Note that equations $\hat{\Delta}_{1,2}F_{F_1,F_2}(x,t) = 0$ represent Fourier law. If $\delta = \alpha/\varepsilon$ and $\kappa = 0$, we have $\Delta_1 = \Delta_2$, and, provided the Fourier Equation $\hat{\Delta}_{1,2}F_F(x,t) = 0$ is fulfilled, we have the particular case of GK equation, which is Equation (37) with $\kappa = 0$, also satisfied by this solution $F(x,t) = F_F(x,t)$. The form of the heat conduction equation Equation (36) with $\kappa \neq 0$ arises in the study of heat propagation in a one-dimensional thin film with account for the phonon transport [108]. From (21), where $\hat{\varepsilon} = \varepsilon - \delta\partial_x^2$, $\hat{D} = \alpha\partial_x^2 + \kappa$, we write the solution of Equation (36) as follows:

$$F(x,t) = e^{-\frac{t}{2}\varepsilon}\frac{t}{4\sqrt{\pi}}\int_0^\infty \frac{d\xi}{\xi\sqrt{\xi}}e^{-\frac{t^2}{16\xi}-\xi(\varepsilon^2+4\kappa)+(\frac{t}{2}\delta+\xi(2\varepsilon\delta-4\alpha))\partial_x^2}e^{-\xi\delta^2\partial_x^4}f(x). \tag{38}$$

By means of the integral presentation $e^{-\xi\delta^2\partial_x^4}f(x) = \int_{-\infty}^{\infty} e^{-\zeta^2+2i\zeta\delta\sqrt{\xi}\partial_x^2}f(x)d\zeta/\sqrt{\pi}$ (see (5)) we obtain after grouping the 2nd order derivative terms in the exponential

$\int_{-\infty}^{\infty} e^{-\zeta^2+(t\delta/2-4\xi\alpha+2\xi\varepsilon\delta+2i\zeta\sqrt{\xi\delta})\partial_x^2} f(x)d\zeta$ the particular bounded solution for the modified hyperbolic heat conduction Equation (36) with initial function $F(x,0)=f(x)$ as follows:

$$F(x,t)=\frac{e^{-\frac{t}{2}\varepsilon t}}{4\pi}\int_0^{\infty}\frac{d\xi}{\xi\sqrt{\xi}}e^{-\frac{t^2}{16\xi}-\xi(\varepsilon^2+4\kappa)}\int_{-\infty}^{\infty}e^{-\zeta^2}\hat{S}f(x)d\zeta, \tag{39}$$

where $\hat{S}=e^{\nu\partial_x^2}$, $\nu=a+ib\zeta$, $a=t\delta/2-4\xi\alpha+2\xi\varepsilon\delta$, $b=2\sqrt{\xi}\delta$. Note, that the obtained integral solution is valid also if the constant term in the r.h.s. of the Equation (36) is negative: $\kappa<0$ and $\varepsilon>2\sqrt{|\kappa|}$, which insures the integral convergence. Equations with negative sign of κ have sense and they arise for ballistic heat propagation in thin films [108].

The application of the above solution is direct: it was demonstrated that heat propagation in thin films obeys the extended form of GK Equation (36). In particular, the following equation for thin films was proposed [108]:

$$\left\{\frac{\partial^2}{\partial t}+2\frac{\partial}{\partial t}-\frac{10Kn_b^2}{3}\frac{\partial^2}{\partial x^2}-3Kn_b^2\frac{\partial^3}{\partial x^2\partial t}+1\right\}F(x,t)=0, \tag{40}$$

where $F(x,t)$ is the ballistic component of the dimensionless energy (or quasi-temperature) and Kn is the Knudsen number, describing the molecular and the boundary effects. Equation (40) is a GK type Equation (36) with the following coefficients: $\alpha=\frac{10}{3}Kn_b^2$, $\varepsilon=2$, $\delta=3Kn_h^2$, $\kappa=-1$.

In what follows, we will consider some examples of heat pulse propagation.

8. Propagation of a Heat Surge with the Guyer-Krumhansl Heat Equation

Consider the example of the initial flash $F(x,0)=\delta(x)$, corresponding to an intense and instant laser point heating at $x=0$ at the moment of time $t=0$. From Equation (16) for $\nu=a+ib\zeta$, $a=t\delta/2-4\xi\alpha+2\xi\varepsilon\delta$, $b=2\sqrt{\xi}\delta$ we obtain $\hat{S}\delta(x)=\frac{e^{-\frac{x^2}{4(a+ib\zeta)}}}{2\sqrt{\pi|a+ib\zeta|}}$. Then from (39) the analytical solution of GK type Equation (36) with $f(x)=\delta(x)$ follows:

$$F(x,t)|_{f(x)=\delta(x)}=\frac{e^{-\frac{t}{2}\varepsilon t}}{8\pi^{3/2}}\int_0^{\infty}\frac{d\xi}{\xi\sqrt{\xi}}e^{-\frac{t^2}{16\xi}-\xi(\varepsilon^2+4\kappa)}\Phi(x;a,b), \tag{41}$$

where we denote the special function

$$\Phi(x;a,b)\equiv\int_{-\infty}^{\infty}\frac{e^{-\zeta^2-\frac{x^2}{4(a+ib\zeta)}}d\zeta}{\sqrt{|a+ib\zeta|}},\quad a=t\delta/2-4\xi\alpha+2\xi\varepsilon\delta\,,\ b=2\delta\sqrt{\xi}. \tag{42}$$

Function $\Phi(x;a,b)$ reduces for some particular values of a and b to the hypergeometric function ${}_0F_2(\{\beta_1,\beta_2\};z)$, this last being the particular case of ${}_pF_q(\{\alpha_1...\alpha_p\};\{\beta_1...\beta_q\};z)$. Moreover, $\Phi(0;0,b)=\frac{1}{\sqrt{|b|}}\Gamma\left(\frac{1}{4}\right)$ and $\int_{-\infty}^{\infty}\frac{e^{-\zeta^2-\frac{x^2}{4b\zeta}}d\zeta}{\sqrt{|\zeta|}}=\left(\begin{array}{c}\Gamma\left(\frac{1}{4}\right){}_0F_2\left(;\left\{\frac{1}{2},\frac{3}{4}\right\};-\frac{x^4}{64b^2}\right)-\\ \sqrt{\frac{\pi}{b}}|x|\,{}_0F_2\left(;\left\{\frac{3}{4},\frac{5}{4}\right\};\ \frac{x^4}{64b^2}\right)\end{array}\right)\text{sgn}b$. The numerical calculation of the double integral (41), taken with care around the point $a=0$, corresponding to the time $t=4\xi\left(2\frac{\alpha}{\delta}-\varepsilon\right)$, yields real values. The example of the solution (41) of the GK type Equation (36) with $\alpha=\delta=\varepsilon=\kappa=1$ for the initial flash $f(x)=\delta(x)$ is shown in Figure 8.

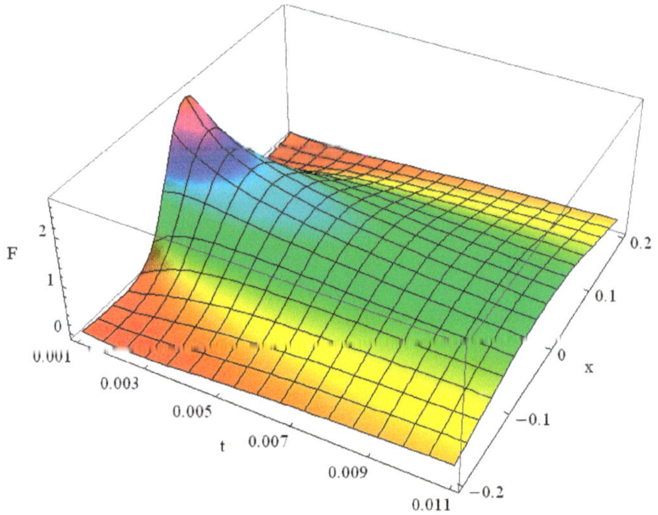

Figure 8. Solution of the Guyer-Krumhansl equation with initial $\delta(x)$ function for $\alpha = \varepsilon = \delta = \kappa = 1$ in the interval of time $t \in [10^{-3}, 10^{-2}]$.

From the comparison of Figure 8 with Figure 1 we conclude that the solution of the Fourier heat Equation (1) (see Figure 1) is more pronounced at short times, it has a higher peak, while the solution of GK Equation (36) spreads faster (see Figure 8) when the additional terms $\varepsilon = \delta = 1$ are comparable with the others order.

In the context of heat conduction in thin films we solved the respective GK equation for $Kn = 0.2$ and for $Kn = 1$. The results for the ballistic dimensionless energy change after a laser pulse heating of a thin film, modelled by the GK type Equation (40), are shown in Figure 9 for Knudsen number $Kn = 1$ and in Figure 10 for Knudsen number $Kn = 0.2$. Their comparison confirms much faster heat propagation in a thin film with $Kn = 1$, where the ballistic transport mechanism contributes noticeably, as compared with the film with small value of $Kn = 0.2$.

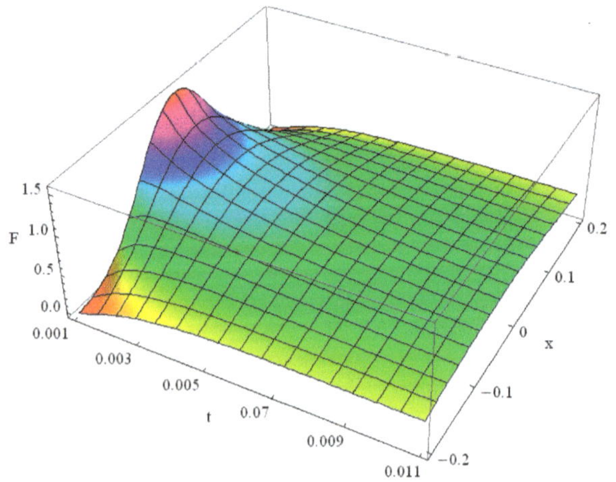

Figure 9. Heat pulse propagation in the Guyer-Krumhansl model with Knudsen number $Kn = 1$.

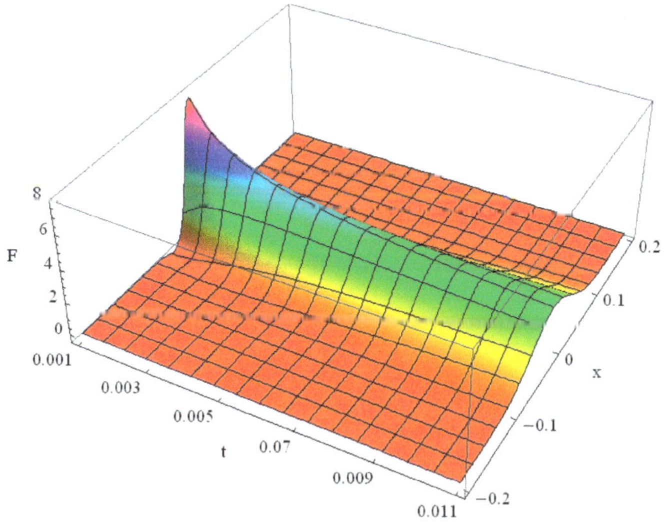

Figure 10. Heat pulse propagation in the Guyer-Krumhansl (GK) model with Knudsen number $Kn = 0.2$.

9. Solution of the Guyer-Krumhansl Equation for the Exponential–Polynomial Initial Function

Interesting features of the heat conduction, governed by the Guyer-Krumhansl equation, arise from consideration of the evolution of the power–exponential initial function $F(x,0) = e^{\gamma x}x^n$. Consideration of such a function is useful not only because it allows better approximation of experimental data than the usual polynomial x^n and $f(x) = e^{\gamma x}x^n$, $\gamma x < 0$, but also as itself it represents a pulse, and practically any solitary space wave or surge is easy to approximate by its sum $f(x) = e^{\gamma x}x^n$. The operational method allows the exact analytical solutions for these functions to be obtained. Indeed, we make use of the operational identity (11) to obtain for $\hat{S} = e^{\nu \partial_x^2}$, $\nu = t\delta/2 - 4\xi\alpha + 2\xi\varepsilon\delta + i2\sqrt{\xi}\delta\zeta$, $\hat{S}\, e^{\gamma x}x^n = e^{\gamma x + \gamma^2 \nu} H_n(x + 2\gamma\nu, \nu)$. The solution for arbitrary values of $n \in$ Integers, $\gamma \in$ Reals can be obtained upon the following integration:

$$F(x,t) = \frac{e^{\frac{t}{2}(\gamma^2\delta - \varepsilon) + \gamma x}t}{4\pi} \int_0^\infty \frac{d\xi}{\xi\sqrt{\xi}} e^{-\frac{t^2}{16\xi} - \xi d} \int_{-\infty}^{\infty} e^{-\zeta^2 + ib\zeta\gamma^2} H_n(x + 2\gamma\nu, \nu)\, d\zeta, \tag{43}$$

where $\nu = a + ib\zeta$, $a = t\delta/2 - 4\xi\alpha + 2\xi\varepsilon\delta$, $b = 2\sqrt{\xi}\delta$, $d = \varepsilon^2 + 4\kappa + 2\gamma^2(2\alpha - \varepsilon\delta)$. The integration can be done in elementary functions if we account for the series presentation $H_n(x + 2\gamma\nu, \nu) = n! \sum_{r=0}^{[n/2]} \frac{(x+2\gamma\nu)^{n-2r}\nu^r}{(n-2r)!r!} = (-i)^n (a + ib\zeta)^{\frac{n}{2}} H_n\left(\frac{i(x+2\gamma(a+ib\zeta))}{2\sqrt{a+ib\zeta}}\right)$. We omit the final result for conciseness. For example, in the simplest case of $n = -\gamma = 1$, i.e., $f(x) = e^{-x}x$ we obtain the following solution of the GK equation:

$$F(x,t) = e^{-x+t(\delta-\varepsilon)/2} e^{-\frac{t}{2}\sqrt{q}} (x - \delta t + (2\alpha + \delta(\delta - \varepsilon))t/\sqrt{q}),\ q = (\delta - \varepsilon)^2 + 4(\kappa + \alpha). \tag{44}$$

Let us now consider the initial smooth function $F(x,0) = (|x| + 2)^2 e^{-|x|}$. By means of the above developed operational technique we obtained the following exact analytical solution of the GK equation in elementary functions:

$$F(x,t) = e^{-|x|-\frac{t}{2}(\sqrt{r}-\delta+\varepsilon)} \times$$
$$\left((2+|x|)^2 - t\delta(3+2|x|) + 2(t\delta)^2 + \frac{4t\eta}{r}\left(\frac{2}{\sqrt{r}}+t\right) + \frac{t\rho}{\sqrt{r}}(3+2|x|-2t\delta)\right), \tag{45}$$
where
$$\rho = 2\alpha + \delta(\delta-\varepsilon), \quad \eta = \alpha^2 - \alpha\delta\varepsilon - \delta^2\kappa, \quad r = 4(\alpha+\kappa) + (\varepsilon-\delta)^2.$$

Direct substitution of the solution satisfies the GK equation. Earlier it was demonstrated that such solutions can show bizarre behavior with local maximums and local minimums even with negative values, occurring in the middle of the domain, so that theoretically for a certain set of parameters of the GK equation the maximum principle can be violated [58]. However, for applications, such as those in the thin films, modelled by the GK type Equation (40), there is no problem with the second law of thermodynamics violation. Indeed, the propagation of the smooth spatial heat distribution $F(x,0) = (|x|+2)^2 e^{-|x|}$ in the range of the values of the Knudsen number between $Kn = 0.05 \div 2$ changes no more than three percent.

The difference between these plots is not distinguishable visually and we present here just one example for $Kn = 0.2$ in Figure 11. It corresponds to the values $\alpha = \frac{2}{15}$, $\delta = 0.12$, $\varepsilon = 2$, $\kappa = -1$. The behavior of the solution surprisingly resembles that for $\alpha \approx \varepsilon \approx \delta \approx 1$, which we omit for brevity, as well as that for $Kn = 2$, for which $\alpha = 13\frac{1}{3}$, $\delta = 12$, $\varepsilon = 2$, $\kappa = -1$.

To contrast it, we demonstrate the behavior of the solution of GK equation (see (36), $\kappa = 0$) for relatively small contribution of the Cattaneo's term: $\alpha = \varepsilon = \delta = 10$, which does not depend on the sign and on the exact value of the constant term κ (we omit proper figures with $\kappa \neq 0$) for conciseness) and it has bizarre non-Fourier behavior, shown in Figure 12.

The initial pulse rapidly decreases and assumes negative values, then it gradually approaches zero. It maintains negative values in the whole space domain, but for the vicinity of $x = 0$. Around $x = 0$ the solution becomes positive: $F(x \approx 0, t > 0.8) > 0$ and then it relaxes to zero: $F(x, t \to \infty) = 0$.

In this case, based on the obtained exact vanishing analytical solution of the GK equation with the initial function $F(x,0) = (|x|+2)^2 e^{-|x|}$ and $F(x,\infty) = 0$, we conclude the minimum of the solution over the time-space domain is in the middle of the domain (see Figure 12). The local maximum of the solution may also occur in the middle of the domain (see [58,59]).

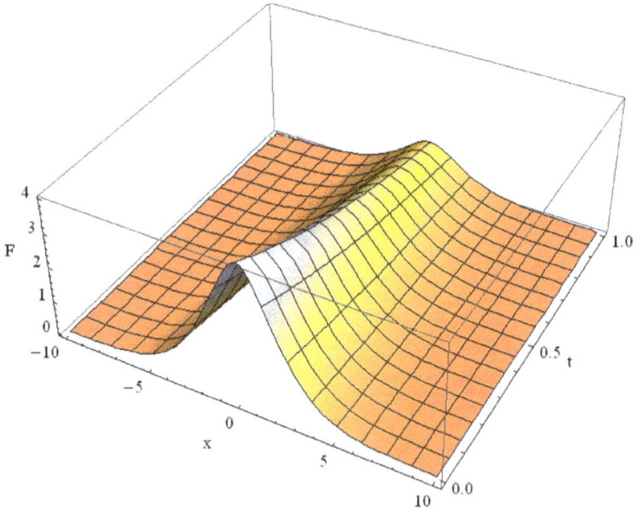

Figure 11. Solution of GK type equation for heat transport in thin films with Knudsen number $Kn = 0.2$ for the initial function $F(x,0) = (|x|+2)^2 e^{-|x|}$.

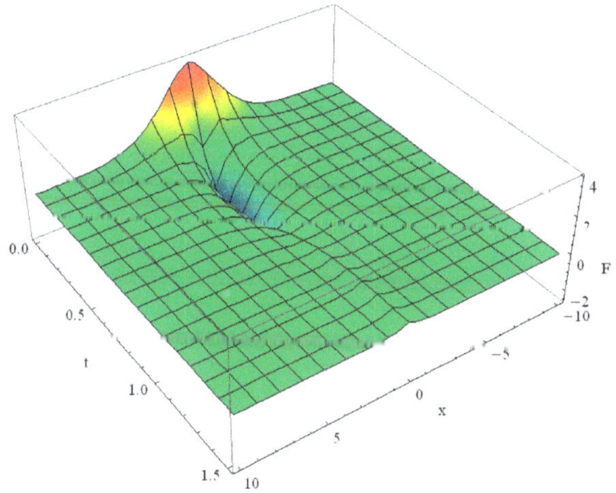

Figure 12. Solution of GK equation with distinct ballistic transport: $\alpha = \varepsilon = \delta = 10$, $\kappa = 0$ for the initial function $F(x,0) = (|x|+2)^2 e^{-|x|}$.

10. Harmonic Solution of Guyer-Krumhansl Equation and Temperature Distribution in Thin Films

The other good example of a given initial temperature distribution, is given by a harmonic function $f(x) = \exp(inx)$, which is necessary to consider approximations or expansion of the initial function into the Fourier series to fit experimental data distributions. The action of the heat conduction operator yields the following bounded at infinite times solution:

$$F(x,t) = e^{inx}\frac{e^{-\frac{t}{2}(\varepsilon+n^2\delta)}t}{4\pi}\int_0^\infty \frac{d\xi}{\xi\sqrt{\xi}}e^{-\frac{t^2}{16\xi}-\xi(\varepsilon^2+4\kappa+2n^2(-2\alpha+\varepsilon\delta))}\int_{-\infty}^{\infty} e^{-\zeta^2}e^{-2i\zeta n^2\delta\sqrt{\xi}}d\zeta. \tag{46}$$

Upon integration we end up with the explicit solution of the GK Equations (36) and (37) for the periodic initial function $f(x) = \exp(inx)$, expressed in elementary functions as follows:

$$F(x,t) = \exp\left(inx - \frac{t}{2}(\varepsilon + n^2\delta + \sqrt{U})\right),\ U = (\varepsilon + n^2\delta)^2 + 4(\kappa - \alpha n^2). \tag{47}$$

Interestingly, the solution (47) of GK type Equation (36) for the initial harmonic function $f(x) = \exp(inx)$ repeats the solution of telegraph Equation (33) upon the substitution $\varepsilon \to \varepsilon + n^2\delta$, where $\delta > 0$, $\varepsilon > 0$, in (33). In other words, the solution (47) of Equation GK type (36) for $f(x) = e^{inx}$ is also the solution of the HHE (22) with the harmonic dependent coefficient $\varepsilon + n^2\delta$ for the first order time derivative:

$$\left(\frac{\partial^2}{\partial t^2} + \left(\varepsilon + n^2\delta\right)\frac{\partial}{\partial t}\right)F(x,t) = \left(\alpha\frac{\partial^2}{\partial x^2} + \kappa\right)F(x,t) \tag{48}$$

Thus, for the initial harmonic function $f(x) = e^{inx}$ and same values of the coefficients $\alpha, \varepsilon, \kappa, \delta$ in Equations (36) and (48), these share identical particular solution (47). For the initial function $\phi(x) = \sum_n c_n \exp(inx)$, the solution of Equations (36) and (48) will be given by the series $\Phi = \sum_n c_n F(x,t)$, where $F(x,t)$ is the solution of (47). In this sense the GK type equation with the harmonic initial function can be viewed as the telegraph equation with the harmonic dependent coefficient for $\partial/\partial t$: $(\varepsilon + n^2\delta)\partial/\partial t$ and the GK equation for $f(x) = e^{inx}$ is in fact the Cattaneo equation for $f(x) = e^{inx}$ with $(\varepsilon + n^2\delta)\partial_t$ instead of $\varepsilon\partial_t$ term. This radically changes the solution behavior. Indeed, the relaxation time for all higher harmonics e^{inx} in the telegraph Equation (33) is the same;

only low harmonics may fade out faster. On the contrary, the solution (47) of the GK type Equation (36) for the harmonic function $f(x) = e^{inx}$ and the solution of the equivalent telegraph equation with $\varepsilon \to \varepsilon + n^2\delta$ (48) for high harmonics show that these harmonics fade out faster than the fundamental one e^{ix} ($n = 1$) and the higher the harmonic number n, the lower is its relaxation time. The examples for $n = 3, 5$ are shown in Figure 13.

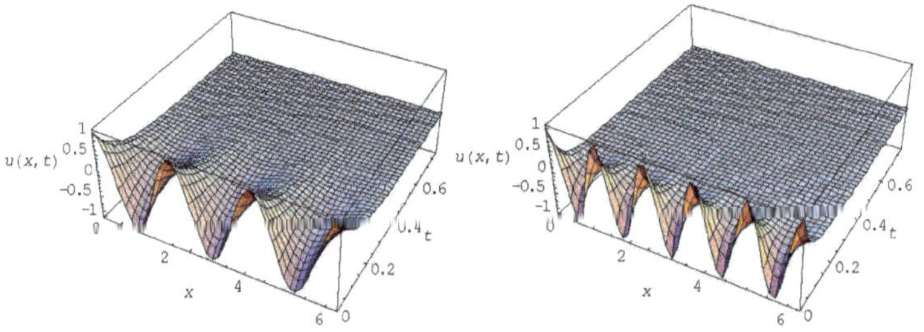

Figure 13. Solution of GK type Equation: $\mathrm{Re}[F(x,t)]$ for $\alpha = 0.5$, $\varepsilon = 1$, $\kappa = 5, \delta = 1$ for $f(x) = e^{3ix}$ ($n = 3$) left, and for $f(x) = e^{5ix}$ ($n = 5$) right.

Compare Figure 13 with the solution of the telegraph Equation (22) (see Figures 6 and 7). High harmonics of the telegraph equation, shown in Figure 7, fade out slower than the low harmonics, shown in Figure 6. For the GK type equation or the HHE with $\varepsilon \to \varepsilon + n^2\delta$ on the contrary, high harmonics fade out faster than the fundamental and the low ones, as evidenced in Figure 13. The effective thermal conductivity in the common telegraph Equation (22) is constant for higher harmonics and in certain cases even for all n. On the contrary, in the GK type Equation (36), which is HHE with $\varepsilon \to \varepsilon + n^2\delta$ for $f(x) = \exp(inx)$ (see (48)), the thermal conductivity for the harmonic function $f(x) = e^{inx}$ rises with increase of the harmonic number n.

In the context of the heat conduction in thin films, for $Kn = 1$ we plotted the bounded solutions of Equation (40), for $f(x) = e^{inx}$, $n = 1, 3$, in Figures 14 and 15 respectively. High harmonics fade out rapidly as follows from their comparison with teach other. For $Kn = 0.2$ we present the solutions in Figures 16 and 17 for $n = 1, 3$ respectively. The comparison with Figure 14, Figure 15 shows that the solutions for $Kn = 0.2$ relax much slower than those for $Kn = 1$.

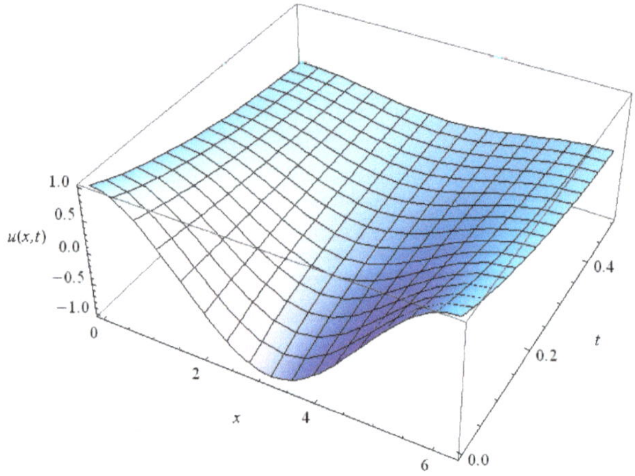

Figure 14. The behavior of the 1st harmonic ($n = 1$) of the GK type Equation (36) solution for $Kn = 1$.

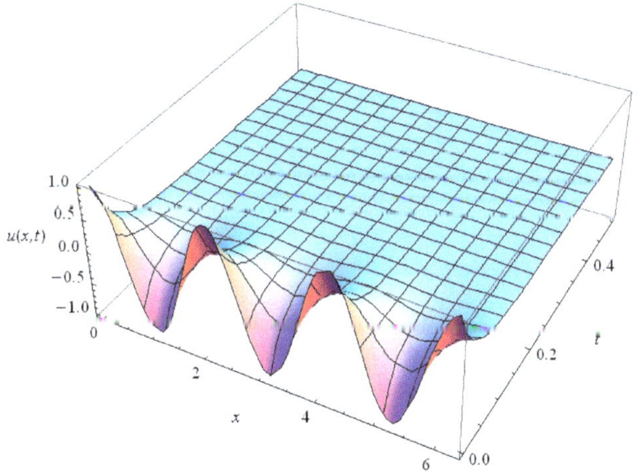

Figure 15. The behavior of the 3rd harmonic ($n = 3$) of the GK type Equation (36) solution for $Kn = 1$.

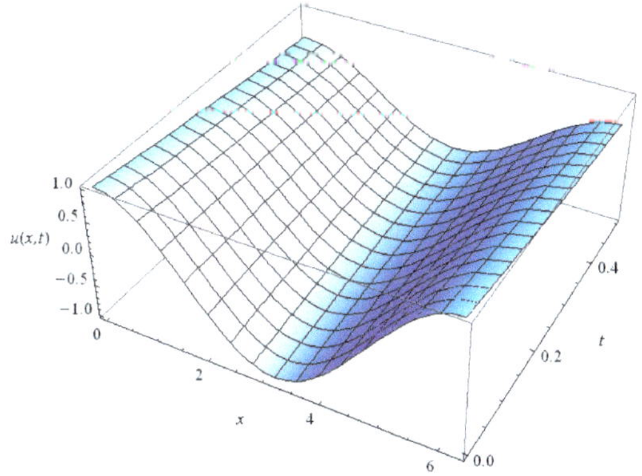

Figure 16. The behavior of the 1st harmonic ($n = 1$) of the GK type Equation (36) solution for $Kn = 0.2$.

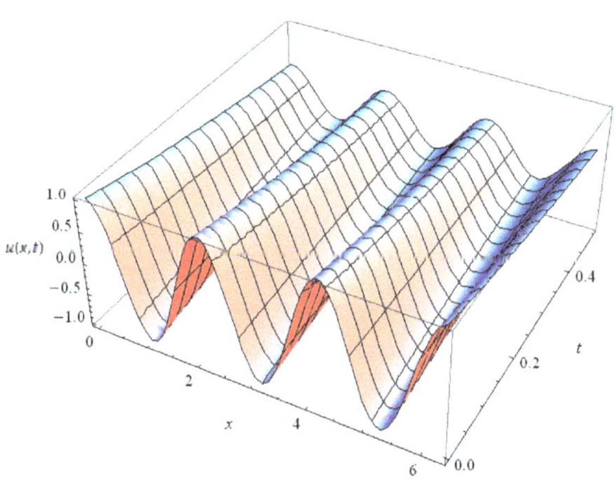

Figure 17. The behavior of the 3rd harmonic ($n = 3$) of the GK type Equation (36) solution for $Kn = 0.2$.

This means that high values of the Knudsen number provide the ballistic transport contribution to the heat conduction, and the latter increases in the GK model, in particular, for high harmonics. The proper relaxation times for the first harmonic $f(x) = e^{ix}$ read as follows: ${}_{n=1}\tau_{Kn=1} \approx 0.5$ and ${}_{n=1}\tau_{Kn=0.2} \approx 2.5$; for the third harmonic $f(x) = e^{3ix}$ the relaxation times are ${}_{n=3}\tau_{Kn=1} \approx 0.2$ and ${}_{n=3}\tau_{Kn=0.2} \approx 3$. The ratios of proper relaxation times are ${}_{n=1}\tau_{Kn=1}/{}_{n=1}\tau_{Kn=0.2} = 1/5$ and ${}_{n=3}\tau_{Kn=1}/{}_{n=3}\tau_{Kn=0.2} = 1/15$. The heat conduction is evidently better for $Kn = 1$ than for $Kn = 0.2$.

11. Conclusions

We obtained exact analytical solutions for heat conduction in the Fourier, Cattaneo, and Guyer-Krumhansl heat transport models. Extended forms of these equations with additional constant terms were solved by the operational method. The obtained solutions directly satisfy the considered equations. The initial functions, modelling various types of pulses and signals were considered. The power–exponential pulse and the flash heat pulse were considered, modelling the laser heat pulse experimental technique, common for thermal conductivity measurements.

The analytical solutions were compared with each other. The finite speed of the heat propagation in the Cattaneo model was demonstrated. By reducing the effect of the second time derivative in the equation for $\alpha = \varepsilon = 10$, maintaining the heat conductivity unchanged, we obtain faster fading of the solution. In this case diffusive heat conduction prevails over the wave-like propagation process. On the contrary, for $\alpha = \varepsilon = 0.1$ the wave-like propagation of the initial function dominates. For $\alpha = \varepsilon = 1$, both heat transport mechanisms contribute; the wave propagates at a constant speed, accompanied by fading.

The exact analytical solution of HHE in three space dimensions for initial flash $f = \delta(x, y, z)$ was obtained in terms of modified Bessel functions. The HHE was solved with the harmonic initial function $f = e^{inx}$. The obtained solution fades in time exponentially and does not spread. The space phase shift of the solution depends on the sign of the linear term in HHE.

The GK type heat equation with linear term was solved operationally; the solution was written in terms of integrals and special functions. The propagation of the initial heat flash was studied and demonstrated for the model of heat conduction in thin films. Small values of Knudsen number result in slower relaxation of the initial heat surge; for $Kn \approx 1$ the ballistic heat transport contributes and the heat surge fades out much faster. An exact solution for the evolution of the exponential-monomial function $F(x, 0) = x^n e^{-\gamma x}$ in the GK equation was obtained. This allows not only the propagation of a surge with power rise, followed by the common exponential fade, to be studied but also the result for the initial function to be obtained, which can be expanded in the series $f(x) = \sum_{n,\gamma} c_n e^{\gamma} x^n$ or approximated by them. The example of the initial smooth heat pulse relaxation in a thin film was considered.

We found that variation of the Knudsen number values in a reasonable range $Kn \in [0.05,\ 2]$ practically does not influence the relaxation time of the initial smooth pulse: the values of the solution of GK type equation vary less than 3% for $0.05 < Kn < 2$. Thus in thin films the conditions on the surface, expressed by the Knudsen number, are particularly important for laser heat flash, but not that important for relaxation of a wide heat spot without a sharp boundary.

For the propagation of the initial laser heat flash $f(x) = \delta(x)$ the relaxation time is ten times smaller for $Kn = 1$ than for $Kn = 0.2$. The relaxation time τ for the harmonic initial function for the first harmonic ($n = 1$) is five times shorter for $Kn = 1$ than for $Kn = 0.2$: ${}_{n=1}\tau_{Kn=1} \approx 0.5$ and ${}_{n=1}\tau_{Kn=0.2} \approx 2.5$; for the third harmonic $f(x) = e^{3ix}$ the relaxation time τ is 15 times shorter for $Kn = 1$ than τ for $Kn = 0.2$: the proper relaxation times are ${}_{n=3}\tau_{Kn=1} \approx 0.2$ and ${}_{n=3}\tau_{Kn=0.2} \approx 3$. The ratios of proper relaxation times are ${}_{n=1}\tau_{Kn=1}/{}_{n=1}\tau_{Kn=0.2} = 1/5$ and ${}_{n=3}\tau_{Kn=1}/{}_{n=3}\tau_{Kn=0.2} = 1/15$.

We obtained exact analytical solutions in terms of the orthogonal polynomials and special functions for three different models of heat conduction: Fourier, Cattaneo, and Guyer-Krumhansl. These solutions allow analytical modelling of the space–time propagation for heat flashes and pulses, which is the established experimental technique. We can also model the relaxation of any initial function, expandable in the Fourier series and in the series of the exponential-polynomial function. This series approximation is applicable practically to any pulse.

In conclusion we stress that the operational technique, used for the solution of hyperbolic heat equations, yields exact analytical solutions. They have clear physical meaning and allow easy understanding of the role of different terms in the equations. The validity of the obtained analytical solutions was checked by their substitution in the equations, the latter were satisfied. Our study not only has theoretical interest, but it also provides practical results of experimental technique and measurements. The obtained exact analytical solutions can be used as a benchmark for numerical solutions of more sophisticated forms of the studied equations.

Conflicts of Interest: The author declares no conflict of interest.

References

1. Von Rosenberg, D.U. *Methods for the Numerical Solution of Partial Differential Equations*; Society of Petroleum Engineers: Richardson, TX, USA, 1969; Volume 6.
2. Smith, G.D. *Numerical Solution of Partial Differential Equations: Finite Difference Methods*; Oxford University Press: Oxford, UK, 1985.
3. Ghia, U.; Ghia, K.N.; Shin, C.T. High-Re solutions for incompressible flow using the Navier-Stokes equations and a multigrid method. *J. Comput. Phys.* **1982**, *48*, 387–411. [CrossRef]
4. Ames, W.F. *Numerical Methods for Partial Differential Equations*; Academic Press: Cambridge, MA, USA, 2014.
5. Johnson, C. *Numerical Solution of Partial Differential Equations by the Finite Element Method*; Courier Corporation: Chelmsford, MA, USA, 2012.
6. Carnahan, B.; Luther, H.A.; Wilkes, J.O. *Applied Numerical Methods*; John Wiley & Sons: Hoboken, NJ, USA, 1969.
7. Spassky, D.; Omelkov, S.; Mägi, H.; Mikhailin, H.; Vasil'ev, A.; Krutyak, N.; Tupitsyna, I.; Dubovik, A.; Yakubovskaya, A.; Belsky, A. Energy transfer in solid solutions $Zn_xMg_{1-x}WO_4$. *Opt. Mater.* **2014**, *36*, 1660–1664. [CrossRef]
8. Kirkin, R.; Mikhailin, V.V.; Vasil'ev, A.N. Recombination of correlated electron-hole Pairs with account of hot capture with emission of optical phonons. *IEEE Trans. Nuclear Sci.* **2015**, *59*, 2057–2064. [CrossRef]
9. Gridin, S.; Belsky, A.; Dujardin, C.; Gektin, A.; Shiran, N.; Vasil'ev, A. Kinetic model of energy relaxation in Csl:A (A = Tl and In) scintillators. *J. Phys. Chem. C* **2015**, *119*, 20578–20590. [CrossRef]
10. Krutyak, N.R.; Mikhailin, V.V.; Vasil'ev, A.N.; Spassky, D.A.; Tupitsyna, I.A.; Dubovik, A.M.; Galashov, E.N.; Shlegel, V.N.; Belsky, A.N. The features of energy transfer to the emission centers in $ZnWO_4$ and $ZnWO_4$:Mo. *J. Lumin.* **2013**, *144*, 105–111. [CrossRef]
11. Savon, A.E.; Spassky, D.A.; Vasil'ev, A.N.; Mikhailin, V.V. Numerical simulation of energy relaxation processes in a $ZnMoO_4$ single crystal. *Opt. Spectrosc.* **2012**, *112*, 72–78. [CrossRef]
12. Denisov, V.I.; Shvilkin, B.N.; Sokolov, V.A.; Vasili'ev, M.I. Pulsar radiation in post-Maxwellian vacuum nonlinear electrodynamics. *Phys. Rev. D* **2016**, *94*, 45021. [CrossRef]
13. Denisov, V.I.; Sokolov, V.A. Analysis of regularizing properties of nonlinear electrodynamics in the Einstein-Born-Infeld theory. *J. Exp. Theor. Phys.* **2011**, *113*, 926–933. [CrossRef]
14. Denisov, V.I.; Sokolov, V.A.; Vasiliev, M.I. Nonlinear vacuum electrodynamics birefringence effect in a pulsar's strong magnetic field. *Phys. Rev. D* **2014**, *90*, 023011. [CrossRef]
15. Denisov, V.I.; Kozar, A.V.; Sharikhin, V.F. Investigation of the trajectories of a magnetized particle in the equatorial plane of a magnetic dipole. *Mosc. Univ. Phys. Bull.* **2010**, *65*, 164–169. [CrossRef]
16. Vladimirova, Y.V.; Zadkov, V.N. Frequency modulation spectroscopy of coherent dark resonances of multi-level atoms in a magnetic field. *Mosc. Univ. Phys. Bull.* **2010**, *65*, 493. [CrossRef]
17. Yanyshev, D.N.; Balykin, V.I.; Vladimirova, Y.V.; Zadkov, V.N. Dynamics of atoms in a femtosecond optical dipole trap. *Phys. Rev. A* **2013**, *87*, 033411. [CrossRef]
18. Bulin, A.-L.; Vasil'ev, A.; Belsky, A.; Amans, D.; Ledoux, G.; Dujardin, C. Modelling energy deposition in nanoscintillators to predict the efficiency of the X-ray-induced photodynamic effect. *Nanoscale* **2015**, *7*, 5744–5751. [CrossRef] [PubMed]

19. Pastukhov, V.M.; Vladimirova, Y.V.; Zadkov, V.N. Photon-number statistics from resonance fluorescence of a two-level atom near a plasmonic nanoparticle. *Phys. Rev. A* **2014**, *90*, 063831. [CrossRef]
20. Gong, L.; Zhao, J.; Huang, S. Numerical study on layout of micro-channel heat sink for thermal management of electronic devices. *Appl. Therm. Eng.* **2015**, *88*, 480–490. [CrossRef]
21. Xia, G.; Ma, D.; Zhai, Y.; Li, Y.; Liu, R.; Du, M. Experimental and numerical study of fluid flow and heat transfer characteristics in microchannel heat sink with complex structure. *Energy Convers. Manag.* **2015**, *105*, 848–857. [CrossRef]
22. Wang, R.; Du, J.; Zhu, Z. Effects of wall slip and nanoparticles' thermophoresis on the convective heat transfer enhancement of nanofluid in a microchannel. *J. Thermal Sci. Technol.* **2016**, *11*, JTST00017. [CrossRef]
23. Roy, R.; Ghosal, S. Homotopy perturbation method for the analysis of heat transfer in an annular fin with temperature-dependent thermal conductivity. *J. Heat Transfer* **2016**, *139*, 022001. [CrossRef]
24. Demiray, S.T.; Bulut, H.; Belgacem, F.B. Sumudu transform method for analytical solutions of fractional type ordinary differential equations. *Math. Probl. Eng.* **2015**, *2015*, 131690.
25. Mathai, A.M. M-convolutions of products and ratios, statistical distributions and fractional calculus. *Analysis* **2016**, *36*, 15–26. [CrossRef]
26. Saxena, R.K.; Mathai, A.M.; Haubold, H.J. Computational solutions of unified fractional reaction-diffusion equations with composite fractional time derivative. *Commun. Nonlinear Sci. Numer. Simul.* **2015**, *27*, 1–11. [CrossRef]
27. Mathai, A.M. Fractional integral operators in the complex matrix variate case. *Linear Algebra Appl.* **2013**, *439*, 2901–2913. [CrossRef]
28. Haubold, H.J.; Mathai, A.M.; Saxena, R.K. Further solutions of fractional reaction-diffusion equations in terms of the H-function. *J. Comput. Appl. Math.* **2011**, *235*, 1311–1316. [CrossRef]
29. Mathai, A.M. Fractional differential operators in the complex matrix-variate case. *Linear Algebra Appl.* **2015**, *478*, 200–217. [CrossRef]
30. Akinlar, M.A.; Kurulay, M. A novel method for analytical solutions of fractional partial differential equations. *Math. Probl. Eng.* **2013**, *2013*, 195708. [CrossRef]
31. Mathai, A.M.; Saxena, R.K.; Haubold, H.J. A certain class of Laplace transforms with applications to reaction and reaction-diffusion equations. *Astrophys. Space Sci.* **2006**, *305*, 283–288. [CrossRef]
32. Zhukovsky, K.V. Operational solution for non-integer ordinary and evolution type partial differential equations. *Axioms* **2016**, in press.
33. Saxena, R.K.; Mathai, A.M.; Haubold, H.J. Solutions of certain fractional kinetic equations and a fractional diffusion equation. *J. Math. Phys.* **2010**, *51*, 103506. [CrossRef]
34. Filobello-Nino, U.; Vazquez-Leal, H.; Benhammouda, B.; Perez-Sesma, A.; Jimenez-Fernandez, V.M.; Cervantes-Perez, J.; Sarmiento-Reyes, A.; Huerta-Chua, J.; Morales-Mendoza, L.J.; Gonzalez-Lee, M.; et al. Analytical solutions for systems of singular partial differential-algebraic equations. *Discret. Dyn. Nat. Soc.* **2015**, *2015*, 752523. [CrossRef]
35. Benhammouda, B.; Vazquez-Leal, H. Analytical solutions for systems of partial differential-algebraic equations. *SpringerPlus* **2014**, *3*, 137. [CrossRef] [PubMed]
36. Hesam, S.; Nazemi, A.R.; Haghbin, A. Analytical solution for the Fokker-Planck equation by differential transform method. *Sci. Iran.* **2012**, *19*, 1140–1145. [CrossRef]
37. Vitanov, N.K.; Dimitrova, Z.I.; Vitanov, K.N. Modified method of simplest equation for obtaining exact analytical solutions of nonlinear partial differential equations: Further development of the methodology with applications. *Appl. Math. Comp.* **2015**, *269*, 363–378. [CrossRef]
38. Bateman, H.; Erdélyi, A.; Magnus, W.; Oberhettinger, F.; Tricomi, F.G. *Higher Transcendental Functions*; McGraw-Hill Book Company: New York, NY, USA; 1953; Volume II.
39. Mathai, M.; Saxena, R.K.; Haubold, H.J. *The H-Function: Theory and Applications*; Springer Science & Business Media: Berlin, Germany, 2009.
40. Mathai, A.M.; Haubold, H.J. *Special Functions for Applied Scientists*; Springer Science & Business Media: Berlin, Germany, 2008.
41. Wolf, K.B. *Integral Transforms in Science and Engineering*; Plenum Press: New York, NY, USA, 1979.
42. Dattoli, G.; Srivastava, H.M.; Zhukovsky, K. A new family of integral transforms and their applications, Integral Transform. *Integral Transform Spec. Funct.* **2006**, *17*, 31–37. [CrossRef]

43. Appèl, P.; de Fériet, J.K. *Fonctions Hypergéométriques et Hypersphériques; Polynômes d'Hermite*; Gauthier-Villars: Paris, France, 1926.
44. Dattoli, G.; Srivastava, H.M.; Zhukovsky, K. Orthogonality properties of the Hermite and related polynomials. *J. Comput. Appl. Math.* **2005**, *182*, 165–172. [CrossRef]
45. Dattoli, G. Generalized polynomials, operational identities and their applications. *J. Comput. Appl. Math.* **2000**, *118*, 111–123. [CrossRef]
46. Dattoli, G.; Srivastava, H.M.; Zhukovsky, K.V. Operational methods and differential equations with applications to Initial-Value problems. *Appl. Math. Comput.* **2007**, *181*, 979–1001. [CrossRef]
47. Dattoli, G.; Mikhailin, V.V.; Zhukovsky, K.V. Influence of a constant magnetic field on the radiation of a planar undulator. *Mosc. Univ. Phys. Bull.* **2009**, *64*, 507–512. [CrossRef]
48. Dattoli, G.; Mikhailin, V.V.; Zhukovsky, K. Undulator radiation in a periodic magnetic field with a constant component. *J. Appl. Phys.* **2008**, *104*, 124507. [CrossRef]
49. Zhukovsky, K. Analytical account for a planar undulator performance in a constant magnetic field. *J. Electromagn. Waves Appl.* **2014**, *28*, 1869–1887. [CrossRef]
50. Zhukovsky, K.V. Harmonic radiation in a double-frequency undulator with account for broadening. *Mosc. Univ. Phys. Bull.* **2015**, *70*, 232–239. [CrossRef]
51. Zhukovsky, K. High harmonic generation in undulators for FEL. *Nucl. Instr. Meth. B* **2015**, *369*, 9–14. [CrossRef]
52. Zhukovsky, K.V. Harmonic generation by ultrarelativistic electrons in a planar undulator and the emission-line broadening. *J. Electromagn. Waves Appl.* **2015**, *29*, 132–142. [CrossRef]
53. Zhukovsky, K. High harmonic generation in the undulators for free electron lasers. *Opt. Commun.* **2015**, *353*, 35–41. [CrossRef]
54. Zhukovsky, K. Emission and tuning of harmonics in a planar two-frequency undulator with account for broadening. *Laser Part. Beams* **2016**, *34*, 447–456. [CrossRef]
55. Haimo, D.T.; Markett, C. A representation theory for solutions of a higher-order heat equation, I. *J. Math. Anal. Appl.* **1992**, *168*, 89–107. [CrossRef]
56. Haimo, D.T.; Markett, C. A representation theory for solutions of a higher-order heat equation, II. *J. Math. Anal. Appl.* **1992**, *168*, 289–305. [CrossRef]
57. Zhukovsky, K.V. Exact solution of Guyer-Krumhansl type heat equation by operational method. *Int. J. Heat Mass Transfer* **2016**, *96*, 132–144. [CrossRef]
58. Zhukovsky, K.V. Violation of the maximum principle and negative solutions with pulse propagation in Guyer-Krumhansl model. *Int. J. Heat Mass Transfer* **2016**, *98*, 523–529. [CrossRef]
59. Zhukovsky, K.V.; Srivastava, H.M. Analytical solutions for heat diffusion beyond Fourier law. *Appl. Math. Comp.* **2007**, *293*, 423–437. [CrossRef]
60. Dattoli, G.; Zhukovsky, K.V. Quark flavour mixing and the exponential form of the Kobayashi-Maskawa matrix. *Eur. Phys. J. C* **2007**, *50*, 817–821. [CrossRef]
61. Dattoli, G.; Zhukovsky, K. Quark mixing in the standard model and the space rotations. *Eur. Phys. J. C* **2007**, *52*, 591–595. [CrossRef]
62. Dattoli, G.; Zhukovsky, K. Quark mixing and the exponential form of the Kobayashi-Maskawa matrix. *Phys. Atom. Nucl.* **2008**, *71*, 1807–1812. [CrossRef]
63. Dattoli, G.; Zhukovsky, K.V. Neutrino Mixing and the exponential form of the Pontecorvo-Maki-Nakagawa-Sakata matrix. *Eur. Phys. J. C* **2008**, *55*, 547–552. [CrossRef]
64. Zhukovsky, K.; Borisov, A. Exponential parameterization of neutrino mixing matrix—Comparative analysis with different data sets and CP violation. *Eur. Phys. J. C* **2016**, in press. [CrossRef]
65. Zhukovsky, K.; Melazzini, F. Exponential parameterization of neutrino mixing matrix with account of CP-violation data. *Eur. Phys. J. C* **2016**, *76*, 462. [CrossRef]
66. Haubold, H.J.; Mathai, A.M.; Saxena, R.K. Analysis of solar neutrino data from Super-Kamiokande I and II. *Entropy* **2014**, *16*, 1414–1425. [CrossRef]
67. Mathai, A.M.; Haubold, H.J. On a generalized entropy measure leading to the pathway model with a preliminary application to solar neutrino data. *Entropy* **2013**, *15*, 4011–4025. [CrossRef]
68. Mathai, A.M.; Saxena, R.K.; Haubold, H.J. Back to the solar neutrino problem. *Space Res. Today* **2012**, *185*, 112–123.

69. Zhukovsky, K. Solution of some types of differential equations: Operational calculus and inverse differential operators. *Sci. World J.* **2014**, *2014*, 454865. [CrossRef] [PubMed]
70. Zhukovsky, K.V. A method of inverse differential operators using ortogonal polynomials and special functions for solving some types of differential equations and physical problems. *Mosc. Univ. Phys. Bull.* **2015**, *70*, 93–100. [CrossRef]
71. Zhukovsky, K.V.; Dattoli, G. Evolution of non-spreading Airy wavepackets in time dependent linear potentials. *Appl. Math. Comp.* **2011**, *217*, 7966–7974. [CrossRef]
72. Zhukovsky, K. Operational solution for some types of second order differential equations and for relevant physical problems. *J. Math. Anal. Appl.* **2017**, *446*, 628–647. [CrossRef]
73. Zhukovsky, K.V. Operational solution of differential equations with derivatives of non-integer order, Black-Scholes type and heat conduction. *Mosc. Univ. Phys. Bull.* **2016**, *71*, 237–244. [CrossRef]
74. Gould, H.W.; Hopper, A.T. Operational formulas connected with two generalizations of Hermite polynomials. *Duke Math. J.* **1962**, *29*, 51–63. [CrossRef]
75. Fourier, J.P.J. *The Analytical Theory of Heat*; Cambridge University Press: London, UK, 1878.
76. Cattaneo, C. Sur une forme de l'equation de la chaleur eliminant le paradoxe d'une propagation instantanee. *Comptes Rendus Hebdomadaires des Séances de l'Académie des Sciences* **1958**, *247*, 431–433.
77. Guyer, R.A.; Krumhansl, J.A. Solution of the linearized phonon Boltzmann equation. *Phys. Rev.* **1966**, *148*, 766–778. [CrossRef]
78. Srivastava, H.M.; Manocha, H.L. *A Treatise on Generating Functions*; Ellis Horwood Limited: Chichester, UK, 1984.
79. Parker, W.J.; Jenkins, R.J.; Butler, C.P.; Abbott, G.L. Flash method of determining thermal diffusivity, heat capacity, and thermal conductivity. *J. Appl. Phys.* **1961**, *32*, 1679. [CrossRef]
80. Onsager, L. Reciprocal Relations in irreversible processes. *Phys Rev.* **1931**, *37*, 119. [CrossRef]
81. Peshkov, V. Second sound in Helium II. *J. Phys. (Mosc.)* **1944**, *8*, 381.
82. Ackerman, C.; Guyer, R.A. Temperature pulses in dielectric solids. *Ann. Phys.* **1968**, *50*, 128–185. [CrossRef]
83. Ackerman, C.; Overton, W.C. Second sound in solid helium-3. *Phys. Rev. Lett.* **1969**, *22*, 764. [CrossRef]
84. McNelly, T.F.; Rogers, S.J.; Channin, D.J.; Rollefson, R.; Goubau, W.M.; Schmidt, G.E.; Krumhansl, J.A.; Pohl, R.O. Heat pulses in NaF: Onset of second sound. *Phys. Rev. Lett.* **1979**, *24*, 100. [CrossRef]
85. Narayanamurti, V.; Dynes, R.D. Observation of second sound in Bismuth. *Phys. Rev. Lett.* **1972**, *26*, 1461–1465. [CrossRef]
86. Frederick Emmons Terman. *Radio Engineers' Handbook*, 1st ed.; McGraw-Hill: New York, NY, USA, 1943.
87. Guyer, R.A.; Krumhansl, J.A. Thermal conductivity, second sound and phonon hydrodynamic phenomena in non-metallic crystals. *Phys. Rev.* **1966**, *148*, 778–788. [CrossRef]
88. Baringhaus, J.; Ruan, M.; Edler, F.; Tejeda, A.; Sicot, M.; Taleb-Ibrahimi, A.; Li, A.; Jiang, Z.; Conrad, E.H.; Berger, C.; et al. Exceptional ballistic transport in epitaxial graphene nanoribbons. *Nature* **2014**, *506*, 349–354. [CrossRef] [PubMed]
89. Hochbaum, A.I.; Chen, R.; Delgado, R.D.; Liang, W.; Garnett, E.C.; Najarian, M.; Majumdar, A.; Yang, P. Enhanced thermoelectric performance of rough silicon nanowires. *Nature (Lond.)* **2008**, *451*, 163–167. [CrossRef] [PubMed]
90. Boukai, I.; Bunimovich, Y.; Tahir-Kheli, J.; Yu, J.-K.; Goddard, W.A.; Heath, J.R. Silicon nanowires as efficient thermoelectric materials. *Nature (Lond.)* **2008**, *451*, 168–171. [CrossRef] [PubMed]
91. Chiritescu, C.; Cahill, D.G.; Nguyen, N.; Johnson, D.; Bodapati, A.; Keblinski, P.; Zschack, P. Ultralow thermal conductivity in disordered, layered WSe_2 crystals. *Science* **2007**, *315*, 351–353. [CrossRef] [PubMed]
92. Maldovan, M. Transition between ballistic and diffusive heat transport regimes in silicon materials. *Appl. Phys. Lett.* **2012**, *101*, 113110. [CrossRef]
93. Casimir, H.B.G. Note on the conduction of heat in crystals. *Physica* **1938**, *5*, 495–500. [CrossRef]
94. Minnich, J.; Johnson, J.A.; Schmidt, A.J.; Esfarjani, K.; Dresselhaus, M.S.; Nelson, K.A.; Chen, G. Thermal conductivity spectroscopy technique to measure phonon mean free paths. *Phys. Rev. Lett.* **2011**, *107*, 095901. [CrossRef] [PubMed]
95. Cahill, G. Thermal conductivity measurement from 30 to 750 K: The 3ω method. *Rev. Sci. Instrum.* **1990**, *61*, 802. [CrossRef]
96. Paddock, A.; Eesley, G.L. Transient thermoreflectance from thin metal films. *J. Appl. Phys.* **1986**, *60*, 285. [CrossRef]

97. Hsiao, T.; Chang, H.; Liou, S.; Chu, M.; Lee, S.; Chang, C. Observation of room-temperature ballistic thermal conduction persisting over 8.3 mm in SiGe nanowires. *Nat. Nanotechnol.* **2013**, *8*, 534–538. [CrossRef] [PubMed]
98. Kaminski, W. Hyperbolic heat conduction equations for materials with a nonhomogeneous inner structure. *J. Heat Transfer* **1990**, *112*, 555–560. [CrossRef]
99. Mitra, K.; Kumar, S.; Vedavarz, A.; Moallemi, M.K. Experimental evidence of hyperbolic heat conduction in processed meat. *J. Heat Transfer* **1995**, *117*, 568–573. [CrossRef]
100. Herwig, H.; Beckert, K. Fourier versus non-Fourier heat conduction in materials with a nonhomogeneous inner structure. *J. Heat Transfer* **2000**, *122*, 363–365. [CrossRef]
101. Roetzel, W.; Putra, N.; Das, S.K. Experiment and analysis for non-Fourier conduction in materials with non-homogeneous inner structure. *Int. J. Thermal Sci.* **2003**, *42*, 541–552. [CrossRef]
102. Scott, P.; Tilahun, M.; Vick, B. The question of thermal waves in heterogeneous and biological materials. *J. Biomech. Eng.* **2009**, *131*, 074518. [CrossRef] [PubMed]
103. Both, S.; Czél, B.; Fülöp, T.; Gróf, G.; Gyenis, Á.; Kovács, R.; Ván, P.; Verhás, J. Deviation from the Fourier law in room-temperature heat pulse experiments. *arXiv* **2015**. [CrossRef]
104. Zhang, Y.; Ye, W. Modified ballistic-diffusive equations for transient non-continuum heat conduction. *Int. J. Heat Mass Transfer* **2015**, *83*, 51–63. [CrossRef]
105. Chen, G. Ballistic-diffusive heat-conduction equations. *Phys. Rev. Lett.* **2001**, *86*, 2297–2300. [CrossRef] [PubMed]
106. Kovacs, R.; Van, P. Generalized heat conduction in heat pulse experiments. *Int. J. Heat Mass Transfer.* **2015**, *83*, 613–620. [CrossRef]
107. Van, P.; Fulop, T. Universality in heat conduction theory: Weakly non-local thermodynamics. *Ann. Phys.* **2012**, *524*, 470–478. [CrossRef]
108. Lebon, G.; Machrafi, H.; Gremela, M.; Dubois, C. An extended thermodynamic model of transient heat conduction at sub-continuum scales. *Proc. R. Soc. A* **2011**, *467*, 3241–3256. [CrossRef]

axioms

MDPI

Article

Fractional Integration and Differentiation of the Generalized Mathieu Series

Ram K. Saxena [1] **and Rakesh K. Parmar** [2,*]

[1] Department of Mathematics and Statistics, Jai Narain Vyas University, Jodhpur-342004, Rajasthan, India; ram.saxena@yahoo.com
[2] Department of Mathematics, Government College of Engineering and Technology, Bikaner-334004, Rajasthan, India
* Correspondence: rakeshparmar27@gmail.com

Academic Editor: Hans J. Haubold
Received: 27 April 2017; Accepted: 22 June 2017; Published: 27 June 2017

Abstract: We aim to present some formulas for the Saigo hypergeometric fractional integral and differential operators involving the generalized Mathieu series $S_\mu(r)$, which are expressed in terms of the Hadamard product of the generalized Mathieu series $S_\mu(r)$ and the Fox–Wright function ${}_p\Psi_q(z)$. Corresponding assertions for the classical Riemann–Liouville and Erdélyi–Kober fractional integral and differential operators are deduced. Further, it is emphasized that the results presented here, which are for a seemingly complicated series, can reveal their involved properties via the series of the two known functions.

Keywords: Mathieu series; generalized Mathieuseries; fractional calculus operators

1. Introduction and Preliminaries

Fractional calculus, which has a long history, is an important branch of mathematical analysis (calculus) where differentiations and integrations can be of arbitrary non-integer order. During the past four decades or so, fractional calculus has been widely and extensively investigated and has gained importance and popularity due mainly to its demonstrated applications in numerous and diverse fields of science and engineering such as turbulence and fluid dynamics, stochastic dynamical system, plasma physics and controlled thermonuclear fusion, nonlinear control theory, image processing, nonlinear biological systems, and astrophysics (see, for detail, [1–5]).

We recall Saigo fractional integral and differential operators involving Gauss's hypergeometric function ${}_2F_1$ as a kernel. Let α, β, $\eta \in \mathbb{C}$, $\Re(\alpha) > 0$ and $x > 0$, then Saigo's fractional integral and differential operators $\left(I_{0+}^{\alpha,\beta,\eta} f\right)(x)$, $\left(I_{-}^{\alpha,\beta,\eta} f\right)(x)$ and $\left(D_{0+}^{\alpha,\beta,\eta} f\right)(x)$, $\left(D_{-}^{\alpha,\beta,\eta} f\right)(x)$ are defined as (see, for example, [1,2,4–6]):

$$\left(I_{0+}^{\alpha,\beta,\eta} f\right)(x) = \frac{x^{-\alpha-\beta}}{\Gamma(\alpha)} \int_0^x (x-t)^{\alpha-1} {}_2F_1\left(\alpha+\beta, -\eta; \alpha; 1-\frac{t}{x}\right) f(t)\, \mathrm{d}t, \tag{1}$$

$$\left(I_{-}^{\alpha,\beta,\eta} f\right)(x) = \frac{1}{\Gamma(\alpha)} \int_x^\infty (t-x)^{\alpha-1} t^{-\alpha-\beta} {}_2F_1\left(\alpha+\beta, -\eta; \alpha; 1-\frac{x}{t}\right) f(t)\, \mathrm{d}t, \tag{2}$$

and

$$\begin{aligned}\left(D_{0+}^{\alpha,\beta,\eta} f\right)(x) &= \left(I_{0+}^{-\alpha,-\beta,\alpha+\eta} f\right)(x) \\ &= \left(\frac{d}{dx}\right)^n \left(I_{0+}^{-\alpha+n,-\beta-n,\alpha+\eta-n} f\right)(x) \quad (n = [\Re(\alpha)] + 1),\end{aligned} \tag{3}$$

$$\left(D_{-}^{\alpha,\beta,\eta} f\right)(x) = \left(I_{-}^{-\alpha,-\beta,\alpha+\eta} f\right)(x)$$
$$= (-1)^n \left(\frac{d}{dx}\right)^n \left(I_{-}^{-\alpha+n,-\beta-n,\alpha+\eta} f\right)(x) \quad (n = [\Re(\alpha)] + 1). \tag{4}$$

Here and in what follows, $[x]$ denotes the greatest integer less than or equal to the real number x. When $\beta = -\alpha$, the operators in (1)–(4) coincide with the classical Riemann–Liouville fractional integrals and derivatives of order $\alpha \in \mathbb{C}$ with $\Re(\alpha) > 0$ and $x > 0$ (see, e.g., [1,4]):

$$\left(I_{0+}^{\alpha,-\alpha,\eta} f\right)(x) = \left(I_{0+}^{\alpha} f\right)(x) \equiv \frac{1}{\Gamma(\alpha)} \int_0^x (x-t)^{\alpha-1} f(t)\, dt, \tag{5}$$

$$\left(I_{-}^{\alpha,-\alpha,\eta} f\right)(x) = \left(I_{-}^{\alpha} f\right)(x) = \frac{1}{\Gamma(\alpha)} \int_x^{\infty} (t-x)^{\alpha-1} f(t)\, dt, \tag{6}$$

and

$$\left(D_{0+}^{\alpha,-\alpha,\eta} f\right)(x) = \left(D_{0+}^{\alpha} f\right)(x) = \left(\frac{d}{dx}\right)^n \frac{1}{\Gamma(n-\alpha)} \int_0^x (x-t)^{n-\alpha-1} f(t)\, dt$$
$$= \left(\frac{d}{dx}\right)^n \left(I_{0+}^{n-\alpha} f\right)(x) \qquad (n = [\Re(\alpha)] + 1), \tag{7}$$

$$\left(D_{-}^{\alpha,-\alpha,\eta} f\right)(x) = \left(D_{-}^{\alpha} f\right)(x) = (-1)^n \left(\frac{d}{dx}\right)^n \frac{1}{\Gamma(n-\alpha)} \int_x^{\infty} (t-y)^{n-\alpha-1} f(t)\, dt$$
$$= (-1)^n \left(\frac{d}{dx}\right)^n \left(I_{-}^{n-\alpha} f\right)(x) \qquad (n = [\Re(\alpha)] + 1). \tag{8}$$

Here and in the following, let $\mathbb{C}$, $\mathbb{R}^+$, and $\mathbb{N}$ be the sets of complex numbers, positive real numbers, and positive integers, respectively, and let $\mathbb{N}_0 := \mathbb{N} \cup \{0\}$.

If $\beta = 0$ in (1)–(4) yields the so-called Erdélyi–Kober fractional integrals and derivatives of order $\alpha \in \mathbb{C}$ with $\Re(\alpha) > 0$ and $x > 0$ (see, e.g., [1,4]):

$$\left(I_{0+}^{\alpha,0,\eta} f\right)(x) = \left(I_{\eta,\alpha}^{+} f\right)(x) = \frac{x^{-\alpha-\eta}}{\Gamma(\alpha)} \int_0^x (x-t)^{\alpha-1} t^{\eta} f(t)\, dt, \tag{9}$$

$$\left(I_{-}^{\alpha,0,\eta} f\right)(x) = \left(K_{\eta,\alpha}^{-} f\right)(x) \equiv \frac{x^{\eta}}{\Gamma(\alpha)} \int_x^{\infty} (t-x)^{\alpha-1} t^{-\alpha-\eta} f(t)\, dt, \tag{10}$$

and

$$\left(D_{0+}^{\alpha,0,\eta} f\right)(x) = \left(D_{\eta,\alpha}^{+} f\right)(x)$$
$$= \left(\frac{d}{dx}\right)^n \left(I_{0+}^{-\alpha+n,-\alpha,\alpha+\eta-n} f\right)(x) \quad (n = [\Re(\alpha)] + 1), \tag{11}$$

$$\left(D_{-}^{\alpha,0,\eta} f\right)(x) = \left(D_{\eta,\alpha}^{-} f\right)(x)$$
$$= (-1)^n \left(\frac{d}{dx}\right)^n \left(I_{-}^{-\alpha+n,-\alpha,\alpha+\eta} f\right)(x) \quad (n = [\Re(\alpha)] + 1), \tag{12}$$

$$\left(D_{\eta,\alpha}^{+} f\right)(x) = x^{-\eta} \left(\frac{d}{dx}\right)^n \frac{1}{\Gamma(n-\alpha)} \int_0^x t^{\alpha+\eta} (x-t)^{n-\alpha-1} f(t)\, dt \quad (n = [\Re(\alpha)] + 1), \tag{13}$$

$$\left(D_{\eta,\alpha}^{-}f\right)(x) = x^{\eta+\alpha}\left(\frac{d}{dx}\right)^{n}\frac{1}{\Gamma(n-\alpha)}\int_{x}^{\infty} t^{-\eta}(t-x)^{n-\alpha-1}f(t)\,\mathrm{d}t \quad (n = [\Re(\alpha)]+1). \tag{14}$$

A detailed account of such operators along with their properties and applications has been considered by several authors (see, for details, [1–5]).

The following familiar infinite series of the form

$$S(r) = \sum_{n\geq 1}\frac{2n}{(n^2+r^2)^2}, \qquad r>0 \tag{15}$$

is known in literature as the Mathieu series. Émile Leonard Mathieu was the first to investigate such a series in 1890 in his book elasticity of solid bodies [7]. An alternative version of (15)

$$\widetilde{S}(r) = \sum_{n\geq 1}(-1)^{n-1}\frac{2n}{(n^2+r^2)^2}, \qquad r>0 \tag{16}$$

was introduced by Pogány et al. [8]. Closed form integral representations for $S(r)$ and $\widetilde{S}(r)$ are given by (see e.g., [8,9])

$$S(r) = \frac{1}{r}\int_0^{\infty}\frac{x\sin(rx)}{\mathrm{e}^x-1}\,\mathrm{d}x \tag{17}$$

and

$$\widetilde{S}(r) = \frac{1}{r}\int_0^{\infty}\frac{x\sin(rx)}{\mathrm{e}^x+1}\,\mathrm{d}x, \tag{18}$$

respectively. Several interesting problems and solutions deal with integral representations and bounds for the following mild generalization of the Mathieu series and its alternative version with a fractional power defined by ([10], p. 2, Equation (16)) (see also, [11], p. 181)

$$S_\mu(r) = \sum_{n\geq 1}\frac{2n}{(n^2+r^2)^{\mu+1}} \qquad (r>0,\ \mu>0) \tag{19}$$

and

$$\widetilde{S}_\mu(r) = \sum_{n\geq 1}(-1)^{n-1}\frac{2n}{(n^2+r^2)^{\mu+1}} \qquad (r>0,\ \mu>0), \tag{20}$$

respectively. Such a series has been widely considered in mathematical literature (see, e.g., papers of Cerone and Lenard [10], Diananda [12] and Pogány et al. [8]). Various applications of the familiar Mathieu series and its generalizations in probability theory with other variants such as trigonometric Mathieu series, harmonic Mathieu series, Fourier–Mathieu series and some other particular forms of the Mathieu series can be found in a recent paper [13].

Recently, Tomovski and Pogány [14] studied the several integral representations of the generalized fractional order Mathieu-type power series

$$S_\mu(r;z) = \sum_{n\geq 1}\frac{2n\,z^n}{(n^2+r^2)^{\mu+1}}, \qquad (\mu>0,\ r\in\mathbb{R},\ |z|<1). \tag{21}$$

Obviously, we have

$$S_\mu(r;1) = S_\mu(r) \quad \text{and} \quad S_\mu(r;-1) = -\widetilde{S}_\mu(r).$$

Various other investigations and generalizations of the Mathieu series with its alternative variants can also be found in [11,14–24], and the references cited therein.

The concept of the Hadamard product (or the convolution) of two analytic functions is useful in our present investigation. It can help us to decompose a newly emerged function into two known functions. If, in particular, one of the power series defines an entire function, then the Hadamard product series defines an entire function, too. Let

$$f(z) := \sum_{n=0}^{\infty} a_n z^n \ (|z| < R_f) \text{ and } g(z) := \sum_{n=0}^{\infty} b_n z^n \ (|z| < R_g)$$

be two power series whose radii of convergence are denoted by R_f and R_g, respectively. Then, their Hadamard product is the power series defined by

$$(f * g)(z) := \sum_{n=0}^{\infty} a_n b_n z^n = (g * f)(z) \ (|z| < R) \tag{22}$$

where

$$R = \lim_{n\to\infty} \left| \frac{a_n b_n}{a_{n+1} b_{n+1}} \right| = \left(\lim_{n\to\infty} \left| \frac{a_n}{a_{n+1}} \right| \right) \cdot \left(\lim_{n\to\infty} \left| \frac{b_n}{b_{n+1}} \right| \right) = R_f \cdot R_g,$$

therefore, in general, we have $R \geq R_f \cdot R_g$ [25,26]. For various other investigations involving the Hadamard product (or the convolution), the interested reader may be referred to several recent papers on the subject (see, for example, [27,28] and the references cited in each of these papers).

In this paper, our aim is to study the compositions of the generalized fractional integration and differentiation operators (1)–(4) with the generalized Mathieu series (21) in terms of the Hadamard product (22) of the generalized Mathieu series and the Fox–Wright function. Further, corresponding assertions for the classical Riemann–Liouville and Erdélyi–Kober fractional integral and differential operators are deduced. The results presented in Theorems together with Corollaries are sure to be new and potentially useful, mainly because they are expressed in terms of the Hadamard product with two known functions. At least, a seemingly complicated resulting series expressed in terms of two known functions *means* that certain properties involved in the complicated resulting series can be revealed via the series of the known functions.

2. Fractional Integration of the Mathieu Series

We first recall the Fox–Wright function ${}_p\Psi_q(z)$ $(p, q \in \mathbb{N}_0)$ with p numerator and q denominator parameters defined for $\alpha_1, \ldots, \alpha_p \in \mathbb{C}$ and $\beta_1, \ldots, \beta_q \in \mathbb{C} \setminus \mathbb{Z}_0^-$ by (see, for details, [1,3]; see also [4,29]):

$${}_p\Psi_q \left[\begin{matrix} (\alpha_1, A_1), \cdots, (\alpha_p, A_p); \\ (\beta_1, B_1), \cdots, (\beta_q, B_q); \end{matrix} \ z \right] = \sum_{n=0}^{\infty} \frac{\Gamma(\alpha_1 + A_1 n) \cdots \Gamma(\alpha_p + A_p n)}{\Gamma(\beta_1 + B_1 n) \cdots \Gamma(\beta_q + B_q n)} \frac{z^n}{n!} \tag{23}$$

$$\left(A_j \in \mathbb{R}^+ \ (j = 1, \ldots, p); \ B_j \in \mathbb{R}^+ \ (j = 1, \ldots, q); \ 1 + \sum_{j=1}^{q} B_j - \sum_{j=1}^{p} A_j \geq 0 \right),$$

where the equality in the convergence condition holds true for

$$|z| < \nabla := \left(\prod_{j=1}^{p} A_j^{-A_j} \right) \cdot \left(\prod_{j=1}^{q} B_j^{B_j} \right).$$

In particular, when $A_j = B_k = 1$ $(j = 1, \ldots, p; \ k = 1, \ldots, q)$, (23) reduces immediately to the generalized hypergeometric function ${}_pF_q$ $(p, q \in \mathbb{N}_0)$ (see, e.g., [29]):

$${}_pF_q \left[\begin{matrix} \alpha_1, \ldots, \alpha_p; \\ \beta_1, \ldots, \beta_q; \end{matrix} \ z \right] = \frac{\Gamma(\beta_1) \cdots \Gamma(\beta_q)}{\Gamma(\alpha_1) \cdots \Gamma(\alpha_p)} \, {}_p\Psi_q \left[\begin{matrix} (\alpha_1, 1), \cdots, (\alpha_p, 1); \\ (\beta_1, 1), \cdots, (\beta_q, 1); \end{matrix} \ z \right]. \tag{24}$$

Lemma 1. *Let* $\alpha,\ \beta,\ \eta \in \mathbb{C}$. *Then, there exists the relation*

(a) *If* $\Re(\alpha) > 0$ *and* $\Re(\sigma) > \max[0, \Re(\beta - \eta)]$, *then*

$$(I_{0+}^{\alpha,\beta,\eta} t^{\sigma-1})(x) = \frac{\Gamma(\sigma)\Gamma(\sigma+\eta-\beta)}{\Gamma(\sigma-\beta)\Gamma(\sigma+\alpha+\eta)} x^{\sigma-\beta-1} \tag{25}$$

In particular, for $x > 0$, *we have*

$$(I_{0+}^{\alpha} t^{\sigma-1})(x) = \frac{\Gamma(\sigma)}{\Gamma(\sigma+\alpha)} x^{\sigma+\alpha-1} \qquad (\Re(\alpha) > 0,\ \Re(\sigma) > 0), \tag{26}$$

$$(I_{\eta,\alpha}^{+} t^{\sigma-1})(x) = \frac{\Gamma(\sigma+\eta)}{\Gamma(\sigma+\alpha+\eta)} x^{\sigma-1} \qquad (\Re(\alpha) > 0,\ \Re(\sigma) > -\Re(\eta)) \tag{27}$$

(b) *If* $\Re(\alpha) > 0$ *and* $\Re(\sigma) < 1 + \min[\Re(\beta), \Re(\eta)]$, *then*

$$(I_{-}^{\alpha,\beta,\eta} t^{\sigma-1})(x) = \frac{\Gamma(1-\sigma+\beta)\Gamma(1-\sigma+\eta)}{\Gamma(1-\sigma)\Gamma(1-\sigma+\alpha+\beta+\eta)} x^{\sigma-\beta-1}. \tag{28}$$

In particular, for $x > 0$, *we have*

$$(I_{-}^{\alpha} t^{\sigma-1})(x) = \frac{\Gamma(1-\alpha-\sigma)}{\Gamma(1-\sigma)} x^{\sigma+\alpha-1} \qquad (0 < \Re(\alpha) < 1 - \Re(\sigma)), \tag{29}$$

$$(K_{\eta,\alpha}^{-} t^{\sigma-1})(x) = \frac{\Gamma(1-\sigma+\eta)}{\Gamma(1-\sigma+\alpha+\eta)} x^{\sigma-1} \qquad (\Re(\sigma) < 1 + \Re(\sigma)). \tag{30}$$

We begin the exposition of the main results by presenting the composition formulas of generalized fractional integrals, (1) and (2), involving the generalized Mathieu series in terms of the Hadamard product (22) of the generalized Mathieu series (21) and the Fox–Wright function (23). It is emphasized that the results presented here, which are for a seemingly complicated series, can reveal their involved properties via the series of the two known functions.

Theorem 1. *Let* $\alpha, \beta, \eta, \sigma \in \mathbb{C}$ *and* $\rho > 0, \mu > 0, r \in \mathbb{R}$ *be such that* $\Re(\alpha) > 0$ *and* $\Re(\sigma) > \max[0, \Re(\beta - \eta)]$. *Then, the following Saigo hypergeometric fractional integral* $I_{0+}^{\alpha,\beta,\eta}$ *of* $S_\mu(r, t^\rho)$ *holds true:*

$$\left(I_{0+}^{\alpha,\beta,\eta}\left\{t^{\sigma-1} S_\mu(r,t^\rho)\right\}\right)(x)$$
$$= x^{\sigma-\beta+\rho-1} S_\mu(r,x^\rho) * {}_3\Psi_2\left[\begin{array}{c}(1,1),(\sigma+\rho,\rho),(\sigma+\eta-\beta+\rho,\rho);\\ (\sigma-\beta+\rho,\rho),(\sigma+\alpha+\eta+\rho,\rho);\end{array} x^\rho\right]. \tag{31}$$

Proof. Using the definitions (1) and (21), by changing the order of integration and applying the relation (25), we find that $x > 0$

$$\left(I_{0+}^{\alpha,\beta,\eta}\left\{t^{\sigma-1} S_\mu(r,t^\rho)\right\}\right)(x) = \sum_{k=1}^{\infty} \frac{2k}{(k^2+r^2)^{\mu+1}} \left(I_{0+}^{\alpha,\beta,\eta} t^{\sigma+\rho k-1}\right)(x)$$
$$= x^{\sigma-\beta-1} \sum_{k=1}^{\infty} \frac{2k}{(k^2+r^2)^{\mu+1}} \frac{\Gamma(\sigma+\rho k)\Gamma(\sigma+\eta-\beta+\rho k)}{\Gamma(\sigma-\beta+\rho k)\Gamma(\sigma+\alpha+\eta+\rho k)} x^{\rho k}. \tag{32}$$

by applying the Hadamard product (22) in (32), which in the view of (21) and (23), yields the desired formula (31). □

Theorem 2. *Let* $\alpha, \beta, \eta, \sigma \in \mathbb{C}$ *and* $\rho > 0, \mu > 0, r \in \mathbb{R}$ *be such that* $\Re(\alpha) > 0$ *and* $\Re(\sigma) < 1 + \min[\Re(\beta), \Re(\eta)]$. *Then, the following Saigo hypergeometric fractional integral* $I_{-}^{\alpha,\beta,\eta}$ *of* $S_\mu\left(r, \frac{1}{t^\rho}\right)$ *holds true:*

$$\left(I_{-}^{\alpha,\beta,\eta}\left\{t^{\sigma-1}S_\mu\left(r,\frac{1}{t^\rho}\right)\right\}\right)(x)$$
$$= x^{\sigma-\rho-\beta-1}S_\mu\left(r,\frac{1}{x^\rho}\right) * {}_3\Psi_2\left[\begin{array}{c}(1,1),(1-\sigma+\beta+\rho,\rho),(1-\sigma+\eta+\rho,\rho);\\(1-\sigma+\rho,\rho),(1-\sigma+\alpha+\beta+\eta+\rho,\rho);\end{array}\frac{1}{x^\rho}\right]. \quad (33)$$

Proof. Using the definitions (2) and (21), by changing the order of integration and applying the relation (28)

$$\left(I_{-}^{\alpha,\beta,\eta}\left\{t^{\sigma-1}S_\mu\left(r,\frac{1}{t^\rho}\right)\right\}\right)(x) = \sum_{k=1}^{\infty}\frac{2k}{(k^2+r^2)^{\mu+1}}\left(I_{-}^{\alpha,\beta,\eta}t^{\sigma-\rho k-1}\right)(x)$$
$$= x^{\sigma-\beta-1}\sum_{k=1}^{\infty}\frac{2k}{(k^2+r^2)^{\mu+1}}\frac{\Gamma(1-\sigma+\beta+\rho k)\Gamma(1-\sigma+\eta+\rho k)}{\Gamma(1-\sigma+\rho k)\Gamma(1-\sigma+\alpha+\beta+\eta+\rho k)}x^{-\rho k}. \quad (34)$$

by applying the Hadamard product (22) in (34), which in the view of (21) and (23), yields the desired formula (33). □

Further, we deduce the fractional integral formulas for the classical Riemann–Liouville and Erdélyi–Kober fractional integral and differential operators by letting $\beta = -\alpha$ and $\beta = 0$ respectively, which are asserted by Corollaries 1–4 below.

Corollary 1. *Let* $\alpha, \sigma \in \mathbb{C}$ *and* $\rho > 0, \mu > 0, r \in \mathbb{R}$ *be such that* $\Re(\alpha) > 0$ *and* $\Re(\sigma) > 0$. *Then, the following Riemann–Liouville fractional integral* I_{0+}^{α} *of* $S_\mu(r, t^\rho)$ *holds true:*

$$\left(I_{0+}^{\alpha}\left\{t^{\sigma-1}S_\mu(r,t^\rho)\right\}\right)(x) = x^{\sigma+\rho+\alpha-1}S_\mu(r,x^\rho) * {}_2\Psi_1\left[\begin{array}{c}(1,1),(\sigma+\rho,\rho);\\(\sigma+\alpha+\rho,\rho);\end{array}x^\rho\right]. \quad (35)$$

Corollary 2. *Let* $\alpha, \eta, \sigma \in \mathbb{C}$ *and* $\rho > 0, \mu > 0, r \in \mathbb{R}$ *be such that* $\Re(\alpha) > 0$ *and* $\Re(\sigma) > -\Re(\eta)$. *Then, the following Erdélyi–Kober fractional integral* $I_{\eta,\alpha}^{+}$ *of* $S_\mu(r, t^\rho)$ *holds true:*

$$\left(I_{\eta,\alpha}^{+}\left\{t^{\sigma-1}S_\mu(r,t^\rho)\right\}\right)(x) = x^{\sigma+\rho-1}S_\mu(r,x^\rho) * {}_2\Psi_1\left[\begin{array}{c}(1,1),(\sigma+\eta+\rho,\rho);\\(\sigma+\alpha+\eta+\rho,\rho);\end{array}x^\rho\right]. \quad (36)$$

Corollary 3. *Let* $\alpha, \sigma \in \mathbb{C}$ *and* $\rho > 0, \mu > 0, r \in \mathbb{R}$ *be such that* $0 < \Re(\alpha) < 1 - \Re(\sigma)$. *Then, the following Riemann–Liouville fractional integral* I_{-}^{α} *of* $S_\mu\left(r, \frac{1}{t^\rho}\right)$ *holds true:*

$$\left(I_{-}^{\alpha}\left\{t^{\sigma-1}S_\mu\left(r,\frac{1}{t^\rho}\right)\right\}\right)(x) = x^{\sigma+\alpha-\rho-1}S_\mu\left(r,\frac{1}{x^\rho}\right) * {}_2\Psi_1\left[\begin{array}{c}(1,1),(1-\sigma-\alpha+\rho,\rho);\\(1-\sigma+\rho,\rho);\end{array}\frac{1}{x^\rho}\right]. \quad (37)$$

Corollary 4. *Let* $\alpha, \eta, \sigma \in \mathbb{C}$ *and* $\rho > 0, \mu > 0, r \in \mathbb{R}$ *be such that* $\Re(\alpha) > 0$ *and* $\Re(\sigma) < 1 + \Re(\eta)$. *Then, the following Erdélyi–Kober fractional integral* $K_{\eta,\alpha}^{-}$ *of* $S_\mu\left(r, \frac{1}{t^\rho}\right)$ *holds true:*

$$\left(K_{\eta,\alpha}^{-}\left\{t^{\sigma-1}S_\mu\left(r,\frac{1}{t^\rho}\right)\right\}\right)(x) = x^{\sigma-\rho-1}S_\mu\left(r,\frac{1}{x^\rho}\right) * {}_2\Psi_1\left[\begin{array}{c}(1,1),(1-\sigma+\eta+\rho,\rho);\\(1-\sigma+\alpha+\eta+\rho,\rho);\end{array}\frac{1}{x^\rho}\right]. \quad (38)$$

The results obtained in this section can be presented in terms of Gauss's hypergeometric functions by taking $\rho = 1$. Here, we present results for the classical Riemann–Liouville fractional integral operators.

Corollary 5. *Let the conditions of Corollary 1 be satisfied, and let $\Re(\sigma) > 0$ and $\Re(\sigma+\alpha+1) > 0$. Then, for $x > 0$, there holds the relation*

$$\left(I_{0+}^{\alpha}\left\{t^{\sigma-1}\,S_\mu(r,t)\right\}\right)(x) = x^{\sigma+\alpha}\frac{\Gamma(\sigma+1)}{\Gamma(\sigma+\alpha+1)}\,S_\mu(r,x) * {}_2F_1\left[\begin{matrix}1,\sigma+1;\\ \sigma+\alpha+1;\end{matrix}\; x\right]. \tag{39}$$

Corollary 6. *Let the conditions of Corollary 3 be satisfied, and let $\Re(1-\sigma) > 0$ and $\Re(2-\sigma-\alpha) > 0$. Then, for $x > 0$, there holds the relation*

$$\left(I_{-}^{\alpha}\left\{t^{\sigma-1}\,S_\mu\left(r,\frac{1}{t}\right)\right\}\right)(x) = x^{\sigma+\alpha-2}\frac{\Gamma(2-\sigma-\alpha)}{\Gamma(2-\sigma)}\,S_\mu\left(r,\frac{1}{x}\right) * {}_2F_1\left[\begin{matrix}1,2-\sigma-\alpha;\\ 2-\sigma;\end{matrix}\; \frac{1}{x}\right]. \tag{40}$$

3. Fractional Differentiation of the Mathieu Series

In this section, we present the composition formulas of generalized fractional derivatives, (3) and (4), involving the generalized Mathieu series in terms of the Hadamard product (22) of the generalized Mathieu series (21) and the Fox–Wright function (23).

Lemma 2. *Let α, β, $\eta \in \mathbb{C}$. Then, there exists the relations*

(a) *If $\Re(\alpha) > 0$ and $\Re(\sigma) > -\min[0, \Re(\alpha+\beta+\eta)]$, then*

$$(D_{0+}^{\alpha,\beta,\eta}t^{\sigma-1})(x) = \frac{\Gamma(\sigma)\Gamma(\sigma+\alpha+\beta+\eta)}{\Gamma(\sigma+\beta)\Gamma(\sigma+\eta)}x^{\sigma+\beta-1} \tag{41}$$

In particular, for $x > 0$, we have

$$(D_{0+}^{\alpha}t^{\sigma-1})(x) = \frac{\Gamma(\sigma)}{\Gamma(\sigma-\alpha)}x^{\sigma-\alpha-1} \qquad (\Re(\alpha) > 0,\ \Re(\sigma) > 0), \tag{42}$$

$$(D_{\eta,\alpha}^{+}t^{\sigma-1})(x) = \frac{\Gamma(\sigma+\alpha+\eta)}{\Gamma(\sigma+\eta)}x^{\sigma-1} \qquad (\Re(\alpha) > 0,\ \Re(\sigma) > -\Re(\alpha+\eta)). \tag{43}$$

(b) *If $\Re(\alpha) > 0$, $\Re(\sigma) < 1+\min[\Re(-\beta-n), \Re(\alpha+\eta)]$ and $n = [\Re(\alpha)]+1$, then*

$$(D_{-}^{\alpha,\beta,\eta}t^{\sigma-1})(x) = \frac{\Gamma(1-\sigma-\beta)\Gamma(1-\sigma+\alpha+\eta)}{\Gamma(1-\sigma)\Gamma(1-\sigma+\eta-\beta)}x^{\sigma+\beta-1}. \tag{44}$$

In particular, for $x > 0$, we have

$$(D_{-}^{\alpha}t^{\sigma-1})(x) = \frac{\Gamma(1-\sigma+\alpha)}{\Gamma(1-\sigma)}x^{\sigma-\alpha-1} \qquad (\Re(\alpha) > 0,\ \Re(\sigma) < 1+\Re(\alpha)-n), \tag{45}$$

$$(D_{\eta,\alpha}^{-}t^{\sigma-1})(x) = \frac{\Gamma(1-\sigma+\alpha+\eta)}{\Gamma(1-\sigma-\eta)}x^{\sigma-1} \qquad (\Re(\alpha) > 0,\ \Re(\sigma) < 1+\Re(\alpha+\eta)-n). \tag{46}$$

Theorem 3. *Let $\alpha, \beta, \eta, \sigma \in \mathbb{C}$ and $\rho > 0, \mu > 0, r \in \mathbb{R}$ be such that $\Re(\alpha) \geq 0$ and $\Re(\sigma) > -\min[0, \Re(\alpha+\beta+\eta)]$. Then, the following Saigo hypergeometric fractional derivative $D_{0+}^{\alpha,\beta,\eta}$ of $S_\mu(r,t^\rho)$ holds true:*

$$\begin{aligned}&\left(D_{0+}^{\alpha,\beta,\eta}\left\{t^{\sigma-1}\,S_\mu(r,t^\rho)\right\}\right)(x)\\ &= x^{\sigma+\beta+\rho-1}\,S_\mu(r,x^\rho) * {}_3\Psi_2\left[\begin{matrix}(1,1),(\sigma+\rho,\rho),(\sigma+\alpha+\beta+\eta+\rho,\rho);\\ (\sigma+\beta+\rho,\rho),(\sigma+\eta+\rho,\rho);\end{matrix}\; x^\rho\right].\end{aligned} \tag{47}$$

Proof. Using the definitions (3) and (21), by changing the order of integration and applying the relation (41), we find for $x > 0$

$$\left(D_{0+}^{\alpha,\beta,\eta}\left\{t^{\sigma-1}S_{\mu}(r,t^{\rho})\right\}\right)(x)=\sum_{k=1}^{\infty}\frac{2k}{(k^2+r^2)^{\mu+1}}\left(D_{0+}^{\alpha,\beta,\eta}t^{\sigma+\rho k-1}\right)(x)$$

$$=x^{\sigma+\beta-1}\sum_{k=1}^{\infty}\frac{2k}{(k^2+r^2)^{\mu+1}}\frac{\Gamma(\sigma+\rho k)\Gamma(\sigma+\alpha+\beta+\eta+\rho k)}{\Gamma(\sigma+\beta+\rho k)\Gamma(\sigma+\eta+\rho k)}x^{\rho k}. \quad (48)$$

by applying the Hadamard product (22) in (48), which in the view of (21) and (23), yields the desired formula (47). □

Theorem 4. *Let $\alpha, \beta, \eta, \sigma \in \mathbb{C}$ and $\rho > 0, \mu > 0, r \in \mathbb{R}$ be such that $\Re(\alpha) \geq 0$ and $\Re(\sigma) < 1+\min[\Re(-\beta-n), \Re(\alpha+\eta)]$, $n = [\Re(\alpha)]+1$. Then, the following Saigo hypergeometric fractional derivative $D_{-}^{\alpha,\beta,\eta}$ of $S_{\mu}\left(r,\frac{1}{t^{\rho}}\right)$ holds true:*

$$\left(D_{-}^{\alpha,\beta,\eta}\left\{t^{\sigma-1}S_{\mu}\left(r,\frac{1}{t^{\rho}}\right)\right\}\right)(x)$$

$$=x^{\sigma-\rho+\beta-1}S_{\mu}\left(r,\frac{1}{x^{\rho}}\right)*{}_3\Psi_2\left[\begin{matrix}(1,1),(1-\sigma-\beta+\rho,\rho),(1-\sigma+\alpha+\eta+\rho,\rho); \\ (1-\sigma+\rho,\rho),(1-\sigma+\eta-\beta+\rho,\rho);\end{matrix}\ \frac{1}{x^{\rho}}\right]. \quad (49)$$

Proof. Using the definitions (4) and (21), by changing the order of integration and applying the relation (44), we find for $x > 0$

$$\left(D_{-}^{\alpha,\beta,\eta}\left\{t^{\sigma-1}S_{\mu}\left(r,\frac{1}{t^{\rho}}\right)\right\}\right)(x)=\sum_{k=1}^{\infty}\frac{2k}{(k^2+r^2)^{\mu+1}}\left(D_{-}^{\alpha,\beta,\eta}t^{\sigma-\rho k-1}\right)(x)$$

$$=x^{\sigma+\rho+\beta-1}\sum_{k=1}^{\infty}\frac{2k}{(k^2+r^2)^{\mu+1}}\frac{\Gamma(1-\sigma-\beta+\rho k)\Gamma(1-\sigma+\alpha+\eta+\rho k)}{\Gamma(1-\sigma+\rho k)\Gamma(1-\sigma+\eta-\beta+\rho k)}x^{-\rho k}. \quad (50)$$

by applying the Hadamard product (22) in (50), which in the view of (21) and (23), yields the desired formula (49). □

Now, we deduce fractional derivative formulas for the classical Riemann–Liouville and Erdélyi–Kober fractional integral and differential operators by letting $\beta = -\alpha$ and $\beta = 0$ respectively, which are asserted by Corollaries 7–10 below.

Corollary 7. *Let $\alpha, \sigma \in \mathbb{C}$ and $\rho > 0, \mu > 0, r \in \mathbb{R}$ be such that $\Re(\alpha) \geq 0$ and $\Re(\sigma) > 0$. Then, the following Riemann–Liouville fractional differentiation D_{0+}^{α} of $S_{\mu}(r,t^{\rho})$ holds true:*

$$\left(D_{0+}^{\alpha}\left\{t^{\sigma-1}S_{\mu}(r,t^{\rho})\right\}\right)(x)=x^{\sigma+\rho-\alpha-1}S_{\mu}(r,x^{\rho})*{}_2\Psi_1\left[\begin{matrix}(1,1),(\sigma+\rho,\rho); \\ (\sigma-\alpha+\rho,\rho);\end{matrix}\ x^{\rho}\right]. \quad (51)$$

Corollary 8. *Let $\alpha, \eta, \sigma \in \mathbb{C}$ and $\rho > 0, \mu > 0, r \in \mathbb{R}$ be such that $\Re(\alpha) \geq 0$ and $\Re(\sigma) > -\Re(\alpha+\eta)$ Then, the following Erdélyi–Kober fractional derivative $D_{\eta,\alpha}^{+}$ of $S_{\mu}(r,t^{\rho})$ holds true:*

$$\left(D_{\eta,\alpha}^{+}\left\{t^{\sigma-1}S_{\mu}(r,t^{\rho})\right\}\right)(x)=x^{\sigma+\rho-1}S_{\mu}(r,x^{\rho})*{}_2\Psi_1\left[\begin{matrix}(1,1),(\sigma+\alpha+\eta+\rho,\rho); \\ (\sigma+\eta+\rho,\rho);\end{matrix}\ x^{\rho}\right]. \quad (52)$$

Corollary 9. *Let $\alpha, \sigma \in \mathbb{C}$ and $\rho > 0$, $\mu > 0$, $r \in \mathbb{R}$ be such that $\Re(\alpha) \geq 0$ and $\Re(\sigma) < \Re(\alpha) - [\Re(\alpha)]$. Then, the following Riemann–Liouville fractional differentiation D_{-}^{α} of $S_\mu\left(r, \frac{1}{t^\rho}\right)$ holds true:*

$$\left(D_{-}^{\alpha}\left\{t^{\sigma-1} S_\mu\left(r, \frac{1}{t^\rho}\right)\right\}\right)(x) = x^{\sigma-\rho-\alpha-1} S_\mu\left(r, \frac{1}{x^\rho}\right) * {}_2\Psi_1\left[\begin{array}{c}(1,1),(1-\sigma+\alpha+\rho,\rho);\\(1-\sigma+\rho,\rho);\end{array} \frac{1}{x^\rho}\right]. \quad (53)$$

Corollary 10. *Let $\alpha, \eta, \sigma \in \mathbb{C}$ and $\rho > 0$, $\mu > 0$, $r \in \mathbb{R}$ be such that $\Re(\alpha) \geq 0$ and $\Re(\sigma) < \Re(\alpha+\eta) - [\Re(\alpha)]$. Then, the following Erdélyi–Kober fractional differentiation $D_{\eta,\alpha}^{-}$ of $S_\mu\left(r, \frac{1}{t^\rho}\right)$ holds true:*

$$\left(D_{\eta,\alpha}^{-}\left\{t^{\sigma-1} S_\mu\left(r, \frac{1}{t^\rho}\right)\right\}\right)(x) = x^{\sigma-\rho-1} S_\mu\left(r, \frac{1}{x^\rho}\right) * {}_2\Psi_1\left[\begin{array}{c}(1,1),(1-\sigma+\alpha+\eta+\rho,\rho);\\(1-\sigma+\eta+\rho,\rho);\end{array} \frac{1}{x^\rho}\right] \quad (54)$$

The results obtained in this section can be presented in terms of Gauss's hypergeometric functions by taking $\rho = 1$. Here, we present results for the classical Riemann–Liouville fractional derivative operators.

Corollary 11. *Let the conditions of Corollary 7 be satisfied, and let $\Re(\sigma+1) > 0$ and $\Re(\sigma-\alpha+1) > 0$. Then, for $x > 0$, there holds the relation*

$$\left(D_{0+}^{\alpha}\left\{t^{\sigma-1} S_\mu(r,t)\right\}\right)(x) = x^{\sigma-\alpha-2}\frac{\Gamma(\sigma+1)}{\Gamma(\sigma-\alpha+1)} S_\mu(r,x) * {}_2F_1\left[\begin{array}{c}1,\sigma+1;\\\sigma-\alpha+1;\end{array} x\right]. \quad (55)$$

Corollary 12. *Let the conditions of Corollary 9 be satisfied, and let $\Re(2-\sigma) > 0$ and $\Re(2-\sigma+\alpha) > 0$. Then, for $x > 0$, there holds the relation*

$$\left(D_{-}^{\alpha}\left\{t^{\sigma-1} S_\mu\left(r, \frac{1}{t}\right)\right\}\right)(x) = x^{\sigma-\alpha-2}\frac{\Gamma(2-\sigma+\alpha)}{\Gamma(2-\sigma)} S_\mu\left(r, \frac{1}{x}\right) * {}_2F_1\left[\begin{array}{c}1,2-\sigma+\alpha;\\2-\sigma;\end{array} \frac{1}{x}\right]. \quad (56)$$

4. Concluding Remarks and Observations

In our present investigation, with the help of the concept of the Hadamard product (or the convolution) of two analytic functions, we have obtained the composition formulas of the generalized fractional integrals, (1) and (2), involving the generalized Mathieu series in terms of the Hadamard product (22) of the generalized Mathieu series (21) and the Fox–Wright function (23). Further, we have also deduced the fractional integral formulas for the classical Riemann–Liouville and the Erdélyi–Kober fractional integral and differential operators by letting $\beta = -\alpha$ and $\beta = 0$, respectively. The results presented here, which are for a seemingly complicated series, can reveal their involved properties via the series of the two known functions.

Acknowledgments: The authors thanks to the anonymous referees for valuable comments and useful suggestions, improving the quality of the article.

Author Contributions: Ram K. Saxena and Rakesh K. Parmar contributed equally to this work.

Conflicts of Interest: The authors declare no conflict of interest.

References

1. Kilbas, A.A.; Srivastava, H.M.; Trujillo, J.J. *Theory and Applications of Fractional Differential Equations, Volume 204*; North-Holland Mathematical Studies; Elsevier (North-Holland) Science Publishers: New York, NY, USA, 2006.
2. Kiryakova, V.S. *Generalized Fractional Calculus and Applications*; Pitman Research Notes in Mathematics Series; Longman Scientific and Technical: Harlow, UK; John Wiley and Sons, Inc.: New York, NY, USA, 1993.

3. Mathai, A.M.; Saxena, R.K.; Haubold, H.J. *The H-Functions: Theory and Applications*; Springer: New York, NY, USA, 2010.
4. Samko, S.G.; Kilbas, A.A.; Marichev, O.I. *Fractional Integrals and Derivatives: Theory and Applications*; Translated from the Russian: Integrals and Derivatives of Fractional Order and Some of Their Applications ("Nauka i Tekhnika", Minsk, 1987); Gordon and Breach Science Publishers: Reading, UK, 1993.
5. Srivastava, H.M.; Saxena, R.K. Operators of fractional integration and their applications. *Appl. Math. Comput.* 2001, 118, 1–52.
6. Saigo, M. A remark on integral operators involving the Gauss hypergeometric functions. *Math. Rep. Kyushu Univ.* **1978**, *11*, 135–143.
7. Mathieu, E.L. *Traité de Physique Mathématique, VI-VII: Théorie de L'élasticité des Corps Solides*; Gauthier-Villars: Paris, France, 1890.
8. Pogány, T.K.; Srivastava, H.M.; Tomovski, Ž. Some families of Mathieu a-series and alternating Mathieu a-series. *Appl. Math. Comput.* **2006**, *173*, 69–108.
9. Emersleben, O. Über die Reihe $\sum_{k=1}^{\infty} \frac{k}{(k^2+r^2)^2}$. *Math. Ann.* **1952**, *125*, 165–171.
10. Cerone, P.; Lenard, C.T. On integral forms of generalized Mathieu series. *JIPAM J. Inequal. Pure Appl. Math.* **2003**, *4*, 1–11.
11. Milovanović, G.V.; Pogány, T.K. New integral forms of generalized Mathieu series and related applications. *Appl. Anal. Discret. Math.* **2013**, *7*, 180–192.
12. Diananda, P.H. Some inequalities related to an inequality of Mathieu. *Math. Ann.* **1980**, *250*, 95–98.
13. Srivastava, H.M.; Tomovski, Ž.; Leškovski, D. Some families of Mathieu type series and Hurwitz-Lerch zeta functions and associated probability distributions. *Appl. Comput. Math.* **2015**, *14*, 349–380.
14. Tomovski, Ž.; Pogány, T.K. Integral expressions for Mathieu-type power series and for the Butzer-Flocke-Hauss Ω-function. *Fract. Calc. Appl. Anal.* **2011**, *14*, 623–634.
15. Baricz, Á.; Butzer, P.L.; Pogány, T.K. Alternating Mathieu series, Hilbert—Eisenstein series and their generalized Omega functions. In *Analytic Number Theory, Approximation Theory, and Special Functions*; Rassias, T., Milovanović, G.V., Eds.; In Honor of Hari M. Srivastava; Springer: New York, NY, USA, 2014; pp. 775–808.
16. Choi, J.; Parmar, R.K.; Pogány, T.K. Mathieu-type series built by (p,q)—Extended Gaussian hypergeometric function. *Bull. Korean Math. Soc.* **2017**, *54*, 789–797.
17. Choi, J.; Srivastava, H.M. Mathieu series and associated sums involving the Zeta functions. *Comput. Math. Appl.* **2010**, *59*, 861–867.
18. Elezović, N.; Srivastava, H.M.; Tomovski, Ž. Integral representations and integral transforms of some families of Mathieu type series. *Integral Transforms Spec. Funct.* **2008**, *19*, 481–495.
19. Pogány, T.K.; Parmar, R.K. On p-extended Mathieu series; submitted manuscript. *arXiv* **2016**, arXiv:1606.08369.
20. Pogány T.K.; Tomovski, Ž. Bounds improvement for alternating Mathieu type series. *J. Math. Inequal.* **2010**, *4*, 315–324.
21. Saxena, R.K.; Pogány, T.K.; Saxena, R. Integral transforms of the generalized Mathieu series. *J. Appl. Math. Stat. Inform.* **2010**, *6*, 5–16.
22. Tomovski, Ž. Integral representations of generalized Mathieu series via Mittag-Leffler type functions. *Fract. Calc. Appl. Anal.* **2007**, *10*, 127–138.
23. Tomovski, Ž. New integral and series representations of the generalized Mathieu series. *Appl. Anal. Discret. Math.* **2008**, *2*, 205–212.
24. Tomovski, Ž.; Pogány, T.K. New upper bounds for Mathieu-type series. *Banach J. Math. Anal.* **2009**, *3*, 9–15.
25. Kiryakova, V. On two Saigo's fractional integral operators in the class of univalent functions. *Fract. Calc. Appl. Anal.* **2006**, *9*, 159–176.
26. Pohlen, T. The Hadamard Product and Universal Power Series. Ph.D. Thesis, Universität Trier, Trier, Germany, 2009.
27. Srivastava, H.M.; Agarwal, R.; Jain, S. Integral transform and fractional derivative formulas involving the extended generalized hypergeometric functions and probability distributions. *Math. Method Appl. Sci.* **2017**, *40*, 255–273.

28. Srivastava, R.; Agarwal, R.; Jain, S. A family of the incomplete hypergeometric functions and associated integral transform and fractional derivative formulas. *Filomat* **2017**, *31*, 125–140.
29. Srivastava, H.M.; Karlsson, P.W. *Multiple Gaussian Hypergeometric Series*; Ellis Horwood Ltd.: Chichester, UK, 1985.

axioms

MDPI

Article

An Evaluation of ARFIMA (Autoregressive Fractional Integral Moving Average) Programs †

Kai Liu [1], YangQuan Chen [2,*] and Xi Zhang [1]

1 School of Mechanical Electronic & Information Engineering, China University of Mining and Technology, Beijing 100083, China; kliu_cmutb@163.com (K.L.); zhangxi6681@163.com (X.Z.)
2 Mechatronics, Embedded Systems and Automation Lab, School of Engineering, University of California, Merced, CA 95343, USA
* Correspondence: ychen53@ucmerced.edu; Tel.: +1-209-228-4672; Fax: +1-209-228-4047
† This paper is an extended version of our paper published in An evaluation of ARFIMA programs. In Proceedings of the International Design Engineering Technical Conferences & Computers & Information in Engineering Conference, Cleveland, OH, USA, 6–9 August 2017; American Society of Mechanical Engineers: New York, NY, USA, 2017; In Press.

Academic Editors: Hans J. Haubold and Javier Fernandez
Received: 13 March 2017; Accepted: 14 June 2017; Published: 17 June 2017

Abstract: Strong coupling between values at different times that exhibit properties of long range dependence, non-stationary, spiky signals cannot be processed by the conventional time series analysis. The autoregressive fractional integral moving average (ARFIMA) model, a fractional order signal processing technique, is the generalization of the conventional integer order models—autoregressive integral moving average (ARIMA) and autoregressive moving average (ARMA) model. Therefore, it has much wider applications since it could capture both short-range dependence and long range dependence. For now, several software programs have been developed to deal with ARFIMA processes. However, it is unfortunate to see that using different numerical tools for time series analysis usually gives quite different and sometimes radically different results. Users are often puzzled about which tool is suitable for a specific application. We performed a comprehensive survey and evaluation of available ARFIMA tools in the literature in the hope of benefiting researchers with different academic backgrounds. In this paper, four aspects of ARFIMA programs concerning simulation, fractional order difference filter, estimation and forecast are compared and evaluated, respectively, in various software platforms. Our informative comments can serve as useful selection guidelines.

Keywords: ARFIMA; long range dependence; fractional order; survey

1. Introduction

Humans are obsessed about their future so much that they worry more about their future more than enjoying the present. Time series modelling and analysis are scientific ways to predict the future. When dealing with empirical time series data, it usually comes to the classic book of Box and Jekin's methodology for time series models in the 1970s, in which it introduced the autoregressive integrated moving average (ARIMA) models to forecast and predict the future behavior [1,2]. However, the ARIMA model as well as Poisson processes, Markov processes, autoregressive (AR), moving average (MA), autoregressive moving average (ARMA) and ARIMA processes, can only capture short-range dependence (SRD). They belong to the conventional integer order models [3].

In time series analysis, another traditional assumption is that the coupling between values at different time instants decreases rapidly as the time difference or distance increases. Long-range dependence (LRD), also called long memory or long-range persistence, is a phenomenon that may arise in the analysis of spatial or time series data. LRD was first highlighted in the hydrological data by the British hydrologist H. E. Hurst, and then the other statistics in econometrics, network traffic, linguistics and the Earth sciences, etc. LRD, which is characterized by the Hurst parameter, means that there is a strong coupling effect between values at different time separations. Thus, LRD also indicates that the decay of the autocorrelation function (ACF) is algebraic and slower than exponential decay so that the area under the function curve is infinite. This behavior can be also called inverse power-law delay. Different from the analytical results of linear integer-order differential equations, which are represented by the combination of exponential functions, the analytical results of the linear fractional-order differential equations are represented by the Mittag–Leffler function, which intrinsically exhibits a power-law asymptotic behavior [4–6].

Due to the increasing demand on modeling and analysis of LRD and self-similarity in time series, such as financial data, communications networks traffic data and underwater noise, the fractional order signal processing (FOSP) technique is becoming a booming research area. Moreover, fractional Fourier transform (FrFT), which is the generalization of the fast Fourier transform (FFT), has become one of the most valuable and frequently used techniques in the frequency domain of the fractional order systems [3].

Compared to the conventional integer order models, the ARFIMA model gives a better fit and result when dealing with the data which possess the LRD property. Sun et al. applied the ARFIMA model to analyze the data and predict the future levels of the elevation of Great Salt Lake (GSL) [7]. The results showed that the prediction results have a better performance compared to the conventional ARMA models. Li et al. examined four models for the GSL water level forecasting: ARMA, ARFIMA, autoregressive conditional heteroskedasticity (GARCH) and fractional integral autoregressive conditional heteroskedasticity (FIGARCH). They found that FIGARCH offers the best performance, indicating that conditional heteroscedasticity should be included in time series with high volatility [8]. Sheng and Chen proposed a new ARFIMA model with stable innovations to analyze the GSL data, and predicted the future levels. They also compared accuracy with previously published results [9]. Contreras-Reyes and Palma developed the statistical tools afmtools package in R for analyzing ARFIMA models. In addition, the implemented methods are illustrated with applications to some numerical examples and tree ring data base [10]. Baillie and Chung considered the estimation of both univariate and multivariate trend-stationary ARFIMA models, which generated a long memory autocorrelated process around a deterministic time trend. The model was found to be remarkably successful at representing annual temperature and width of tree ring time series data [11]. OxMetrics is an econometric software including the Ox programming language for econometrics and statistics, developed by Doornik and Hendry. Several papers and manuals are available for the ARFIMA model with OxMetrics [12–14].

Nowadays, there are lots of numerical tools available for the analysis of the ARFIMA processes since these applications are developed by different groups based on different algorithms and definitions of accuracies and procedures. As a consequence, the estimation and prediction results may be different or even conflicting with others. For the scholars or engineers who are going to do the modeling work of the ARFIMA processes, they might get confused as to which tool is more suitable to choose. Thus, we have evaluated techniques concerning the ARFIMA process so as to provide some guidelines when choosing appropriate methods to do the analysis. With this motivation, this paper briefly introduces their usage and algorithms, evaluates the accuracy, compares the performance, and provides informative comments for selection. Through such efforts, it is hoped that informative guidelines are provided to the readers when they face the problem of selecting a numerical tool for a specific application.

For one thing, many publications about fractional systems dynamics use their novel fractional order calculus ideas to represent with encouraging results [15]. However, in reality, when it comes to engineers with zero background, they do not even know which tool to start to use.

The rest of the paper is organized as the follows: Section 2 introduces the basic mathematics of LRD and the ARFIMA model. Section 3 gives a brief review and description on the software commonly used for the analysis of the ARFIMA processes. In Section 3, the quantitative performances of the tools are evaluated and compared in four primary categories—simulation, processing, estimation and prediction in the ARFIMA process. Conclusions are given in Sections 4 and 5.

2. LRD and ARFIMA Model

When the hydrologist H.E. Hurst spent many years analyzing the records of elevation of the Nile river in the 1950s, he found a strange phenomena: the long-range recording of the elevation of the Nile river has much stronger coupling, and the autocorrelation function (ACF) decays slower than exponentially [16]. In order to quantify the level of coupling, the rescaled range (R/S) analysis method was provided to estimate the coupling level, which is now called the Hurst parameter. Furthermore, many valuable Hurst parameter estimators were provided to more accurately characterize the LRD time series [17]. Since then, the LRD or long memory phenomenon has attracted numerous research studies. Based on Hurst's analysis, more suitable models, such as ARFIMA and fractional integral generalized autoregressive conditional heteroscedasticity (FIGARCH) were built to accurately analyze LRD processes.

The rescaled range (R/S) method is one of the time-domain analysis of Hurst parameter defined as follows [16]:

$$E[\frac{R(n)}{S(n)}]_{n\to\infty} = Cn^H, \tag{1}$$

where $E(\cdot)$ denotes the expected value of the observations, $R(n)$ is the range of the first n values, $S(n)$ is their standard deviation, and C is a constant. Whittle's Maximum Likelihood Estimator (MLE) and wavelet analysis using periodogram based analysis in the frequency domain [18].

Autocorrelation function (ACF) analysis is one of the useful techniques for identifying trends and periodicities in the data, in a manner that is often more precise than can be obtained with simple visual inspections. In addition, LRD or long memory property can be defined by ACF.

Let $\{X(t); t \in (-\infty, +\infty)\}$ and the ACF $\rho(k)$ is defined as:

$$\rho(k) = \frac{Cov(X_t, X_{t-k})}{Var(X_t)}, \tag{2}$$

where $Cov(\cdot)$ is the covariance and $Var(\cdot)$ is the variance.

A stationary time series defined over $t = 0, 1, 2, 3\cdots$ is said to be long memory if $\sum_{k=0}^{\infty} |\rho_{(k)}|$ diverges, where $\rho_{(k)}$ is the ACF of the process. Otherwise, the time series is said to be short memory or SRD. Another definition of long memory if for some frequency, $f \in [0, 0.5]$, the power spectrum $P(f)$, becomes unbounded.

The power spectrum $P(f)$ is defined by:

$$P(f) = \int_{-\infty}^{\infty} e^{-2\pi i f k}\rho(k)dk, \tag{3}$$

where $-\infty < f < \infty$, $i = \sqrt{-1}$ and $\rho(k)$ is the ACF.

The spectral density $S(f)$ is a normalized form of $P(f)$, defined by:

$$S(f) = \frac{P(f)}{\sigma^2} = \int_{-\infty}^{\infty} e^{-2\pi i f k}\rho(k)dk. \tag{4}$$

If the spectrum becomes unbounded, then the ACF are not absolutely summable [19]. Therefore, ACF is defined as time domain analysis, while power spectrum density (PSD) is used for the frequency domain analysis.

The ACF of the stationary SRD stochastic models, such as the ARMA processes and Markov processes, is absolutely summable, while the correlations function ρ_k is not absolutely summable for the processes with long-range dependence [19]. Signals with long-range correlations, which are characterized by inverse power-law decaying autocorrelation function, occur ubiquitously in nature and many man-made systems. Because of the strong coupling and the slow decaying autocorrelation, these processes are also said to be long memory processes. Typical examples of LRD signals include financial time series, underwater noise, electroencephalography (EEG) signal, etc. The level of the dependence or coupling of LRD processes can be indicated or measured by the estimated Hurst parameter, or the Hurst exponent [16]. The value of the Hurst Exponent varies between 0 and 1. If $H = 0.5$, the time series has no statistical dependence. If $H < 0.5$, the time series is a negatively correlated process or an anti-persistent process. If $H > 0.5$, the time series is a positively correlated process [20]. The LRD processes are also closely related to fractional calculus. In order to capture the property of coupling or hyperbolic decaying autocorrelation, fractional calculus based LRD models have been suggested, such as ARFIMA and FIGARCH models [21,22]. The ARFIMA model is a generalization of ARMA model, which is a typical fractional order system.

2.1. Autoregressive (AR) Model

The notation AR(p) refers to the autoregressive model of order p. The AR(p) model is written as [2]:

$$X_t = c + \sum_{i=1}^{p} \phi_i X_{t-i} + \varepsilon_t, \tag{5}$$

where $\phi_1, \cdots, \phi_p$ are autoregressive parameters, c is a constant, and the random variable ε_t is the white noise. Some constraints are necessary on the values of the parameters so that the model remains stationary. For example, processes in the AR(1) model with $|\phi_1| \geq 1$ are not stationary. In statistics and signal processing, an autoregressive (AR) model is a representation of a type of random process; as such, it describes certain time-varying processes in nature, economics, etc.

2.2. Moving Average (MA) Model

The notation MA(q) refers to the moving average model of order q [2]:

$$X_t = \mu + \sum_{i=1}^{q} \theta_i \varepsilon_{t-i} + \varepsilon_t, \tag{6}$$

where the $\theta_1, \cdots, \theta_q$ are the moving average parameters of the model, μ is the expectation of X_t (often assumed to equal 0), and the ϵ_t , ϵ_{t-1} ,... are again, white noise error terms. The moving average (MA) smooths a time series, which can produce cyclic and a trend like plots even when the original data are themselves independent random events with fixed mean. This characteristic lessens its usefulness as a control mechanism.

2.3. ARIMA and ARFIMA Model

The above AR and MA models can be generalized as follows [2]:

$$(1 - \sum_{i=1}^{p} \phi_i B^i)(1 - B)^d (X_t - \mu) = (1 + \sum_{i=1}^{q} \theta_i B^i)\epsilon_t. \tag{7}$$

The above $(1 - B)^d$ is called a difference operator ∇^d. The ARMA or ARIMA models can only capture the SRD property, since d is confined in the range of integer order. Therefore, in order to capture the LRD property of the fractional systems, the ARFIMA(p,d,q) model is thereby proposed accordingly. In fact, the operator can be defined in a natural way by using binomial expansion for any real number d with Gamma function:

$$(1 - B)^d = \sum_{k=0}^{\infty} \binom{d}{k} (-B)^k = \sum_{k=0}^{\infty} \frac{\Gamma(d+1)}{\Gamma(k+1)\Gamma(d+1-k)} (-B)^k. \tag{8}$$

Many authors suggested the use of the fractionally ARIMA model by using a fractional difference operator rather than an integer one could better take into account this phenomenon of LRD [23]. Hosking et al. defined an extension of the ARIMA model, which allows for the possibility of stationary long-memory models [24]. Thus, the general form of ARIMA(p,q,d) process X_t in Equation (7)—the ARFIMA(p,d,q) process is defined as:

$$\Phi(B)(1-B)^d X_t = \Theta(B)\varepsilon_t, \tag{9}$$

where $d \in (-0.5, 0.5)$, and $(1-B)^d$ is defined as the fractional difference operator in Equation (8). ARFIMA(p,d,q) processes are widely used in modeling LRD time series, where p is the autoregressive order, q is the moving average order and d is the level of differencing [25]. The larger the value of d, the more closely it approximates a simple integrated series, and it may approximate a general integrated series better than a mixed fractional difference and ARMA model.

Figure 1 presents the discrete ARFIMA process that can be described as the output of the fractional-order system driven by a discrete white Gaussian noise (wGn). The ARFIMA(p,d,q) process is the natural generalization of the standard ARIMA or ARMA processes. In a fractionally differenced model, the difference coefficient d is a parameter to be estimated first [26]. The intensity of self-similar of ARFIMA is measured by a parameter d. For the finite variance process with fractional Gaussian noise, d has a closed relation with Hurst parameter H [3,26,27]:

$$d = H - 1/2. \tag{10}$$

In addition, for the infinite variance process with fractional α-stable noise, d is related with Hurst and characteristic exponent α [18,22]:

$$d = H - 1/\alpha. \tag{11}$$

In this way, the parameter d may be chosen to model long-time effects, whereas p and q may be selected to model relatively short-time effects.

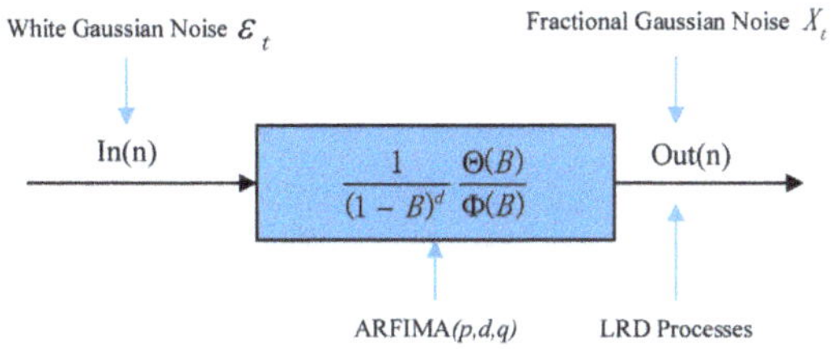

Figure 1. ARFIMA model.

3. Review and Evaluation

ARFIMA(p,d,q) processes are widely used in modeling LRD time series, especially for the high frequency trading data, network traffic and hydrology dataset, etc. In practice, several time series exhibit LRD in their observations, leading to the development of a number of estimation and prediction methodologies to account for the slowly decaying autocorrelations. The ARFIMA process is one of the best-known classes of long-memory models. As introduced in Section 1, most statistical analysis programs have been embedded with ARFIMA models. A summary of the current software dealing with ARFIMA model analysis is as follows:

1. MATLAB applications
   MATLAB® (Matrix Laboratory) is a multi-paradigm numerical computing environment and

fourth-generation programming language developed by MathWorks (Natick, MA 01760-2098, USA). The MATLAB applications are interactive applications written to perform technical computing tasks with the MATLAB scripting language from MATLAB File Exchange, through additional MATLAB products, and by users.

2. SAS software
   SAS (Statistical Analysis System) is a software suite developed by SAS Institute (Cary, NC 27513-2414, USA) for advanced analytics, multivariate analyses, business intelligence, data management, and predictive analytics.
3. R packages
   R packages and projects are contributed by RStudio (Boston, MA 02210, USA) team on CRAN (Comprehensive R Archive Network). R users are doing some of the most innovative and important work in science, education, and industry. It is a daily inspiration and challenge to keep up with the community and all it is accomplishing.
4. OxMetrics
   Ox^{TM} is an object-oriented matrix language with a comprehensive mathematical and statistical function library developed by Timberlake Consultants Limited (Richmond, Surrey TW9 3GA, UK). Many packages were written for Ox including software mainly for econometric modelling. The Ox packages for time series analysis and forecasting.

MATLAB codes are open-source applications where we could download, view and revise the codes if possible while other three are packaged and embedded in the software modules. In the following evaluation parts, we could clearly see the differences between them even with the same inputs. Four primary embedded functions concerning simulation, fractional difference filter, parameter estimation and forecast, are tested and evaluated for the ARFIMA processes in Table 1. It should be noted that the first two functions can be regarded as the forward problem solving systems, while the latter two are developed for the backward problem solving systems which are much more significant. In view of the above, this section can be divided into four parts.

Table 1. Numerical tools for the ARFIMA process.

Procedures	MATLAB	R	SAS	OxMetrics
Simulation	✔*	✔	✔	✗
Fractional Difference	✔	✔	✔	✔
Parameter Estimation	✔	✔	✔	✔
Forecast	✔	✔	✗	✔

✔* means it can simulate ARFIMA processes, but cannot choose or define the initial seeds.

3.1. Simulation

On the website of MATLAB Central, there are two files that can simulate ARFIMA processes. They are developed by Fatichi [28] and Caballero [29]. However, users cannot choose initial random seeds, that is, it can only simulate one certain series of ARFIMA process. The ARFIMA(p,d,q) estimator is developed by Inzelt, which is used for a linear stationary ARFIMA(p,d,q) process [30].

R is a freely available language and environment for statistical computing and graphics, which provides a wide variety of statistical and graphical techniques: linear and nonlinear modelling, statistical tests, time series analysis, classification, clustering, etc. Like the MATLAB Central, CRAN is a platform that stores identical, up-to-date, versions of code and documentation for R. There are several major packages concerning ARFIMA process according to the authors' survey in Table 2.

Table 2. Comparison of ARFIMA packages in R.

Package	Author	Release Date	Typical Functions	Requirements
fractal	William Constantine et al. [31]	2016-05-21	hurstSpec	R (≥ 3.0.2)
fracdiff	Martin Maechler et al. [32]	2012-12-02	fracdiff	longmemo, urca R (≥ 2.6.0), polynom
afmtools	Javier E. Contreras-Reyes et al. [33]	2012-12-28	arfima.whittle	fracdiff, hypergeo, sandwich, longmemo
ArfimaMLM	Patrick Kraft et al. [34]	2015-01-21	arfimaMLM	R (≥ 3.0.0), fractal
arfima	Justin Q. Veenstra et al. [35]	2015-12-31	arfima	R (≥ 2.14.0), ltsa

The first two packages are used for the processing of ARFIMA processes, including Hurst fitting, calculation and fractional order difference and so on, while the latter two are mainly used for the parameter estimation of ARFIMA. The last package arfima is the most comprehensive tool that could simulate, estimate and predict the results of ARFIMA processes. In the paper, we use the last one package to compare with the other software.

SAS and R could also generate the ARFIMA process by defining the order of AR(p) and MA(q), setting the parameters ϕ, θ and d, respectively. In addition, the number of the initial random seeds could/should be set for the stochastic process. Random seeds are defined by the internal algorithms, which make the initial stochastic process a difference. Therefore, it may be a big difference if picking arbitrary seeds. In order to illustrate the above problems, we have generated the ARFIMA$(1, 0.4, 1)$ process with $d = 0.4$, $\phi = 0.5$, $\theta = -0.1$ and $\sigma = 1$. Then, we set 100 different initial random seeds with 3000 observations and do the same estimation. It should be kept in mind that, even with the same simulation software that generates the processes, the estimation results could be a big difference in Figure 2. However, from the perspective of the sample-path analysis for the stochastic processes, this could be the advantage compared to the MATLAB ARFIMA applications, which can only generate one certain series (path). Furthermore, we have also found that the SAS software is somewhat better or "conservative", while R software is more "aggressive" in Figure 2. We could check the comparisons below with dashed lines showing true values of parameters.

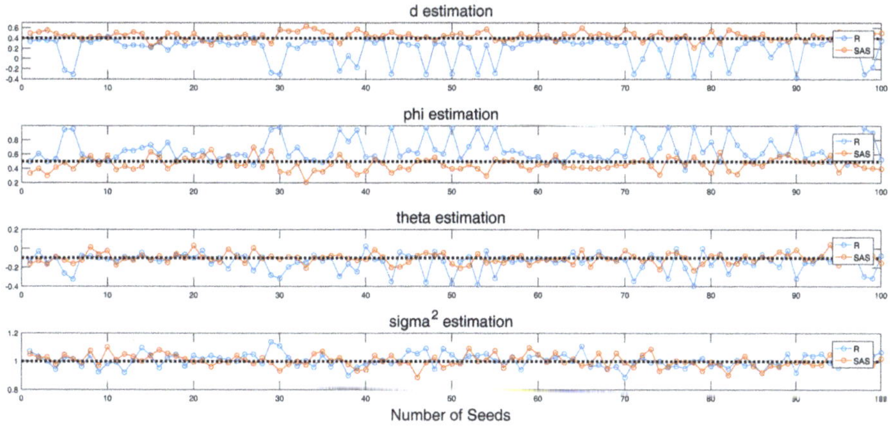

Figure 2. Estimation results of SAS and R.

It should be also noted that, even with a certain series of the initial starting random seeds, the estimation results could also have quite a few variations. For example, we have set the fractional order d from 0 to 0.5, and do the simulation and estimation accordingly in MATLAB. It can be seen that the estimation $\hat{d}$ is jumping up and down around the true values (red line) in Figure 3.

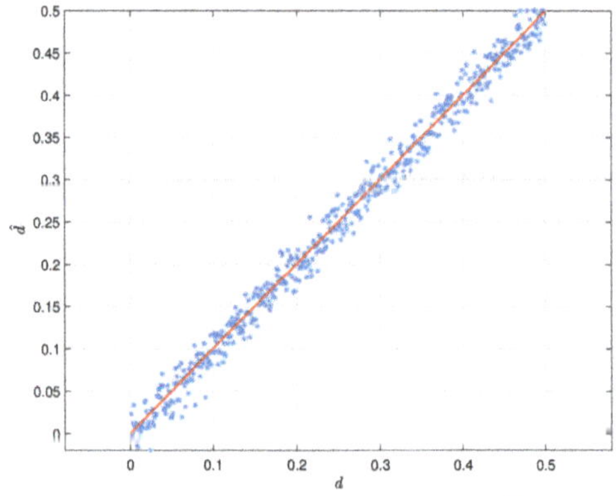

Figure 3. Comparison of $\hat{d}$ and d.

Here are some comments of this subsection:

1. Estimation results also depend on the initial random seeds, even the series that are from their own simulations.
2. The test results may be different if not enough points/observations are generated. More than 300 points are preferred.
3. Estimation results may not be accurate if they only use one method. R should be more desirable to try first.

3.2. *Fractional Order Difference Filter*

Many time series signals contain trends, i.e., they are non-stationary. It is usually preferable to specify and remove the trends explicitly to get the smoothed or stationary data for the further analysis and modeling. According to the theory of Box–Jekins, an ARIMA model can be viewed as a "filter" that tries to separate the signal from the noise, and the signal is then extrapolated into the future to obtain forecasts [2]. Since the beginning of the 1980s, the long memory ARFIMA model has been introduced and investigated by many scholars especially for the parameter estimation problems. Shumway and Stoffer gave a brief overview of "long memory ARMA" models in [36]. This type of long memory model might be considered to use when the ACF of the series tapers slowly to 0 and spectral densities are unbounded at $f = 0$. Jensen et al. derived an algorithm for the calculation of fractional differences based on circular convolutions method in [37]. In fact, there are a lot of estimation methods concerning fractional difference algorithms.

In some instances, however, we may see a persistent pattern of non-zero correlations that begins with a first lag correlation that is not close to 1. In these cases, models that incorporate "fractional differencing" may be useful. Therefore, differencing the time series data by using the approximated binomial expression of the long-memory filter is a prerequisite to estimates of the memory parameter in the ARFIMA(p, d, q) model. The user should not only set numeric vector of p and q, but also specify the order of the fractional difference filter. By passing through fractional order difference filter, the ARFIMA series will yield residuals that are uncorrelated and normally distributed with constant variance in Figure 4.

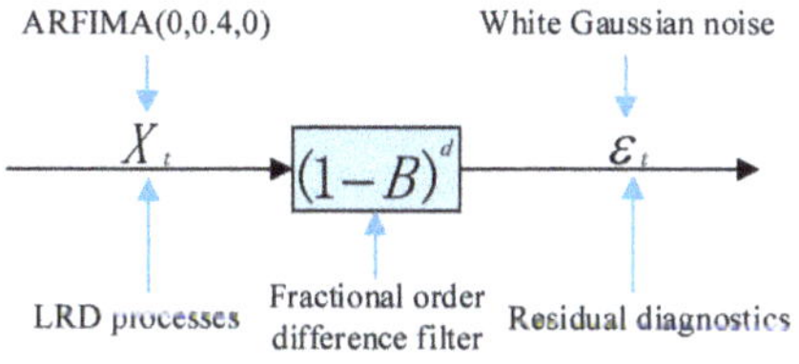

Figure 4. Fractional order difference of the ARFIMA(0, 0.4, 0) process.

The sample ACF and partial autocorrelation function (PACF) are useful qualitative tools to assess the presence of autocorrelation at individual lags. The Ljung-Box Q-test is a more quantitative way to test for autocorrelation at multiple lags jointly [38]. The Ljung-Box test statistic is given by:

$$Q(L) = N(N+2)\sum_{k=1}^{L}\frac{\hat{\rho}_k^2}{N-k}, \tag{12}$$

where N is the sample size, L is the number of autocorrelation lags, and $\rho(k)$ is the sample autocorrelation at lag k. Under the null hypothesis, the asymptotic distribution of Q is chi-square with L degrees of freedom. If we use lbqtest function in the MATLAB Econometrics Toolbox, it returns the rejection decision and p-value for the hypothesis test. Similar functions Box test of stats and ljung.wge of tswge are also available in the R package. p-values indicate the strength at which the test rejects the null hypothesis. If all of the p-values are larger than 0.01, there is strong evidence to accept the hypothesis that the residuals are not autocorrelated.

Thus, we have generated an ARFIMA(0,0.4,0) process in Figure 5 and use fractional order difference filter with the order $d = 0.4$ to filter the LRD property in Figure 6. It is obvious that, by passing through the fractional order difference filter, the slowly decaying property of LRD has been eliminated.

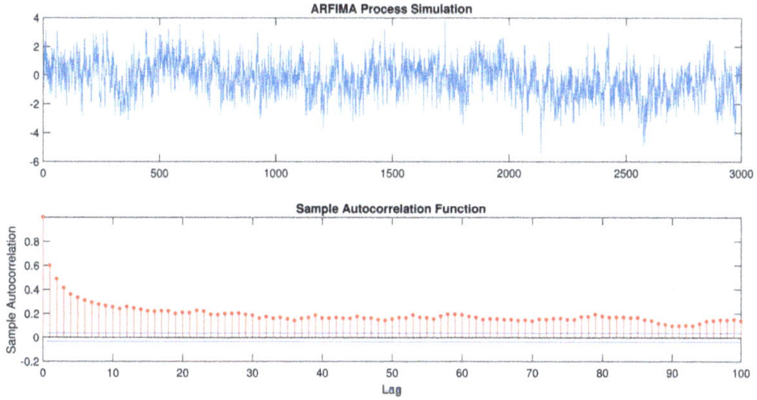

Figure 5. Simulation of ARFIMA(0, 0.4, 0) process.

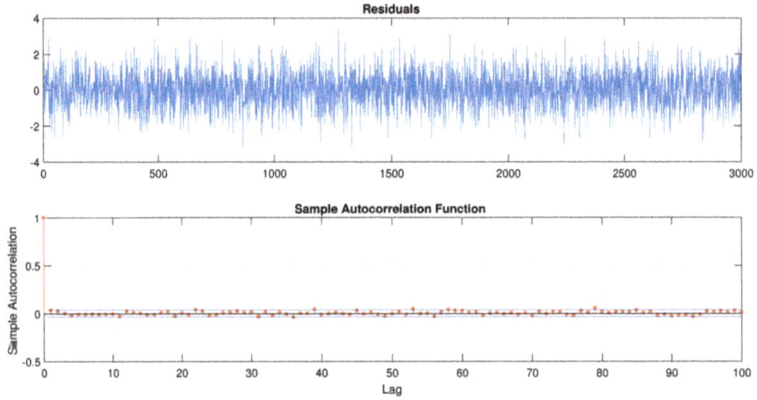

Figure 6. ARFIMA(0,0.4,0) process passing through the fractional difference filter.

In order to evaluate residuals, p-values are used to quantify the goodness of fitting in Table 3.

Table 3. Fractional filters.

Software	MATLAB	R	SAS	Ox
Function	d_filter	diffseries	fdif	fracdiff
p-values with 1,5,10,15 lags	0.0710 0.2253 0.5850 0.5330	0.09998 0.2395 0.5320 0.4571	0.1062 0.2414 0.5198 0.4473	0.0862 0.2114 0.5898 0.5473

Here are some comments of this subsection: all of the four programs above have fractional order operators to filter the LRD process successfully. In general, d is the parameter to be estimated first [26]. If we use the calculation defined by the Hurst method in Equations (10) and (11), d could probably be the fractional one. Therefore, the fractional order filter would be the primary tool to eliminate the LRD property or the heavy-tailedness in order to get the stationary series.

Meanwhile, however, the fractional order d is closely related to the Hurst parameter in Equations (10) and (11). There are more than ten methods to estimate Hurst parameters, R/S method, aggregated variance method, absolute value method, periodogram method, whittle method, Higuchi's method, etc. These methods are mainly useful as simple diagnostic tools for LRD time series. These Hurst estimators have been introduced to analyze the LRD time series in [17,39,40]. Therefore, the results of Hurst estimators can be different if applying different methods. In addition, from Equation (8), it is interesting to note that there are infinite factorial series in the expansion of binomial expansion. In practice, we usually take the first three factorials for approximation. That is to say, the accuracy of differencing is also determined by how many factorials are used for approximation. Consequently, these different methods make the subsequent estimations differ from each other in the following sections.

3.3. Parameter Estimation

From Figure 2, we could see that, even though R and SAS can both simulate the ARFIMA processes, the properties of these processes are not the same mainly because of the distinctive random seeds defined by different software. Therefore, when reviewing and evaluating the accuracy of above software, the same ARFIMA series should be guaranteed first. Herein, we have proposed to use the following steps to compare the results in Figure 7. In addition, OxMetrics is the software that cannot generate ARFIMA simulation, but it can estimate and forecast ARFIMA-FIGARCH processes. MATLAB cannot generate multiple ARFIMA series for the same parameter combinations. We have thus used R and SAS to provide the ARFIMA series for the inputs of estimations.

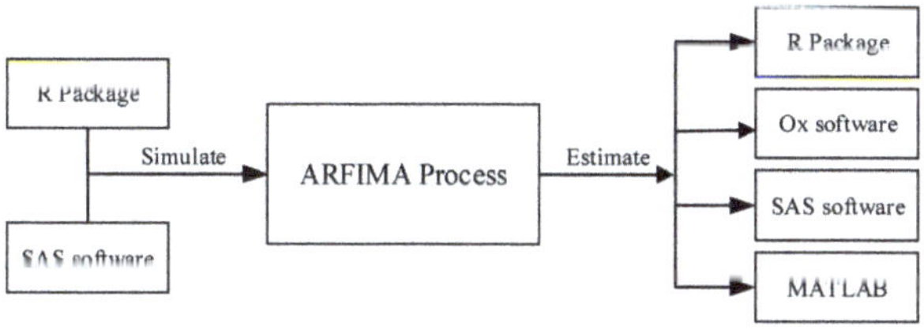

Figure 7. Simulation and estimation of the ARFIMA process.

Since we have received simulation results, the parameters of ARFIMA processes can be estimated and compared with true values (parameter setting values). First, we have used the simulation data from R software and have then used these three programs to do the estimation in Figure 8. Second, we have used SAS to do the same simulation and have then used the other three to do the estimation in Figure 9. Without loss of generality, we pick 10 groups of 3000 observations to see who could capture the accuracy.

Here are some comments from this subsection: from the above plots, it is very interesting to find that the estimation results of ARFIMA simulations are relatively accurate when they come from the same simulation data set. However, OxMetrics and MATLAB estimate the negative values of θ.

In order to further test whether the Ox and MATLAB can only return the negative values of θ or if they just return the inverse values. The parameters are set to the inverse values, accordingly, with $\phi = -0.5$ and $\theta = 0.2$; while, in the previous test, they are $\phi = 0.5$ and $\theta = -0.1$, respectively. The result presented in Figure 10 validates the comments above.

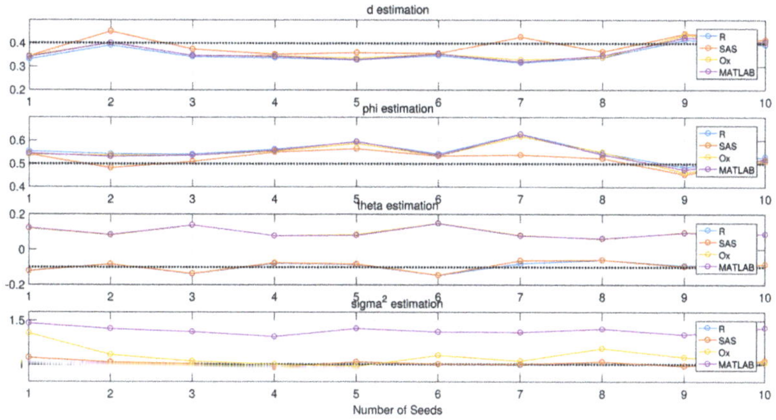

Figure 8. Parameter estimation with different methods (R simulation inputs).

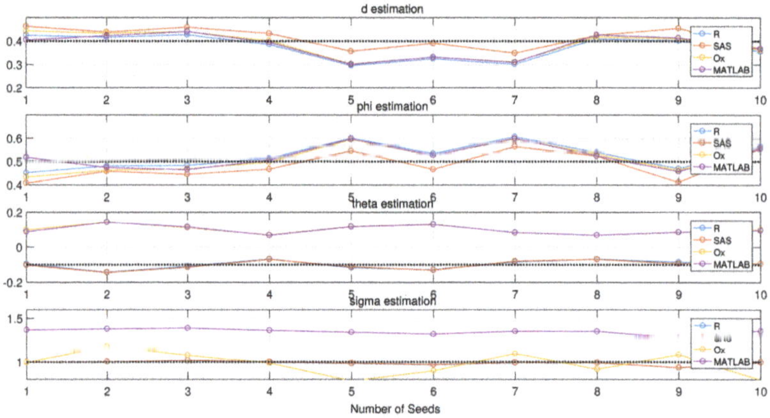

Figure 9. Parameter estimation with different methods (SAS simulation inputs).

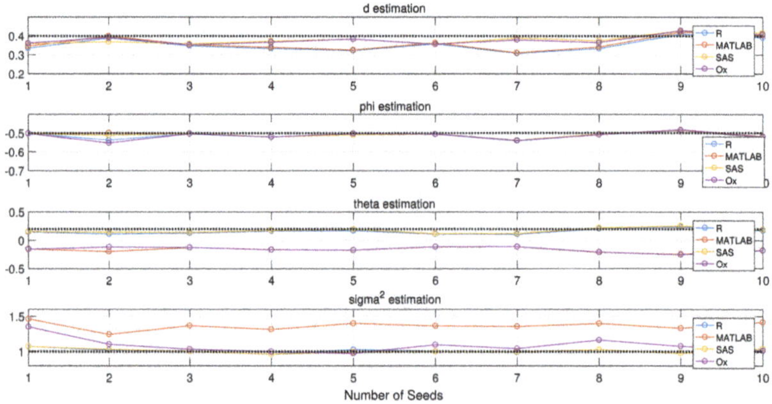

Figure 10. Parameter estimation with different methods (R simulation inputs).

3.4. Forecast

Simulation data could only be the auxiliary part of these software, since it can never be a powerful and useful tool for the ARFIMA process analysis if it cannot retrieve the estimation parameters from the real data with LRD. Moreover, the last and the most significant part of the ARFIMA process is to forecast and thereby predict the future behavior. Therefore, mean absolute percentage error (MAPE) values are used for the evaluation of the forecast results for the data from real life. The error square of the prediction results from different methods with the increasing number of predictions are illustrated in Figure 11:

$$\sigma_t^2 = \sum_{t=1}^{n} (\hat{y}_t - y_t)^2. \tag{13}$$

Data description: Centered annual pinus longaeva tree ring width measurements at Mammoth Creek, Utah, USA from 0 A.D (anno Domini) to 1989 A.D with 1990 sampling points in time series [41,42]. The data can be divided into two parts: the first part with 1900 observations are used to estimate ARFIMA parameters, and the second part with 90 points are used for comparison with the prediction results from the fitted ARFIMA models. Finally, the results with the implemented methods that are applied to real-life time series are summarized in Table 4.

Here are some comments of this subsection:

1. d is the parameter to be estimated first when doing ARFIMA model fitting. Therefore, if the estimation of d is different for a certain time series, the following estimations for AR(Φ) and MA(Θ) will be different.
2. The ideal length (horizon) of predictions is within 30 steps. With the increasing steps of forecast, prediction errors are adding up. If a long range prediction series is required, R and MATLAB should be priorities for their smaller prediction errors.
3. Compared with other forecast results with true values in Table 4, R produces the minimum prediction errors and MAPE.

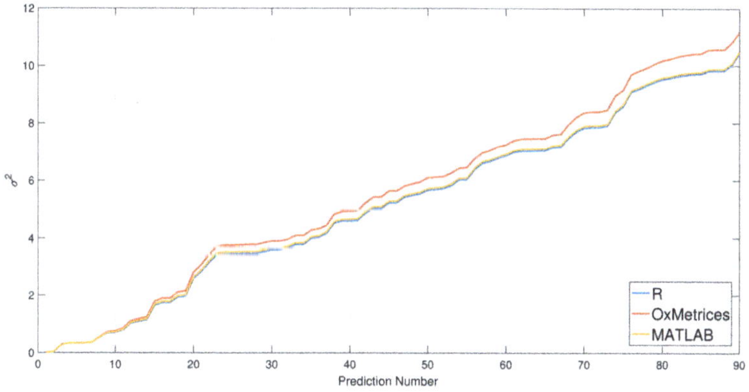

Figure 11. Prediction comparison with different methods.

Table 4. Parameter estimations and forecast comparisons.

Number	Parameters	R	SAS	OxMetrices	MATLAB
1	mu	0.9833	N/A (Not Applicable)	0.98799	0.9878
2	d	0.1670	0.1479624	0.282087	0.2313
3	ar	0.9070119	0.8939677	−0.254265	0.6473
4	ma	0.8603811	0.8318787	0.18698	0.6393
5	sigma	0.1078173	0.1073417	0.1246	0.1163
6	p value Lag1	0.9195	N/A	0.7709458	0.9101
7	p value Lag5	0.6369	N/A	0.341324	0.6959
8	p value Lag10	0.8659	N/A	0.4367925	0.9037
9	p value Lag15	0.6491	N/A	0.6229542	0.6776
10	LogLikelihood	2117.224	1851.5512	−570.599	1162.527
11	MAPE	28.95	N/A	29.36	29.02

4. Summary of Selection Guidelines

Qualitative analysis as well as quantitative evaluations of the selected ARFIMA tools have been conducted in the previous sections. In order to make it easier for researchers from different backgrounds, we summarize the selection guidelines for the ARFIMA process modeling and analysis.

1. R and SAS software are priorities for the simulation of ARFIMA process, since they could define the initial seeds. R is one of the desirable tools for the estimation of ARFIMA process, since it has more than five packages including Hurst estimators, ACF plot, Quantile-Quantile (QQ) plot, white noise test and some LRD examples.

2. Estimation results of the ARFIMA process may be different if the number of observations is not large enough. Therefore, more than one estimation method should be used in order to guarantee the accuracy.
3. d is the parameter to be estimated first. All of this software could use fractional difference functions to filter the trend and thereafter stationarize time series data.
4. The ideal length (horizon) of predictions is within 30 steps. If a long range prediction series is required, R and MATLAB are the priorities for their smaller prediction errors.

5. Conclusions

Compared to the conventional integer order models that can only capture SRD, the ARFIMA model gives a better fitting result, especially for the data with the LRD property. Nowadays, some programs have been integrated with ARFIMA solutions. However, the final results of estimation and prediction could be different or even conflicting if choosing different methods. Therefore, a comprehensive review and evaluation of the numerical tools for the ARFIMA process is presented in the paper so as to provide some guidelines when choosing appropriate methods to do the time series analysis of LRD data. Through such efforts, it is hoped that an informative guidance is provided to the readers when they face the problem of selecting a numerical tool for a specific application.

Supplementary Materials: The MATLAB code, SAS code and R code for Sections 3.1–3.4, and Ox code are available online at www.mdpi.com/2075-1680/6/2/16/s1.

Acknowledgments: We would like to thank the anonymous reviewers for their insightful comments on the paper.

Author Contributions: YangQuan Chen designed the experiment and Xi Zhang analyzed the data. Kai Liu performed the experiment and wrote the paper.

Conflicts of Interest: The authors declare no conflict of interest.

References

1. Box, G.E.; Hunter, W.G.; Hunter, J.S. *Statistics for Experimenters: An Introduction to Design, Data Analysis, and Model Building*; John Wiley & Sons: Hoboken, NJ, USA, 1978; Volume 1.
2. Box, G.E.; Jenkins, G.M.; Reinsel, G.C.; Ljung, G.M. *Time Series Analysis: Forecasting and Control*; John Wiley & Sons: Hoboken, NJ, USA, 2015.
3. Sheng, H.; Chen, Y.; Qiu, T. *Fractional Processes and Fractional-Order Signal Processing: Techniques and Applications*; Springer Science & Business Media: New York, NY, USA, 2011.
4. Mathai, A.M.; Saxena, R.K. *The H-Function with Applications in Statistics and Other Disciplines*; John Wiley & Sons: Hoboken, NJ, USA, 1978.
5. Saxena, R.; Mathai, A.; Haubold, H. On fractional kinetic equations. *Astrophys. Space Sci.* **2002**, *282*, 281–287.
6. Saxena, R.; Mathai, A.; Haubold, H. On generalized fractional kinetic equations. *Phys. A Stat. Mech. Its Appl.* **2004**, *344*, 657–664.
7. Sun, R.; Chen, Y.; Li, Q. Modeling and prediction of Great Salt Lake elevation time series based on ARFIMA. In Proceedings of the International Design Engineering Technical Conferences and Computers and Information in Engineering Conference, Las Vegas, NV, USA, 4–7 September 2007; American Society of Mechanical Engineers: New York, NY, USA , 2007; pp. 1349–1359.
8. Li, Q.; Tricaud, C.; Sun, R.; Chen, Y. Great Salt Lake surface level forecasting using FIGARCH model. In Proceedings of the International Design Engineering Technical Conferences and Computers and Information in Engineering Conference, Las Vegas, NV, USA, 4–7 September 2007; American Society of Mechanical Engineers: New York, NY, USA, 2007; pp. 1361–1370.
9. Sheng, H.; Chen, Y. The modeling of Great Salt Lake elevation time series based on ARFIMA with stable innovations. In Proceedings of the International Design Engineering Technical Conferences and Computers and Information in Engineering Conference, San Diego, CA, USA, 30 August–2 September 2009; American Society of Mechanical Engineers: New York, NY, USA, 2009; pp. 1137–1145.
10. Contreras-Reyes, J.E.; Palma, W. Statistical analysis of autoregressive fractionally integrated moving average models in R. *Comput. Stat.* **2013**, *28*, 2309–2331.

11. Baillie, R.T.; Chung, S. Modeling and forecasting from trend-stationary long memory models with applications to climatology. *Int. J. Forecast.* **2002**, *18*, 215–226.
12. Doornik, J.A.; Ooms, M. Computational aspects of maximum likelihood estimation of autoregressive fractionally integrated moving average models. *Comput. Stat. Data Anal.* **2003**, *42*, 333–348.
13. Doornik, J.A.; Ooms, M. *A package for estimating, forecasting and simulating ARFIMA models: Arfima package 1.0 for Ox.* Erasmus University: Rotterdam, The Netherlands, 1999.
14. Doornik, J.A.; Ooms, M. Inference and forecasting for ARFIMA models with an application to US and UK inflation. *Stud. Nonlinear Dyn. Econ.* **2004**, *8*, 1208–1218.
15. Burnecki, K. *Identification, Validation and Prediction of Fractional Dynamical Systems*; Oficyna Wydawnicza Politechniki Wrocławskiej: Wroclaw, Poland, 2012.
16. Hurst, H.E. Long-term storage capacity of reservoirs. *Trans. Am. Soc. Civ. Eng.* **1951**, *116*, 770–808.
17. Ye, X.; Xia, X.; Zhang, J.; Chen, Y. Effects of trends and seasonalities on robustness of the Hurst parameter estimators. *IET Signal Process.* **2012**, *6*, 849–856.
18. Samorodnitsky, G.; Taqqu, M.S. *Stable Non-Gaussian Random Processes: Stochastic Models with Infinite Variance*; CRC Press: Boca Raton, FL, USA, 1994; Volume 1.
19. Woodward, W.A.; Gray, H.L.; Elliott, A.C. *Applied Time Series Analysis with R*, 2nd ed.; CRC Press: Boca Raton, FL, USA, 2016.
20. Kale, M.; Butar, F.B. Fractal analysis of time series and distribution properties of Hurst exponent. *J. Math. Sci. Math. Educ.* **2011**, *5*, 8–19.
21. Karasaridis, A.; Hatzinakos, D. Network heavy traffic modeling using α-stable self-similar processes. *IEEE Trans. Commun.* **2001**, *49*, 1203–1214.
22. Sheng, H.; Chen, Y. FARIMA with stable innovations model of Great Salt Lake elevation time series. *Signal Process.* **2011**, *91*, 553–561.
23. Granger, C.W.; Joyeux, R. An introduction to long-memory time series models and fractional differencing. *J. Time Ser. Anal.* **1980**, *1*, 15–29.
24. Hosking, J.R. Fractional differencing. *Biometrika* **1981**, *68*, 165–176.
25. Brockwell, P.J.; Davis, R.A. *Time Series: Theory and Methods*; Springer Science & Business Media: New York, NY, USA, 2013.
26. Reisen, V.; Abraham, B.; Lopes, S. Estimation of parameters in ARFIMA processes: A simulation study. *Commun. Stat.-Simul. Comput.* **2001**, *30*, 787–803.
27. Reisen, V.A. Estimation of the fractional difference parameter in the ARIMA (p, d, q) model using the smoothed periodogram. *J. Time Ser. Anal.* **1994**, *15*, 335–350.
28. Fatichi, S. ARFIMA Simulations. Available online: https://www.mathworks.com/matlabcentral/fileexchange/25611-arfima-simulations/content/ARFIMA_SIM.m (accessed on 15 June 2017).
29. Caballero, C.V.R. ARFIMA(p, d, q). Available online: https://www.mathworks.com/matlabcentral/fileexchange/53301-arfima-p-d-q-/content/dgp_arfima.m (accessed on 15 June 2017).
30. Inzelt, G. ARFIMA(p, d, q) Estimator. Available online: https://www.mathworks.com/matlabcentral/fileexchange/30238-arfima-p-d-q--estimator (accessed on 15 June 2017).
31. Constantine, W.; Percival, D.; Constantine, M.W.; Percival, D.B. The Fractal Package for R. Available online: https://cran.r-project.org/web/packages/fractal/fractal.png (accessed on 15 June 2017).
32. Maechler, M.; Fraley, C.; Leisch, F. The Fracdiff Package for R. Available online: https://cran.r-project.org/web/packages/fracdiff/fracdiff.png (accessed on 15 June 2017).
33. Contreras-Reyes, J.E.; Goerg, G.M.; Palma, W. The Afmtools Package for R. Available online: http://www2.uaem.mx/r-mirror/web/packages/afmtools/afmtools.png (accessed on 15 June 2017).
34. Kraft, P.; Weber, C.; Lebo, M. The ArfimaMLM Package for R. Available online: https://cran.r-project.org/web/packages/ArfimaMLM/ArfimaMLM.png (accessed on 15 June 2017).
35. Veenstra, J.Q.; McLeod, A. The Arfima Package for R. Available online: https://cran.r-project.org/web/packages/arfima/arfima.png (accessed on 15 June 2017).
36. Shumway, R.H.; Stoffer, D.S. *Time Series Analysis and Its Applications: With R Examples*; Springer Science & Business Media: New York, NY, USA, 2010.
37. Jensen, A.N.; Nielsen, M.Ø. A fast fractional difference algorithm. *J. Time Ser. Anal.* **2014**, *35*, 428–436.
38. Ljung, G.M.; Box, G.E. On a measure of lack of fit in time series models. *Biometrika* **1978**, *65*, 297–303.
39. Sheng, H.; Chen, Y.; Qiu, T. On the robustness of Hurst estimators. *IET Signal Process.* **2011**, *5*, 209–225.

40. Chen, Y.; Sun, R.; Zhou, A. An improved Hurst parameter estimator based on fractional Fourier transform. In Proceedings of the International Design Engineering Technical Conferences and Computers and Information in Engineering Conference, Las Vegas, NV, USA, 4–7 September 2007; American Society of Mechanical Engineers: New York, NY, USA, 2007; pp. 1223–1233.
41. Palma, W. *Long-Memory Time Series: Theory and Methods*; John Wiley & Sons: Hoboken, NJ, USA, 2007; Volume 662.
42. Palma, W.; Olea, R. An efficient estimator for locally stationary Gaussian long-memory processes. *Ann. Stat.* **2010**, *38*, 2958–2997.

axioms

MDPI

Article

Applications of Skew Models Using Generalized Logistic Distribution

Pushpa Narayan Rathie [1,*,†], Paulo Silva [2,†] and Gabriela Olinto [3,†]

1 Department of Statistics and Applied Mathematics, Federal University of Ceara, Fortaleza, CE 60020-181, Brazil
2 DICRE, Credit Risk Management, Bank of Brazil—BB, Brasilia 70073-901, Brazil; paulodourado@bb.com.br
3 Department of Statistics, University of Brasilia, Brasilia 70910-900, Brazil; gabrielaolinto@hotmail.com
* Correspondence: pushpanrathie@yahoo.com; Tel.: +55-85-3366-9447
† These authors contributed equally to this work.

Academic Editors: Humberto Bustince and Hans J. Haubold
Received: 12 February 2016; Accepted: 7 April 2016; Published: 15 April 2016

Abstract: We use the skew distribution generation procedure proposed by Azzalini [*Scand. J. Stat.*, **1985**, *12*, 171–178] to create three new probability distribution functions. These models make use of normal, student-*t* and generalized logistic distribution, see Rathie and Swamee [Technical Research Report No. 07/2006. Department of Statistics, University of Brasilia: Brasilia, Brazil, 2006]. Expressions for the moments about origin are derived. Graphical illustrations are also provided. The distributions derived in this paper can be seen as generalizations of the distributions given by Nadarajah and Kotz [*Acta Appl. Math.*, **2006**, *91*, 1–37]. Applications with unimodal and bimodal data are given to illustrate the applicability of the results derived in this paper. The applications include the analysis of the following data sets: (a) spending on public education in various countries in 2003; (b) total expenditure on health in 2009 in various countries and (c) waiting time between eruptions of the Old Faithful Geyser in the Yellow Stone National Park, Wyoming, USA. We compare the fit of the distributions introduced in this paper with the distributions given by Nadarajah and Kotz [*Acta Appl. Math.*, **2006**, *91*, 1–37]. The results show that our distributions, in general, fit better the data sets. The general *R* codes for fitting the distributions introduced in this paper are given in Appendix A.

Keywords: generalized logistic distribution; normal distribution; Student-*t* distribution; skew distributions

MSC: 60E05; 62B15; 33C60; 60E10

---

1. Introduction

The skew symmetric models have been considered by several researchers. Skew normal distribution is a classical example. Abtahi *et al.* [1] constructed skew student-*t* and skew Cauchy distributions. Recently, Rathie *et al.* [2,3] introduced a system of univariate skew distributions by utilizing Rathie and Swamee [4] generalized logistic distribution. For certain values of the parameters, this skew distribution approximates nicely the skew normal distribution. Recently, Gupta and Kundu [5] defined and studied two generalizations of the logistic distribution by introducing skewness parameters. In this paper we use Azzalini's formula to generate new asymmetric distributions by using the Generalized Logistic, Normal and Student-*t* distributions. The distributions are Skew Normal-Generalized Logistic (Skew Normal-GL), Skew Generalized Logistic-Normal (Skew GL-Normal), Skew Student-*t*-Generalized Logistic (Skew t-GL) and Skew Generalized Logistic-Student-*t* (Skew GL-*t*). The models Skew GL-t and Skew GL-Normal generated by using $f(.)$

as define by the Equation (1) are bimodal for values of a close to 0, it is important to note that the values of the parameters determining uni/bi modal shapes are yet to be investigated.

We apply these distributions to three real data sets (expenditure on education, expenditure on health and waiting time between eruptions of the Old Faithful Geyser. We compare the fit of the distributions introduced in this paper with the distributions given by Nadarajah and Kotz [6], the results show that: (1) our distributions, in general, fit better the data sets; (2) The Skew GL-Normal, Skew GL-t, Skew Normal-GL and Skew t-GL distributions can be used to model symmetrical and asymmetrical unimodal data; (3) The Skew GL-Normal and Skew GL-t distributions can be used to adjust bimodal symmetrical and asymmetrical data, offering good fits, showing a high flexibility which is not common in the literature on probability distributions, which are mostly unimodal. This may be very important in practical applications; (4) The distributions are robust to numerical calculations in practical applications. The general R codes for fitting the distributions introduced in this paper are given in Appendix A.

The paper is organized as follows. In Section 2, we introduce the Skew Normal-GL, Skew GL-Normal, Skew t-GL and Skew GL-t distributions and we obtained the mathematical expressions for the moments, respectively. In Section 3, we apply the new distributions in three real data sets. Finally, in Section 4, we point out some final considerations of the results obtained in this paper.

We conclude this introduction section with some results which will be useful in the subsequent sections of this paper.

1.1. Generalized Logistic Distribution

We start by defining the symmetric generalized logistic density function and its cumulative distribution function studied recently by Rathie and Swamee [4]:

$$f(x) = \frac{[a + b(1+p)|x|^p]\exp\left[-x\left(a + b|x|^p\right)\right]}{\{\exp\left[-x\left(a+b|x|^p\right)\right] + 1\}^2}, \tag{1}$$

$$F(x) = \{\exp\left[-x\left(a+b|x|^p\right)\right] + 1\}^{-1}, \tag{2}$$

where $x \in \mathbb{R}, a \geq 0, b \geq 0, p \geq 0$ (with a and b are not zero simultaneously), and $\mathbb{R}$ is the set of real numbers. For the values $a = 1.59413$, $b = 0.07443$ and $p = 1.939$, this distribution approximates very well the normal distribution with a maximum error of 4×10^{-4} at $x = 0$ for the density function and 7.757×10^{-5} at $x = 2.81$ for the distribution function. For approximations to Student-t distribution, see Rathie *et al.* [3]. The case $a = 0$ was studied, and applied to a civil engineering problem by Swamee and Rathie [7]. In the recent review article on univariate normal distribution, Rathie [8] pointed out that the generalized logistic distribution defined in (1) and (2) is invertible and that the approximation to the normal distribution is important for practical applications.

1.2. Azzalini's Skew Distribution

Azzalini [9] obtained the following skew density function:

$$h(x) = 2f(x)G(w(x)) \quad (-\infty < x < \infty), \tag{3}$$

where $f(x)$ is a symmetric probability density function about the origin, $G(x)$ is the cumulative distribution function of a symmetric density function about the origin, and $w(x)$ is an odd function of x. In this paper, we take $w(x) = cx$, $c \in \mathbb{R}$.

1.3. Moments

It is easy to calculate the n-th moments of $h(x)$ given in (3) with $w(x) = cx$, $c \in \mathbb{R}$, which are

$$E(X^n) = 2\int_0^\infty x^n f(x)dx, \tag{4}$$

when n is even, and

$$E(X^n) = 4\int_0^\infty x^n f(x)G(cx)dx - 2\int_0^\infty x^n f(x)dx, \tag{5}$$

when n is odd.

1.4. Generalized Hypergeometric Function

The H-function, which is a generalization of Meijer's G function, is given below

$$\begin{aligned} &H_{p,q}^{m,n}\left[x \,\middle|\, \begin{matrix} (a_1, A_1), ..., (a_n, A_n), (a_{n+1}, A_{n+1}), ..., (a_p, A_p) \\ (b_1, B_1), ..., (b_n, B_n), (b_{n+1}, B_{n+1}), ..., (b_p, B_p) \end{matrix}\right] = \\ &= \frac{1}{2\pi i}\int_L \frac{\prod_{j=1}^m \Gamma(b_j - B_j s)\prod_{j=1}^n \Gamma(1 - a_j + A_j s)}{\prod_{j=m+1}^q \Gamma(1 - b_j + B_j s)\prod_{j=n+1}^p \Gamma(a_j - A_j s)} x^s ds. \end{aligned} \tag{6}$$

As a special case, we have

$$H_{p,q}^{m,n}\left[x \,\middle|\, \begin{matrix} (a_1, 1), ..., (a_p, 1) \\ (b_1, 1), ..., (b_p, 1) \end{matrix}\right] = G_{p,q}^{m,n}\left[x \,\middle|\, \begin{matrix} a_1, ..., a_p \\ b_1, ..., b_p \end{matrix}\right]. \tag{7}$$

For details, see Luke [10], Springer [11], or Mathai *et al.* [12].

2. Skew Distributions

2.1. Skew Normal-Generalized Logistic Distribution

The skew normal-generalized logistic distribution (Skew Normal-GL), using (3), with $f(x)$ standard normal and $G(x)$ given in Equation (2) is defined by

$$h(x) = \sqrt{\frac{2}{\pi}}\frac{\exp[\frac{-x^2}{2}]}{\{\exp\left[-cx\left(a + b|cx|^p\right)\right] + 1\}} \quad (-\infty < x < \infty). \tag{8}$$

In order that (8) is identifiable, we can rewrite it in the following form:

$$h(x) = \sqrt{\frac{2}{\pi}}\frac{\exp[\frac{-x^2}{2}]}{\{\exp\left[-x\left(A + B|x|^p\right)\right] + 1\}} \quad (-\infty < x < \infty), \tag{9}$$

where $A = ac \in \mathbb{R}$ and $B = bc|c|^p \in \mathbb{R}$. Plots for probability density function (9), varying some values of A, B and p, to show different forms of the Skew Normal-GL distribution are illustrated in Figure 1. We can see that the density has symmetric, asymmetric to the left and asymmetric to the right behavior, which may be important for practical purposes.

In the next subsection, we obtain n-th moments. For $B = 0$, our results give alternative expressions for the results obtained earlier by Nadarajah and Kotz [6].

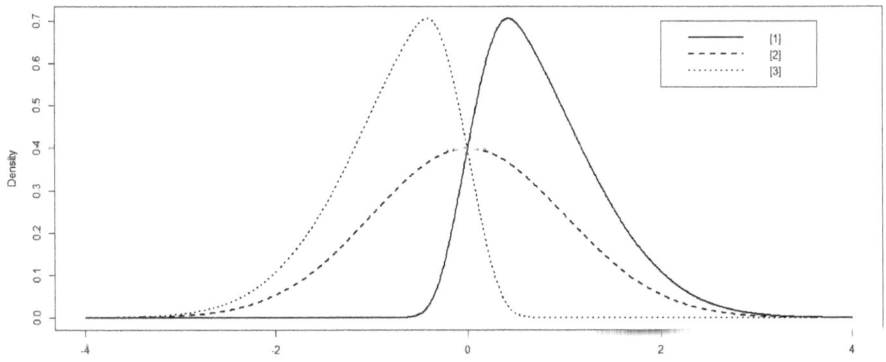

Figure 1. The skew normal-generalized logistic distribution (Skew Normal-GL) using the following parameter values: [1] $A = 6$, $B = 10.8$ and $p = 2$; [2] $A = 0$, $B = 0$ and $p = 3$; [3] $A = -6$, $B = -10.8$ and $p = 2$.

Moments

The n-th moments of (9) are

$$E(X^n) = \sqrt{\frac{2^n}{\pi}}\,\Gamma\left(\frac{n+1}{2}\right), \tag{10}$$

for even values of n. For odd values of n, on using $(1+y)^{-1} = \sum_{r=0}^{\infty}(-y)^r$, one gets

$$\begin{aligned} E(X^n) = {} & \frac{4}{\sqrt{2\pi}}\sum_{r=1}^{\infty}(-1)^r \sum_{k=0}^{\infty}\frac{(-rB)^k}{k!}\frac{(rA)^{-(p+1)k-n-1}}{2}\times \\ & H_{1,1}^{1,1}\left[\frac{1}{rA\sqrt{2}}\middle|\begin{matrix}(-(p+1)k-n,1)\\(0,\frac{1}{2})\end{matrix}\right] + \left(2-\sqrt{2\pi}\right)\sqrt{\frac{2^n}{\pi}}\Gamma\left(\frac{n+1}{2}\right). \end{aligned} \tag{11}$$

2.2. *Skew Generalized Logistic-Normal Distribution*

This section deals with the skew generalized logistic-normal distribution (Skew GL-Normal) defined below in (13). Moments about origin are obtained. The skew generalized logistic-normal distribution, using (3), with $f(x)$ given in (1) and

$$G(x) = \frac{1}{2}\left\{1 + \frac{x}{\sqrt{2\pi}}\,G_{1,2}^{1,1}\left[\frac{x^2}{2}\middle|\begin{matrix}\frac{1}{2}\\0,\frac{-1}{2}\end{matrix}\right]\right\} \quad (-\infty < x < \infty), \tag{12}$$

is defined by

$$h(x) = 2\frac{[a + b(1+p)|x|^p]\exp\left[-x\left(a + b|x|^p\right)\right]}{\{\exp\left[-x\left(a+b|x|^p\right)\right]+1\}^2}G(cx) \quad (-\infty < x < \infty). \tag{13}$$

Plots for probability density function (13), varying some values of a, b, p and c, to show different forms of the Skew GL-Normal distribution are illustrated in Figure 2. As in Skew Normal-GL distribution, the density has symmetric, asymmetric to the left and asymmetric to the right behavior. It is interesting to note that, for values of parameter a near to zero, the Skew GL-Normal distribution has a bimodal shape, which may be very important in practical applications.

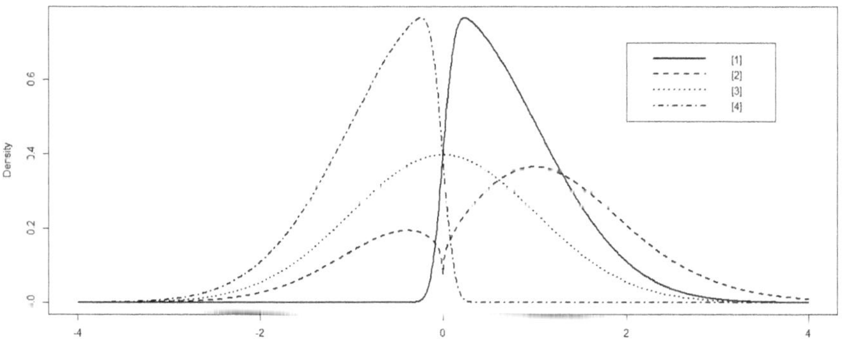

Figure 2. The Skew GL-Normal distribution using the following parameter values: [1] $a = 1.59$, $b = 0.0727$, $p = 1.962$ and $c = 10$; [2] $a = 0.3$, $b = 0.7$, $p = 0.4$ and $c = 0.6$; [3] $a = 1.59$, $b = 0.0727$, $p = 1.962$ and $c = 0$; [4] $a = 1.59$, $b = 0.0727$, $p = 1.962$ and $c = -10$.

In the next subsection, we obtain the moments of Skew GL-Normal distribution . For $b = 0$, the results of this section give alternative expressions for the results obtained earlier by Nadarajah and Kotz [6].

Moments

Using (4) the n-th moments of (13) for even values of n has been calculated earlier by Rathie and Swamee [4], and are given by

$$E(X^n) = 2\sum_{r=0}^{\infty}(-1)^r(1+r)\left[aI_{h,r} + b(1+p)I_{h+p,r}\right], \tag{14}$$

where

$$\begin{aligned} I_{\alpha,r} &= \int_0^{\infty} x^{\alpha}\exp\left[-(1+r)x\left(a+bx^p\right)\right]dx \\ &= [a(1+r)]^{-\alpha-1}H_{1,1}^{1,1}\left[\frac{a^{p+1}(1+r)^p}{b}\middle| \begin{array}{c}(1,1)\\(\alpha+1,p+1)\end{array}\right]. \end{aligned} \tag{15}$$

For odd values of n, on using (7),

$$E(X^n) = \frac{4c}{\sqrt{2\pi}}\sum_{r=0}^{\infty}(-1)^r(1+r)\left[aL_{n+1,a(1+r)} + b(1+p)L_{n+p+1,a(1+r)}\right], \tag{16}$$

where

$$\begin{aligned} L_{\alpha,\theta} &= \frac{2^{\alpha}}{\sqrt{\pi}\theta^{\alpha+1}}\sum_{k=0}^{\infty}\left(\frac{-2^{p+1}(1+r)b}{\theta^{p+1}}\right)^k \times \\ &\frac{1}{k!}G_{3,2}^{1,3}\left[\frac{2c^2}{\theta^2}\middle| \begin{array}{c}-\frac{1}{2}, \frac{1-\alpha-k(p+1)}{2}, \frac{-\alpha-k(p+1)}{2}\\ 0, -\frac{1}{2}\end{array}\right]. \end{aligned} \tag{17}$$

2.3. *Skew Student-t-Generalized Logistic Distribution*

The skew student-t-generalized logistic distribution (Skew t-GL), using Azzalini's formula, is defined by

$$h(x) = \frac{2}{\sqrt{v}B\left(v/2,\, 1/2\right)}\left(1+\frac{x^2}{v}\right)^{\frac{-(1+v)}{2}}\left\{1+exp\left[-cx\left(a+b|cx|^p\right)\right]\right\}^{-1}, \tag{18}$$

for $-\infty < x < \infty$ and $v > 0$. As before, taking $A_1 = ac \in \mathbb{R}$ and $B_1 = bc|c|^p \in \mathbb{R}$, (18) can be rewritten as

$$h(x) = \frac{2}{\sqrt{v}B\left(v/2,\, 1/2\right)}\left(1+\frac{x^2}{v}\right)^{\frac{-(1+v)}{2}}\left\{1+exp\left[-x\left(A_1+B_1|x|^p\right)\right]\right\}^{-1}, \tag{19}$$

for $-\infty < x < \infty$, $v > 0$ and $B(.)$ is the Beta function, defined by $B(a,b) = \frac{\Gamma(a)\Gamma(b)}{\Gamma(a+b)}$. Plots for probability density function (19), varying some values of A_1, B_1, p and v, showing different unimodal forms of the Skew t-GL distribution are illustrated in Figure 3. We can see that the density also has symmetric, asymmetric to the left and asymmetric to the right behavior. However, the Skew t-GL distribution has heavy tails.

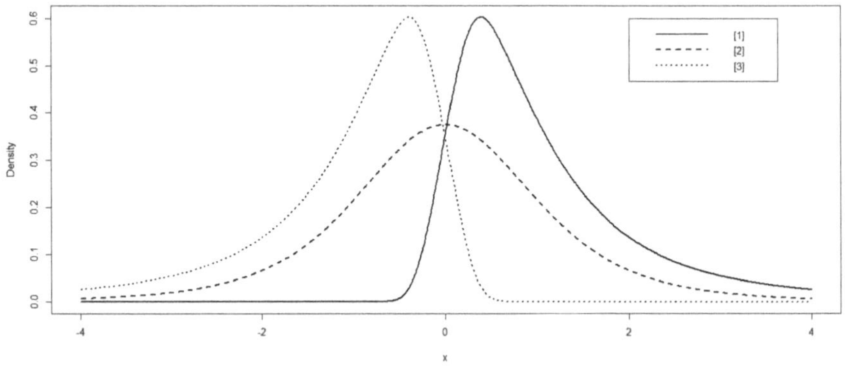

Figure 3. The Skew t-GL distribution using the following parameter values: [1] $A_1 = 6$, $B_1 = 10.8$, $v = 2$ and $p = 2$; [2] $A_1 = 0$, $B_1 = 0$, $v = 4$ and $p = 3$; [3] $A_1 = -6$, $B_1 = -10.8$, $v = 2$ and $p = 2$.

In the next subsection, moments of the Skew t-GL distribution (19) are obtained.

Moments

The n-th moments of (19), using (4), are given by

$$E(X^n) = \frac{1}{\sqrt{\pi}\Gamma(v/2)}\left[\Gamma\left(\frac{v-n}{2}\right)\Gamma\left(\frac{1+n}{2}\right)v^{n/2}\right], \tag{20}$$

if n is an even integer, and $0 < n < v$. For odd integer n, and using the equation (3.389.2) of Gradshteyn *et al.* [13] or Prudnikov *et al.* [14], the moments are given by

$$\begin{aligned} E(X^n) = {} & \frac{v^{n/2}}{\pi\Gamma\left(\frac{v}{2}\right)}\sum_{r=0}^{\infty}\sum_{k=0}^{\infty}\frac{(-1)^{r+k}\left(B_1 r\right)^k v^{\frac{k(p+1)}{2}}}{k!}S^{v,p,k}_{n,\frac{v+1}{2}}\left(A_1 r\right) - \\ & \frac{v^{n/2}}{\pi\Gamma\left(\frac{v}{2}\right)}\sum_{r=0}^{\infty}\sum_{k=0}^{\infty}\frac{(-1)^{r+k}\left(B_1(1+r)\right)^k v^{\frac{k(p+1)}{2}}}{k!}S^{v,p,k}_{n,\frac{v+1}{2}}\left(A_1(1+r)\right), \end{aligned} \tag{21}$$

where

$$S^{\gamma,p,k}_{\alpha,\beta}(\theta) = G^{1,3}_{3,1}\left[\frac{\theta^2\gamma}{4}\,\middle|\, \begin{matrix} \frac{1-\alpha-k(p+1)}{2} \\ \beta - \left(\frac{\alpha+1+k(p+1)}{2}\right),\ 0,\ \frac{1}{2} \end{matrix}\right]. \tag{22}$$

For $B_1 = 0$, (21) reduces to the result obtained earlier by Nadarajah and Kotz [6].

2.4. Skew Generalized Logistic-Student-t Distribution

The skew generalized logistic-student-t distribution (Skew GL-t), using (3), with $f(x)$ given in (1) and

$$G(x) = \frac{1}{2}\left\{1 + \frac{x}{\sqrt{v\pi}\Gamma\left(\frac{v}{2}\right)} G_{2,2}^{1,2}\left[\frac{x^2}{v}\middle| \begin{array}{c} \frac{1-v}{2}, \frac{1}{2} \\ 0, \frac{-1}{2} \end{array}\right]\right\} \quad (-\infty < x < \infty), \tag{23}$$

is defined by

$$h(x) = 2\frac{[a + b(1+p)|x|^p]\exp\left[-x\left(a + b|x|^p\right)\right]}{\{\exp\left[-x\left(a + b|x|^p\right)\right] + 1\}^2} G(cx) \quad (-\infty < x < \infty). \tag{24}$$

Plots for probability density function (24), for different values of values of a, b, p, c and v, showing different forms of the Skew GL-t distribution are illustrated in Figure 4. As in Skew GL-Normal distribution, the density also has symmetric, asymmetric to the left and asymmetric to the right behavior. Again, it is interesting to note that, for values of the parameter a near to zero, the Skew GL-t distribution has a bimodal shape and heavy tails.

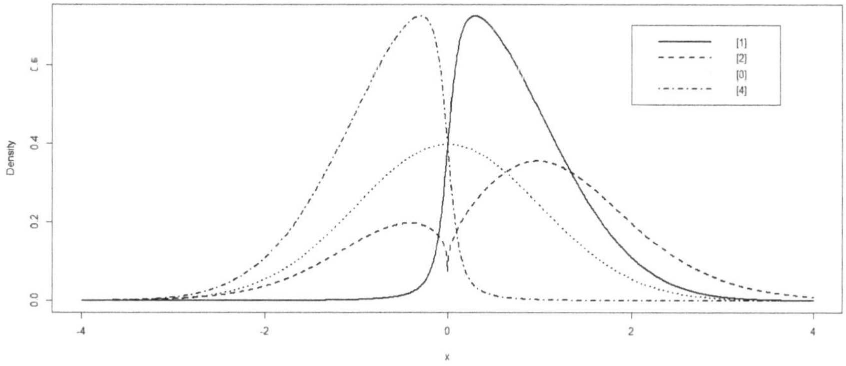

Figure 4. The Skew GL-t distribution using the following parameter values: [1] $a = 1.59$, $b = 0.0727$, $p = 1.962$, $v = 2$ and $c = 10$; [2] $a = 0.3$, $b = 0.7$, $p = 0.4$, $v = 3$ and $c = 0.6$; [3] $a = 1.59$, $b = 0.0727$, $p = 1.962$, $v = 3$ and $c = 0$; [4] $a = 1.59$, $b = 0.0727$, $p = 1.962$, $v = 2$ and $c = -10$.

In the next subsection, we obtain the moments of Skew GL-t distribution. For $b = 0$, the results of this section give alternative expressions for what have been given earlier by Nadarajah and Kotz [6].

Moments

Using (4), the n-th moments of (24), for even values of n, has been calculated earlier by Rathie and Swamee [4], and is given by

$$E(X^n) = 2\sum_{r=0}^{\infty}(-1)^r(1+r)\left[aI_{h,r} + b(1+p)I_{h+p,r}\right], \tag{25}$$

where

$$\begin{aligned} I_{\alpha,r} &= \int_0^{\infty} x^{\alpha}\exp\left[-(1+r)x\left(a + bx^p\right)\right]dx \\ &= [a(1+r)]^{-\alpha-1}H_{1,1}^{1,1}\left[\frac{a^{p+1}(1+r)^p}{b}\middle| \begin{array}{c} (1,1) \\ (\alpha+1, p+1) \end{array}\right]. \end{aligned} \tag{26}$$

For odd values of n, on using (7),

$$E(X^n) = \frac{2c}{\sqrt{v\pi}\Gamma\left(\frac{v}{2}\right)} \sum_{r=0}^{\infty} (-1)^r (1+r) \left[aR_{n+1,a(1+r)} + b(1+p)R_{n+p+1,a(1+r)} \right], \tag{27}$$

where

$$\begin{aligned} R_{\alpha,\theta} = {} & \frac{2^{\alpha}}{\sqrt{\pi}\theta^{\alpha+1}} \sum_{k=0}^{\infty} \left(\frac{-2^{p+1}(1+r)b}{\theta^{p+1}} \right)^k \times \\ & \frac{1}{k!} G^{1,4}_{4,2} \left[\frac{4c^2}{v\theta^2} \middle| \begin{matrix} \frac{1-v}{2}, \frac{1}{2}, \frac{1-\alpha-k(p+1)}{2}, \frac{-\alpha-k(p+1)}{2} \\ 0, -\frac{1}{2} \end{matrix} \right]. \end{aligned} \tag{28}$$

3. Applications to Real Data

In this section, we apply all the distributions introduced in this paper to three real data sets. The first data set is related to the total spending on public education (% of GDP—Gross Domestic Product) in various countries in 2003, which is unimodal and asymmetrical. The second data set relates to the total expenditure, in 2009, on health (% of GDP—Gross Domestic Product) in various countries, which has a bimodal and asymmetric behavior, even if not so evident. And finally, the third data set is related to the waiting time between eruptions of the Old Faithful Geyser in the Yellow Stone National Park, Wyoming, USA, which is clearly bimodal and asymmetrical.

The performance of the models was then compared by using the Akaike criterion (AIC), Bayesian criterion (BIC), Modified Akaike criterion (AICC) and Komogorov-Sminorv test (KS-Test). The information criterion AIC, BIC and AICC are given by

$$\begin{aligned} AIC &= -2\log(f(\mathbf{x}|\boldsymbol{\theta})) + 2p; \\ BIC &= -2\log(f(\mathbf{x}|\boldsymbol{\theta})) + p\log(n); \\ AICC &= -2\log(f(\mathbf{x}|\boldsymbol{\theta})) + 2\frac{p(p+1)}{n-p-1}, \end{aligned} \tag{29}$$

where $\log(f(\mathbf{x}|\boldsymbol{\theta}))$ is the log-likelihood function, p is then number of parameters of models and n is the sample size. The models that have lowest AIC, BIC and AICC values are better.

The accuracy of the models was then compared by using Mean Square Error (MSE), Mean Deviation Absolute (MDA) and Max Deviation (MaxD). The MSE, MDA and MaxD are given by

$$\begin{aligned} MSE &= \frac{\sum_{i=1}^{n} (F_e(x_i) - \hat{F}(x_i))^2}{n} \\ MAD &= \frac{\sum_{i=1}^{n} |F_e(x_i) \quad \hat{F}(x_i)|}{n} \\ MaxD &= \max(|F_e(x_i) - \hat{F}(x_i)|), \ i = 1, \ldots, n, \end{aligned} \tag{30}$$

where $F_e(x_i)$ is the empirical cumulative distribution and $\hat{F}(x_i)$ is the fitted cumulative distribution of the data. The models that have minimum values of MSE, MAD and MaxD (close to zero) are better. In Appendix A, we give a general *R* code for fitting the distributions introduced in this paper for practical purposes.

3.1. Application 1: Expenditure on Education

We use the data of total spending on public education (% of GDP- Gross Domestic Product) in various countries in 2003. These data were obtained from [15]. Expenditure on public education includes the current and capital spending by private and government agencies on educational institutions (both public and private), educational administration and subsidies to private (student/family) entities.

To adjust this data set, we modify the models by introducing a location parameter μ and a scale parameter σ by changing x to $(x-\mu)/\sigma$ everywhere in the density function divided by σ. The software *R* was used to calculate the estimates of the parameters through maximum likelihood method and the *R* function *constrOptim* [16] was used to maximize the log-likelihood function (Appendix A). The reason for using the *R* function *constrOptim* is to guarantee that the estimated parameters are consistent within their respective parametric space. The maximum likelihood estimates for the parameters of the models are given by:

- Skew Normal-GL: $\hat{A} = -0.000227$, $\hat{B} = -8.83 \times 10^{-14}$, $\hat{p} = 2.29$, $\hat{\mu} = 4.83$ and $\hat{\sigma} = 1.51$;
- Skew GL-Normal: $\hat{a} = 22.76$, $\hat{b} = 2.67$, $\hat{p} = 26.40$, $\hat{c} = 43.94$, $\hat{\mu} = 2.60$ and $\hat{\sigma} = 38.21$;
- Skew *t*-GL: $\hat{A}_1 = 0.167$, $\hat{B}_1 = 0.005$, $\hat{p} = 5.19$, $\hat{v} = 3.44$, $\hat{\mu} = 4.21$ and $\hat{\sigma} = 1.40$;
- Skew GL-*t*: $\hat{a} = 22.75$, $\hat{b} = 2.55$, $\hat{p} = 26.66$, $\hat{v} = 1968.59$, $\hat{c} = 43.92$, $\hat{\mu} = 2.60$ and $\hat{\sigma} = 38.18$.

We compare the results of our distribution with the corresponding distribution (special cases) introduced by Nadarajah and Kotz [6] (Skew Normal-Logistic, Skew Logistic-Normal, Skew *t*-Logistic and Skew Logistic-*t* distributions). The Figure 5 illustrates the fit of the distributions introduced in this paper. The Figure B1 (Appendix B) illustrates the fit of Nadarajah and Kotz [6] distributions. The Figure 6 illustrates the $pp - plot$ of all distributions. The performance of the all fitted distributions are given in Table 1. Observing the results in Table 1 we can see that, looking the p-value of the KS test, all distributions can be used to model the data. According to the accuracy, the Skew GL-Normal, Skew GL-*t* and Skew Logistic-Normal distributions indicated better results with similar values. However, Skew GL-Normal and Skew GL-*t* distributions presented smaller values of AIC, BIC and AICC compared to the Skew Logistic-Normal distribution. In contrast, Skew Normal-Logistic distribution showed the worst results followed by Skew Normal-GL distribution.

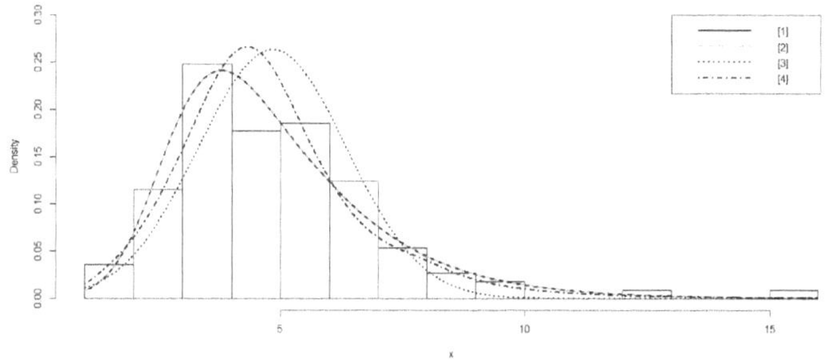

(**a**)

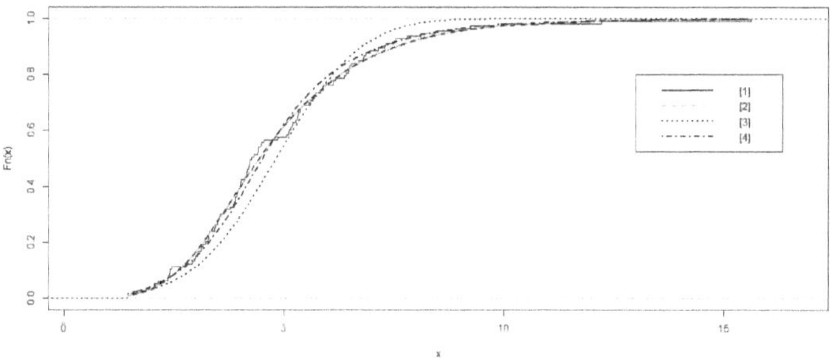

(**b**)

Figure 5. Education data - Fitted distributions. [1] Skew GL-Normal distribution; [2] Skew GL-*t* distribution; [3] Skew Normal-GL distribution; [4] Skew *t*-GL distribution. (**a**) Probability density function; (**b**) Cumulative distribution.

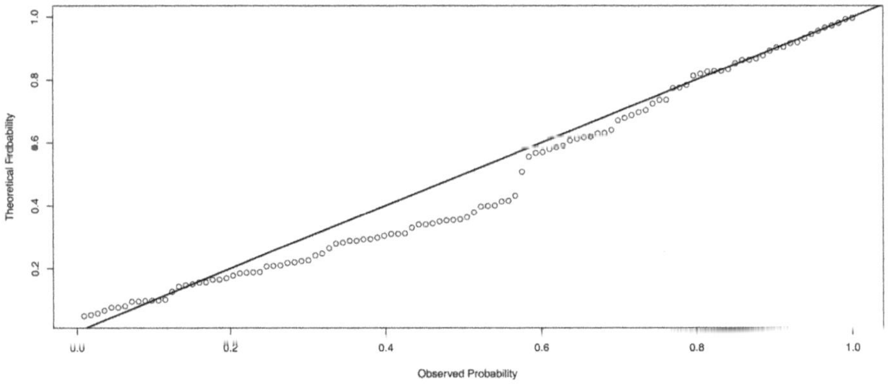

(**a**)

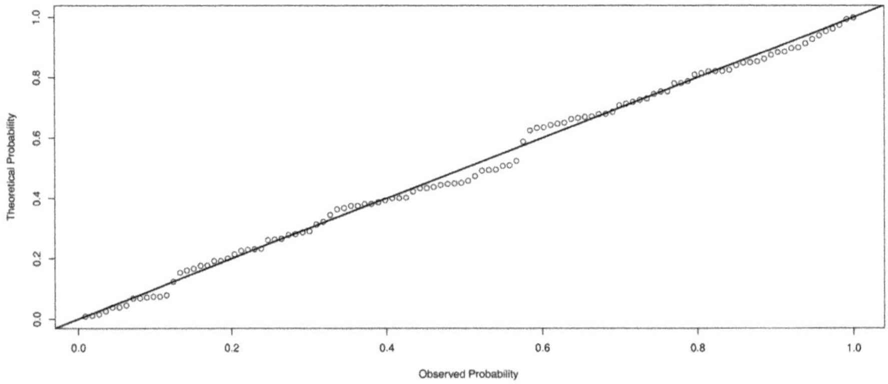

(**b**)

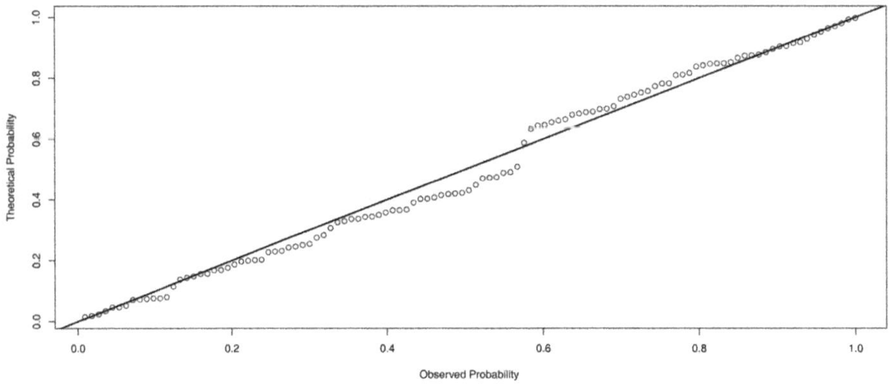

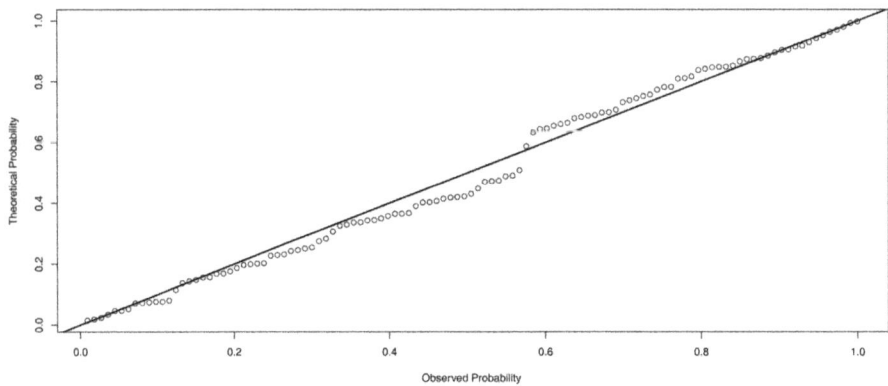

(**c**)

Figure 6. *Cont.*

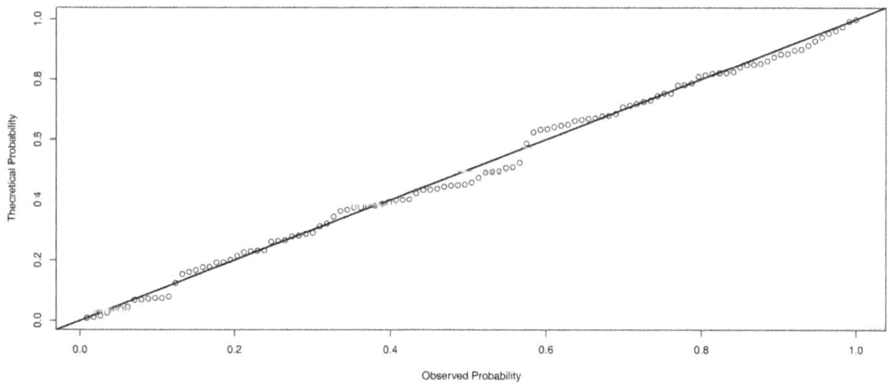

(d)

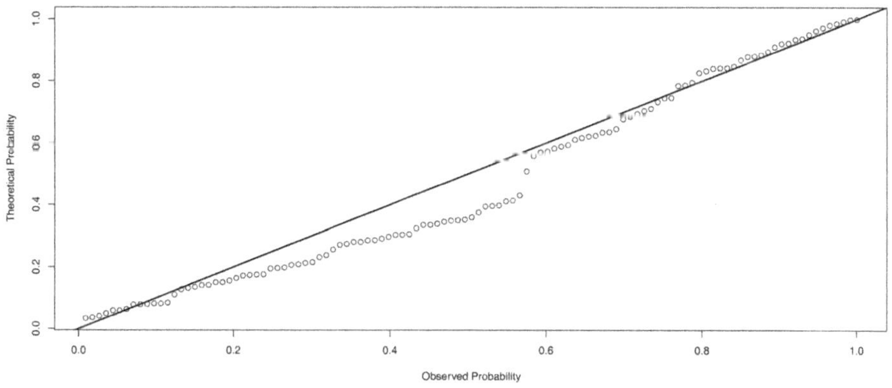

(e)

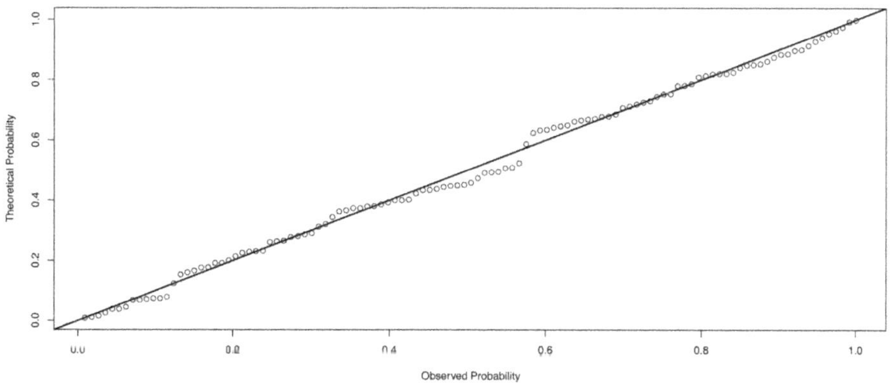

(f)

Figure 6. *Cont.*

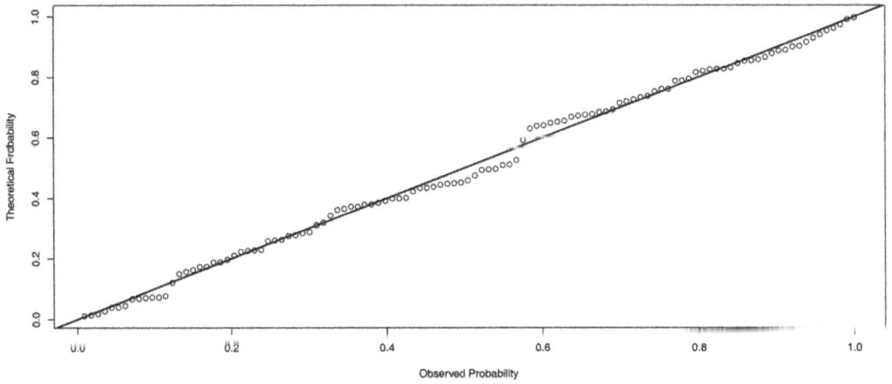

(g)

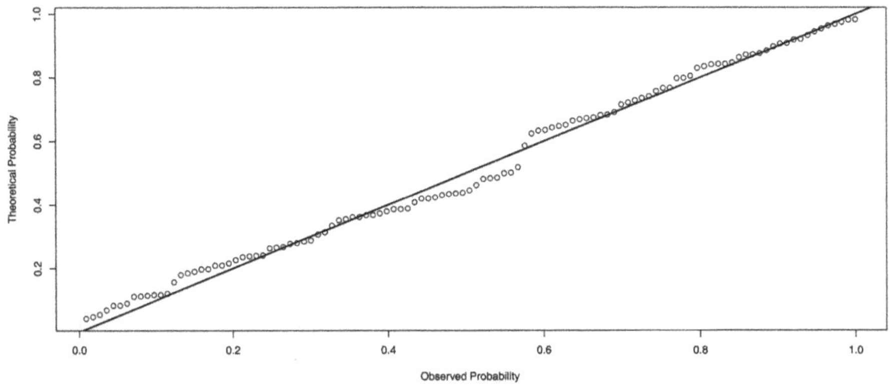

(h)

Figure 6. Education data - PP-Plot. (**a**) Skew Normal-GL; (**b**) Skew GL-Normal; (**c**) Skew t-GL; (**d**) Skew GL-t; (**e**) Skew Normal-Logistic; (**f**) Skew Logistic-Normal; (**g**) Skew t-Logistic; (**h**) Skew Logistic-t.

Table 1. Performance and accuracy of the distributions.

Model	AIC	BIC	AICC	KS-Test (p-Value)	MSE	MAD	MaxD
Skew GL-Normal	448.29	429.20	461.22	0.9973	0.000387	0.0153	0.0481
Skew GL-t	448.29	429.20	461.23	0.9973	0.000387	0.0153	0.0481
Skew Normal-GL	609.70	593.33	620.91	0.1549	0.003777	0.0443	0.1417
Skew t-GL	441.69	422.60	454.62	0.866	0.001086	0.0266	0.0736
Skew Logistic-Normal	452.29	438.65	461.73	0.9973	0.000387	0.0153	0.0481
Skew Logistic-t	465.02	448.66	476.23	0.9818	0.000683	0.0210	0.0588
Skew Normal-Logistic	611.70	598.06	621.14	0.1549	0.004160	0.0581	0.1449
Skew t-Logistic	450.39	434.02	461.59	0.9973	0.000389	0.0154	0.0459

Comparing the proposed distributions in this paper with their corresponding distributions given by Nadarajah and Kotz [6], we can see that: (1) Skew GL-t and Skew Normal-GL distributions have lower values of AIC, BIC, AICC , MSE , MAD and MaxD compared to the Skew Logistic-t and Skew Normal-Logistic distributions, respectively; (2) The Skew t-Logistic gave better accuracy compared to Skew distribution t-GL (Smaller values of MSE, MAD and MaxD), however, Skew t-GL distribution indicated better performance than Skew t-Logistic distribution (Smaller values of AIC, BIC and AICC).

Thus, for this application, in general, the distributions introduced in this paper fit better the data and the Skew GL-Normal and Skew GL-t distributions are preferable to fit this data because they present better and similar results (smaller values of AIC, AICC, BIC, MSE, MAD and MaxD).

3.2. Application 2: Expenditure on Health

We use the data of total expenditure, in 2009, on health (% of GDP—Gross Domestic Product) in various countries. These data are obtained from [17]. Total health expenditure is the sum of expenses with public and private health. It covers the provision of health services (preventive and curative), family planning activities, nutrition activities and emergency aid designated for health but does not include water supply and sanitation.

Again, to adjust this data set we introduced a location parameter μ and a scale parameter σ. For the estimates of model parameters the maximum likelihood method is used. The software R was used to calculate estimates of the parameters by using the R function *constrOptim* [16] to maximize the log-likelihood function (Appendix A). The maximum likelihood estimates for the parameters of the models are given by:

- Skew Normal-GL: $\hat{A} = 9.51 \times 10^{-8}$, $\hat{B} = 0.0005$, $\hat{p} = 7.28$, $\hat{\mu} = 6.62$ and $\hat{\sigma} = 1.99$;
- Skew GL-Normal: $\hat{a} = 0.38$, $\hat{b} = 3.30$, $\hat{p} = 0.54$, $\hat{c} = -1.39$, $\hat{\mu} = 8.61$ and $\hat{\sigma} = 5.98$;
- Skew t-GL: $\hat{A}_1 = 8.86 \times 10^{-7}$, $\hat{B}_1 = 0.069$, $\hat{p} = 4.41$, $\hat{v} = 4.61$, $\hat{\mu} = 5.84$ and $\hat{\sigma} = 2.28$;
- Skew GL-t: $\hat{a} = 0.37$, $\hat{b} = 4.07$, $\hat{p} = 0.53$, $\hat{v} = 0.72$, $\hat{c} = -4.02$, $\hat{\mu} = 8.60$ and $\hat{\sigma} = 6.78$.

We compare the results of our distribution with the corresponding distribution (special cases) introduced by Nadarajah and Kotz [6] (Skew Normal-Logistic, Skew Logistic-Normal, Skew t-Logistic and Skew Logistic-t distributions). The Figure 7 illustrates the fit of the distributions introduced in this paper. The Figure B2 (Appendix B) illustrates the fit of Nadarajah and Kotz [6] distributions. The Figure 8 illustrates the $pp - plot$ of all distributions. The performance of the all fitted distributions are included in Table 2. Observing the Figure 7 we can see that the data presents a bimodal and asymmetric behavior, even if it is not very evident.

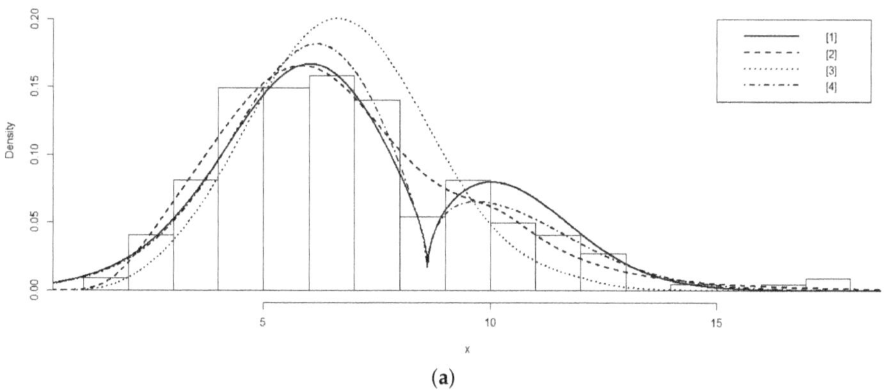

(a)

Figure 7. *Cont.*

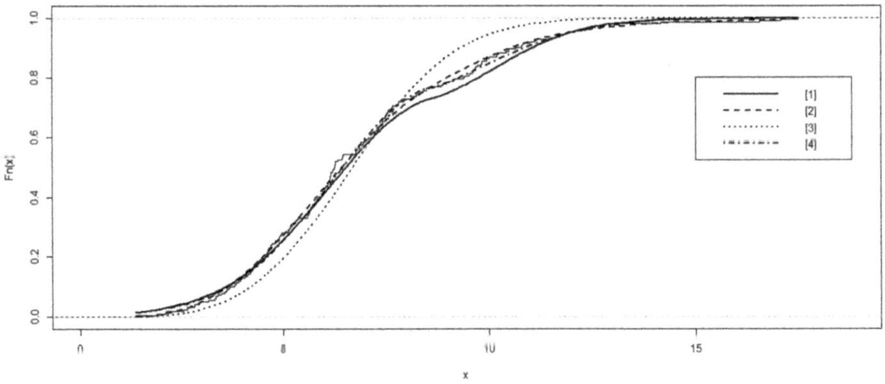

(b)

Figure 7. Health data - Fitted distributions. [1] Skew GL-Normal distribution; [2] Skew GL-*t* distribution; [3] Skew Normal-GL distribution; [4] Skew *t*-GL distribution. (**a**) Probability density function; (**b**) Cumulative distribution.

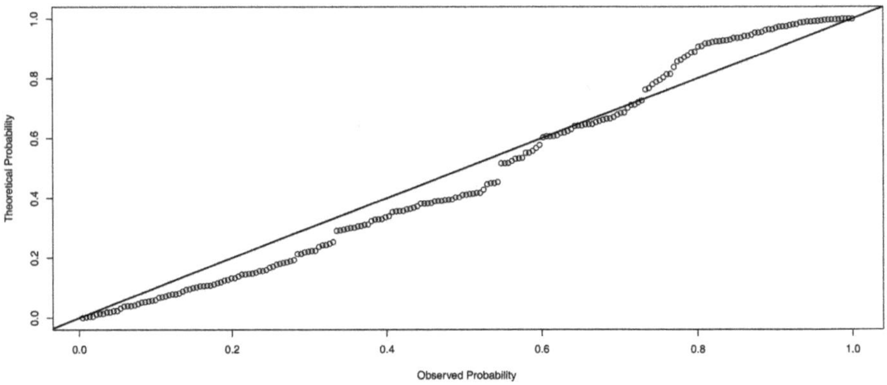

(a)

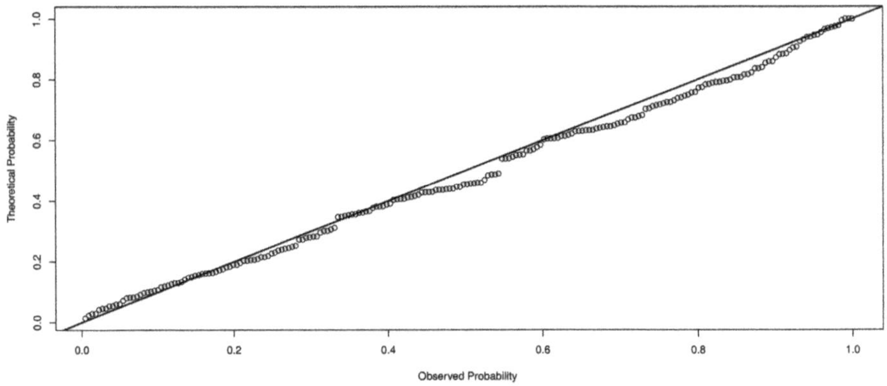

(b)

Figure 8. *Cont.*

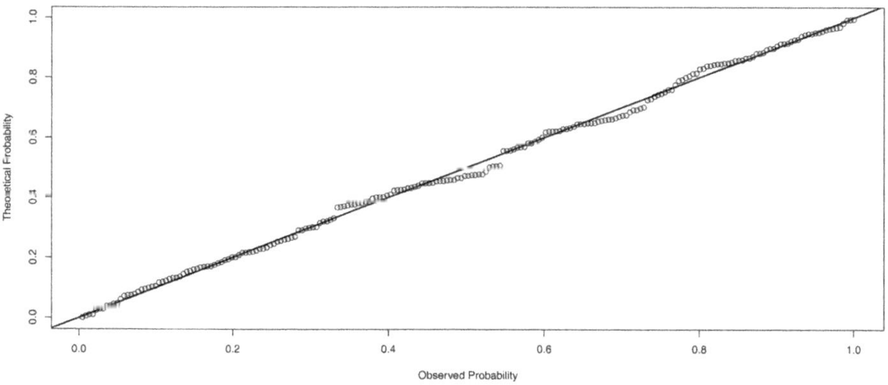

(c)

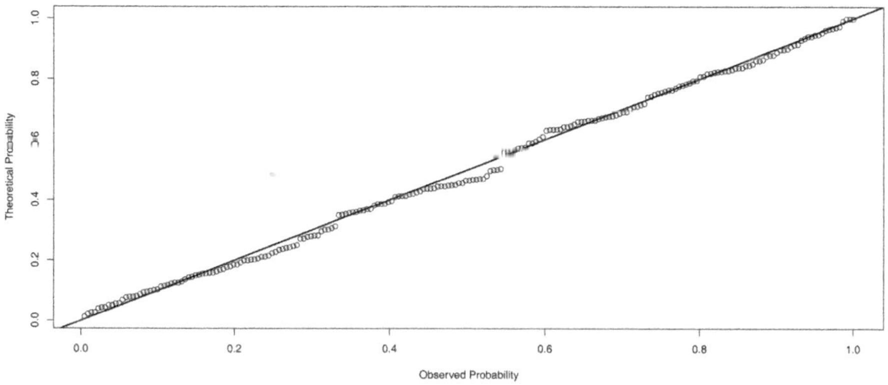

(d)

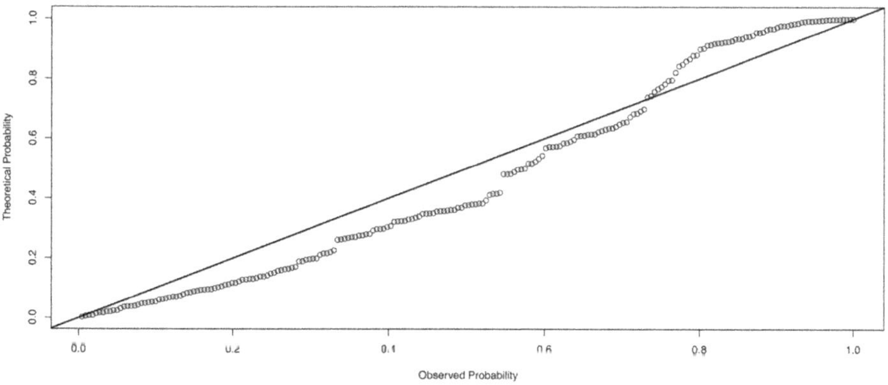

(e)

Figure 8. *Cont.*

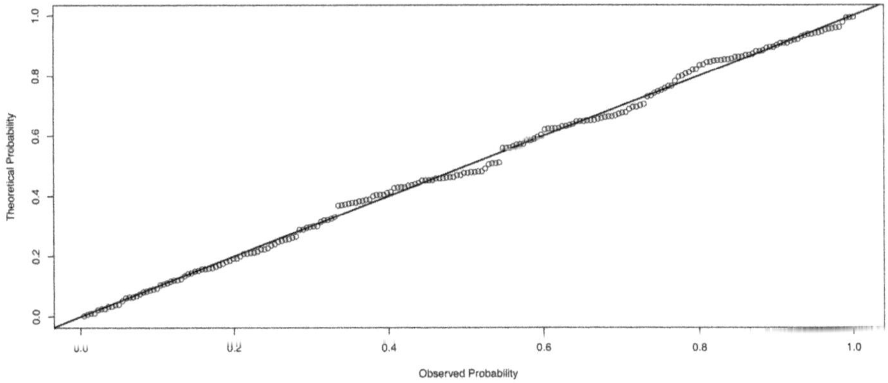

(**f**)

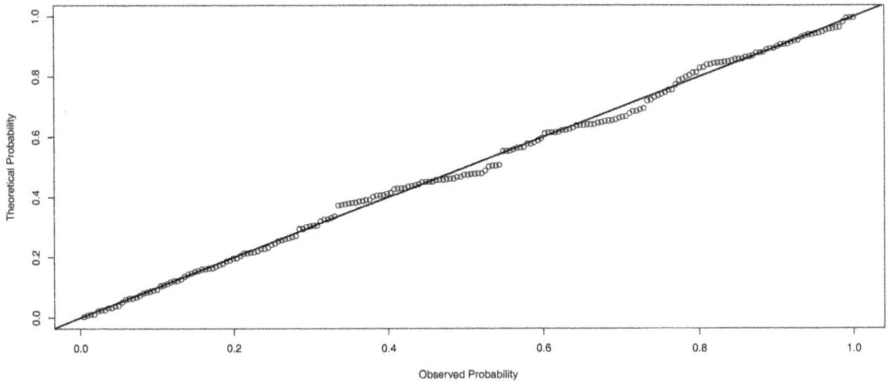

(**g**)

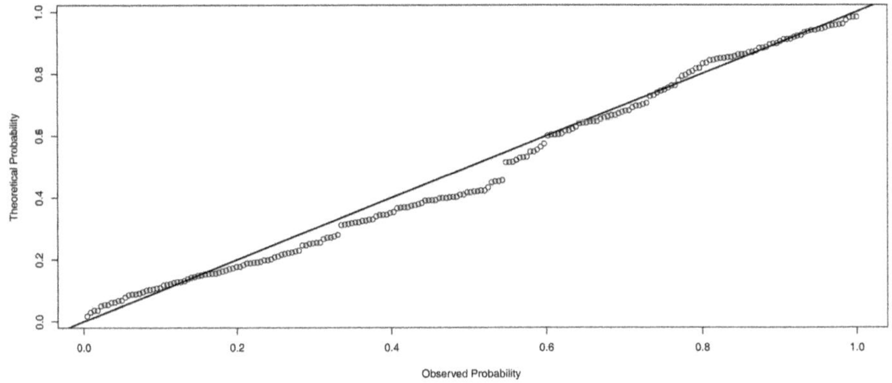

(**h**)

Figure 8. Health data - PP-Plot. (**a**) Skew Normal-GL; (**b**) Skew GL-Normal; (**c**) Skew t-GL; (**d**) Skew GL-t; (**e**) Skew Normal-Logistic; (**f**) Skew Logistic-Normal; (**g**) Skew t-Logistic; (**h**) Skew Logistic-t.

Table 2. Performance and accuracy of the distributions.

Model	AIC	BIC	AICC	KS-Test (p-Value)	MSE	MAD	MaxD
Skew GL-Normal	1040.04	1016.26	1050.52	0.993	0.000169	0.0104	0.0379
Skew GL-t	1038.21	1017.54	1050.42	0.9904	0.000152	0.0103	0.0378
Skew Normal-GL	1426.81	1406.42	1438.42	0.05759	0.003472	0.0497	0.1226
Skew t-GL	1048.29	1024.50	1061.76	0.9774	0.000196	0.0103	0.0402
Skew Logistic-Normal	1041.70	1024.71	1051.42	0.993	0.000211	0.0110	0.0401
Skew Logistic-t	1057.58	1037.19	1069.18	0.2235	0.001299	0.0279	0.0975
Skew Normal-Logistic	1442.35	1425.36	1452.07	0.02585	0.005780	0.0686	0.1371
Skew t-Logistic	1040.60	1019.22	1054.21	0.9774	0.000233	0.0111	0.0407

Observing the results in Table 2 we can see that, through the p-value of the KS test, only the Skew-Normal Logistic distribution could not be used to model the data. The Skew GL-t distribution had the smallest values of AIC, BIC, AICC, MSE, MAD and MaxD followed by Skew GL-Normal and Skew t-Logistics distributions. In contrast, Skew-Normal Logistic distribution showed the worst results followed by Skew Normal-GL distribution.

Comparing the distributions proposed in this paper with their corresponding distributions given by Nadarajah and Kotz [6], we can see that: (1) the Skew GL-Normal Skew GL-t and Skew Normal-GL distributions has smaller values of AIC, BIC, AICC, MSE, MAD and MaxD compared to the Skew Logistic-Normal Skew Logistic-t and Skew Normal-Logistic distributions, respectively; (2) The Skew t-Logistic distribution resulted in better performance when compared to the Skew t-GL distribution (Smaller values of AIC, BIC and AICC), however, the Skew t-GL distribution obtained better accuracy than Skew t-Logistic distribution (Smaller values of MSE, MAD and MaxD).

Again, for this application, in general, the distributions proposed in this paper fit better the data and the Skew GL-t distribution is preferred to fit this data presenting better results (smaller values of AIC, AICC, BIC, MSE, MAD and MaxD).

3.3. Application 3: Waiting Time between Eruptions of Old Faithful Geyser

This application shows the versatility of the Skew GL-Normal and the Skew GL-t distributions. Using data available in the free statistical software R we see the shape of bimodal distribution. Among the variables available, the waiting time between eruptions of Old Faithful Geyser in Yellow Stone National Park, Wyoming, USA was used. The data has 272 observations given in minutes.

Once more, to adjust this data set we modify the models by introducing a location parameter μ and a scale parameter σ. The software R was used to calculate the estimates of the parameters through maximum likelihood method and the R function *constrOptim* [16] was used to maximize the log-likelihood function (Appendix A). The maximum likelihood estimates for the parameters of the models are given by:

- Skew Normal-GL: $\hat{A} = -8.23 \times 10^{-8}$, $\hat{B} = -0.010$, $\hat{p} = 5.94$, $\hat{\mu} = 73.59$ and $\hat{\sigma} = 9.76$;
- Skew GL-Normal: $\hat{a} = 0.42$, $\hat{b} = 1.59$, $\hat{p} = 1.57$, $\hat{c} = 0.48$, $\hat{\mu} = 66.59$ and $\hat{\sigma} = 17.31$;
- Skew t-GL: $\hat{A}_1 = -3.33 \times 10^{-8}$, $\hat{B}_1 = -0.222$, $\hat{p} = 6.37$, $\hat{v} = 10714.65$, $\hat{\mu} = 75.58$ and $\hat{\sigma} = 14.37$;
- Skew GL-t: $\hat{a} = 0.89$, $\hat{b} = 10.89$, $\hat{p} = 1.57$, $\hat{v} = 35.10$, $\hat{c} = 1.02$, $\hat{\mu} = 66.59$ and $\hat{\sigma} = 36.51$.

We now compare the results of our distribution with the corresponding distribution (special cases) introduced by Nadarajah and Kotz [6] (Skew Normal-Logistic, Skew Logistic-Normal, Skew t-Logistic and Skew Logistic-t distributions). It is interesting to note that, when we try to adjust the distributions given by Nadarajah and Kotz [6] to the data, which have a marked bimodal behavior, we had numerical problems when trying to calculate their cumulative distribution functions, which did not happen with our distributions. Thus, it was not possible to calculate the p-value of the KS test, MSE, MAD and MaxD for the Skew Normal-Logistic, Skew Logistic-Normal, Skew t-Logistic and Skew Logistic-t distributions.

The Figure 9 illustrates the fit of the distributions introduced in this paper. The Figure 10 illustrates the fit of the density function of the distributions introduced by Nadarajah and Kotz [6]. The Figure 11 illustrates the $pp - plot$ of our distributions. The performance of the all fitted distributions are given in Table 3. Observing the results of the Table 3 we can see that, only the Skew GL-Normal and Skew GL-t distributions adjusted well to the data with similar accuracies. However, observing the AIC, BIC and AICC values, the Skew GL-t distribution had a slightly better result. In contrast, Skew Normal-GL and Skew t-GL distributions, even having no numerical problems, are not indicated to model these data showing poor results. Finally, the Skew Logistic-Normal Skew Logistic-t, Skew Normal-Logistic and Skew t-Logistic distributions presented numerical problems when calculating the cumulative distribution functions, showing that they are not flexible enough to model bimodal data. So, for this application, the Skew GL-t distribution is preferred to fit this data presenting better results (smaller values of AIC, AICC, BIC, MSE, MAD and MaxD).

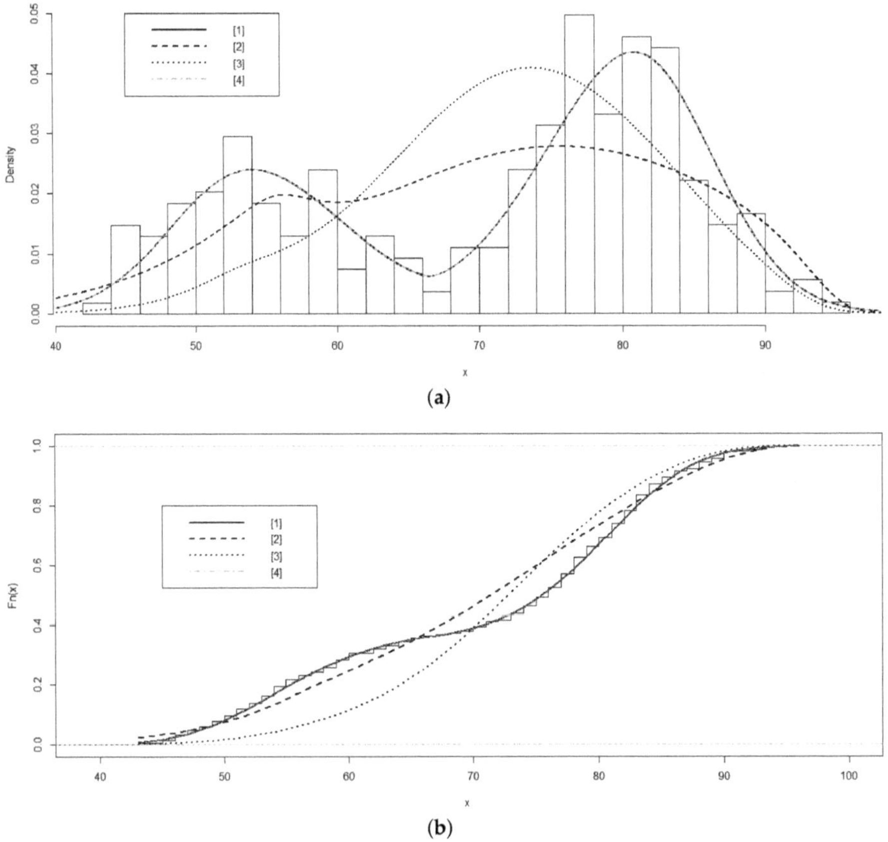

Figure 9. Faithful data - Fitted distributions. [1] Skew GL-Normal distribution; [2] Skew GL-t distribution; [3] Skew Normal-GL distribution; [4] Skew t-GL distribution. (**a**) Probability density function; (**b**) Cumulative distribution.

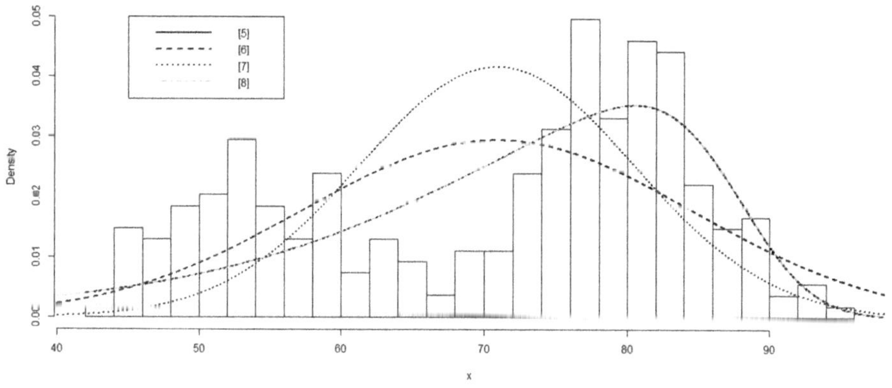

Figure 10. Faithful data - Fitted Nadarajah and Kotz [6] distributions. [5] Skew Logistic-Normal distribution; [6] Skew Logistic-*t* distribution; [7] Skew Normal-Logistic distribution; [8] Skew *t*-Logistic distribution.

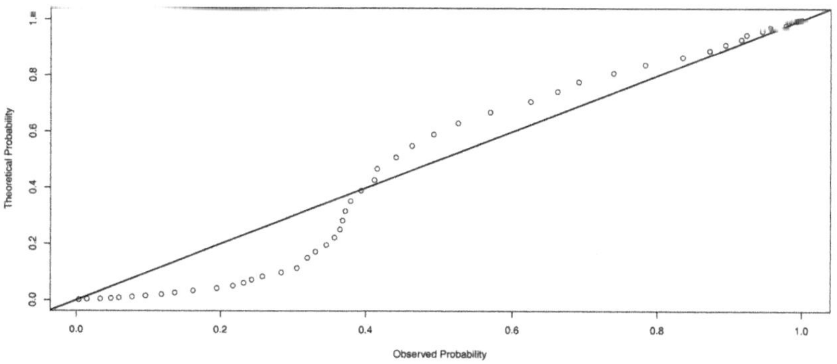

(**a**)

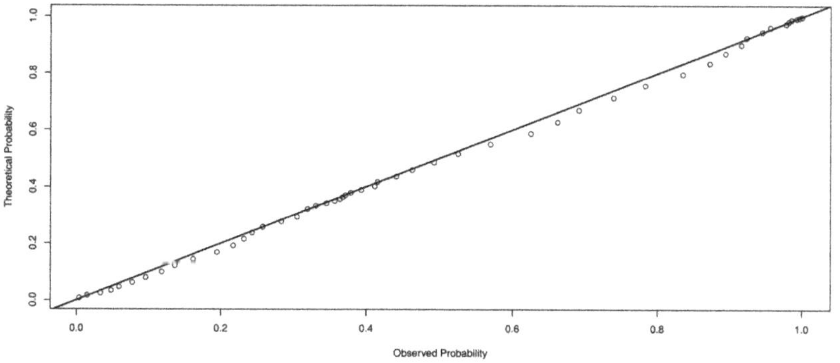

(**b**)

Figure 11. *Cont.*

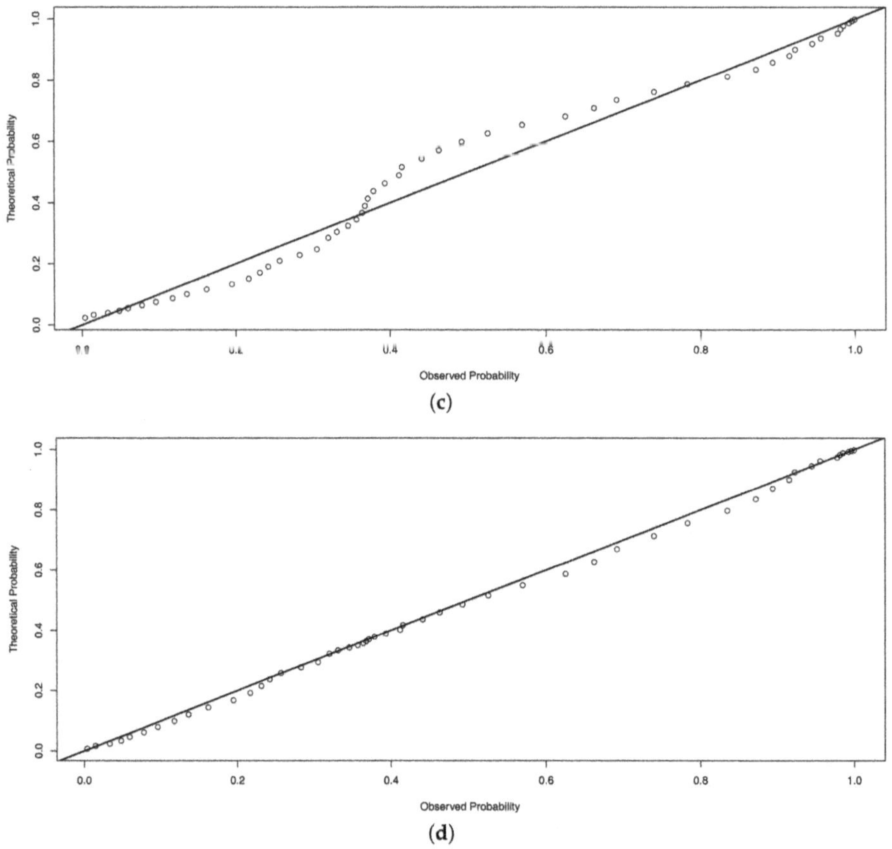

Figure 11. Faithful data - PP-Plot. (**a**) Skew Normal-GL; (**b**) Skew GL-Normal; (**c**) Skew *t*-GL; (**d**) Skew GL-*t*.

Table 3. Performance and accuracy of the distributions.

Model	AIC	BIC	AICC	KS-Test (*p*-Value)	MSE	MAD	MaxD
Skew GL-Normal	2055.26	2033.63	2066.95	0.7344	0.000439	0.0171	0.0378
Skew GL-*t*	2053.27	2028.02	2066.84	0.7344	0.000439	0.0171	0.0378
Skew Normal-GL	3454.74	3433.11	3466.43	<0.0001	0.009393	0.0805	0.1915
Skew *t*-GL	2120.98	2095.74	2134.55	0.01705	0.002896	0.0448	0.1077
Skew Logistic-Normal	2149.71	2131.68	2159.49	-	-	-	-
Skew Logistic-*t*	2147.71	2126.08	2159.40	-	-	-	-
Skew Normal-Logistic	3494.18	3476.15	3503.95	-	-	-	-
Skew *t*-Logistic	2178.58	2156.95	2190.26	-	-	-	-

From the results of Sections 3.1, 3.2 and 3.3, we can see that: (1) in general, our distributions adjusted the data better than the distributions given by Nadarajah and Kotz [6]; (2) The Skew GL-Normal, Skew GL-*t*, Skew *t*-GL and Skew Normal-GL distributions can be used to model symmetrical and asymmetrical unimodal data; (3) Skew GL-*t* and Skew GL-Normal distributions can be used to adjust bimodal symmetrical and asymmetrical data, showing high flexibility which is not common in the literature on probability distributions, and this can be very important in practical applications; (4) For application 1, the Skew GL-Normal and Skew GL-*t* distributions are preferable to fit this data because they present better and similar results (smaller values of AIC, AICC, BIC, MSE, MAD and MaxD), and, for applications 2 and 3, the Skew GL-t distribution is preferred to fit this data presenting better results. Finally, the distributions introduced in this paper are robust to numerical computation.

4. Conclusions

In this paper, we proposed new skew probability density functions using the Azzalini's formula $2f(x)G(cx)$, where f is a symmetric density about zero, and G is a distribution function of a symmetric density about zero. The expressions for f and G are taken from normal, student-t and generalized logistic distributions. We derived expressions for the n-th moments in terms of the H and Meijer G functions [12].

We apply new distributions to three data sets. One application for unimodal data is provided for total expenditure on education in various countries in 2003. Two applications of bimodal data are given for the total expenditure on health in various countries in 2009 and waiting time between eruptions of the Old Faithful Geyser. We conclude that:

1. In general, the distributions introduced in this paper fit better the data when compared with the Skew Logistic-Normal, Skew Logistic-t, Skew Normal-Logistic and Skew t-Logistic distributions, introduced by Nadarajah and Kotz [6];
2. The Skew GL-Normal, Skew GL-t, Skew Normal-GL and Skew t-GL distributions can be used to model symmetrical and asymmetrical unimodal data;
3. The Skew GL-Normal and Skew GL-t distributions can be used to adjust bimodal symmetrical and asymmetrical data, offering good fits, showing a high flexibility which is not common in the literature on probability distributions, and this can be very important in practical applications;
4. For application 1, the Skew GL-Normal and Skew GL-t distributions are preferable to fit this data because they present better and similar results (smaller values of AIC, AICC, BIC, MSE, MAD and MaxD), and, for applications 2 and 3, the Skew GL-t distribution is preferred to fit this data presenting better results;
5. The distributions proposed in this paper apply to all applications without presenting numerical problems, unlike the proposed distributions by Nadarajah and Kotz [6] which had serious numerical problems to adjust bimodal data. (Section 3.3).

Thus, the proposed distributions in this paper are flexible to adjust symmetric and asymmetric data, with unimodal and bimodal behavior, and are robust to numerical computation in practical applications.

Acknowledgments: Gabriela Olinto thanks CNPq (National Counsel of Technological and Scientific Development) for granting a scholarship during 2011–2012 to participate in this research work and Pushpa Narayan Rathie thanks CAPES, Brazil, for supporting his Senior National Visiting Professorship. We wish to thank three anonymous referees for their valuable comments, which improved substantially the presentation of this manuscript.

Author Contributions: Pushpa Narayan Rathie, Paulo Silva and Gabriela Olinto contributed equally to this work.

Conflicts of Interest: The authors declare no conflict of interest.

Appendix General *R* Codes

In this appendix we give the general *R* code for fitting the distributions introduced in this paper for practical purpose.

Appendix A.1 Skew Normal-GL Distribution

```
#READ THE DATA
data <- read.csv(file.choose(), header=T, stringsAsFactor=F, sep=';')

#SKEW NORMAL GENERALIZED LOGISTIC DISTRIBUTION - DENSITY
dnormgen <- function(x, a, b, p, c, mu, sigma){
```

```
(2/sigma)*dnorm(x=x, mean=mu, sd=sigma)/(1+exp(-c*(x-mu)/sigma*(a+b*abs(c*(x-mu)/sigma)**p)))
}

#GIVE THE INITIAL PARAMETERS HERE

theta <- theta0

#LOG-LIKELIHOOD FUNCTION

loglik <- function(pars){
a <- pars[1]
b <- pars[2]
p <- pars[3]
c <- pars[4]
mu <- pars[5]
sigma <- pars[6]
logl <-sum(log(dnormgen(x, a=a, b=b, p=p, c=c, mu=mu, sigma=sigma)))
return(-logl)
}

#FIT

fit=constrOptim(theta=theta, f=loglik, ui=rbind(c(1, 0, 0, 0, 0, 0),
c(0, 1, 0, 0, 0, 0),
c(0, 0, 1, 0, 0, 0),
c(0, 0, 0, 0, 1, 0),
c(0, 0, 0, 0, 0, 1)), ci=c(0, 0, 0, 0, 0)
, method="Nelder-Mead", outer.iterations=300)
```

Appendix A.2 Skew GL-Normal Distribution

```
#READ THE DATA
data <- read.csv(file.choose(), header=T, stringsAsFactor=F, sep=';')

#SKEW GENERALIZED LOGISTIC NORMAL DISTRIBUTION - DENSITY
dglnorm <- function(x, a, b, p, c, mu, sigma){
2/sigma*{(a+b*(1+p)*abs((x-mu)/sigma)**p)*exp(-(x-mu)/sigma*(a+b*abs((x-mu)/sigma)**p))/
(1+exp(-(x-mu)/sigma*(a+b*abs((x-mu)/sigma)**p)))**2}*pnorm(c*(x-mu)/sigma, mean=0,sd=1)
}

#GIVE THE INITIAL PARAMETERS HERE

theta <- theta0

#LOG-LIKELIHOOD FUNCTION

loglik <- function(pars){
a <- pars[1]
b <- pars[2]
p <- pars[3]
c <- pars[4]
mu <- pars[5]
sigma <- pars[6]
logl <-sum(log(dglnorm(x, a=a, b=b, p=p, c=c, mu=mu, sigma=sigma)))
return(-logl)
}

#FIT

fit=constrOptim(theta=theta, f=loglik, ui=rbind(c(1, 0, 0, 0, 0, 0),
c(0, 1, 0, 0, 0, 0),
c(0, 0, 1, 0, 0, 0),
c(0, 0, 0, 0, 1, 0),
c(0, 0, 0, 0, 0, 1)), ci=c(0, 0, 0, 0, 0)
```

```
, method="Nelder-Mead", outer.iterations=300)
```

Appendix A.3 Skew t-GL Distribution

```
#READ THE DATA
data <- read.csv(file.choose(), header=T, stringsAsFactor=F, sep=';')

#SKEW T GENERALIZED LOGISTIC DISTRIBUTION - DENSITY
dtgen <- function(x, a, b, p, c, v, mu, sigma){
(2/sigma)*dt((x-mu)/sigma, df=v)/(1+exp(-c*(x-mu)/sigma*(a+b*abs(c*(x-mu)/sigma)**p)))
}

#GIVE THE INITIAL PARAMETERS HERE

theta <- theta0

#LOG-LIKELIHOOD FUNCTION

loglik  <- function(pars){
a <- pars[1]
b <- pars[2]
p <- pars[3]
c <- pars[4]
v <- pars[5]
mu <- pars[6]
sigma <- pars[7]
logl <-sum(log(dtgen(x, a=a, b=b, p=p, c=c, v=v, mu=mu, sigma=sigma)))
return(-logl)
}

#FIT

fit=constrOptim(theta=theta, f=loglik, ui=rbind(c(1, 0, 0, 0, 0, 0, 0),
c(0, 1, 0, 0, 0, 0, 0),
c(0, 0, 1, 0, 0, 0, 0),
c(0, 0, 0, 0, 1, 0, 0),
c(0, 0, 0, 0, 0, 1, 0)), ci=c(0, 0, 0, 0, 0)
, method="Nelder-Mead", outer.iterations=300)
```

Appendix A.4 Skew GL-t Distribution

```
#READ THE DATA
data <- read.csv(file.choose(), header=T, stringsAsFactor=F, sep=';')

#SKEW GENERALIZED LOGISTIC T DISTRIBUTION - DENSITY
dglt <- function(x, a, b, p, c, v, mu, sigma){
2/sigma*{(a+b*(1+p)*abs((x-mu)/sigma)**p)*exp(-(x-mu)/sigma*(a+b*abs((x-mu)/sigma)**p))/
(1+exp(-(x-mu)/sigma*(a+b*abs((x-mu)/sigma)**p)))**2}*pt(c*(x-mu)/sigma, df=v)
}

#GIVE THE INITIAL PARAMETERS HERE

theta <- theta0

#LOG-LIKELIHOOD FUNCTION

loglik <- function(pars){
a <- pars[1]
b <- pars[2]
p <- pars[3]
c <- pars[4]
v <- pars[5]
mu <- pars[6]
sigma <- pars[7]
```

```
logl <-sum(log(dglt(x, a=a, b=b, p=p, c=c, v=v, mu=mu, sigma=sigma)))
return(-logl)
}

#FIT

fit=constrOptim(theta=theta, f=loglik, ui=rbind(c(1, 0, 0, 0, 0, 0, 0),
c(0, 1, 0, 0, 0, 0, 0),
c(0, 0, 1, 0, 0, 0, 0),
c(0, 0, 0, 0, 1, 0, 0),
c(0, 0, 0, 0, 0, 1, 0)), ci=c(0, 0, 0, 0, 0)
, method="Nelder-Mead", outer.iterations=300)
```

Appendix Density Plots

In this appendix we give the plots of Nadarajah and Kotz [6] estimated densities given in Sections 3.1 and 3.2.

Appendix B.1 Application 1: Expenditure on Education

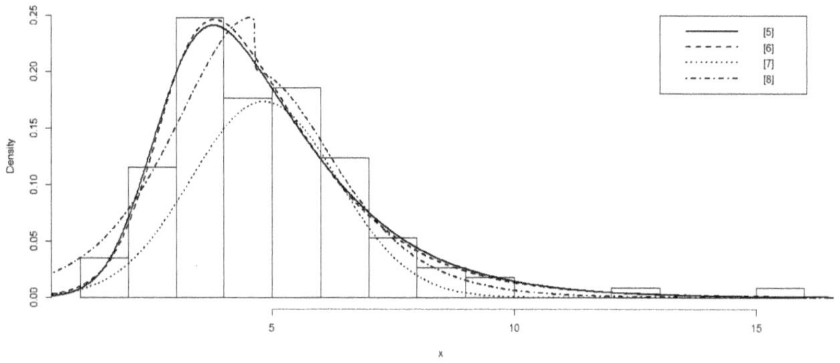

(**a**)

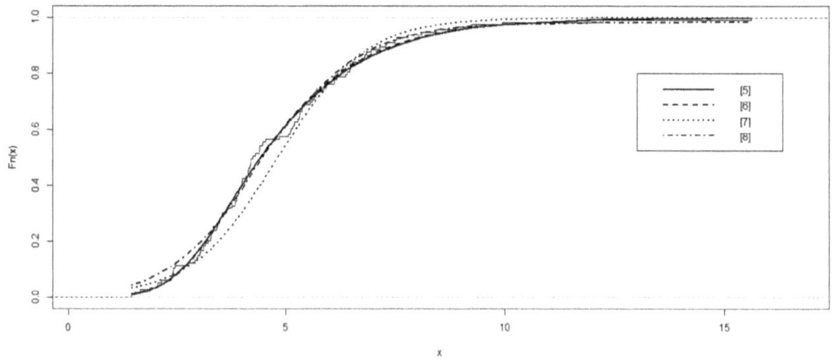

(**b**)

Figure B1. Education data - Fitted Nadarajah and Kotz [6] distributions. [5] Skew Logistic-Normal distribution; [6] Skew Logistic-*t* distribution; [7] Skew Normal-Logistic distribution; [8] Skew *t*-Logistic distribution. (**a**) Probability density function; (**b**) Cumulative distribution.

Application 2: Expenditure on Health

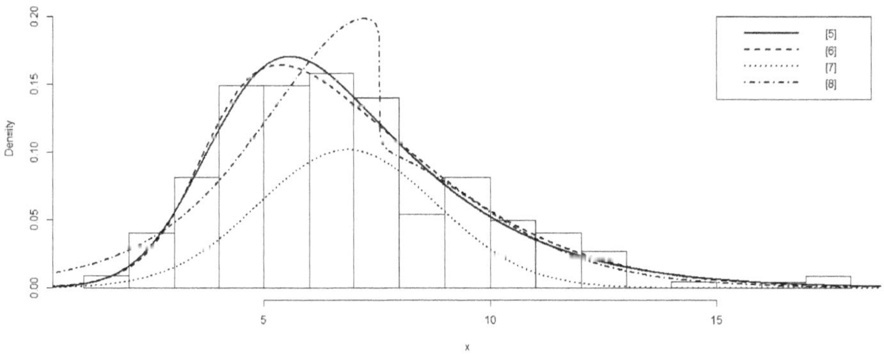

(a)

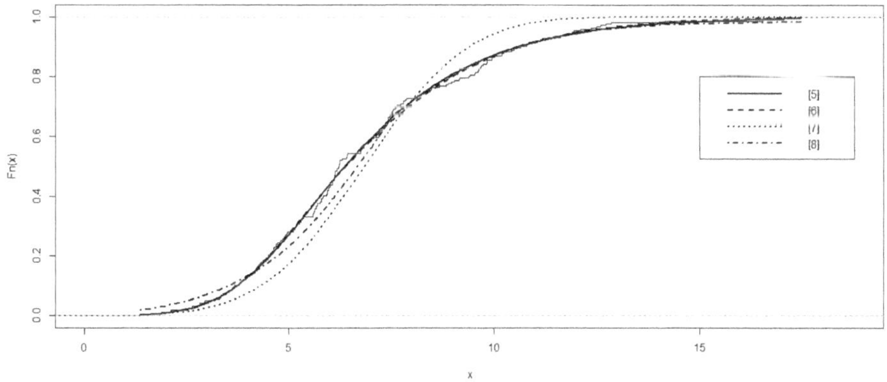

(b)

Figure B2. Health data - Fitted Nadarajah and Kotz [6] distributions. [5] Skew Logistic-Normal distribution; [6] Skew Logistic-*t* distribution; [7] Skew Normal-Logistic distribution; [8] Skew *t*-Logistic distribution. (**a**) Probability density function; (**b**) Cumulative distribution.

References

1. Abtahi, A.; Behboodian, J.; Shafari, M.A. General class of univariate skew distributions considering Stein's lemma and infinite divisibility. *Metrika* **2010**, *75*, 193–206.
2. Rathie, P.N.; Coutinho, M. A new skew generalized logistic distribution and approximations to skew normal distribution. *Aligarh J. Stat.* **2011**, *31*, 1–12.
3. Rathie, P.N.; Swamee, P.K., Matos, G.C.; Coutinho, M.; Carrijo, T.B. H-function and statistical distributions. *Ganita* **2008**, *59*, 23–37.
4. Rathie, P.N.; Swamee, P.K. *On a New Invertible Generalized Logistic Distribution Approximation to Normal Distribution*; Technical Research Report No. 07/2006; Department of Statistics, University of Brasilia: Brasilia, Brazil, 2006.
5. Gupta, R.D.; Kundu, D. Generalized logistic distributions. *J. Appl. Stat. Sci.* **2010**, *18*, 51–66.
6. Nadarajah, S.; Kotz, S. Skew distributions generated from different families. *Acta Appl. Math.* **2006**, *91*, 1–37.
7. Swamee, P.K.; Rathie, P.N. Invertible alternatives to normal and lognormal distributions. *J. Hydrol. Eng. ASCE* **2007**, *12*, 218–221.
8. Rathie, P.N. *Normal Distribution, Univariate*; Springer: Berlin, Germany, 2011; pp. 1012–1013.

9. Azzalini, A. A class of distributions which includes the normal ones. *Scand. J. Stat.* **1985**, *12*, 171–178.
10. Luke, Y.L. *The Special Functions and Their Approximations*; Academic Press: New York, NY, USA, 1969.
11. Springer, M.D. *Algebra of Random Variables*; John Wiley: New York, NY, USA, 1979.
12. Mathai, A.M.; Saxena, R.K.; Haubold, H.J. *The H-Function: Theory and Applications*; Springer: New York, NY, USA, 2010.
13. Gradshteyn, I.S.; Ryzhik, I.M. *Table of Integrals, Series and Products*; Academic Press: San Diego, CA, USA, 2000.
14. Prudnikov, A.P.; Brychkov, Y.A.; Marichev, O.I. *Integrals and Series*; Breach Science: Amsterdam, The Netherlands, 1986.
15. The World Bank: Working for a World Free of Poverty. Government expenditure on education as % of GDP (%). Available online: http://data.worldbank.org/indicator/SE.XPD.TOTL.GD.ZS (accessed on 1 February 2016).
16. R Core Team. *R: A Language and Environment for Statistical Computing*; R Foundation for Statistical Computing: Vienna, Austria, 2015. Available online: http://www.R-project.org/ (accessed on 1 February 2016).
17. The World Bank: Working for a World Free of Poverty. Health expenditure, total (% of GDP). Available online: http://data.worldbank.org/indicator/SH.XPD.TOTL.ZS (accessed on 1 February 2016).

axioms

MDPI

Article

Closed-Form Representations of the Density Function and Integer Moments of the Sample Correlation Coefficient

Serge B. Provost

Department of Statistical & Actuarial Sciences, The University of Western Ontario,
London, ON N6A 5B7, Canada; provost@stats.uwo.ca

Academic Editors: Angel Garrido and Hans J. Haubold
Received: 7 May 2015; Accepted: 18 June 2015; Published: 20 July 2015

Abstract: This paper provides a simplified representation of the exact density function of $\mathcal{R}$, the sample correlation coefficient. The odd and even moments of $\mathcal{R}$ are also obtained in closed forms. Being expressed in terms of generalized hypergeometric functions, the resulting representations are readily computable. Some numerical examples corroborate the validity of the results derived herein.

Keywords: sample correlation coefficient; hypergeometric function; density function; moments

1. Introduction

Given $\{(X_i, Y_i), i = 1, \ldots, n\}$, a simple random sample of size n from a bivariate normal distribution, the sample correlation coefficient,

$$\mathcal{R} = \frac{1}{n}\sum_{i=1}^{n}\left(\frac{X_i - \overline{X}}{S_X}\right)\left(\frac{Y_i - \overline{Y}}{S_Y}\right) \tag{1}$$

where $\overline{X} = \sum_{i=1}^{n} X_i/n$, $\overline{Y} = \sum_{i=1}^{n} Y_i/n$, $S_X^2 = \sum_{i=1}^{n}(X_i - \overline{X})^2/n$ and $S_Y^2 = \sum_{i=1}^{n}(Y_i - \overline{Y})^2/n$, is the maximum likelihood estimator of $\rho_{X,Y}$, Pearson's product-moment correlation coefficient. Fisher [1] obtained the following series representation of the density function of $\mathcal{R}$:

$$f_{\mathcal{R}}(r) = \frac{2^{n-3}}{\pi(n-3)!}\left(1-\rho^2\right)^{\frac{n-1}{2}}\left(1-r^2\right)^{\frac{n-4}{2}}\sum_{i=0}^{\infty}\Gamma^2\left(\frac{n+i-1}{2}\right)\frac{(2\rho r)^i}{i!} \tag{2}$$

which converges for $-1 < \rho r < 1$.

Closed-form representations of the exact density of $\mathcal{R}$ are derived in Section 2. They are given in terms of the generalized hypergeometric function,

$${}_pF_q(a_1, \ldots, a_p; b_1, \ldots, b_q; z) = \sum_{k=0}^{\infty}\frac{(a_1)_k \cdots (a_p)_k}{(b_1)_k \cdots (b_q)_k}\frac{z^k}{k!} \tag{3}$$

where, for example, $(a_1)_k = \Gamma(a_1 + k)/\Gamma(a_1)$. More specifically, it will be shown that the exact density of $\mathcal{R}$ can be expressed as

$$\begin{aligned} g(r) \;=\; & \frac{2^{n-3}}{\pi(n-3)!}\left(1-\rho^2\right)^{\frac{n-1}{2}}\left(1-r^2\right)^{\frac{n-4}{2}} \\ & \times\left[\Gamma^2\left(\tfrac{n-1}{2}\right){}_2F_1\left(\tfrac{n-1}{2}, \tfrac{n-1}{2}; \tfrac{1}{2}; \rho^2 r^2\right) + 2\rho r\,\Gamma^2\left(\tfrac{n}{2}\right){}_2F_1\left(\tfrac{n}{2}, \tfrac{n}{2}; \tfrac{3}{2}; \rho^2 r^2\right)\right] \end{aligned} \tag{4}$$

for $-1 < \rho r < 1$, which simplifies to

$$g(r) = \kappa(n,\rho)(1-r^2)^{\frac{n}{2}-2}_{2}F_1(n-1,\, n-1;\, n-1/2;\, (1+\rho r)/2) \tag{5}$$

where $\kappa(n,\rho) = [(n-2)\,\mathcal{B}^2\left(\frac{n-1}{2},\frac{n}{2}\right)(1-\rho^2)^{\frac{n-1}{2}}]/[\pi\, 2^{n+1}\,\mathcal{B}(n-1,n)]$, $\mathcal{B}(a,b) = \Gamma(a)\Gamma(b)/\Gamma(a+b)$ denoting the beta function. For various results on the hypergeometric function ${}_2F_1(a,b;c,z)$ and its main properties, the reader is referred to Olver *et al.* [2], Chapter 15. Closed-form representations of the odd and even moments of $\mathcal{R}$ are provided in Section 3 and some numerical examples are included in Section 4.

Fisher's $\mathcal{Z}$-transform is a well-known transformation of $\mathcal{R}$ whose associated approximate normal distribution is known to present some shortcomings, especially when the sample size is small and $|\rho|$ is large, in which case the distribution of $\mathcal{R}$ is markedly skewed. Winterbottom [3] showed that the normal approximation requires large sample sizes to be valid. It is also known that, in the bivariate normal case, the asymptotic variance of Fisher's $\mathcal{Z}$ statistic does not depend on ρ. Furthermore, as pointed out by Hotelling [4], the variance of $\mathcal{R}$ changes with the mean. The density and moment expressions derived in this paper remain accurate for any values of ρ and n.

2. The Exact Density $\mathcal{R}$

It should be noted that the series representation of the density function of $\mathcal{R}$ given in Equation (2) converges very slowly. It was indeed observed that, in certain instances, more than 1000 terms may be necessary to reach convergence. Closed-form representations of the exact density function of $\mathcal{R}$ are derived in this section.

First, we note that the identity,

$$\frac{\Gamma[1/2]}{k!\,\Gamma[1/2+k]} = \frac{2^{2k}}{(2k)!} \tag{6}$$

can be established by re-expressing the Legendre duplication formula,

$$\Gamma(2k) = \pi^{-1/2}\, 2^{2k-1}\Gamma(k)\,\Gamma(k+1/2) \tag{7}$$

as

$$[2k\,\Gamma(2k)] = (\Gamma(1/2))^{-1}\, 2^{2k}\,[k\,\Gamma(k)]\,\Gamma(1/2\,+\,k)$$

Moreover, since $\Gamma(3/2+k) = (1/2+k)\,\Gamma(1/2+k) = (1/2)\,(2k+1)\,\Gamma(1/2+k)$ and $\Gamma(3/2) = (1/2)\,\Gamma(1/2)$, it follows from Equation (6) that

$$\frac{\Gamma(3/2)}{k!\,\Gamma(3/2+k)} = \frac{2^{2k}}{(2k+1)!} \tag{8}$$

In order to prove that the representation of the density function of given in Equation (4) is equivalent to the series representation (2), it suffices to show that

$$\begin{aligned}\sum_{k=0}^{\infty}\frac{(2r\rho)^k}{k!}\Gamma^2[(k+n-1)/2] &= \Gamma^2\left(\tfrac{n}{2}-\tfrac{1}{2}\right){}_2F_1\left(\tfrac{n}{2}-\tfrac{1}{2},\tfrac{n}{2}-\tfrac{1}{2};\tfrac{1}{2};r^2\rho^2\right)\\ &\quad +2r\rho\,\Gamma^2\left(\tfrac{n}{2}\right){}_2F_1\left(\tfrac{n}{2},\tfrac{n}{2};\tfrac{3}{2};r^2\rho^2\right)\end{aligned} \tag{9}$$

Now, letting $k = 2j+1$, we establish that when k odd,

$$2r\rho\sum_{j=0}^{\infty}\frac{(2r\rho)^{2j}}{(2j+1)!}\Gamma^2[(2j\,+\,n)/2] = 2r\rho\,\Gamma^2\left(\tfrac{n}{2}\right){}_2F_1\left(\tfrac{n}{2},\tfrac{n}{2};\tfrac{3}{2};r^2\rho^2\right) \tag{10}$$

Note that

$$\begin{aligned} 2r\rho \sum_{j=0}^{\infty} \frac{(2r\rho)^{2j}}{(2j+1)!}\Gamma^2(j+n/2) &= 2r\rho \sum_{j=0}^{\infty} \frac{(2r\rho)^{2j}}{(2j+1)!}\Gamma^2(j+n/2) \\ &= 2r\rho \sum_{j=0}^{\infty} (r\rho)^{2j}\,\Gamma(j+n/2)\,\Gamma(j+n/2)\frac{2^{2j}}{(2j+1)!} \end{aligned}$$

However,

$$\begin{aligned} 2r\rho\,\Gamma^2\left(\tfrac{n}{2}\right){}_2F_1\left(\tfrac{n}{2},\tfrac{n}{2};\tfrac{3}{2};r^2\rho^2\right) &= 2r\rho\,\Gamma^2\left(\tfrac{n}{2}\right)\sum_{j=0}^{\infty} \frac{\Gamma(\frac{n}{2}+j)\Gamma(\frac{n}{2}+j)\Gamma(\frac{3}{2})}{\Gamma(\frac{n}{2})\Gamma(\frac{n}{2})\Gamma(\frac{3}{2}+j)}\frac{(r^2\rho^2)^j}{j!} \\ &= 2r\rho \sum_{j=0}^{\infty} \Gamma(n/2+j)\,\Gamma(n/2+j)\,(r^2\rho^2)^j\left(\frac{\Gamma(3/2)}{j!\,\Gamma(3/2+j)}\right) \end{aligned}$$

which, in view of Equation (8), proves the result.

We now show that when $k = 2i$,

$$\sum_{i=0}^{\infty} \frac{(2r\rho)^{2i}}{(2i)!}\Gamma^2((2i+n-1)/2) = {}_2F_1\left(\frac{n}{2}-\frac{1}{2},\frac{n}{2}-\frac{1}{2};\frac{1}{2};r^2\rho^2\right)\Gamma^2\left(\frac{n}{2}-\frac{1}{2}\right) \tag{11}$$

First, note that

$${}_2F_1\left(\frac{n}{2}-\frac{1}{2},\frac{n}{2}-\frac{1}{2};\frac{1}{2};r^2\rho^2\right)\Gamma^2\left(\frac{n}{2}-\frac{1}{2}\right) = \Gamma^2\left(\frac{n}{2}-\frac{1}{2}\right)\sum_{i=0}^{\infty}\frac{\Gamma\left(\frac{n}{2}-\frac{1}{2}+i\right)\Gamma\left(\frac{n}{2}-\frac{1}{2}+i\right)\Gamma\left(\frac{1}{2}\right)}{\Gamma\left(\frac{n}{2}-\frac{1}{2}\right)\Gamma\left(\frac{n}{2}-\frac{1}{2}\right)\Gamma\left(\frac{1}{2}+i\right)}\frac{(r^2\rho^2)^i}{i!}$$

The result is established by applying identity (7) wherein k is replaced by $k-1$. Thus, one has the following closed-form representation of the exact density function of $\mathcal{R}$:

$$\begin{aligned} g_1(r) &= \frac{1}{\pi(n-3)!}\,2^{n-3}(1-r^2)^{\frac{n-4}{2}}(1-\rho^2)^{\frac{n-1}{2}} \\ &\times\left[\Gamma^2\left(\tfrac{n}{2}-\tfrac{1}{2}\right){}_2F_1\left(\tfrac{n}{2}-\tfrac{1}{2},\tfrac{n}{2}-\tfrac{1}{2};\tfrac{1}{2};\rho^2 r^2\right) + 2\rho\, r\,\Gamma^2\left(\tfrac{n}{2}\right){}_2F_1\left(\tfrac{n}{2},\tfrac{n}{2};\tfrac{3}{2};\rho^2 r^2\right)\right] \end{aligned} \tag{12}$$

A simplified representation of this expression can be obtained by making use of the following identity listed under "Quadratic transformations with fixed a, b, z" on the Wolfram website, http://functions.wolfram.com/HypergeometricFunctions/Hypergeometric2F1/17/02/10/ :

$$\begin{aligned} {}_2F_1\left(a,b;\tfrac{a+b+1}{2};z\right) &= \frac{\sqrt{\pi}\,\Gamma\left(\frac{a+b+1}{2}\right)}{\Gamma\left(\frac{a+1}{2}\right)\Gamma\left(\frac{b+1}{2}\right)}\,{}_2F_1\left(\tfrac{a}{2},\tfrac{b}{2};\tfrac{1}{2};(2z-1)^2\right) \\ &+\frac{2\sqrt{\pi}\,(2z-1)\Gamma\left(\frac{a+b+1}{2}\right)}{\Gamma\left(\frac{a}{2}\right)\Gamma\left(\frac{b}{2}\right)}\,{}_2F_1\left(\tfrac{a+1}{2},\tfrac{b+1}{2};\tfrac{3}{2};(2z-1)^2\right) \end{aligned} \tag{13}$$

which, on making the substitutions, $a \to n-1, b \to n-1 \text{and} z \to (1+\rho\, r)/2$, becomes

$$\begin{aligned} {}_2F_1\left(n-1,n-1;n-\tfrac{1}{2};\tfrac{1+\rho r}{2}\right) &= \frac{\sqrt{\pi}\,\Gamma\left(n-\frac{1}{2}\right)}{\Gamma\left(\frac{n}{2}\right)\Gamma\left(\frac{n}{2}\right)}\,{}_2F_1\left(\tfrac{n-1}{2},\tfrac{n-1}{2};\tfrac{1}{2};\rho^2 r^2\right) \\ &+\frac{2\,r\,\rho\sqrt{\pi}\,\Gamma\left(n-\frac{1}{2}\right)}{\Gamma\left(\frac{n-1}{2}\right)\Gamma\left(\frac{n-1}{2}\right)}\,{}_2F_1\left(\tfrac{n}{2},\tfrac{n}{2};\tfrac{3}{2};\rho^2 r^2\right) \end{aligned} \tag{14}$$

Multiplying both sides by $\Gamma^2\left(\frac{n-1}{2}\right)\Gamma^2\left(\frac{n}{2}\right)/\left\{\Gamma\left(n-\frac{1}{2}\right)\sqrt{\pi}\right\}$ then yields

$$\begin{aligned} \frac{\Gamma^2\left(\frac{n-1}{2}\right)\Gamma\left(\frac{n}{2}\right)^2}{\Gamma\left(n-\frac{1}{2}\right)\sqrt{\pi}}\,{}_2F_1\left(n-1,n-1;n-\tfrac{1}{2};\tfrac{1+\rho r}{2}\right) &= \Gamma^2\left(\tfrac{n-1}{2}\right){}_2F_1\left(\tfrac{n-1}{2},\tfrac{n-1}{2};\tfrac{1}{2};\rho^2 r^2\right) \\ &+2\,\rho\, r\,\Gamma^2\left(\tfrac{n}{2}\right){}_2F_1\left(\tfrac{n}{2},\tfrac{n}{2};\tfrac{3}{2};\rho^2 r^2\right) \end{aligned} \tag{15}$$

Hence, the following form of the exact density function of $\mathcal{R}$:

$$\frac{2^{n-3}\Gamma^2(\frac{n-1}{2})\Gamma^2(\frac{n}{2})(1-\rho^2)^{\frac{n-1}{2}}}{\pi^{3/2}\Gamma(n-\frac{1}{2})(n-3)!}(1-r^2)_2^{\frac{n-4}{2}}F_1\left(n-1,n-1;n-\tfrac{1}{2};\tfrac{1}{2}(1+r\rho)\right)$$
$$= \frac{2^{n-3}\mathcal{B}^2(\frac{n-1}{2},\frac{n}{2})\Gamma(n-\frac{1}{2})(1-\rho^2)^{\frac{n-1}{2}}}{\pi^{3/2}(n-3)!}(1-r^2)^{\frac{n-4}{2}}{}_2F_1\left(n-1,n-1;n-\tfrac{1}{2};\tfrac{1}{2}(1+\rho r)\right)$$

which, on letting $k = n-1$ in Equation (6), gives

$$\frac{\mathcal{B}^2\left(\frac{n-1}{2},\frac{n}{2}\right)(2n-2)!\,(1-\rho^2)^{\frac{n-1}{2}}}{2^{n+1}\,\pi(n-3)!(n-1)!}\left(1-r^2\right)^{\frac{n-4}{2}}{}_2F_1\left(n-1,n-1;n-\frac{1}{2};\frac{1}{2}(1+\rho r)\right)$$

Finally the following representation of the density function of $\mathcal{R}$ is obtained on writing $(2n-2)!/[(n-3)!(n-1)!]$ as $(n-2)\Gamma(2n-1)/[\Gamma((n-1)\Gamma(n)] = (n-2)/\mathcal{B}(n-1,n)$:

$$g(r) = \frac{(n-2)\mathcal{B}^2(\frac{n-1}{2},\frac{n}{2})(1-\rho^2)^{\frac{n-1}{2}}}{2^{n+1}\mathcal{B}(n-1,n)\,\pi}(1-r^2)_2^{\frac{n}{2}-2}F_1\left(n-1\,,\,n-1\,;\,n-\tfrac{1}{2}\,;\,\tfrac{1+\rho r}{2}\right) \tag{16}$$

Incidentally, this expression is more compact than that proposed by Hotelling [4].

3. Closed Forms for the Moments of $\mathcal{R}$

It is shown in this section that the moments of $\mathcal{R}$ can also be expressed in closed forms. The following moment expressions are available in Anderson [5] pp. 151–152:

$$E\left(\mathcal{R}^k\right) = \frac{(1-\rho^2)^{\frac{n-1}{2}}}{\sqrt{\pi}\,\Gamma(\frac{n-1}{2})}\sum_{i=0}^{\infty}\frac{(2\rho)^{2i+1}}{(2i+1)!}\frac{\Gamma(\frac{3}{2}+\frac{k-1}{2}+i)\,\Gamma^2(\frac{n}{2}+i)}{\Gamma(\frac{n+1}{2}+\frac{k-1}{2}+i)}\ \text{for}k\text{odd} \tag{17}$$

and

$$E\left(\mathcal{R}^k\right) = \frac{(1-\rho^2)^{\frac{n-1}{2}}}{\sqrt{\pi}\,\Gamma(\frac{n-1}{2})}\sum_{i=0}^{\infty}\frac{(2\rho)^{2i}}{(2i)!}\frac{\Gamma(\frac{1}{2}+\frac{k}{2}+i)\,\Gamma^2(\frac{n}{2}-\frac{1}{2}+i)}{\Gamma(\frac{n-1}{2}+\frac{k}{2}+i)}\ \text{for}k\text{even} \tag{18}$$

We will show that when k is odd,

$$E\left(\mathcal{R}^k\right) = \frac{2\rho\,(1-\rho^2)^{\frac{n-1}{2}}\,\Gamma(\frac{k}{2}+1)\,\Gamma^2(\frac{n}{2})}{\sqrt{\pi}\,\Gamma(\frac{n-1}{2})\,\Gamma(\frac{k+n}{2})}\,{}_3F_2\left(\tfrac{k}{2}+1,\tfrac{n}{2},\tfrac{n}{2};\tfrac{3}{2},\tfrac{k}{2}+\tfrac{n}{2};\rho^2\right) \tag{19}$$

and when k is even,

$$E\left(\mathcal{R}^k\right) = \frac{(1-\rho^2)^{\frac{n-1}{2}}\,\Gamma(\frac{k+1}{2})\,\Gamma(\frac{n-1}{2})}{\sqrt{\pi}\,\Gamma(\frac{k+n-1}{2})}\,{}_3F_2\left(\tfrac{k}{2}+\tfrac{1}{2},\tfrac{n}{2}-\tfrac{1}{2},\tfrac{n}{2}-\tfrac{1}{2};\tfrac{1}{2},\tfrac{k}{2}+\tfrac{n}{2}-\tfrac{1}{2};\rho^2\right) \tag{20}$$

where the generalized hypergeometric function, ${}_pF_q(a_1,\ldots,a_p;b_1,\ldots,b_q;z)$, is as defined in Equation (3).

Since

$${}_3F_2(n_1,n_2,n_3;d_1,d_2;v) = \sum_{k=0}^{\infty}\frac{\Gamma(n_1+k)\Gamma(n_2+k)\Gamma(n_3+k)\Gamma(d_1)\Gamma(d_2)\,v^k}{\Gamma(n_1)\,\Gamma(n_2)\,\Gamma(n_3)\,\Gamma(d_1+k)\,\Gamma(d_2+k)\,k!}$$

then, according to Equation (19), when k is odd, one has

$$E\left(\mathcal{R}^k\right) = \frac{(1-\rho^2)^{\frac{n-1}{2}}}{\sqrt{\pi}\,\Gamma(\frac{n-1}{2})}\frac{2\rho\,\Gamma(\frac{k}{2}+1)\Gamma^2(\frac{n}{2})}{\Gamma(\frac{k+n}{2})}\sum_{i=0}^{\infty}\frac{\Gamma(\frac{k}{2}+1+i)\Gamma^2(\frac{n}{2}+i)\Gamma(\frac{3}{2})\Gamma(\frac{k}{2}+\frac{n}{2})\,\rho^{2i}}{\Gamma(\frac{k}{2}+1)\Gamma^2(\frac{n}{2})\Gamma(\frac{3}{2}+i)\Gamma(\frac{k}{2}+\frac{n}{2}+i)\,i!}$$
$$= \frac{(1-\rho^2)^{\frac{n-1}{2}}}{\sqrt{\pi}\,\Gamma(\frac{n-1}{2})}\sum_{i=0}^{\infty}\frac{2\,\Gamma(\frac{k}{2}+1+i)\,\Gamma^2(\frac{n}{2}+i)\,\Gamma(\frac{3}{2})\,\rho^{2i+1}}{\Gamma(\frac{k}{2}+\frac{n}{2}+i)\,\Gamma(\frac{3}{2}+i)\,i!}$$

which, in light of Equation (8), that is, $\frac{\Gamma(3/2)}{\Gamma(3/2+i)\,i!} = \frac{2^{2i}}{(2i+1)!}$, is seen to be equal to the expression given in Equation (17).

Now, when k is even, according to Equation (20), one has

$$\begin{aligned} E\left(\mathcal{R}^k\right) &= \frac{(1-\rho^2)^{\frac{n-1}{2}}}{\sqrt{\pi}} \frac{\Gamma\left(\frac{k+1}{2}\right)\Gamma\left(\frac{n-1}{2}\right)}{\Gamma\left(\frac{1}{2}(k+n-1)\right)} \sum_{i=0}^{\infty} \frac{\Gamma\left(\frac{k}{2}+\frac{1}{2}+i\right)\Gamma^2\left(\frac{n}{2}-\frac{1}{2}+i\right)\Gamma\left(\frac{1}{2}\right)\Gamma\left(\frac{k}{2}+\frac{n}{2}-\frac{1}{2}\right)\rho^{2i}}{\Gamma\left(\frac{k}{2}+\frac{1}{2}\right)\Gamma^2\left(\frac{n}{2}-\frac{1}{2}\right)\Gamma\left(\frac{1}{2}+i\right)\Gamma\left(\frac{k}{2}+\frac{n}{2}-\frac{1}{2}+i\right)i!} \\ &= \frac{(1-\rho^2)^{\frac{n-1}{2}}}{\sqrt{\pi}} \sum_{i=0}^{\infty} \frac{\Gamma\left(\frac{k}{2}+\frac{1}{2}+i\right)\Gamma^2\left(\frac{n}{2}-\frac{1}{2}+i\right)\Gamma\left(\frac{1}{2}\right)\rho^{2i}}{\Gamma\left(\frac{n}{2}-\frac{1}{2}\right)\Gamma\left(\frac{k}{2}+\frac{n}{2}-\frac{1}{2}+i\right)\Gamma\left(\frac{1}{2}+i\right)i!} \end{aligned}$$

which turns out to be equal to the right-hand side of Equation (18) on noting that, as proved earlier, $\frac{\Gamma\left(\frac{1}{2}\right)}{\Gamma\left(\frac{1}{2}+i\right)i!} = \frac{2^{2i}}{(2i)!}$

4. Numerical Examples

When the series representations of the density function or the moments of $\mathcal{R}$ are utilized, the number of terms required to achieve convergence depends on the length of the observation vector, the underlying correlation coefficient and the point at which the density function is evaluated in the former case or the order of the required moment in the latter. In certain instances, even 1000 terms turn out to be insufficient. The proposed closed-form expressions, which for all intents and purposes produce exact numerical results, can be evaluated much more quickly.

Consider for example the case, $n = 10$ and $\rho = -0.97$. Table 1 reports the values of the probability density function (PDF) of $\mathcal{R}$, first determined from $f(r)$ as specified by Equation (2), truncated to 500 and 1000 terms, and then, from $g(r)$, the exact closed-form representation given in Equation (16), for $r = -0.99, -0.25, 0.05, 0.25, 0.95$.

Table 1. PDF of $\mathcal{R}$ as evaluated from $f(r)$ truncated to m terms and $g(r)$.

r	$f(r)[m = 500]$	$f(r)[m = 1000]$	$g(r)(Closedform)$
-0.99	21.0839	21.1043	21.1043
-0.25	0.0000284304	0.0000284304	0.0000284304
0.05	2.15111×10^{-6}	2.15111×10^{-6}	2.15111×10^{-6}
0.25	4.20668×10^{-7}	4.20668×10^{-7}	4.20668×10^{-7}
0.95	4.61344×10^{-11}	1.1523×10^{-11}	1.15232×10^{-11}

Similarly, when $n = 75$ and $\rho = 0.80$, one obtains the numerical results appearing in Table 2.

Table 2. PDF of $\mathcal{R}$ as evaluated from $f(r)$ truncated to m terms and $g(r)$.

r	$f(r)[m = 500]$	$f(r)[m = 1000]$	$g(r)(Closedform)$
-0.90	1.08277×10^{-18}	1.07281×10^{-18}	1.57819×10^{-59}
-0.60	4.50675×10^{-19}	4.50675×10^{-19}	5.23693×10^{-36}
0.60	0.0128167	0.0128167	0.0128167
0.95	6.01144×10^{-7}	6.01144×10^{-7}	6.01144×10^{-7}

Certain moments of $\mathcal{R}$ are included Table 3 for some values of k, n and ρ, along with the computing times associated with the evaluation of the truncated series representations of the moments given in Equations (17) and (18) and the closed-form representations specified by Equations (19) and (20). We observed that the computing times can be significantly reduced by making use of the closed-form expressions. All the calculations were carried out with the symbolic computing software *Mathematica*, the code being available from the author upon request.

Table 3. Certain moments of $\mathcal{R}$ and associated computing times in seconds.

Formula	(n, ρ, k)	k^{th} moment	Timing
(17) *1000 terms*	$(800, 0.75, 7)$	0.134421	0.468
(19) *closed-form*	$(800, 0.75, 7)$	0.134421	0.032
(18) *1000 terms*	$(200, -0.91, 12)$	0.324631	0.577
(20) *closed-form*	$(200, -0.91, 12)$	0.324631	0.047
(17) *1000 terms*	$(8, 0.255, 23)$	0.001752	0.327
(19) *closed-form*	$(8, 0.255, 23)$	0.001752	5.72459×10^{-16}
(18) *1000 terms*	$(60, 0.051, 36)$	1.16476×10^{-13}	0.514
(20) *closed-form*	$(60, 0.051, 36)$	1.16476×10^{-13}	6.67869×10^{-16}

Acknowledgments: The financial support of the Natural Sciences and Engineering Research Council of Canada is gratefully acknowledged. Thanks are also due to two referees for their valuable comments and suggestions.

Conflicts of Interest: The author declares no conflict of interest.

References

1. Fisher, R.A. Distribution of the values of the correlation coefficient in samples from an indefinitely large population. *Biometrika* **1915**, *10*, 507–521. [CrossRef]
2. Olver, F.W.J.; Lozier, D.W.; Boisvert, R.; Clark, C.W. *NIST Handbook on Mathematical Functions;* Cambridge University Press: Cambridge, UK, 2010.
3. Winterbottom, A. A note on the derivation of Fisher's transformation of the correlation coefficient. *Am. Stat.* **1979**, *33*, 142–143.
4. Hotelling, H. New light on the correlation coefficient and its transforms. *J. R. Stat. Soc. Ser. B* **1953**, *15*, 193–232.
5. Anderson, T.W. *An Introduction to Multivariate Statistical Analysis;* Wiley: New York, NY, USA, 1984.

axioms

MDPI

Article

On some Integral Representations of Certain G-Functions

Seemon Thomas

Department of Statistics, St Thomas College Pala, Kerala 686574, India; palaseemon@gmail.com; Tel.: +91-949-532-5341

Academic Editor: Hans J. Haubold
Received: 3 November 2015; Accepted: 23 December 2015; Published: 31 December 2015

Abstract: This is a brief exposition of some statistical techniques utilized to obtain several useful integral equations involving G-functions.

Keywords: generalized type-1 Dirichlet model; marginal density; Meijer's G-function; moments

1. Introduction

Thomas and George [1] introduced a generalized type-1 Dirichlet model having several mathematical and statistical properties. The density has been derived from the property of the following ratios:
Let $x_1, ..., x_k$ be such that $0 < x_i < 1,\ i = 1, ..., k,\ 0 < x_1 + \cdots + x_k < 1$ and let

$$\frac{x_1}{x_1+x_2}, \frac{x_1+x_2}{x_1+x_2+x_3}, ..., \frac{x_1+\cdots+x_{k-1}}{x_1+\cdots+x_k}, x_1+\cdots+x_k$$

be independently distributed as type-1 beta with parameters $(\alpha_1, \alpha_2), (\alpha_1+\alpha_2+\beta_2, \alpha_3), (\alpha_1+\alpha_2+\alpha_3+\beta_2+\beta_3, \alpha_4), ..., (\alpha_1+\cdots+\alpha_k+\beta_2+\cdots+\beta_k, \alpha_{k+1})$ respectively. Then $(x_1, ..., x_k)$ has the density function of the following form:

$$f(x_1, ..., x_k) = c_k x_1^{\alpha_1-1} \ldots x_k^{\alpha_k-1}(x_1+x_2)^{\beta_2}...(x_1+\cdots+x_k)^{\beta_k}(1-x_1-\cdots-x_k)^{\alpha_{k+1}-1}. \quad (1)$$

Obviously it is a generalization of type-1 Dirichlet probability model. The normalizing constant c_k can be evaluated as

$$\begin{aligned} c_k &= \frac{\Gamma(\alpha_1+\alpha_2)}{\Gamma(\alpha_1)\Gamma(\alpha_2)...\Gamma(\alpha_{k+1})} \frac{\Gamma(\alpha_1+\alpha_2+\alpha_3+\beta_2)}{\Gamma(\alpha_1+\alpha_2+\beta_2)}... \\ &\times \frac{\Gamma(\alpha_1+\cdots+\alpha_k+\beta_2+\cdots+\beta_{k-1})}{\Gamma(\alpha_1+\cdots+\alpha_{k-1}+\beta_2+\cdots+\beta_{k-1})} \frac{\Gamma(\alpha_1+\cdots+\alpha_{k+1}+\beta_2+\cdots+\beta_k)}{\Gamma(\alpha_1+\cdots+\alpha_k+\beta_2+\cdots+\beta_k)} \end{aligned}$$

for $\Re(\alpha_j) > 0,\ j = 1, ..., k+1,\ \Re(\alpha_1+\cdots+\alpha_j+\beta_2+\cdots+\beta_j) > 0,\ j = 2, ..., k$, where $\Re$ denotes the real part of $(\cdot)$. For more properties of the Model (1) one may refer Thomas and George [1]. Note that

$$x_1 = \frac{x_1}{x_1+x_2} \frac{x_1+x_2}{x_1+x_2+x_3} ... \frac{x_1+\cdots+x_{k-1}}{x_1+\cdots+x_k}(x_1+\cdots+x_k)$$

is structurally a product of k independent real variables and its density can be written in terms of a G-function of the type $G_{k,k}^{k,0}(\cdot)$. The majority of the established special functions can be represented in terms of the G-function. A notable property of G-functions is the closure property. The closure property implies that whenever a function is expressible as a G-function of a constant multiple of some constant power of the function argument, the derivative and the antiderivative of this function are expressible so too. A general G-function is defined as the following Mellin-Barnes integral:

$$\begin{aligned} G_{p,q}^{m,n}(z) &= G_{p,q}^{m,n}\left(z\Big|_{b_1,\ldots,b_q}^{a_1,\ldots,a_p}\right) \\ &= \frac{1}{2\pi i}\int_{\mathcal{L}} \frac{\left\{\prod_{j=1}^{m}\Gamma(b_j+s)\right\}\left\{\prod_{j=1}^{n}\Gamma(1-a_j-s)\right\}}{\left\{\prod_{j=m+1}^{q}\Gamma(1-b_j-s)\right\}\left\{\prod_{j=n+1}^{p}\Gamma(a_j+s)\right\}} z^{-s}ds \end{aligned} \quad (2)$$

where $i=\sqrt{(-1)}$ and $\mathcal{L}$ is a suitable contour.

The existence of different types of contours, properties and applications of G-functions are available in Mathai and Haubold [2].

2. Integral Representations

All the random variables considered above take values in $[0,1]$ and hence the density functions can be uniquely determined by their moments. For arbitary t, we have

$$E(x_1^t) = c_k' \prod_{j=1}^{k} \frac{\Gamma(\alpha_1+\cdots+\alpha_j+\beta_2+\cdots+\beta_j+t)}{\Gamma(\alpha_1+\cdots+\alpha_{j+1}+\beta_2+\cdots+\beta_j+t)} \quad (3)$$

where

$$c_k' = \prod_{j=1}^{k} \frac{\Gamma(\alpha_1+\cdots+\alpha_{j+1}+\beta_2+\cdots+\beta_j)}{\Gamma(\alpha_1+\cdots+\alpha_j+\beta_2+\cdots+\beta_j)}.$$

Note that the moments of the product of independent random variables are the products of the respective moments. Treating Equation (3) as a Mellin transform of the density of x_1, the density is available by the inverse Mellin transform. Thus, the density of x_1 is the following:

$$\begin{aligned} g(x_1) &= c_k' x_1^{-1} \frac{1}{2\pi i}\int_{\mathcal{L}} \frac{\Gamma(\alpha_1+t)}{\Gamma(\alpha_1+\alpha_2+t)} \cdots \frac{\Gamma(\alpha_1+\cdots+\alpha_k+\beta_2+\cdots+\beta_k+t)}{\Gamma(\alpha_1+\cdots+\alpha_{k+1}+\beta_2+\cdots+\beta_k+t)} x_1^{-t}dt \\ &= c_k' x_1^{-1} G_{k,k}^{k,0}\left[x_1\Big|_{\alpha_1,\ldots,\alpha_1+\cdots+\alpha_k+\beta_2+\cdots+\beta_k}^{\alpha_1+\alpha_2,\ldots,\alpha_1+\cdots+\alpha_{k+1}+\beta_2+\cdots+\beta_k}\right] \end{aligned} \quad (4)$$

for $0 < x_1 < 1$ and zero elsewhere.

Proposition 1.

$$G_{k,k}^{k,0}\left[x_1\Big|_{\alpha_1,\ldots,\alpha_1+\cdots+\alpha_k+\beta_2+\cdots+\beta_k}^{\alpha_1+\alpha_2,\ldots,\alpha_1+\cdots+\alpha_{k+1}+\beta_2+\cdots+\beta_k}\right] =$$

$$\frac{x_1}{\Gamma(\alpha_2)\Gamma(\alpha_3)\ldots\Gamma(\alpha_{k+1})} \int_0^{1-x_1}\int_0^{1-x_1-x_2}\cdots\int_0^{1-x_1-\cdots-x_{k-1}} x_1^{\alpha_1-1}\ldots x_k^{\alpha_{k-1}}$$
$$\times (x_1+x_2)^{\beta_2}\ldots(x_1+\cdots+x_k)^{\beta_k}(1-x_1-\cdots-x_k)^{\alpha_{k+1}^{-1}}dx_k dx_{k-1}\ldots dx_2.$$

Proof. The result follows by equating Equation (4) with the marginal density of x_1 obtained by integrating out $x_2,\ldots,x_k$ from the joint density of $x_1,\ldots,x_k$ given in Equation (1). □

Let X_j, $j=1,\ldots,k$ be an ordered set of points in the Euclidean n-space $\mathcal{R}^n$, $n \geq k$. Let O denotes the origin of a rectangular coordinate system. Now the $1\times n$ vector X_j can be considered as a point in $\mathcal{R}^n$. If $X_1,\ldots,X_k$ are linearly independent then the convex hull generated by these k-points almost surely determine a k-parallelotope in $\mathcal{R}^n$ with the sides $\overrightarrow{OX_1}, \ldots, \overrightarrow{OX_k}$. The random volume or k-content $\nabla_{k,n}$ of this random parallelotope is given by

$$\nabla_{k,n} = |XX'|^{1/2}$$

where

$$X = \begin{pmatrix} X_1 \\ \vdots \\ X_k \end{pmatrix}$$

is a matrix of order $k \times n$, X' is the transpose of X and $|(\cdot)|$ denotes the determinant of $(\cdot)$. The classical approach to random points and random volumes consists in looking at independently distributed isotropic random points and dealing with random geometric configurations with the help of techniques from differential and integral geometry. Mathai [3] looked into random volumes under a more general structure by deleting the assumptions of independence and isotropy. Mathai [3] has shown that if the $k \times n$, $n \geq k$, real random matrix X of full rank k has the density:

$$f(X) = C|XX'|^{\alpha}|I - XX'|^{\beta - \frac{k+1}{2}}$$

for $0 < XX' < I$, then the probability distribution of $\nabla^2_{k,n} = |XX'|$ has the following structure:

$$\nabla^2_{k,n} \stackrel{d}{=} \prod_{j=1}^{k} \text{type-1 beta}[\alpha + \frac{1}{2}(n+1-j), \beta].$$

Thus, it is possible to express the density of $\nabla^2_{k,n}$ as a marginal density of x_1 obtained from the joint density given in Equation (1) with specific set of parameters. The notion $\nabla^2_{k,n}$ has application in the study of variance of multivariate distributions. More details on random volumes may be seen from Mathai [3]. Thomas and Mathai [4] expressed the density of $\nabla^2_{k,n}$ as a marginal density of x_1 in the Model (1) with parameters as

$$\alpha_1 = \alpha + \tfrac{n}{2}, \alpha_2 = \alpha_3 = \cdots = \alpha_{k+1} = \beta, \quad \beta_2 = \cdots = \beta_k = -(\beta + \tfrac{1}{2}).$$

Now let us consider the Gaussian or ordinary hypergeometric function ${}_2F_1(a,b;c;z)$ which is a special function represented by the hypergeometric series:

$${}_2F_1(a,b;c;z) = \sum_{r=0}^{\infty} \frac{(a)_r (b)_r}{(c)_r} \frac{z^r}{r!}$$

where

$$(a)_r = (a+r-1)(a+r-2)\cdots(a) = \frac{\Gamma(a+r)}{\Gamma(a)}; \ (a)_0 = 1, \ a \neq 0$$

when $\Gamma(a)$ is defined.

Proposition 2.

$${}_2F_1(\beta, \beta + \frac{1}{2}; 2\beta; 1-x) = 2^{2\beta-1} x^{-\frac{1}{2}} (1 + x^{1/2})^{1-2\beta}; \ 0 < x < 1.$$

Proof. Let us consider the model (1) for the case when $k = 2$ and take the parameters as $\alpha_1 = \alpha + \frac{n}{2}, \alpha_2 = \alpha_3 = \beta$ and $\beta_2 = -(\beta + \frac{1}{2})$. Now the density of $\nabla^2_{2,n}$ as a marginal density of x_1, can be obtained as the following:

$$g(x_1) = \frac{1}{2^{2\beta} B(2\alpha + n - 1, 2\beta)} x_1^{\alpha + \frac{n}{2} - 1} (1 - x_1)^{2\beta - 1} {}_2F_1(\beta, \beta + \frac{1}{2}; 2\beta; 1 - x_1); \tag{5}$$

for $0 < x_1 < 1$ and zero elsewhere, where $B(\alpha, \beta)$ is the beta function. □

Alternatively, we can obtain the density of $\nabla^2_{2,n}$ by using Meijer's G-function given in Equation (4). Then the density function obtained has the form:

$$g(x_1) = \frac{1}{2B(2\alpha + n - 1, 2\beta)}(x_1^{1/2})^{2\alpha+n-3}(1 - x_1^{1/2})^{2\beta-1};\ 0 < x_1 < 1 \quad (6)$$

and zero elsewhere.

Since the density function is unique, Equations (5) and (6) must be equal. Hence the result follows.

Since Equation (5) is a probability density function we obtain the following relation:

Proposition 3.

$$\int_0^1 x^{[(2\alpha+n-1)-1]/2}(1-x)^{2\beta-1}{}_2F_1(\beta, \beta + \frac{1}{2}; 2\beta; 1-x)\mathrm{d}x = 2^{2\beta}B(2\alpha + n - 1, 2\beta).$$

Many multivariate procedures based on random samples from multivariate normal populations can be interpreted as the study of the distribution of $\nabla^2_{k,n}$. The exact distribution of likelihood ratio criteria for testing hypothesis in MANOVA, MANCOVA, multivariate regression analysis etc can be obtained as a special case of distribution of $\nabla^2_{k,n}$. Thomas and Thannippara [5,6] expressed the density of the above mentioned likelihood ratio criteria in terms of the marginal distribution of the generalized type-1 Dirichlet model given in Equation (1) with specific set of parameter values. The density of the likelihood ratio criterion U_k, m, n for $k = 4$ is obtained to be the following:

$$g(x) = \frac{\Gamma(n+m-1)\Gamma(n+m-3)}{2\Gamma(n-1)\Gamma(n-3)\Gamma(2m)}(x^{1/2})^{n-3}(1-x^{1/2})^{2m-1}{}_2F_1(m+2; m; 2m; 1-x^{1/2}) \quad (7)$$

for $0 < x < 1, n \geq 4$ and zero elsewhere. Since Equation (7) is a probability density function we obtain the following relation:

Proposition 4.

$$\int_0^1 (x^{1/2})^{n-3}(1-x^{1/2})^{2m-1}{}_2F_1(m+2; m; 2m; 1-x^{1/2})\mathrm{d}x = \frac{2\Gamma(n-1)\Gamma(n-3)\Gamma(2m)}{\Gamma(n+m-1)\Gamma(n+m-3)}.$$

Note that the evaluation of a G-function involves evaluation of residues at poles of different orders. Hence in such cases we may end up with psi, gamma or zeta functions. The above results are useful in evaluating the definite integrals involving G-functions of the type $G^{k,0}_{k,k}(\cdot)$.

Conflicts of Interest: The authors declare no conflict of interest.

References

1. Thomas, S.; George, S. A review of Dirichlet distribution and its generalizations. *J. Indian Soc. Probab. Stat.* **2004**, *8*, 72–91.
2. Mathai, A.M.; Haubold, H.J. *Special Functions for Applied Scientists*; Springer: New York, NY, USA, 2008.
3. Mathai, A.M. *An Introduction to Geometrical Probability: Distributional Aspects with Applications*; Gordon and Breach Science Publishers: Amsterdam, The Netherlands, 1999.
4. Thomas, S.; Mathai, A.M. p-Content of a p-parallelotope and its connection to likelihood ratio statistic. *Sankhya A* **2009**, *71*, 49–63.

5. Thomas, S.; Thannippara, A. Distribution of the LR criterion $U_{p,m,n}$ as a marginal distribution of a generalized Dirichlet model. *Statistica* **2008**, *68*, 375–390.
6. Thomas, S.; Thannippara, A. Distribution of Λ-criterion for sphericity test and its connection to a generalized Dirichlet model. *Commun. Stat. Simul. Comput.* **2008**, *37*, 1384–1394.

axiomS

MDPI

Article

On Elliptic and Hyperbolic Modular Functions and the Corresponding Gudermann Peeta Functions

Thomas Ernst

Department of Mathematics, Uppsala University, P.O. Box 480, Uppsala SE-751 06, Sweden; thomas@math.uu.se; Tel.: +46-1826-1924; Fax: +46-1847-1321

Academic Editor: Hans J. Haubold
Received: 18 May 2015; Accepted: 19 June 2015; Published: 8 July 2015

Abstract: In this article, we move back almost 200 years to Christoph Gudermann, the great expert on elliptic functions, who successfully put the twelve Jacobi functions in a didactic setting. We prove the second hyperbolic series expansions for elliptic functions again, and express generalizations of many of Gudermann's formulas in Carlson's modern notation. The transformations between squares of elliptic functions can be expressed as general Möbius transformations, and a conjecture of twelve formulas, extending a Gudermannian formula, is presented. In the second part of the paper, we consider the corresponding formulas for hyperbolic modular functions, and show that these Möbius transformations can be used to prove integral formulas for the inverses of hyperbolic modular functions, which are in fact Schwarz-Christoffel transformations. Finally, we present the simplest formulas for the Gudermann Peeta functions, variations of the Jacobi theta functions. 2010 Mathematics Subject Classification: Primary 33E05; Secondary 33D15.

Keywords: hyperbolic series expansion; Carlson's modern notation; hyperbolic modular function; Möbius transformation; Schwarz-Christoffel transformation; Peeta function

1. Introduction

The elliptic integrals were first classified by Euler and Legendre, and then Gauss, Jacobi and Abel started to study their inverses, the elliptic functions. Starting in the 1830s, Gudermann published a series of papers in German and Latin, with the aim of presenting these functions in a didactic way, and to introduce a short notation for them. This notation, with a small modification, has survided until the present day. Jacobi, in 1829, had found quickly converging Fourier series expansions for most of the twelve elliptic functions, which have been put in q-hypergeometric form in the authors article [1]. As Gudermann [2] showed, there are second series expansions for the twelve elliptic functions, starting from the imaginary period, which are not so quickly converging for all values of the variables; these expansions were also found, without proof, by Glaisher [3]. Since these hyperbolic expansions are virtually unknown today, we prove them again, and also put them into q-hypergeometric form in section two. There are many series expansions for squareroots of rational functions of elliptic functions; as a bonus we also prove some of these. However, before this, we introduce the q-hypergeometric notation in this first section, this can also be found in the book [4].

In section three, we generalize many of Gudermanns formulas to the very general Carlson [5] notation, where many formulas can be put into one single equation by using a clever code, and the symmetry of these functions. This notation has been known for many years, but was only recently published; by coincidence, the author saw it when he was asked to review this article by Carlson. In particular, a formula with squareroots, stated without proof by Gudermann, is generalized to a conjecture of two formulas with squareroots, or twelve elliptic function formulas, which generalize four formulas with squareroots for trigonometric and hyperbolic functions. Gudermann was the first to point out the close relationship between trigonometric and hyperbolic functions. We also state four Möbius transformations in Carlson's notation, and generalize Gudermanns formulas for artanh.

In section four, we come to the hyperbolic modular functions, which have not yet appeared in the English literature; the function $\mathcal{SN}(u)$ is the inverse of an hyperbolic integral, which is formed by changing two minuses to plus in the elliptic integral of the first kind. We calculate the poles, periods, Möbius transformations for squares, and special values of the hyperbolic modular functions. Finally, we compute several addition formulas using a short notation for these functions.

In section five, we consider the Peeta functions, which are theta functions with imaginary function value. We show that the hyperbolic modular functions can be expressed as quotients of Peeta functions, and that the four Peeta functions are solutions of a certain heat equation with the variable q as parameter.

Before presenting the q-series formulas in the next section, we present the necessary definitions.

An elliptic integral is given by

$$F(z) = \int_0^z \frac{dx}{\sqrt[2]{(1-x^2)(1-(kx)^2)}} \tag{1}$$

where $0 < k < 1$.

Abel and Jacobi, inspired by Gauss, discovered that inverting $F(z)$ gave the doubly periodic elliptic function

$$F^{-1}(\omega) = \text{sn}(\omega) \tag{2}$$

In connection with elliptic functions k always denotes the modulus.

Definition 1. Let $\delta > 0$ be an arbitrarily small number. We will always use the following branch of the logarithm: $-\pi + \delta < \text{Im}(\log q) \leq \pi + \delta$. This defines a simply connected space in the complex plane.

The power function is defined by

$$q^a \equiv e^{a \log(q)} \tag{3}$$

Definition 2. The q-factorials and the tilde operator are defined by

$$\langle a;q\rangle_n \equiv \begin{cases} 1, & n = 0; \\ \prod_{m=0}^{n-1}(1-q^{a+m}) & n = 1,2,\ldots \end{cases} \tag{4}$$

$$\langle \widetilde{a};q\rangle_n \equiv \prod_{m=0}^{n-1}(1+q^{a+m}) \tag{5}$$

Definition 3. The q-hypergeometric series is defined by

$${}_2\phi_1(\hat{a},\hat{b};\hat{c}|q;z) \equiv \sum_{n=0}^{\infty} \frac{\langle \hat{a};q\rangle_n \langle \hat{b};q\rangle_n}{\langle 1;q\rangle_n \langle \hat{c};q\rangle_n} z^n \tag{6}$$

where

$$\hat{a} \equiv a \vee \widetilde{a} \tag{7}$$

It is assumed that the denominator contains no zero factors.

By $\sqrt{z}$ we mean the branch $|z|^{\frac{1}{2}} \exp(i\frac{1}{2}\arg z)$. Everywhere we have $y \equiv \frac{\pi u}{2K'}$. To maintain a symmetrical form, we put according to Jacobi and Glaisher $q' \equiv e^{-\pi\frac{K}{K'}}$. The following lemma will be used in the proofs.

Lemma 1.1. *A Fourier series for the logarithmic potential [6].*

$$\log(1 - 2q^{2k}\cos(2x) + q^{4k}) = -\sum_{n=1}^{\infty} \frac{2q^{2kn}}{n} \cos(2nx) \tag{8}$$

2. Hyperbolic series expansions

The following series were published for the first time by Gudermann in [2], see also [7].

Theorem 2.1.

$$\frac{\mathrm{cn}u\,\mathrm{dn}u}{\mathrm{sn}u} = \frac{\pi}{2K'}\left[\frac{2}{\sinh(2y)} - 8\sum_{m=1}^{\infty}\frac{q'^{4m-2}}{1-q'^{4m-2}}\sinh((4m-2)y)\right] \tag{9}$$

$$\frac{\mathrm{sn}u}{\mathrm{cn}u\,\mathrm{dn}u} = \frac{4\pi}{K'k'^2}\sum_{m=1}^{\infty}\frac{q'^{2m-1}}{1-q'^{4m-2}}\sinh((4m-2)y) \tag{10}$$

$$\frac{\mathrm{sn}u\,\mathrm{dn}u}{\mathrm{cn}u} = \frac{\pi}{2K'}\left(\tanh y + 4\sum_{m=1}^{\infty}\frac{q'^m}{1+(-q')^m}\sinh(2my)\right) \tag{11}$$

$$\frac{\mathrm{cn}u}{\mathrm{sn}u\,\mathrm{dn}u} = \frac{\pi}{2K'}\left(\coth y + 4\sum_{m=1}^{\infty}\frac{(-q')^m}{1+(-q')^m}\sinh(2my)\right) \tag{12}$$

$$\frac{\mathrm{sn}u\,\mathrm{cn}u}{\mathrm{dn}u} = \frac{\pi}{2k^2K'}\left[\tanh y + 4\sum_{m=1}^{\infty}\frac{(-q')^m}{1+q'^m}\sinh(2my)\right] \tag{13}$$

$$\frac{\mathrm{dn}u}{\mathrm{sn}u\,\mathrm{cn}u} = \frac{\pi}{2K'}\left[\coth y + 4\sum_{m=1}^{\infty}\frac{q'^m}{1+q'^m}\sinh(2my)\right] \tag{14}$$

Proof. We only prove Equation (9), the other formulas are proved similarly.

$$\begin{aligned}\log \mathrm{sn}u &= \log\left(\frac{2q^{\frac{1}{4}}\sin x}{k^{\frac{1}{2}}}\right)\\ &+ \sum_{n=1}^{\infty}\log(1-2q^{2n}\cos 2x + q^{4n}) - \log(1-2q^{2n-1}\cos 2x + q^{4n-2})\\ &\overset{\text{by}(8)}{=} \log\left(\frac{2q^{\frac{1}{4}}\sin x}{k^{\frac{1}{2}}}\right) + \sum_{m,n=1}^{\infty}\frac{2\cos 2mx(q^{m(2n-1)} - q^{2mn})}{m}\\ &= \log\left(\frac{2q^{\frac{1}{4}}\sin x}{k^{\frac{1}{2}}}\right) + \sum_{m=1}^{\infty}\frac{2q^m\cos 2mx}{m(1+q^m)}\end{aligned} \tag{15}$$

The derivative with respect to u finally gives (9).

Theorem 2.2. *Hyperbolic series for* $\sqrt{\frac{1\pm t}{1\mp t}}, t \in \{cdu, cnu, dnu\}$ *[2]. The comodulus* k' *is small.*

$$k'\sqrt{\frac{1+\mathrm{cd}u}{1-\mathrm{cd}u}} = \frac{\pi}{K'}\left[\frac{1}{\sinh y} - 4\sum_{m=1}^{\infty}\frac{q'^{4m-2}}{1-q'^{2m-1}}\sinh((4m-2)y)\right] \tag{16}$$

$$k'\sqrt{\frac{1-\mathrm{cd}u}{1+\mathrm{cd}u}} = \frac{4\pi}{K'}\sum_{m=1}^{\infty}\frac{q'^{2m-1}}{1-q'^{4m-2}}\sinh((2m-1)y) \tag{17}$$

$$\sqrt{\tfrac{1-\mathrm{cn}u}{1+\mathrm{cn}u}} = \tfrac{\pi}{2K'}\left(\tanh\tfrac{y}{2} + 4\sum_{m=1}^{\infty}\tfrac{q'^m}{1+(-q')^m}\sinh(my)\right) \tag{18}$$

$$\sqrt{\tfrac{1+\mathrm{cn}u}{1-\mathrm{cn}u}} = \tfrac{\pi}{2K'}\left(\coth\tfrac{y}{2} + 4\sum_{m=1}^{\infty}\tfrac{(-q')^m}{1+(-q')^m}\sinh(my)\right) \tag{19}$$

$$k\sqrt{\tfrac{1-\mathrm{dn}u}{1+\mathrm{dn}u}} = \tfrac{\pi}{2K'}\left[\tanh\tfrac{y}{2} + 4\sum_{m=1}^{\infty}\tfrac{(-q')^m}{1+q'^m}\sinh(my)\right] \tag{20}$$

$$k\sqrt{\frac{1+\mathrm{dn}u}{1-\mathrm{dn}u}} = \frac{\pi}{2K'}\left[\coth\frac{y}{2} + 4\sum_{m=1}^{\infty}\frac{q'^m}{1+q'^m}\sinh(my)\right] \tag{21}$$

Proof. All formulas are proved with the help of the previous theorem. We first observe that

$$k'\frac{\mathrm{sn}(\frac{u}{2})}{\mathrm{cn}(\frac{u}{2})\mathrm{dn}(\frac{u}{2})} = \sqrt{\frac{1-\mathrm{cd}u}{1+\mathrm{cd}u}} \tag{22}$$

$$\frac{\mathrm{sn}(\frac{u}{2})\mathrm{dn}(\frac{u}{2})}{\mathrm{cn}(\frac{u}{2})} = \sqrt{\frac{1-\mathrm{cn}u}{1+\mathrm{cn}u}} \tag{23}$$

$$k\frac{\mathrm{sn}(\frac{u}{2})\mathrm{cn}(\frac{u}{2})}{\mathrm{dn}(\frac{u}{2})} = \sqrt{\frac{1-\mathrm{dn}u}{1+\mathrm{dn}u}} \tag{24}$$

The Formulas (16) and (17) follow from Formula (22), the Formulas (18) and (19) follow from Formula (23) and finally, Formulas (20) and (21) follow from Formula (24).

Theorem 2.3. *The following 12 series, found by Gudermann [8] and Glaisher [3], define the second series expansions of the corresponding elliptic functions.*

$$[8, \text{p.}366(6), \text{p.}367(3)]\mathrm{sn}u = \tfrac{\pi}{2K'k}\left[\tanh y + 4\sum_{m=1}^{\infty}\tfrac{(-1)^m q'^{2m}}{1+q'^{2m}}\cosh(2my)\right] \tag{25}$$

$$[8, \text{p.}366(4), \text{p.}368(11)]\mathrm{cn}u = \tfrac{\pi}{2K'k}\left[\tfrac{1}{\cosh y} + 4\sum_{m=1}^{\infty}\tfrac{(-1)^m q'^{2m-1}}{1+q'^{2m-1}}\cosh((2m-1)y)\right] \tag{26}$$

$$[8, \text{p.}366(5), \text{p.}368(19)]\mathrm{dn}u = \tfrac{\pi}{2K'}\left[\tfrac{1}{\cosh y} - 4\sum_{m=1}^{\infty}\tfrac{(-1)^m q'^{2m-1}}{1-q'^{2m-1}}\sinh((4m-2)y)\right] \tag{27}$$

$$[8, \text{p.}366(3), \text{p.}367(7)]\mathrm{ns}u = \tfrac{\pi}{2K'}\left[\coth y + 4\sum_{m=1}^{\infty}\tfrac{q'^{2m}}{1+q'^{2m}}\sinh(2my)\right] \tag{28}$$

$$[8, \text{p.}366(7), \text{p.}368(15)]\mathrm{nc}u = \tfrac{2\pi}{K'k'}\sum_{m=1}^{\infty}\tfrac{q'^{\frac{2m-1}{2}}}{1+q'^{2m-1}}\sinh((2m-1)y) \tag{29}$$

$$[8, \text{p.}367(11), \text{p.}368(20)]\mathrm{nd}u = \tfrac{2\pi}{K'k'}\sum_{m=1}^{\infty}\tfrac{(-1)^{m+1}q'^{\frac{2m-1}{2}}}{1-q'^{2m-1}}\cosh((2m-1)y) \tag{30}$$

$$[8, \text{p.}366(8), \text{p.}368(23)]\mathrm{sc}u = \frac{2\pi}{K'k'}\sum_{m=1}^{\infty}\frac{q'^{\frac{2m-1}{2}}}{1-q'^{2m-1}}\sinh((2m-1)y) \tag{31}$$

$$[8,\text{p.}368(24)]\text{cs}u = \frac{\pi}{2K'}\left[\frac{1}{\text{sinh}y} - 4\sum_{m=1}^{\infty}\frac{q'^{2m-1}}{1-q'^{2m-1}}\sinh\left((2m-1)y\right)\right] \tag{32}$$

$$[8,\text{p.}367(10),\text{p.}368(12)]\text{sd}u = \frac{2\pi}{K'kk'}\sum_{m=1}^{\infty}\frac{(-1)^{n-1}q'^{\frac{2m-1}{2}}}{1+q'^{2m-1}}\sinh\left((2m-1)y\right) \tag{33}$$

$$[8,\text{p.}365(1),\text{p.}368(16)]\text{ds}u = \frac{\pi}{2K'}\left[\frac{1}{\text{sinh}y} + 4\sum_{m=1}^{\infty}\frac{q'^{2m-1}}{1+q'^{2m-1}}\sinh\left((2m-1)y\right)\right] \tag{34}$$

$$[8,(\text{p.}367(4)),\text{p.}367(12)]\text{cd}u = \frac{\pi}{2K'k}\left[1 + 4\sum_{m=1}^{\infty}\frac{(-q')^{m}}{1+q'^{2m}}\sinh\left(2my\right)\right] \tag{35}$$

$$[8,\text{p.}366(9),\text{p.}367(8)]\text{dc}u = \frac{\pi}{2K'}\left[1 + 4\sum_{m=1}^{\infty}\frac{q'^{m}}{1+q'^{2m}}\cosh\left(2my\right)\right] \tag{36}$$

Proof. By addition and subtraction of the Formulas (76) and (77) we obtain Formulas (37)–(40):

$$\text{ds}u + \text{cs}u = \frac{\pi}{2K'}\left[\frac{2}{\sinh y} - 8\sum_{m=1}^{\infty}\frac{q'^{4m-2}}{1-q'^{4m-2}}\sinh\left((2m-1)y\right)\right] \tag{37}$$

$$\text{ds}u - \text{cs}u = \frac{4\pi}{K'}\sum_{m=1}^{\infty}\frac{q'^{2m-1}}{1-q'^{4m-2}}\sinh\left((2m-1)y\right) \tag{38}$$

$$\text{ns}u + \text{cs}u = \frac{\pi}{2K'}\left(\tanh\frac{y}{2} + 4\sum_{m=1}^{\infty}\frac{q'^{m}}{1+(-q')^{m}}\sinh\left(my\right)\right) \tag{39}$$

$$\text{ns}u - \text{cs}u = \frac{\pi}{2K'}\left(\coth\frac{y}{2} + 4\sum_{m=1}^{\infty}\frac{(-q')^{m}}{1+(-q')^{m}}\sinh\left(my\right)\right) \tag{40}$$

New additions and subtractions give Formulas (34), (32) and (28). The substitution $u \mapsto u + iK'$ gives Formulas (26), (27) and (25).

Theorem 2.4. *According to Heine, these 12 series can be written as follows.*

$$\text{sn}u = \frac{\pi}{2K'k}\left[\tanh y - i\text{Im}\Big(-2 + 2{}_2\phi_1(1,\widetilde{0};\widetilde{1}|q'^2;-q'^2e^{-2y})\Big)\right] \tag{41}$$

$$\text{cn}u = \frac{\pi}{2K'k}\left[\frac{1}{\cosh y} - \text{Re}\left[e^{-y}\frac{4q'}{1+q'}{}_2\phi_1(1,\widetilde{\tfrac{1}{2}};\widetilde{\tfrac{3}{2}}|q'^2;-q'^2e^{-2y})\right]\right] \tag{42}$$

$$\text{dn}u = \frac{\pi}{2K'}\left[\frac{1}{\cosh y} + \text{Re}\Big(\frac{4q'e^{-y}}{1-q'}{}_2\phi_1(1,\tfrac{1}{2};\tfrac{3}{2}|-q'^2;-q'^2e^{-2y})\Big)\right] \tag{43}$$

$$\text{ns}u = \frac{\pi}{2K'}\left[\frac{1}{\tanh y} - i\text{Im}\Big(-2 + 2{}_2\phi_1(1,\widetilde{0};\widetilde{1}|q'^2;q'^2e^{-2y})\Big)\right] \tag{44}$$

$$\text{nc}u = \frac{2\pi}{K'k'}\text{Re}\left[\frac{q'^{\frac{1}{2}}e^{-y}}{1+q'}{}_2\phi_1(1,\widetilde{\tfrac{1}{2}};\widetilde{\tfrac{3}{2}}|q'^2;q'e^{-2y})\right] \tag{45}$$

$$\text{nd}u = \frac{2\pi}{K'k'}\text{Re}\left[e^{-y}\frac{q'^{\frac{1}{2}}}{1-q'}{}_2\phi_1(1,\tfrac{1}{2};\tfrac{3}{2}|q'^2;-q'e^{-2y})\right] \tag{46}$$

$$\text{sc}u = \frac{-2\pi i}{K'k'}\text{Im}\left[\frac{q'^{\frac{1}{2}}e^{-y}}{1-q'}{}_2\phi_1(1,\tfrac{1}{2};\tfrac{3}{2}|q'^2;q'e^{-2y})\right] \tag{47}$$

$$\mathrm{cs}u = \frac{\pi}{2K'}\left[\frac{1}{\sinh y} - i\mathrm{Im}\left[e^{-y}\frac{4q'}{1-q'}{}_2\phi_1(1,\tfrac{1}{2};\tfrac{3}{2}|q'^2;q'^2e^{-2y})\right]\right] \tag{48}$$

$$\mathrm{sd}u = \frac{-2\pi i}{K'kk'}\mathrm{Im}\left[e^{-y}\frac{q'^{\frac{1}{2}}}{1+q'}{}_2\phi_1(1,\widetilde{\tfrac{1}{2}};\widetilde{\tfrac{3}{2}}|q'^2;-q'e^{-2y})\right] \tag{49}$$

$$\mathrm{ds}u = \frac{\pi k'}{2K'k}\left[\frac{1}{\sinh y} - i\mathrm{Im}\left[e^{-y}\frac{4q'}{1+q'}{}_2\phi_1(1,\widetilde{\tfrac{1}{2}};\widetilde{\tfrac{3}{2}}|q'^2;q'^2e^{-2y})\right]\right] \tag{50}$$

$$\mathrm{cd}u = \frac{\pi}{2K'k}\left[-1 + 2\mathrm{Re}_2\phi_1(1,\widetilde{0};\widetilde{1}|q'^2;-q'e^{-2y})\right] \tag{51}$$

$$\mathrm{dc}u = \frac{\pi}{2K'}\left[-1 + 2\mathrm{Re}_2\phi_1(1,\widetilde{0};\widetilde{1}|q'^2;q'e^{-2y})\right] \tag{52}$$

3. Some New Elliptic Function Formulas in Carlsons Notation

Bille Carlson (1924–2013) [5] managed to simplify the great number of elliptic function formulas into a series of very general formulas. First put

$$\{\mathrm{p},\mathrm{q},\mathrm{r}\} \equiv \{\mathrm{c},\mathrm{d},\mathrm{n}\} \tag{53}$$

and use Glaisher's abbreviations for Jacobis elliptic functions. Thus q is not a q-analogue in this section.

Furthermore, we put

$$\triangle\,(\mathrm{p},\mathrm{q}) \equiv \mathrm{ps}^2 - \mathrm{qs}^2, \mathrm{p},\mathrm{q} \in \{\mathrm{c},\mathrm{d},\mathrm{n}\} \tag{54}$$

which implies that

$$\begin{gathered}\triangle\,(\mathrm{n},\mathrm{c}) = -\triangle\,(\mathrm{c},\mathrm{n}) = 1, \triangle\,(\mathrm{n},\mathrm{d}) = -\triangle\,(\mathrm{d},\mathrm{n}) = k^2\\ \triangle\,(\mathrm{d},\mathrm{c}) = -\triangle\,(\mathrm{c},\mathrm{d}) = k'^2\end{gathered} \tag{55}$$

The default function values are u,k. All formulas apply for $u \in \Sigma$ (Riemann sphere). It is well-known that

$$\begin{gathered}\lim_{k\to0^+}\mathrm{sn}x = \lim_{k\to0^+}\mathrm{sd}x = \sin x,\ \lim_{k\to0^+}\mathrm{cn}x = \cos x,\ \lim_{k\to0^+}\mathrm{dn}x = 1\\ \lim_{k\to0^+}\mathrm{sc}x = \tan x,\ \lim_{k\to1^-}\mathrm{sn}x = \tanh x,\ \lim_{k\to1^-}\mathrm{cn}x = \lim_{k\to1^-}\mathrm{dn}x = \frac{1}{\cosh x}\\ \lim_{k\to1^-}\mathrm{sc}x = \lim_{k\to1^-}\mathrm{sd}x = \sinh x\end{gathered} \tag{56}$$

All formulas in this section lie between these two limits, *i.e.*, for all limits in k, we get known trigonometric and hyperbolic function (or trivial) formulas. We first give one of Carlsons results; all other formulas are presumably new.

Theorem 3.1. *Addition formulas [5]. Put* $\mathrm{ps}_i \equiv \mathrm{ps}(u_i,k), i = 1,2$, *and similar notation for the other functions. Then*

$$\mathrm{ps}(u_1+u_2,k) = \frac{\mathrm{ps}_1\mathrm{qs}_2\mathrm{rs}_2 - \mathrm{ps}_2\mathrm{qs}_1\mathrm{rs}_1}{\mathrm{ps}_2^2 - \mathrm{ps}_1^2} = \frac{\mathrm{ps}_1^2\mathrm{ps}_2^2 - \triangle\,(\mathrm{p},\mathrm{q})\triangle\,(\mathrm{p},\mathrm{r})}{\mathrm{ps}_1\mathrm{qs}_2\mathrm{rs}_2 + \mathrm{ps}_2\mathrm{qs}_1\mathrm{rs}_1} \tag{57}$$

$$\mathrm{sp}(u_1+u_2,k) = \frac{\mathrm{sp}_1^2 - \mathrm{sp}_2^2}{\mathrm{sp}_1\mathrm{qp}_2\mathrm{rp}_2 - \mathrm{sp}_2\mathrm{qp}_1\mathrm{rp}_1} = \frac{\mathrm{sp}_1\mathrm{qp}_2\mathrm{rp}_2 + \mathrm{sp}_2\mathrm{qp}_1\mathrm{rp}_1}{1 - \triangle\,(\mathrm{p},\mathrm{q})\triangle\,(\mathrm{p},\mathrm{r})\mathrm{sp}_1^2\mathrm{sp}_2^2} \tag{58}$$

$$\mathrm{pq}(u_1+u_2,k) = \frac{\mathrm{ps}_1\mathrm{qs}_2\mathrm{rs}_2 - \mathrm{ps}_2\mathrm{qs}_1\mathrm{rs}_1}{\mathrm{qs}_1\mathrm{ps}_2\mathrm{rs}_2 - \mathrm{qs}_2\mathrm{ps}_1\mathrm{rs}_1} = \frac{\mathrm{ps}_1\mathrm{qs}_1\mathrm{ps}_2\mathrm{qs}_2 + \triangle\,(\mathrm{p},\mathrm{q})\mathrm{rs}_1\mathrm{rs}_2}{\mathrm{qs}_1^2\mathrm{qs}_2^2 + \triangle\,(\mathrm{p},\mathrm{q})\triangle\,(\mathrm{q},\mathrm{r})} \tag{59}$$

$$\mathrm{pq}(u_1+u_2,k) = \frac{\mathrm{pq}_1\mathrm{sq}_1\mathrm{rq}_2 - \mathrm{pq}_2\mathrm{sq}_2\mathrm{rq}_1}{\mathrm{pq}_2\mathrm{sq}_1\mathrm{rq}_2 - \mathrm{pq}_1\mathrm{sq}_2\mathrm{rq}_1} = \frac{\mathrm{pq}_1\mathrm{pq}_2 + \triangle\,(\mathrm{p},\mathrm{q})\mathrm{sq}_1\mathrm{rq}_1\mathrm{sq}_2\mathrm{rq}_2}{1 + \triangle\,(\mathrm{p},\mathrm{q})\triangle\,(\mathrm{q},\mathrm{r})\mathrm{sq}_1^2\mathrm{sq}_2^2} \tag{60}$$

Remark 1. A special case of Formula (58) was first given by Gudermann 1838 [9]. Two special cases of Formula (60) were first given by Gudermann 1838 [9].

Put $\mathrm{ps}_i \equiv \mathrm{ps}(u_i, k), i = 1, 2$, and similar notation for the other functions.

Theorem 3.2. *Formulas for elliptic functions corresponding to product formulas for trigonometric functions.*

$$\begin{aligned} \mathrm{ps}(u_1 + u_2, k) + \mathrm{ps}(u_1 - u_2, k) &= \tfrac{2\mathrm{ps}_1\mathrm{qs}_2\mathrm{rs}_2}{\mathrm{ps}_2^2 - \mathrm{ps}_1^2} \\ \mathrm{ps}(u_1 - u_2, k) - \mathrm{ps}(u_1 + u_2, k) &= \tfrac{2\mathrm{ps}_2\mathrm{qs}_1\mathrm{rs}_1}{\mathrm{ps}_2^2 - \mathrm{ps}_1^2} \end{aligned} \tag{61}$$

Proof. Use Formulas (57).

A special case of Formula (61) was first given by Gudermann 1838 [10].

Theorem 3.3. *Formulas for elliptic functions corresponding to product formulas for trigonometric functions.*

$$\begin{aligned} \mathrm{sp}(u_1 + u_2, k) + \mathrm{sp}(u_1 - u_2, k) &= \tfrac{2\mathrm{sp}_1\mathrm{qp}_2\mathrm{rp}_2}{1 - \triangle(\mathrm{p,q})\triangle(\mathrm{p,r})\mathrm{sp}_1^2\mathrm{sp}_2^2} \\ \mathrm{sp}(u_1 + u_2, k) - \mathrm{sp}(u_1 - u_2, k) &= \tfrac{2\mathrm{sp}_2\mathrm{qp}_1\mathrm{rp}_1}{1 - \triangle(\mathrm{p,q})\triangle(\mathrm{p,r})\mathrm{sp}_1^2\mathrm{sp}_2^2} \end{aligned} \tag{62}$$

Proof. Use Formulas (58).

Special cases of Formula (62) were first given by Legendre [11] 1828 , Jacobi 1829 [12], Laurent [13] and by Gudermann 1838 [10].

Theorem 3.4. *Formulas for elliptic functions corresponding to product formulas for trigonometric functions.*

$$\begin{aligned} \mathrm{pq}(u_1 + u_2, k) + \mathrm{pq}(u_1 - u_2, k) &= \tfrac{2\mathrm{pq}_1\mathrm{pq}_2}{1 + \triangle(\mathrm{p,q})\triangle(\mathrm{q,r})\mathrm{sq}_1^2\mathrm{sq}_2^2} \\ \mathrm{pq}(u_1 + u_2, k) - \mathrm{pq}(u_1 - u_2, k) &= \tfrac{2\triangle(\mathrm{p,q})\mathrm{sq}_1\mathrm{rq}_1\mathrm{sq}_2\mathrm{rq}_2}{1 + \triangle(\mathrm{p,q})\triangle(\mathrm{q,r})\mathrm{sq}_1^2\mathrm{sq}_2^2} \end{aligned} \tag{63}$$

Proof. Use formula Equation (60).

Special cases of Formula (63) were first given by Gudermann 1838 [10].

Theorem 3.5.

$$\mathrm{sp}(u_1 + u_2, k)\mathrm{sp}(u_1 - u_2, k) = \frac{\mathrm{sp}_1^2 - \mathrm{sp}_2^2}{1 - \triangle(\mathrm{p,q}) \triangle(\mathrm{p,r})\mathrm{sp}_1^2\mathrm{sp}_2^2} \tag{64}$$

$$\mathrm{pq}(u_1 + u_2, k)\mathrm{pq}(u_1 - u_2, k) = \frac{1 + \triangle(\mathrm{q,p}) \triangle(\mathrm{p,r})\mathrm{sp}_1^2\mathrm{sp}_2^2}{1 + \triangle(\mathrm{p,q}) \triangle(\mathrm{q,r})\mathrm{sq}_1^2\mathrm{sq}_2^2} \tag{65}$$

Proof. Use Formulas (61), (62), and (63).

Special cases of Formulas (64) and (65) were first given by Jacobi 1829 [12] and by Gudermann 1838 [10].

Theorem 3.6. *Put $f_i = f(2u_i), i = 1, 2$. Then we have*

$$\mathrm{pq}(u_1 + u_2, k)\mathrm{pq}(u_1 - u_2, k) = \frac{\mathrm{pq}_1\mathrm{rq}_2 + \mathrm{rq}_1\mathrm{pq}_2}{\mathrm{rq}_1 + \mathrm{rq}_2} \tag{66}$$

$$\frac{\mathrm{pq}(u_1 - u_2, k)}{\mathrm{pq}(u_1 + u_2, k)} = \frac{\mathrm{sq}_1 + \mathrm{sq}_2}{\mathrm{sq}_1\mathrm{pq}_2 + \mathrm{sq}_2\mathrm{pq}_1} = \frac{\mathrm{sq}_1\mathrm{pq}_2 + \mathrm{sq}_2\mathrm{pq}_1}{\mathrm{sq}_1 + \mathrm{sq}_2} \tag{67}$$

Special cases of Formula (66) were given by Gudermann 1838 [10]. Special cases of Formula (67) were given by Gudermann 1838 [10].

Theorem 3.7. *For $p = n$, Formula (68) holds unaltered. In the two other cases we only use one of the two factors qn, rn in either numerator or denominator. This gives the six formulas*

$$\frac{\mathrm{sp}(u_1 - u_2, k)}{\mathrm{sp}(u_1 + u_2, k)} = \frac{\mathrm{sn}(2u_1)\mathrm{qn}(2u_2) - \mathrm{sn}(2u_2)\mathrm{qn}(2u_1)}{\mathrm{sn}(2u_1)\mathrm{rn}(2u_2) + \mathrm{sn}(2u_2)\mathrm{rn}(2u_1)} \tag{68}$$

Special case of Formula (68) were given by Gudermann 1838 [10].

Theorem 3.8. *Bisection*

$$\mathrm{sp}^2\left(\frac{u}{2}, k\right) = \frac{1}{\triangle\,(\mathrm{p}, \mathrm{q})}\frac{1 - \mathrm{qp}}{1 + \mathrm{rp}} \tag{69}$$

Special cases of the following formulas were given by Gudermann 1838 [10].

Theorem 3.9. *Addition formulas [5]. Put $ps_i = ps(u_i, k), i = 1, 2$, and similar notation for the other functions.*

$$1 + \mathrm{pq}(u_1 \pm u_2, k) = \frac{(\mathrm{pq}_1 + \mathrm{pq}_2)(\mathrm{sr}_1 \mp \mathrm{sr}_2)}{\mathrm{sr}_1\mathrm{pq}_2 \mp \mathrm{sr}_2\mathrm{pq}_1} \tag{70}$$

$$1 - \mathrm{pq}(u_1 \pm u_2, k) = \frac{(\mathrm{pq}_1 - \mathrm{pq}_2)(\mathrm{sr}_1 \pm \mathrm{sr}_2)}{\pm\mathrm{sr}_2\mathrm{pq}_1 - \mathrm{sr}_1\mathrm{pq}_2} \tag{71}$$

Half of the following conjecture was given in [2]. We have the well-known formulas

$$\frac{1}{2}\sqrt{\frac{1 + \cos x}{1 - \cos x}} + \frac{1}{2}\sqrt{\frac{1 - \cos x}{1 + \cos x}} = \frac{1}{\sin x} \tag{72}$$

$$\frac{1}{2}\sqrt{\frac{1 + \cos x}{1 - \cos x}} - \frac{1}{2}\sqrt{\frac{1 - \cos x}{1 + \cos x}} = \cot x \tag{73}$$

$$\frac{1}{2}\sqrt{\frac{1 + \cosh x}{\cosh x - 1}} - \frac{1}{2}\sqrt{\frac{\cosh x - 1}{1 + \cosh x}} = \frac{1}{\sinh x} \tag{74}$$

$$\frac{1}{2}\sqrt{\frac{1 + \cosh x}{\cosh x - 1}} + \frac{1}{2}\sqrt{\frac{\cosh x - 1}{1 + \cosh x}} = \coth x \tag{75}$$

Conjecture 3.10. We have the twelve formulas

$$\frac{\sqrt{\triangle\,(\mathrm{p}, \mathrm{q})}}{2}\sqrt{\frac{\mathrm{pq} + 1}{\mathrm{pq} - 1}} + \frac{\sqrt{\triangle\,(\mathrm{p}, \mathrm{q})}}{2}\sqrt{\frac{\mathrm{pq} - 1}{\mathrm{pq} + 1}} = f \tag{76}$$

$$\frac{\sqrt{\triangle\,(\mathrm{p}, \mathrm{q})}}{2}\sqrt{\frac{\mathrm{pq} + 1}{\mathrm{pq} - 1}} - \frac{\sqrt{\triangle\,(\mathrm{p}, \mathrm{q})}}{2}\sqrt{\frac{\mathrm{pq} - 1}{\mathrm{pq} + 1}} = g \tag{77}$$

where $f, g \in \{\mathrm{qs}, \mathrm{ps}\}$. We choose the right hand side that has the correct limits for $\lim_{k \to 0^+}$ and $\lim_{k \to 1^-}$ in Formulas (72)–(75). Ten of the formulas have one limit among these four formulas, and the remaining two (with $\mathrm{nc}u$) have two limits.

To be able to compute the inverses of the elliptic functions, we must first prove a number of Möbius transformations between squares of elliptic functions which govern their transformations. Most of these can be summarized in the formulas

$$\mathrm{pq}^2 = \frac{\triangle\,(\mathrm{p},\mathrm{r}) + \mathrm{rs}^2}{\triangle\,(\mathrm{q},\mathrm{r}) + \mathrm{rs}^2} \tag{78}$$

$$\mathrm{sr}^2 = \frac{\mathrm{pq}^2 - 1}{\triangle\,(\mathrm{p},\mathrm{r}) - \triangle\,(\mathrm{q},\mathrm{r})\mathrm{pq}^2} \tag{79}$$

$$\mathrm{sp}^2 = \frac{1 - \mathrm{pq}^2}{\triangle\,(\mathrm{q},\mathrm{p})\mathrm{pq}^2} \tag{80}$$

$$\mathrm{pq}^2 = \frac{\triangle\,(\mathrm{p},\mathrm{r})\mathrm{qr}^2 + \triangle\,(\mathrm{q},\mathrm{p})}{\triangle\,(\mathrm{q},\mathrm{r})\mathrm{qr}^2} \tag{81}$$

Formula (81) is its own inverse. With these formulas we can prove integral formulas for the twelve inverse elliptic functions like in [14] and [15]. The integral formulas for the inverses of the elliptic functions are Schwarz-Christoffel mappings from the periodic rectangle of each elliptic function. The Formulas (78) to (81) do not map each periodic rectangle to the next, but if we take all rectangles in the vicinity of the origin and agree to start each equation solving with the prerequisite $\Re(z) > 0$, the formulas give correct values on the Riemann sphere.

We conclude with a few formulas with the function $\mathrm{artanh}(x)$, which again generalize Gudermanns results.

Theorem 3.11. *We have the eighteen formulas*

$$\log\sqrt{\frac{1+\mathrm{pq}(u_1 \pm u_2,k)}{1-\mathrm{pq}(u_1 \pm u_2,k)}} = \mathrm{artanh}\left(\frac{\mathrm{pq}_2}{\mathrm{pq}_1}\right) \pm \mathrm{artanh}\left(\frac{\mathrm{sr}_2}{\mathrm{sr}_1}\right) \tag{82}$$

$$\log\sqrt{\frac{1+\mathrm{pq}(u_1 - u_2,k)}{1+\mathrm{pq}(u_1 + u_2,k)}} = \mathrm{artanh}\left(\frac{\mathrm{sr}_2}{\mathrm{sr}_1}\right) - \mathrm{artanh}\left(\frac{\mathrm{sr}_2\mathrm{pq}_1}{\mathrm{sr}_1\mathrm{pq}_2}\right) \tag{83}$$

$$\log\sqrt{\frac{1-\mathrm{pq}(u_1 + u_2,k)}{1-\mathrm{pq}(u_1 - u_2,k)}} = \mathrm{artanh}\left(\frac{\mathrm{sr}_2}{\mathrm{sr}_1}\right) + \mathrm{artanh}\left(\frac{\mathrm{sr}_2\mathrm{pq}_1}{\mathrm{sr}_1\mathrm{pq}_2}\right) \tag{84}$$

Proof. Use Formulas (70) and (71).

Special cases were given by Gudermann 1838 [10].

Theorem 3.12. *Assume that $p \neq n$ and $q = n$. Then we have the two formulas*

$$\log\sqrt{\frac{\mathrm{sp}(a+b)}{\mathrm{sp}(a-b)}} = \frac{1}{2}\mathrm{artanh}\left(\frac{\mathrm{sn}(2b)}{\mathrm{sn}(2a)}\right) + \frac{1}{2}\mathrm{artanh}\left(\frac{\mathrm{sr}(2b)}{\mathrm{sr}(2a)}\right) \tag{85}$$

Proof. Use Formula (68).

Special cases of Formula (85) were given by Gudermann 1838 [10].

Theorem 3.13. *We have the six formulas*

$$\log\sqrt{\frac{\mathrm{pq}(a+b)}{\mathrm{pq}(a-b)}} = \frac{1}{2}\mathrm{artanh}\left(\frac{\mathrm{sq}(2b)\mathrm{pq}(2a)}{\mathrm{sq}(2a)\mathrm{pq}(2b)}\right) - \frac{1}{2}\mathrm{artanh}\left(\frac{\mathrm{sq}(2b)}{\mathrm{sq}(2a)}\right) \tag{86}$$

Proof. Use Formula (67).

Special cases of Formula (86) were given by Gudermann 1838 [10].

4. Hyperbolic Modular Functions

We will again consider two inverse functions.

Definition 4. Let $0 < k < 1$ and consider the following hyperbolic integral.

$$u \equiv F(x) \equiv \int_0^x \frac{dt}{\sqrt{1+t^2}\sqrt{1+k^2t^2}} \tag{87}$$

Now put $x = \mathcal{SN}(u) \equiv F^{-1}(u)$, the hyperbolic modular sine for the module k. Then we further define

$$\mathcal{CN}(u) \equiv \sqrt{1+x^2} \tag{88}$$

the hyperbolic modular cosine for the module k

$$\mathcal{SC}(u) \equiv \frac{x}{\sqrt{1+x^2}} \tag{89}$$

the hyperbolic modular tangent for the module k

$$\mathcal{DN}(u) \equiv \sqrt{1+k^2x^2} \tag{90}$$

the hyperbolic difference for the module k.

Definition 5. Just like for the elliptic functions, we use the Glaischer notation as follows:

$$\mathcal{NS}u \equiv \frac{1}{\mathcal{SN}u}, \mathcal{NC}u \equiv \frac{1}{\mathcal{CN}u}, \mathcal{ND}u \equiv \frac{1}{\mathcal{DN}u}, \mathcal{CD}u \equiv \frac{\mathcal{CN}u}{\mathcal{DN}u}, etc. \tag{91}$$

We find that

$$\begin{gathered}\lim_{k\to 0^+} \mathcal{SN}x = \lim_{k\to 0^+} \mathcal{SD}x = \mathrm{sinh}x,\ \lim_{k\to 0^+} \mathcal{SN}x = \cosh x,\ \lim_{k\to 0^+} \mathcal{DN}x = 1 \\ \lim_{k\to 0^+} \mathcal{SC}x = \tanh x\ \lim_{k\to 1^-} \mathcal{SN}x = \tan x \\ \lim_{k\to 1^-} \mathcal{CN}x = \lim_{k\to 1^-} \mathcal{DN}x = \frac{1}{\cos x},\ \lim_{k\to 1^-} \mathcal{SC}x = \lim_{k\to 1^-} \mathcal{SD}x = \sin x\end{gathered} \tag{92}$$

Definition 6. We put

$$\{\mathcal{P},\mathcal{Q},\mathcal{R}\} \equiv \{\mathcal{C},\mathcal{D},\mathcal{N}\} \tag{93}$$

Furthermore, we put

$$\triangle\,(\mathcal{P},\mathcal{Q}) \equiv \mathcal{PS}^2 - \mathcal{QS}^2 \tag{94}$$

Definition 7. The Gudermannian function $l(x)$ relates the circular functions and hyperbolic functions without using complex numbers. It is given by

$$\begin{aligned} l(x) &\equiv \int_0^x \frac{dt}{\cosh t} = \arcsin(\tanh x) = \arctan(\sinh x) \\ &= 2\arctan\left[\tanh\left(\tfrac{1}{2}x\right)\right] = 2\arctan\left(e^x\right) - \tfrac{1}{2}\pi \end{aligned} \tag{95}$$

The inverse function or the Mercator function is given by

$$\begin{aligned} \mathcal{L}(x) &\equiv \int_0^x \frac{dt}{\cos t} = \log\frac{1+\sin x}{\cos x} = \log\sqrt{\frac{1+\sin x}{1-\sin x}} \\ &= \log[\tan x + \sec x] = \log\left[\tan\left(\tfrac{1}{4}\pi + \tfrac{1}{2}x\right)\right] \\ &= \operatorname{artanh}(\sin x) = \operatorname{arsinh}(\tan x) \end{aligned} \tag{96}$$

The function $\mathcal{L}(x)$ is the inverse of $l(x)$. Legendre calculated tables for this function. Since $\mathcal{L}\varphi > \varphi$ it follows that [9]

$$\sinh u < \mathcal{SN}u < \sinh\mathcal{L}u \tag{97}$$

We also find that

$$\cosh u < \mathcal{CN}u < \cosh\mathcal{L}u, \tanh u < \mathcal{SC}u < \tanh\mathcal{L}u \tag{98}$$

The hyperbolic elliptic functions can also be transformed to the hyperbolic potential functions by putting

$$\mathcal{SN}u = \sinh\psi, \mathcal{CN}u = \cosh\psi \text{and} \mathcal{SC}u = \tanh\psi \tag{99}$$

The arc ψ is is called the *hyperbolic amplitude* of the argument u for the module kt; or $\psi = \mathcal{A}mu$, and vice versa $u = \mathcal{A}rg\mathcal{A}m(\psi)$.

Next we have [9]

$$D\mathcal{SN}u = \mathcal{CN}u\mathcal{DN}u, D\mathcal{CN}u = \mathcal{SN}u\mathcal{DN}u \tag{100}$$

$$D\mathcal{SC}u = \frac{\mathcal{DN}u}{\mathcal{CN}^2u}, D\mathcal{DN}u = k^2\mathcal{SN}u\mathcal{CN}u \tag{101}$$

Below is a list of the inverse of the four hyperbolic modular functions:

When $t = \mathcal{SN}u$, so $u = \mathcal{A}rg\mathcal{SN}t$;
When $t = \mathcal{CN}u$, so $u = \mathcal{A}rg\mathcal{CN}t$;
When $t = \mathcal{SC}u$, so $u = \mathcal{A}rg\mathcal{SC}t$;
When $t = \mathcal{DN}u$, so $u = \mathcal{A}rg\mathcal{DN}t$;

We have

$$D\mathcal{A}mu = \mathcal{DN}u \tag{102}$$

Formula (87) is equivalent to

$$\mathcal{A}rg\mathcal{A}m(t) = \int_0^{\operatorname{arsinh} t} \frac{d\psi}{\sqrt{1 + k^2 \sinh^2(\psi)}} \tag{103}$$

The poles and periods are shown in the following table:

half period	polesiK	poles$K'+iK$	polesK'	poles0	periods
iK	$\mathcal{SC}$	$\mathcal{ND}$	$\mathcal{DN}$	$\mathcal{CS}$	$2iK, 4K'+4iK, 4K'$
$K'+iK$	$\mathcal{NC}$	$\mathcal{SD}$	$\mathcal{CN}$	$\mathcal{DS}$	$4iK, 2K'+2iK, 4K'$
K'	$\mathcal{DC}$	$\mathcal{CD}$	$\mathcal{SN}$	$\mathcal{NS}$	$4iK, 4K'+4iK, 2K'$

We have the following special values for the twelve hyperbolic modular functions:

u	$\mathcal{SN}$	$\mathcal{CN}$	$\mathcal{DN}$	$\mathcal{SC}$	$\mathcal{SD}$	$\mathcal{ND}$	$\mathcal{CD}$	$\mathcal{CS}$	$\mathcal{DS}$	$\mathcal{NC}$	$\mathcal{DC}$	$\mathcal{NS}$
0	0	1	1	0	0	1	1	∞	∞	1	1	∞
iK	i	0	k'	∞	$\frac{i}{k'}$	$\frac{1}{k'}$	0	0	$-ik'$	∞	∞	$-i$
K'	∞	∞	∞	1	$\frac{1}{k}$	0	$\frac{1}{k}$	1	k	0	k	0
$2K'$	0	-1	-1	0	0	-1	1	∞	∞	-1	1	∞
$2K'+2iK$	0	-1	-1	0	0	-1	-1	∞	∞	-1	-1	∞
$2iK$	0	-1	1	0	0	1	1	∞	∞	-1	-1	∞
$K'+iK$	$\frac{i}{k}$	$\frac{ik'}{k}$	0	$\frac{1}{k'}$	∞	∞	∞	k'	0	$\frac{-ik}{k'}$	0	ik

Just like the trigonometric and hyperbolic potential functions can be transformed to each other by multiplication by i, the trigonometric and hyperbolic modular functions can also be mapped to each other with utter avoidance of imaginary forms [9]. These transformations look like this (we use the Glaisher notation and $'$ means the module k'):

$$\begin{cases} \mathcal{SN}ui = i\mathrm{sn}u, \mathcal{CN}ui = \mathrm{cn}u \\ \mathcal{SC}ui = i\mathrm{sc}u, \mathcal{DN}ui = \mathrm{dn}u \end{cases} \tag{104}$$

$$\begin{cases} \mathrm{sn}ui = i\mathcal{SN}u, \mathrm{cn}ui = \mathcal{CN}u \\ \mathrm{sc}ui = i\mathcal{SC}u, \mathrm{dn}ui = \mathcal{DN}u \end{cases} \tag{105}$$

Then we have by the Jacobi imaginary transformation [12], compare with [16].

$$\begin{cases} \mathcal{SN}u = \mathrm{sc}'u \\ \mathcal{CN}u = \mathrm{nc}'u \\ \mathcal{SC}u = \mathrm{sn}'u \\ \mathcal{DN}u = \mathrm{dc}'u \end{cases} \tag{106}$$

This implies

$$\mathcal{A}\mathrm{m}ui = i\mathrm{am}u, \mathrm{am}ui = i\mathcal{A}\mathrm{m}u \tag{107}$$

To be able to compute the inverses of the hyperbolic modular functions, we must first prove a large number < 132 of Möbius transformations between squares of hyperbolic modular functions which govern their transformations. Most of these can be summarized in the formulas

$$\mathcal{PQ}^2 = \frac{\triangle(\mathcal{R},\mathcal{P}) + \mathcal{RS}^2}{\triangle(\mathcal{R},\mathcal{Q}) + \mathcal{RS}^2} \tag{108}$$

$$\mathcal{SR}^2 = \frac{\mathcal{PQ}^2 - 1}{\triangle(\mathcal{R},\mathcal{P}) - \triangle(\mathcal{R},\mathcal{Q})\mathcal{PQ}^2} \tag{109}$$

$$\mathcal{SP}^2 = \frac{1 - \mathcal{PQ}^2}{\triangle\,(\mathcal{P},\mathcal{Q})\mathcal{PQ}^2} \tag{110}$$

$$\mathcal{PQ}^2 = \frac{\triangle\,(\mathcal{P},\mathcal{R})\mathcal{QR}^2 + \triangle\,(\mathcal{Q},\mathcal{P})}{\triangle\,(\mathcal{Q},\mathcal{R})\mathcal{QR}^2} \tag{111}$$

Formulas (108) and (109) are inverse to each other. Formula (111) is its own inverse. It is the same as the previous Formula (81).

We can now easily prove the following integral formulas:

$$\mathcal{SN}^{-1}(x) = \int_0^x \frac{dt}{\sqrt{1+t^2}\sqrt{1+k^2t^2}}, 0 < x \le \infty \tag{112}$$

$$\mathcal{CN}^{-1}(x) = \int_1^x \frac{dt}{\sqrt{t^2-1}\sqrt{k'^2+k^2t^2}}, \infty \ge x > 1 \tag{113}$$

$$\mathcal{SC}^{-1}(x) = \int_0^x \frac{dt}{\sqrt{1-t^2}\sqrt{1-k'^2t^2}}, 0 < x \le 1 \tag{114}$$

$$\mathcal{DN}^{-1}(x) = \int_1^x \frac{dt}{\sqrt{t^2-1}\sqrt{t^2-k'^2}}, \infty \ge x > 1 \tag{115}$$

$$\mathcal{SD}^{-1}(x) = \int_0^x \frac{dt}{\sqrt{1+k'^2t^2}\sqrt{1-k^2t^2}}, \frac{1}{k} \ge x > 0 \tag{116}$$

$$\mathcal{DC}^{-1}(x) = \int_x^1 \frac{dt}{\sqrt{t^2-k^2}\sqrt{1-t^2}}, k \le x < 1 \tag{117}$$

$$\mathcal{DS}^{-1}(x) = \int_x^\infty \frac{dt}{\sqrt{k'^2+t^2}\sqrt{t^2-k^2}}, x \ge k > 0 \tag{118}$$

$$\mathcal{CD}^{-1}(x) = \int_1^x \frac{dt}{\sqrt{t^2-1}\sqrt{1-k^2t^2}}, \frac{1}{k} \le x < 1 \tag{119}$$

$$\mathcal{CS}^{-1}(x) = \int_x^1 \frac{dt}{\sqrt{1-t^2}\sqrt{k'^2-t^2}}, -\infty \le x < 1 \tag{120}$$

$$\mathcal{NS}^{-1}(x) = \int_x^\infty \frac{dt}{\sqrt{1+t^2}\sqrt{k^2+t^2}}, \infty > x \ge 0 \tag{121}$$

$$\mathcal{NC}^{-1}(x) = \int_x^1 \frac{dt}{\sqrt{1-t^2}\sqrt{k^2+k'^2t^2}}, 0 \le x < 1 \tag{122}$$

$$\mathcal{ND}^{-1}(x) = \int_x^1 \frac{dt}{\sqrt{1-t^2}\sqrt{1-k'^2t^2}}, 0 \le x < 1 \tag{123}$$

Formulas (112)–(115) can be found in [9].

The addition formulas for hyperbolic modular functions are

Theorem 4.1. *Put* $\mathcal{PS}_i \equiv \mathcal{PS}(u_i,k), i = 1,2,$ *and similar notation for the other functions.*

$$\mathcal{PS}(u_1+u_2,k) = \frac{\mathcal{PS}_1\mathcal{QS}_2\mathcal{RS}_2 - \mathcal{PS}_2\mathcal{QS}_1\mathcal{RS}_1}{\mathcal{PS}_2^2 - \mathcal{PS}_1^2} = \frac{\triangle\,(\mathcal{P},\mathcal{Q})\,\triangle\,(\mathcal{P},\mathcal{R}) - \mathcal{PS}_1^2\mathcal{PS}_2^2}{\mathcal{PS}_1\mathcal{QS}_2\mathcal{RS}_2 + \mathcal{PS}_2\mathcal{QS}_1\mathcal{RS}_1} \tag{124}$$

$$\begin{aligned}\mathcal{SP}(u_1+u_2,k) &= \frac{\mathcal{SP}_1^2 - \mathcal{SP}_2^2}{\mathcal{SP}_1\mathcal{QP}_2\mathcal{RP}_2 - \mathcal{SP}_2\mathcal{QP}_1\mathcal{RP}_1} \\ &= \frac{\mathcal{SP}_1\mathcal{QP}_2\mathcal{RP}_2 + \mathcal{SP}_2\mathcal{QP}_1\mathcal{RP}_1}{1 - \triangle(\mathcal{P},\mathcal{Q})\triangle(\mathcal{P},\mathcal{R})\mathcal{SP}_1^2\mathcal{SP}_2^2}\end{aligned} \tag{125}$$

$$\begin{aligned}\mathcal{PQ}(u_1+u_2,k) &= \frac{\mathcal{PS}_1\mathcal{QS}_2\mathcal{RS}_2 - \mathcal{PS}_2\mathcal{QS}_1\mathcal{RS}_1}{\mathcal{QS}_1\mathcal{PS}ps_2\mathcal{RS}_2 - \mathcal{QS}_2\mathcal{PS}_1\mathcal{RS}_1} \\ &= \frac{\mathcal{PS}_1\mathcal{QS}_1\mathcal{PS}_2\mathcal{QS}_2 - \triangle(\mathcal{P},\mathcal{Q})\mathcal{RS}_1\mathcal{RS}_2}{\mathcal{QS}_1^2\mathcal{QS}_2^2 + \triangle(\mathcal{P},\mathcal{Q})\triangle(\mathcal{Q},\mathcal{R})}\end{aligned} \tag{126}$$

$$\mathcal{PQ}(u_1+u_2,k)=\frac{\mathcal{PQ}_1\mathcal{SQ}_1\mathcal{RQ}_2-\mathcal{PQ}_2\mathcal{SQ}_2\mathcal{RQ}_1}{\mathcal{PQ}_2\mathcal{SQ}_1\mathcal{RQ}_2-\mathcal{PQ}_1\mathcal{SQ}_2\mathcal{RQ}_1}$$
$$=\frac{\mathcal{PQ}_1\mathcal{PQ}_2-\Delta(\mathcal{P},\mathcal{Q})\mathcal{SQ}_1\mathcal{RQ}_1\mathcal{SQ}_2\mathcal{RQ}_2}{1+\Delta(\mathcal{P},\mathcal{Q})\Delta(\mathcal{Q},\mathcal{R})\mathcal{SQ}_1^2\mathcal{SQ}_2^2} \tag{127}$$

Proof. Use Formula (105).

Special cases of Formulas (124) to (127) were given in [17]. Formulas (124) and (125) are inverse to each other.

Theorem 4.2. *Addition formulas for complex arguments. Put* $\mathrm{ps}_1\equiv\mathrm{ps}(u_1,k),\mathcal{PS}\equiv\mathcal{PS}(u_2,k)$, *and similar notation for the other functions.*

$$\mathrm{ps}(u_1+iu_2,k)=\frac{\mathrm{ps}_1\mathcal{QSRS}-i\mathcal{PS}\mathrm{qs}_1\mathrm{rs}_1}{\mathcal{PS}^2+\mathrm{ps}_1^2}=\frac{\mathcal{PS}^2\mathrm{ps}_2^2+\Delta(\mathrm{p},\mathrm{q})\Delta(\mathrm{p},\mathrm{r})}{\mathrm{ps}_1\mathcal{QSRS}+i\mathcal{PS}\mathrm{qs}_1\mathrm{rs}_1} \tag{128}$$

$$\mathrm{sp}(u_1+iu_2,k)=\frac{\mathcal{SP}^2+\mathrm{sp}_1^2}{\mathrm{sp}_1\mathcal{QPRP}-i\mathcal{SP}\mathrm{qp}_1\mathrm{rp}_1}=\frac{\mathrm{sp}_1\mathcal{QPRP}+i\mathcal{SP}\mathrm{qp}_1\mathrm{rp}_1}{1+\Delta(\mathrm{p},\mathrm{q})\Delta(\mathrm{p},\mathrm{r})\mathrm{sp}_1^2\mathcal{SP}^2} \tag{129}$$

$$\mathrm{pq}(u_1+iu_2,k)=\frac{\mathrm{ps}_1\mathcal{QSRS}-i\mathcal{PS}\mathrm{qs}_1\mathrm{rs}_1}{\mathrm{qs}_1\mathcal{PSRS}-i\mathcal{QS}\mathrm{ps}_1\mathrm{rs}_1}=\frac{\mathrm{ps}_1\mathrm{qs}_1\mathcal{PSQS}+i\Delta(\mathrm{p},\mathrm{q})\mathrm{rs}_1\mathcal{RS}}{\mathrm{qs}_1^2\mathcal{QS}^2-\Delta(\mathrm{p},\mathrm{q})\Delta(\mathrm{q},\mathrm{r})} \tag{130}$$

$$\mathrm{pq}(u_1+iu_2,k)=\frac{\mathrm{pq}_1\mathrm{sq}_1\mathcal{RQ}-i\mathcal{PQSQ}\mathrm{rq}_1}{\mathcal{PQRQ}\mathrm{sq}_1-i\mathrm{pq}_1\mathcal{SQ}\mathrm{rq}_1}=\frac{\mathrm{pq}_1\mathcal{PQ}+i\Delta(\mathrm{p},\mathrm{q})\mathrm{sq}_1\mathrm{rq}_1\mathcal{SQRQ}}{1-\Delta(\mathrm{p},\mathrm{q})\Delta(\mathrm{q},\mathrm{r})\mathrm{sq}_1^2\mathcal{SQ}^2} \tag{131}$$

Proof. Use Formula (105).

Special cases of Formulas (128) to (131) were first given by Gudermann 1838 [9].

5. The Peeta Functions

The Peeta functions were first introduced by Gudermann [18], to be able to express the hyperbolic modular functions by theta functions in a manner similar to the elliptic functions.

Eagle [7] has also spoken of the great importance of these functions.

The four Peeta functions are defined as follows:

Definition 8.

$$\psi_1(z,q)\equiv-i\theta_1(iz,q);\psi_2(z,q)\equiv\theta_2(iz,q);\psi_3(z,q)\equiv\theta_3(iz,q);\psi_4(z,q)\equiv\theta_4(iz,q) \tag{132}$$

This is equivalent to

$$\begin{aligned}\psi_1(z,q)&\equiv 2\sum_{n=0}^{\infty}(-1)^n\mathrm{QE}((n+\tfrac{1}{2})^2)\sinh{(2n+1)z}\\ \psi_2(z,q)&\equiv 2\sum_{n=0}^{\infty}\mathrm{QE}((n+\tfrac{1}{2})^2)\cosh{(2n+1)z}\\ \psi_3(z,q)&\equiv 1+2\sum_{n=1}^{\infty}\mathrm{QE}(n^2)\cosh{(2nz)}\\ \psi_4(z,q)&\equiv 1+2\sum_{n=1}^{\infty}(-1)^n\mathrm{QE}(n^2)\cosh{(2nz)}\end{aligned} \tag{133}$$

where $\mathrm{QE}(x)\equiv q^x, q\equiv\exp(\pi it), t\in\mathcal{U}$.

Another definition is

$$
\begin{aligned}
\psi_1(z,q) &= 2q^{\frac{1}{4}} \sinh(x) \prod_{k=1}^{\infty} (1-q^{2k})(1-2q^{2k}\cosh(2z)+q^{4k}) \\
\psi_2(z,q) &= 2q^{\frac{1}{4}} \cosh(z) \prod_{k=1}^{\infty} (1-q^{2k})(1+2q^{2k}\cosh(2z)+q^{4k}) \\
\psi_3(z,q) &= \prod_{k=1}^{\infty} (1-q^{2k})(1+2q^{2k-1}\cosh(2z)+q^{4k-2}) \\
\psi_4(z,q) &= \prod_{k=1}^{\infty} (1-q^{2k})(1-2q^{2k-1}\cosh(2z)+q^{4k-2})
\end{aligned}
\tag{134}
$$

This implies [18]

$$
\begin{aligned}
\log\psi_1(z,q) &= f_1(q) + \log\sinh(z) - \sum_{m=1}^{\infty} \tfrac{2q^{2m}}{m(1-q^{2m})} \cosh(2mz) \\
\log\psi_2(z,q) &= f_2(q) + \log\cosh(z) + \sum_{m=1}^{\infty} \tfrac{2(-1)^{m+1}q^{2m}}{m(1-q^{2m})} \cosh(2mz) \\
\log\psi_3(z,q) &= f_3(q) + \sum_{m=1}^{\infty} \tfrac{2(-1)^{m+1}q^{m}}{m(1-q^{2m})} \cosh(2mz) \\
\log\psi_4(z,q) &= f_4(q) - \sum_{m=1}^{\infty} \tfrac{2q^{m}}{m(1-q^{2m})} \cosh(2mz)
\end{aligned}
\tag{135}
$$

The hyperbolic modular functions can be expressed in terms of Peeta functions as follows [18]:

Theorem 5.1.

$$
\begin{aligned}
\mathcal{SN}(u) &= k^{-\frac{1}{2}} \tfrac{\psi_1(z,q)}{\psi_4(z,q)} \\
\mathcal{CN}(u) &= \left(\tfrac{k'}{k}\right)^{\frac{1}{2}} \tfrac{\psi_2(z,q)}{\psi_4(z,q)} \\
\mathcal{DN}(u) &= k'^{\frac{1}{2}} \tfrac{\psi_3(z,q)}{\psi_4(z,q)}
\end{aligned}
\tag{136}
$$

The Peeta functions have the following periodic properties:

$y=$	z	$z+\frac{1}{2}\log q$	$z+\log q$	$z+\frac{\pi i}{2}$	$z+\pi i$
$\psi_1(y)$	$\psi_1(z)$	$q^{-\frac{1}{4}}e^{z}\psi_4(z)$	$-q^{-1}e^{2z}\psi_1(z)$	$-i\psi_2(z)$	$-\psi_1(z)$
$\psi_2(y)$	$\psi_2(z)$	$q^{-\frac{1}{4}}e^{z}\psi_3(z)$	$q^{-1}e^{2z}\psi_2(z)$	$-i\psi_1(z)$	$-\psi_2(z)$
$\psi_3(y)$	$\psi_3(z)$	$q^{-\frac{1}{4}}e^{z}\psi_2(z)$	$q^{-1}e^{2z}\psi_3(z)$	$\psi_4(z)$	$\psi_3(z)$
$\psi_4(y)$	$\psi_4(z)$	$-q^{-\frac{1}{4}}e^{z}\psi_1(z)$	$-q^{-1}e^{2z}\psi_4(z)$	$\psi_3(z)$	$\psi_4(z)$

Theorem 5.2. *The Peeta functions have the following zeros:*

$$
\begin{aligned}
\psi_1(m\pi i + n\log q, q) &= 0 \\
\psi_2(\tfrac{\pi i}{2} + m\pi i + n\log q, q) &= 0 \\
\psi_3(\tfrac{\pi i}{2} + \tfrac{1}{2}\log q + m\pi i + n\log q, q) &= 0 \\
\psi_4(\tfrac{1}{2}\log q + m\pi i + n\log q, q) &= 0
\end{aligned}
\tag{137}
$$

where $m, n \in \mathbb{Z}$.

Theorem 5.3. *The four Peeta functions satisfy the following heat equation*

$$
\frac{\partial^2\psi_i(z,q)}{\partial z^2} = 4q\frac{\partial\psi_i(z,q)}{\partial q}, i = 1,\ldots,4 \tag{138}
$$

Corollary 5.4. *Generalization of heat equation to n variables. Put*

$$\triangle_N \equiv \sum_{i=1}^{N} \frac{\partial^2}{\partial x_i^2}, u(x_1, \ldots, x_N, q) \equiv \prod_{i=1}^{N} \psi_{k(i)}(x_i, q), k(i) \in \{1, \ldots, 4\} \tag{139}$$

Then

$$\triangle_N\, u = 4q\frac{\partial u}{\partial q} \tag{140}$$

Proof. Use [19].

By scaling in q, we can transform Formula (140) to other heat equations.

Example 1. Heat transfer in friction-free (non-viscous) fluid flow.

$$a^2(\frac{\partial^2 u}{\partial x^2} + \frac{\partial^2 u}{\partial y^2}) = \frac{\partial u}{\partial t} \tag{141}$$

Several formulas for theta functions, like logarithmic derivative, immediately transfer to Peeta functions.

Conflicts of Interest: The author declare no conflict of interest.

References

1. Ernst, T. Die Jacobi-Gudermann-Glaisherschen elliptischen Funktionen nach Heine. *Hadronic. J.* **2010**, *33*, 273–302.
2. Gudermann, C. Theorie der Modular-Functionen and der Modular-Integrale: Reihen for die Logarithmen der Modular-Functionen, welche nach den Cosinus der Vielfachen von ηu and $\eta' u$ fortschreiten. *J. Reine Angew. Math.* **1840**, *20*, 103–167. [CrossRef]
3. Glaisher, J.W.L. On the series which represent the twelve elliptic and the four Zeta functions. *Mess. Math.* **1888**, *18*, 1–84.
4. Ernst, T. *A Comprehensive Treatment of q-Calculus*; Springer: Basel, Switzerland, 2012.
5. Carlson, B.C. Symmetry in c, d, n of Jacobian elliptic functions. *J. Math. Anal. Appl.* **2004**, *299*, 242–253. [CrossRef]
6. Enneper, A. *Elliptische Funktionen. Theorie and Geschichte. Zweite Auflage. Neu bearbeitet und herausgegeben von Felix Müller*, L. Nebert. 1890.
7. Eagle, A. *The Elliptic Functions as They Should Be: An Account, with Applications, of the Functions in a New Canonical Form*; Galloway and Porter, Ltd.: Cambridge, England, UK, 1958.
8. Gudermann, C. *Theorie der Modular-Functionen und der Modular-Integrale*; G. Reimer Publisher: Berlin, Germany, 1844.
9. Gudermann, C. Theorie der Modular-Functionen and der Modular-Integrale: Allgemeiner Character der Potenzial-Funktionen. *J. Reine Angew. Math.* **1838**, *18*, 1–54. [CrossRef]
10. Gudermann, C. Theorie der Modular-Functionen and der Modular-Integrale: Die Modular-Functionen von $\frac{iK'}{2}$ and $\frac{K \pm iK'}{2}$. *J. Reine Angew. Math.* **1838**, *18*, 142–175. [CrossRef]
11. Legendre, A.M. *Traité des fonctions elliptiques et des intégrales eulériennes*; Gauthier-Villars: Paris, France, 1828.
12. Jacobi, C.G.J. *Fundamenta Nova*; Königsberg, 1829.
13. Laurent, H. *Théorie élémentaire des fonctions elliptiques*; Gauthier-Villars: Paris, France, 1880.
14. Neville, E. *Elliptic Functions: A Primer*; Pergamon Press: Oxford, UK, 1971.
15. Olver, F.W.J. *NIST Handbook of Mathematical Functions*; Cambridge University Press: Cambridge, UK, 2010.
16. Abramowitz, M.; Stegun, I. *Handbook of Mathematical Funktionen with Formulas, Graphs, and Mathematical Tables*; U.S. Department of Commerce: Washington, DC, USA, 1970.
17. Gudermann, C. Theorie der Modular-Functionen and der Modular-Integrale: Ausdruck von u and el$u$ durch am$u$ in Reihen, welche nach Potenzen des Moduls k fortschreiten. *J. Reine Angew. Math.* **1839**, *19*, 45–83. [CrossRef]

18. Gudermann, C. Theorie der Modular-Functionen und der Modular-Integrale: Die Modular-Functionen, dargestellt als Producte unendlich vieler Factoren. *J. Reine Angew. Math.* **1840**, *20*, 62–87. [CrossRef]
19. Widder, D.V. Series expansions of solutions of the heat equation in n dimensions. *Ann. Mat. Pura Appl.* **1961**, *55*, 389–409. [CrossRef]

axioms

MDPI

Article

Some Aspects of Extended Kinetic Equation

Dilip Kumar

Department of Basic Sciences, Amal Jyothi College of Engineering, Koovapally P.O., Kanjirapally, Kerala 686518, India; dilipkumar.cms@gmail.com or dilipkumar@amaljyothi.ac.in; Tel.: +91-9446195433

Academic Editor: Hans J. Haubold
Received: 20 May 2015; Accepted: 31 August 2015; Published: 18 September 2015

Abstract: Motivated by the pathway model of Mathai introduced in 2005 [*Linear Algebra and Its Applications*, 396, 317–328] we extend the standard kinetic equations. Connection of the extended kinetic equation with fractional calculus operator is established. The solution of the general form of the fractional kinetic equation is obtained through Laplace transform. The results for the standard kinetic equation are obtained as the limiting case.

Keywords: kinetic equation; fractional derivatives and integrals; Mellin convolution; *G*-function; Mittag-Lefler function

MSC: 35K57; 26A33; 44A35; 33C60; 33E12

---

1. Introduction

The chemical evolution of a star like sun could be effectively explained by kinetic equations. The kinetic equations explain the rate of change of chemical composition of a star in terms of the thermonuclear reaction rates for destruction and production of the species involved. An arbitrary reaction is characterized by the rate of change $\frac{dN}{dt}$ of a time dependent quantity $N(t)$ between the destruction rate d and production rate p. Here the destruction or production at time t depends not only on $N(t)$ but also on the past history $N(\tau), \tau < t$ of the variable N. This may be formally represented by following [1,2]

$$\frac{dN(t)}{dt} = -d(N_t) + p(N_t) \tag{1}$$

where N_t denotes the function defined by $N_t(t^*) = N(t - t^*), t^* > 0$. It should be noted that d and p are functionals and Equation (1) represents a functional-differential equation. If we consider a simplified form of Equation (1) we could consider the decay rate of a radio-active substance which is given by a homogeneous differential equation

$$\frac{dN}{dt} = -\lambda N \tag{2}$$

where N is the number density of the radio-active substance and λ is the decay constant. The solution of this differential equation with initial condition $N = N_0$ at $t = 0$ is

$$N(t) = N_0 e^{-\lambda t} \tag{3}$$

If we consider a more general form of the differential Equation (2) for the decay rate of a radio-active substance as

$$\frac{dN}{dt} = -\lambda N^{\alpha} \tag{4}$$

we have the solution of the form

$$N_\alpha(t) = N_0[1 + a(\alpha - 1)t]^{-\frac{1}{\alpha-1}} \tag{5}$$

where a is a constant. One may get the solution in Equation (3) from Equation (5) as $\alpha \to 1$. These types of problems arise in many experimental situations where one needs to switch from one family of functions to another family. In 2005, Mathai [3,4] introduced the pathway model by which one can switch among three different families of functions, say, type-1 beta families, type-2 beta families and gamma families. We get three different functional forms by varying the pathway parameter α. The pathway model in the real scalar case is defined as

$$f(x) = \begin{cases} c_1|x|^\gamma[1 - a(1-\alpha)|x|^\delta]^{\frac{\eta}{1-\alpha}}, 1 - a(1-\alpha)|x|^\delta > 0, \alpha < 1 \\ c_2|x|^\gamma[1 + a(\alpha-1)|x|^\delta]^{-\frac{\eta}{\alpha-1}}, -\infty < x < \infty, \alpha > 1 \\ c_3|x|^\gamma e^{-a\eta|x|^\delta}, -\infty < x < \infty, \alpha \to 1 \end{cases} \tag{6}$$

where $a > 0, \delta > 0, \gamma > 0, \eta > 0$. c_1, c_2 and c_3 are the normalizing constants when we consider the functions as statistical densities. The three different functional forms are respectively generalized type-1 beta, generalized type-2 beta and generalized gamma forms. By writing $1 - \alpha = -(\alpha - 1)$, the generalized type-2 beta form can be obtained from generalized type-1 beta form. Both generalized type-1 beta form and generalized type-2 beta form reduce to generalized gamma form as $\alpha \to 1$.

Due to this switching property, the pathway model has been widely used in many areas. In this paper, we use the pathway model to extend kinetic equations. The present paper is organized as follows: In the next section we discuss the extended kinetic equation and its solution with a brief description of the extended reaction rate probability integral. Connection of the extended kinetic equation with fractional calculus is examined in Section 3. In Section 4 we try to solve fractional kinetic equations and their various generalizations. Concluding remarks are given in Section 5.

2. Extended Kinetic Equations

The following discussion is based on [1,2]. If we consider a production and destruction of nuclei in the proton-proton chain reaction, we can describe it by the equation

$$\frac{dN_i}{dt} = -\sum_j N_i N_j \langle\sigma v\rangle_{ij} + \sum_{k,l\neq i} N_k N_l \langle\sigma v\rangle_{kl} \tag{7}$$

where N_i is the number density of the species i over time. Here the summation is taken over all reactions, productions or destructions of the species i. The number density N_i of the species i can be expressed by the relation $N_i = \rho N_A \frac{X_i}{A_i}$ where ρ is the mass density, X_i is the mass abundance, N_A is the Avogadro number and A_i is the mass of species i in mass units. The mean life time $\tau_j(i)$ of species i for destruction by species j is given by the relation [2]

$$\lambda_j(i) = \frac{1}{\tau_j(i)} = N_j\langle\sigma v\rangle_{ij} = \rho N_A \frac{X_j}{A_j}\langle\sigma v\rangle_{ij} \tag{8}$$

where $\lambda_j(i)$ is the decay rate of i for interaction with j. $\langle\sigma v\rangle_{ij}$ denotes the reaction probability for an interaction involving species i and j defined as

$$\langle\sigma v\rangle_{ij} = \sqrt{\frac{2}{\mu}} \int_0^\infty E^{\frac{1}{2}} f(E)\sigma(E) dE \tag{9}$$

where μ is the reduced mass of the particles given by $\mu = \frac{m_1 m_2}{m_1+m_2}$, $E = \frac{\mu v^2}{2}$ is the kinetic energy of the particles in the center of mass system. Consider the cross section $\sigma(E)$ for low-energy non-resonant reactions given by

$$\sigma(E) = \sum_{\nu=0}^{2} \frac{S^{(\nu)}(0)}{\nu!} E^{\nu-1} e^{-2\pi(\frac{\mu}{2})^{\frac{1}{2}} \frac{Z_i Z_j e^2}{\hbar E^{\frac{1}{2}}}} \left(\frac{E}{B} \ll 1\right) \tag{10}$$

where Z_i and Z_j are the atomic numbers of the nuclei i and j, e is the quantum of electric charge, $\hbar$ is the Planck's quantum of action, B the nuclear barrier height, $S(E)$ is the cross-section factor which is a slowly varying function of energy over a limited energy range and which can be characterized depending on the nuclear reaction. The density function of the relative velocities of the nuclei for a non degenerate and non-relativistic gas is assumed to be Maxwell-Boltzmann given as

$$f_{MBD}(E)\mathrm{d}E = 2\pi\left(\frac{1}{\pi kT}\right)^{\frac{3}{2}} e^{-\frac{E}{kT}} \sqrt{E}\mathrm{d}E \tag{11}$$

By substituting Equation (11) and Equation (10) in Equation (9) the reaction probability $\langle\sigma v\rangle_{ij}$ is obtained as

$$\langle\sigma v\rangle_{ij} = \left(\frac{8}{\pi\mu}\right)^{\frac{1}{2}} \left(\frac{1}{kT}\right)^{\frac{3}{2}} \sum_{\nu=0}^{2} \frac{S^{(\nu)}(0)}{\nu!} \int_0^\infty E^\nu e^{-\frac{E}{kT} - 2\pi(\frac{\mu}{2})^{\frac{1}{2}} \frac{Z_i Z_j e^2}{\hbar E^{\frac{1}{2}}}} \mathrm{d}E \tag{12}$$

If a deviation from the thermodynamic equilibrium with regard to their velocities is considered then it results in a deviation from the Maxwell-Boltzmann velocity. In this context, we consider a more general density function than Maxwell-Boltzmann density function by using the pathway model defined in Equation (6). The pathway energy density function has the form

$$f_{PD}(E)\mathrm{d}E = \frac{2\pi(\alpha-1)^{\frac{3}{2}}}{(\pi kT)^{\frac{3}{2}}} \frac{\Gamma\left(\frac{1}{\alpha-1}\right)}{\Gamma\left(\frac{1}{\alpha-1} - \frac{3}{2}\right)} \sqrt{E}\left[1 + (\alpha-1)\frac{E}{kT}\right]^{-\frac{1}{\alpha-1}} \mathrm{d}E \tag{13}$$

for $\alpha > 1, \frac{1}{\alpha-1} - \frac{3}{2} > 0$. Replacing the Maxwell-Boltzmann density Function Equation (11) by the pathway energy density Equation (13), we get the extended thermonuclear reaction probability integral in the form

$$\begin{aligned}\langle\sigma v\rangle_{ij} = & \left(\frac{8}{\pi\mu}\right)^{\frac{1}{2}} \left(\frac{\alpha-1}{kT}\right)^{\frac{3}{2}} \frac{\Gamma\left(\frac{1}{\alpha-1}\right)}{\Gamma\left(\frac{1}{\alpha-1} - \frac{3}{2}\right)} \sum_{\nu=0}^{2} \frac{S^{(\nu)}(0)}{\nu!} \\ & \times \int_0^\infty E^\nu \left[1 + (\alpha-1)\frac{E}{kT}\right]^{-\frac{1}{\alpha-1}} \exp\left[-2\pi(\tfrac{\mu}{2})^{\frac{1}{2}} \frac{Z_i Z_j e^2}{\hbar E^{\frac{1}{2}}}\right] \mathrm{d}E\end{aligned} \tag{14}$$

Putting $y = \frac{E}{kT}$ and $x = 2\pi(\frac{\mu}{2kT})^{\frac{1}{2}} \frac{Z_i Z_j e^2}{\hbar}$ we get

$$\langle\sigma v\rangle_{ij} = \left(\frac{8}{\pi\mu}\right)^{\frac{1}{2}} (\alpha-1)^{\frac{3}{2}} \frac{\Gamma\left(\frac{1}{\alpha-1}\right)}{\Gamma\left(\frac{1}{\alpha-1} - \frac{3}{2}\right)} \sum_{\nu=0}^{2} \left(\frac{1}{kT}\right)^{-\nu+\frac{1}{2}} \frac{S^{(\nu)}(0)}{\nu!} I_{1\alpha}\left(\nu, 1, x, \frac{1}{2}\right) \tag{15}$$

where

$$I_{1\alpha}\left(\nu, 1, x, \frac{1}{2}\right) = \int_0^\infty y^\nu [1 + (\alpha-1)y]^{-\frac{1}{\alpha-1}} e^{-xy^{-\frac{1}{2}}} \mathrm{d}y \tag{16}$$

Following [5], by taking the Mellin transform of Equation (16) and simplifying, we get

$$M_{I_{1\alpha}}(s) = \frac{\Gamma(s)\Gamma(\nu + 1 + \frac{s}{2})\Gamma\left(\frac{1}{\alpha-1} - \nu - 1 - \frac{s}{2}\right)}{(\alpha-1)^{\nu+1+\frac{s}{2}}\Gamma\left(\frac{1}{\alpha-1}\right)} \tag{17}$$

where $\Re(s) > 0, \Re(\nu+1+\frac{s}{2}) > 0, \Re\left(\frac{1}{\alpha-1} - \nu - 1 - \frac{s}{2}\right) > 0$. By taking the inverse Mellin transform we get,

$$I_{1\alpha}\left(\nu,1,x,\frac{1}{2}\right) = \frac{1}{(\alpha-1)^{\nu+1}\Gamma\left(\frac{1}{\alpha-1}\right)} \frac{1}{2\pi i}\int_L \Gamma(s)\Gamma\left(\nu+1+\frac{s}{2}\right)$$
$$\times\Gamma\left(\tfrac{1}{\alpha-1} - \nu - 1 - \tfrac{s}{2}\right)\left[x(\alpha-1)^{\frac{1}{2}}\right]^{-s} ds \tag{18}$$

where L is a suitable contour which separates the poles of $\Gamma(s)$ and $\Gamma(\nu+1+\frac{s}{2})$ from the poles of $\Gamma\left(\frac{1}{\alpha-1} - \nu - 1 - \frac{s}{2}\right)$. Putting $s = 2s'$ and using Legendre's duplication formula [6]

$$\Gamma(2z) = \pi^{-\frac{1}{2}}2^{2z-1}\Gamma(z)\Gamma\left(z+\frac{1}{2}\right), z \in \mathbf{C} \tag{19}$$

we get

$$I_{1\alpha}\left(\nu,1,x,\frac{1}{2}\right) = \frac{\pi^{-\frac{1}{2}}}{(\alpha-1)^{\nu+1}\Gamma\left(\frac{1}{\alpha-1}\right)} \frac{1}{2\pi i}\int_{L'} \Gamma(s')\Gamma\left(\frac{1}{2}+s'\right)\Gamma(\nu+1+s')$$
$$\times\Gamma\left(\tfrac{1}{\alpha-1} - \nu - 1 - s'\right)\left[\tfrac{(\alpha-1)x^2}{4}\right]^{-s'} ds' \tag{20}$$
$$= \frac{\pi^{-\frac{1}{2}}}{(\alpha-1)^{\nu+1}\Gamma\left(\frac{1}{\alpha-1}\right)} G^{3,1}_{1,3}\left(\frac{(\alpha-1)x^2}{4}\Big|^{2-\frac{1}{\alpha-1}+\nu}_{0,\frac{1}{2},\nu+1}\right) \tag{21}$$

where $G^{3,1}_{1,3}(.)$ is the G-function originally introduced by C.S. Meijer in 1936, see [5,7,8]. The $G^{3,1}_{1,3}(.)$ used in Equation (21) converges for all $\frac{(\alpha-1)x^2}{4}, x \neq 0$. The contour line L' appearing in the integral in Equation (20) is $c - i\infty$ to $c + i\infty$ for $0 < c < \frac{1}{\alpha-1} - \nu - 1$ so that all the poles of $\Gamma(s'), \Gamma\left(\frac{1}{2}+s'\right)$ and $\Gamma(\nu+1+s')$ lie to the left and all the poles of $\Gamma\left(\frac{1}{\alpha-1} - \nu - 1 - s'\right)$ lie to the right. $G^{3,1}_{1,3}(.)$ is evaluated as the sum of the residues at the poles of $\Gamma(s'), \Gamma\left(\frac{1}{2}+s'\right)$ and $\Gamma(\nu+1+s')$.

In most of the cases the nuclear factor $S^{(\nu)}(0)$ used in Equation (15) is approximately constant across the fusion window. Hence taking $S^{(\nu)}(0) = 0$ for $\nu = 1$ and $\nu = 2$ and taking $S^0(0) = S(0)$ we get

$$\langle\sigma v\rangle_{ij} = \left(\frac{8(\alpha-1)}{\mu kT}\right)^{\frac{1}{2}} \frac{1}{\pi\Gamma\left(\frac{1}{\alpha-1} - \frac{3}{2}\right)} S(0) G^{3,1}_{1,3}\left(\frac{(\alpha-1)x^2}{4}\Big|^{2-\frac{1}{\alpha-1}}_{0,\frac{1}{2},1}\right) \tag{22}$$

The following derivations are adapted from [9]. From the Mellin-Barnes representation of the G-function, $G^{3,1}_{1,3}\left(\frac{(\alpha-1)x^2}{4}\right)$ appearing in Equation (20) with $\nu = 0$, the poles of $\Gamma(s')$ are $s' = 0, -1, -2, \ldots$; the poles of $\Gamma\left(\frac{1}{2}+s'\right)$ are $s' = -\frac{1}{2}, -\frac{3}{2}, -\frac{5}{2}, \ldots$; and the poles of $\Gamma(1+s')$ are $s' = -1, -2, -3, \ldots$. Here the poles of $\Gamma(s')$ and $\Gamma(1+s')$ will coincide at all points except at $s' = 0$ and hence the pole $s' = 0$ is a pole of order 1, $s' = -\frac{1}{2}, -\frac{3}{2}, -\frac{5}{2}, \ldots$ are each of order 1 and $s' = -1, -2, -3, \ldots$ are each of order 2. The sum of residues corresponding to the pole $s' = 0$ is given by

$$R_1 = \sqrt{\pi}\Gamma\left(\frac{1}{\alpha-1} - 1\right) \tag{23}$$

The sum of the residues corresponding to the poles $s' = -\frac{1}{2}, -\frac{3}{2}, -\frac{5}{2}, \ldots$ is

$$R_2 = \sum_{r=0}^{\infty} \frac{(-1)^r}{r!}\Gamma\left(-\frac{1}{2} - r\right)\Gamma\left(\frac{1}{2} - r\right)\Gamma\left(\frac{1}{\alpha-1} - \frac{1}{2} + r\right)\left[\frac{(\alpha-1)x^2}{4}\right]^{-\frac{1}{2}+r}$$

$$= -2\pi\Gamma\left(\tfrac{1}{\alpha-1}-\tfrac{1}{2}\right)\left[\tfrac{(\alpha-1)x^2}{4}\right]^{\frac{1}{2}}{}_1F_2\left(\tfrac{1}{\alpha-1}-\tfrac{1}{2};\tfrac{3}{2},\tfrac{1}{2};-\tfrac{(\alpha-1)x^2}{4}\right) \tag{24}$$

where ${}_1F_2$ is the hypergeometric function defined by

$${}_1F_2(a;b,c;x)=\sum_{r=0}^{\infty}\frac{(a)_r}{(b)_r(c)_r}\frac{x^r}{r!}$$

where

$$(a)_r=\begin{cases} a(a+1)\cdots(a+r-1) & \text{if} r\geq 1, a\neq 0\\ 1 & \text{if} r=0\end{cases}$$

The sum of the residues corresponding to poles $s'=-1,-2,-3,\ldots$ of order 2 can be obtained as follows:

$$\begin{aligned} R_3 = & \sum_{r=0}^{\infty}\lim_{s'\to-1-r}\frac{\partial}{\partial s'}\Big[(s'+1+r)^2\Gamma(1+s')\Gamma(s')\Gamma\left(\tfrac{1}{2}+s'\right)\\ & \times\Gamma\left(\tfrac{1}{\alpha-1}-1-s'\right)\left(\tfrac{(\alpha-1)x^2}{4}\right)^{-s'}\Big]\\ = & \sum_{r=0}^{\infty}\lim_{s'\to-1-r}\frac{\partial}{\partial s'}\left[\frac{\Gamma^2(2+s'+r)\Gamma\left(\frac{1}{2}+s'\right)\Gamma\left(\frac{1}{\alpha-1}-1-s'\right)}{(s'+r)^2(s'+r-1)^2\cdots(s'+1)^2s'}\left(\frac{(\alpha-1)x^2}{4}\right)^{-s'}\right]\\ = & \sum_{r=0}^{\infty}\lim_{s'\to-1-r}\frac{\partial}{\partial s'}\Phi(s') \end{aligned} \tag{25}$$

where

$$\Phi(s')=\frac{\Gamma^2(2+s'+r)\Gamma\left(\frac{1}{2}+s'\right)\Gamma\left(\frac{1}{\alpha-1}-1-s'\right)}{(s'+r)^2(s'+r-1)^2\cdots(s'+1)^2s'}\left(\frac{(\alpha-1)x^2}{4}\right)^{-s'}$$

We have

$$\frac{\partial}{\partial s'}\Phi(s')=\Phi(s')\frac{\partial}{\partial s'}[\ln(\Phi(s'))]$$

$$\begin{aligned} \ln\Phi(s') &= 2\ln[\Gamma(2+s'+r)]+\ln\left[\Gamma\left(\tfrac{1}{2}+s'\right)\right]+\ln\left[\Gamma\left(\tfrac{1}{\alpha-1}-1-s'\right)\right]\\ &\quad -s'\ln\left(\tfrac{(\alpha-1)x^2}{4}\right)-2\ln(s'+r)-\cdots-2\ln(s'+1)-\ln(s')\\ \tfrac{\partial}{\partial s'}[\ln(\Phi(s'))] &= 2\Psi(2+s'+r)+\Psi\left(\tfrac{1}{2}+s'\right)+\Psi\left(\tfrac{1}{\alpha-1}-1-s'\right)\\ &\quad -\ln\left(\tfrac{(\alpha-1)x^2}{4}\right)-\tfrac{2}{s'+r}-\tfrac{2}{s'+r-1}-\cdots-\tfrac{2}{s'+1}-\tfrac{1}{s'} \end{aligned}$$

$$\begin{aligned} \lim_{s'\to-1-r}\left\{\frac{\partial}{\partial s'}\ln[\Phi(s')]\right\} &= \Psi\left(-\frac{1}{2}-r\right)+\Psi\left(\frac{1}{\alpha-1}+r\right)+\Psi(1+r)\\ &\quad +\Psi(2+r)-\ln\left(\tfrac{(\alpha-1)x^2}{4}\right) \end{aligned} \tag{26}$$

where $\Psi(z)$ is a Psi function or digamma function (see Mathai [5]) and $\Psi(1)=-\gamma,\gamma=0.5772156649\ldots$ is Euler's constant. Now

$$\lim_{s'\to-1-r}\Phi(s)=\frac{(-1)^{1+r}2\sqrt{\pi}\Gamma\left(\frac{1}{\alpha-1}+r\right)}{\left(\frac{3}{2}\right)_r r!(1+r)!}\left(\frac{(\alpha-1)x^2}{4}\right)^{1+r} \tag{27}$$

Then by using Equation (25)–Equation (27) we get,

$$R_3=\sum_{r=0}^{\infty}\frac{(-1)^{1+r}2\sqrt{\pi}\Gamma\left(\frac{1}{\alpha-1}+r\right)}{\left(\frac{3}{2}\right)_r r!(1+r)!}\left(\frac{(\alpha-1)x^2}{4}\right)^{1+r}$$

$$\times\left[\Psi\left(-\tfrac{1}{2}-r\right)+\Psi\left(\tfrac{1}{\alpha-1}+r\right)+\Psi(1+r)+\Psi(2+r)-\ln\left(\tfrac{(\alpha-1)x^2}{4}\right)\right]$$

$$= \left(\frac{2\sqrt{\pi}(\alpha-1)x^2}{4}\right)\sum_{r=0}^{\infty}\left(\frac{(\alpha-1)x^2}{4}\right)^r\left[A_r - \ln\left(\frac{(\alpha-1)x^2}{4}\right)\right]B_r \tag{28}$$

where

$$A_r = \Psi\left(-\frac{1}{2}-r\right) + \Psi\left(\frac{1}{\alpha-1}+r\right) + \Psi(1+r) + \Psi(2+r) \tag{29}$$

and

$$B_r = \frac{(-1)^r\Gamma\left(\frac{1}{\alpha-1}+r\right)}{\left(\frac{3}{2}\right)_r r!(1+r)!} \tag{30}$$

Thus the series representation for the reaction probability is

$$\langle\sigma v\rangle_{ij} = \left[\frac{8(\alpha-1)}{\mu kT}\right]^{\frac{1}{2}}\frac{1}{\Gamma\left(\frac{1}{\alpha-1}-\frac{3}{2}\right)}S(U)\left\{\frac{1}{\sqrt{\pi}}\Gamma\left(\frac{1}{\alpha-1}-1\right)\right.$$
$$-2\Gamma\left(\frac{1}{\alpha-1}-\frac{1}{2}\right)\left[\frac{(\alpha-1)x^2}{4}\right]^{\frac{1}{2}}{}_1F_2\left(\frac{1}{\alpha-1}-\frac{1}{2};\frac{3}{2},\frac{1}{2};-\frac{(\alpha-1)x^2}{4}\right)$$
$$\left.+\frac{2(\alpha-1)x^2}{4\sqrt{\pi}}\sum_{r=0}^{\infty}\left(\frac{(\alpha-1)x^2}{4}\right)^r\left[A_r - \ln\left(\frac{(\alpha-1)x^2}{4}\right)\right]B_r\right\} \tag{31}$$

where A_r and B_r are as defined in Equations (29) and (30). For detailed theory of extended reaction rates and its series representations see Haubold and Kumar [10,11], Kumar and Haubold [12].

The following discussion is adapted from [1]. The solution of the differential Equation (4) with initial condition $N_i(t) = 1$ when $t = 0$ is

$$N_i^{\alpha}(t)\mathrm{d}t = [1+(\alpha-1)c_i t]^{-\frac{1}{\alpha-1}}\mathrm{d}t \tag{32}$$

When c_i in Equation (32) is a constant, the total number of reactions in the time interval $0 \le t \le t_0$ is obtained as

$$\int_0^{t_0} N_i^{\alpha}(t)\mathrm{d}t = \int_0^{t_0}[1+(\alpha-1)c_i t]^{-\frac{1}{\alpha-1}}\mathrm{d}t = \frac{1}{c_i(2-\alpha)}\left\{[1+(\alpha-1)c_i t_0]^{-\frac{2-\alpha}{\alpha-1}} - 1\right\} \tag{33}$$

Now $[1+(\alpha-1)c_i t_0]^{-\frac{2-\alpha}{\alpha-1}} - 1$ is the probability that the lifetime of species i is $\le t_0$ when t follows a distribution with density function

$$f(t) = c_i(2-\alpha)[1+(\alpha-1)c_i t]^{-\frac{1}{\alpha-1}}, 0 \le t < \infty, c_i > 0, 1 < \alpha < 2 \tag{34}$$

or we have

$$N_i(t) = \frac{f(t)}{c_i(2-\alpha)} \tag{35}$$

Equation (34) is the Tsallis statistics for $\alpha > 1$, see [13,14] which can also be seen as a particular case of the pathway model Equation (6) for $\alpha > 1$. If c_i in Equation (32) is a function of time, say $c_i(t)$, then it should be replaced by $\int c_i(t)\mathrm{d}t$. When $c_i = c_i(t) = d_i t$ where $d_i > 0$ is independent of t, then in this case $\int c_i(t)\mathrm{d}t = \frac{d_i t^2}{2}$, then

$$N_i(t) = \frac{\Gamma\left(\frac{1}{\alpha-1}-\frac{1}{2}\right)}{\Gamma\left(\frac{1}{\alpha-1}\right)}\left[\frac{\pi}{2d_i(\alpha-1)}\right]^{\frac{1}{2}} g(t) \tag{36}$$

where

$$g(t) = \frac{\Gamma\left(\frac{1}{\alpha-1}\right)}{\Gamma\left(\frac{1}{\alpha-1}-\frac{1}{2}\right)}\left[\frac{2d_i(\alpha-1)}{\pi}\right]^{\frac{1}{2}}[1+(\alpha-1)\frac{d_i t^2}{2}]^{-\frac{1}{\alpha-1}}, 0 \le t < \infty, d_i > 0, 1 < \alpha < 2 \tag{37}$$

The density in Equation (34) is the lifetime density of the destruction of the species i, with the expected mean value

$$\mathbf{E}(t) = \frac{1}{c_i(3-2\alpha)} \tag{38}$$

where $\mathbf{E}(\cdot)$ is the expected value of $(\cdot)$. The mean value of the lifetime density function given in Equation (37) is

$$\mathbf{E}(t) = \frac{\Gamma\left(\frac{1}{\alpha-1}\right)}{\Gamma\left(\frac{1}{\alpha-1}-\frac{1}{2}\right)} \left[\frac{2(\alpha-1)}{\pi d_i}\right]^{\frac{1}{2}} \frac{1}{2-\alpha} \tag{39}$$

Now as $\alpha \to 1$ we get the expected mean lifetime

$$\mathbf{E}(t) = \left(\frac{2}{\pi d_i}\right)^{\frac{1}{2}} \tag{40}$$

of the lifetime density function

$$g^*(t) = \left(\frac{2d_i}{\pi}\right)^{\frac{1}{2}} \mathrm{e}^{-\frac{d_i t^2}{2}}, 0 \le t < \infty, d_i > 0 \tag{41}$$

considered by Haubold and Mathai [1].

From the lifetime density function given in Equation (34) and the mean lifetime Equation (38) we can infer that

1. the expected lifetime of the species depends on the value of c_i and α. As $c_i(3-2\alpha)$ increases the expected lifetime decreases and vice versa.
2. $\frac{1}{c_i(2-\alpha)} f(t)\Delta(t)$ can be interpreted as the amount of the net destruction in a small time interval $\Delta(t)$. As the net destruction is faster the lifetime becomes smaller.

3. Connection of Extended Kinetic Equation to Fractional Calculus

Let $\Phi_1(t_1)$ be an integrable function as defined in Equation (34) and $\Phi_2(t_2) = \theta(t)$ be any integrable function, then by the Mellin convolution property, we have

$$\Phi(u) = \int_t \frac{1}{t} \Phi_1(t) \Phi_2\left(\frac{u}{t}\right) \mathrm{d}t = \int_t \frac{1}{t} \Phi_1\left(\frac{u}{t}\right) \Phi_2(t) \mathrm{d}t \tag{42}$$

Then $\Phi(u)$, after substituting $\Phi_1(t_1)$ and $\Phi_2(t_2)$, takes the form

$$\Phi(u) = c_i(2-\alpha) \int_{t=0}^{\infty} \frac{1}{t} \left[1 + (\alpha-1) c_i \frac{u}{t}\right]^{-\frac{1}{\alpha-1}} \theta(t) \mathrm{d}t \tag{43}$$

for $\alpha > 1$ which can be considered as a generalized fractional Kober type operator of an integrable function $\theta(t)$. Here as $\alpha \to 1_-$ we have

$$\Phi(u) = c_i(2-\alpha) \int_{t=0}^{\infty} t^{-1} \mathrm{e}^{-c_i \frac{u}{t}} \theta(t) \mathrm{d}t \tag{44}$$

More general cases can be seen in the paper Mathai and Haubold [15].

4. Fractional Kinetic Equation and Its Solution

The following discussion is based on [16,17]. In Equation (4), if instead of an ordinary classical integral we use a fractional integral, we get the reaction equation as

$$N(t) - N_0 = -c^{\nu}{}_0 D_t{}^{-\nu} N(t), \nu > 0 \tag{45}$$

where ${}_0D_t{}^{-\nu}N(t)$ represents the Riemann-Liouville fractional integral defined as

$$
{}_0D_t{}^{-\nu}f(t) = \frac{1}{\Gamma(\nu)}\int_0^t (t-u)^{\nu-1}f(u)\mathrm{d}u, \nu > 0 \tag{46}
$$

with ${}_0D_t^0 f(t) = f(t)$. Taking Laplace transform, simplifying and then taking the inverse Laplace transform one gets

$$
N(t) = N_0 \sum_{k=0}^{\infty} \frac{(-1)^k (ct)^{k\nu}}{\Gamma(1+k\nu)} = N_0 E_\nu(-c^\nu t^\nu) \tag{47}
$$

where $E_\nu(-c^\nu t^\nu)$ is the Mittag-Leffler function, introduced by M. G. Mittag-Leffler in 1903 [18] as

$$
E_\beta(z) = \sum_{k=0}^{\infty} \frac{z^k}{\Gamma(\beta k+1)}, \beta \in \mathbf{C}, \Re(\beta) > 0, z \in \mathbf{C} \tag{48}
$$

If we consider a generalization of the fractional kinetic equation considered by Mathai, Haubold and Saxena [16,17] in the form

$$
N(t) - N_0 t^{\omega-1} E^{\gamma}_{\nu,\omega}(-c^\nu t^\nu) = -c^\nu {}_0D_t{}^{-\nu}N(t), \nu > 0, \omega > 0, \gamma > 0 \tag{49}
$$

where $E^{\gamma}_{\nu,\omega}(.)$ is the generalized three parameter Mittag-Leffler function introduced by Prabhakar [19] defined as

$$
E^{\gamma}_{\beta,\rho}(z) = \sum_{k=0}^{\infty} \frac{(\gamma)_k z^k}{\Gamma(\beta k+\rho)k!}, \beta, \rho, \gamma \in \mathbf{C}, \Re(\beta) > 0, \Re(\rho) > 0, z \in \mathbf{C} \tag{50}
$$

then the solution of the fractional kinetic Equation (49) is

$$
N(t) = N_0 t^{\omega-1} E^{\gamma+1}_{\nu,\omega}(-c^\nu t^\nu) \tag{51}
$$

5. Conclusions

The linear and non-linear kinetic equations establish a connection between the Boltzmann-Gibbs statistical mechanics and Tsallis non-extensive statistical mechanics. The pathway parameter α plays a key role in switching between these two cases. The theory of extended reaction rates and its closed form solutions can be seen in Haubold and Kumar [10] and Kumar and Haubold [12]. Further, the fractional diffusion equation and its solution help us to understand the connection with the classical Laplace transform. In 2013, the author has solved the fractional kinetic equations discussed here by P_α-transform [20]. Various fractional differential equations and their solution by various transforms are studied by many authors, see [21,22]. It should be noted that the Mittag-Leffler function arises in the solution of a fractional diffusion equation whereas the exponential function arises in the solution of its classical counterpart. A possible connection of the extended kinetic equation to fractional calculus can be established through the procedure adopted here.

Acknowledgments: The author acknowledges gratefully the encouragements given by Professor A. M. Mathai, Centre for Mathematical and Statistical Sciences, KFRI, Peechi, Kerala, and Hans J. Haubold, Chief Scientist, Office of Outer Space Affairs, Vienna, Austria in this research work. The author would also like to thank the unknown referees for their valuable suggestions which enhanced the paper in the present form.

Conflicts of Interest: The author declares no conflict of interest.

References

1. Haubold, H.J.; Mathai, A.M. A heurisitc remark on the periodic variation in the number of solar neutrinos detected on Earth. *Astrophys. Space Sci.* **1995**, *228*, 113–134. [CrossRef]
2. Haubold, H.J.; Mathai, A.M. The fractional kinetic equation and thermonuclear functions. *Astrophys. Space Sci.* **2000**, *273*, 53–63. [CrossRef]

3. Mathai, A.M. A Pathway to matrix-variate gamma and normal densities. *Linear Algebra Appl.* **2005**, *396*, 317–328. [CrossRef]
4. Mathai, A.M.; Haubold, H.J. Pathway model, superstatistics, Tsallis statistics and a generalized measure of entropy. *Phys. A* **2007**, *375*, 110–122. [CrossRef]
5. Mathai, A.M. *A Handbook of Generalized Special Functions for Statistical and Physical Sciences*; Clarendo Press: Oxford, UK, 1993.
6. Rainville, E.D. *Special Functions*; Chelsea Publishing Company: New York, NY, USA, 1960.
7. Mathai, A.M.; Saxena, R.K.; Haubold, H.J. *The H-function: Theory and Applications*; Springer: New York, NY, USA, 2010.
8. Meijer, C.S. Uber Whittakersche bzw. Besselsche Funktionen und deren Produkte. *Nieuw Arch. Wiskd.* **1936**, *18*, 10–39.
9. Haubold, H.J.; Mathai, A.M. The Maxwell-Boltzmannian Approach to the Nuclear Reaction Rate Theory. *Fortschr. Phys.* **1985**, *33*, 623–644. [CrossRef]
10. Haubold, H.J.; Kumar, D. Extension of thermonuclear functions through the pathway model including Maxwell-Boltzmann and Tsallis distributions. *Astropart. Phys.* **2008**, *29*, 70–76. [CrossRef]
11. Haubold, H.J.; Kumar, D. Fusion yield: Guderley modeland Tsallis statistics. *J. Plasma Phys.* **2011**, *77*, 1–14. [CrossRef]
12. Kumar, D.; Haubold, H.J. On extended thermonuclear functions through pathway model. *Adv. Space Res.* **2009**, *45*, 698–708. [CrossRef]
13. Tsallis, C. Possible generalization of Boltzmann-Gibbs statistics. *J. Stat. Phys.* **1988**, *52*, 479–487. [CrossRef]
14. Tsallis, C. *Introduction to Nonextensive Statistical Mechanics: Approaching a Complex World*; Springer: New York, NY, USA, 2009.
15. Mathai, A.M.; Haubold, H.J. Erdélyi-Kober Fractional Integral Operators from a Statistical Perspective-I. 2013; arXiv:1303.3978.
16. Saxena, R.K.; Mathai, A.M.; Haubold, H.J. On fractional kinetic equations. *Astrophys. Space Sci.* **2002**, *282*, 281–287. [CrossRef]
17. Saxena, R.K.; Mathai, A.M.; Haubold, H.J. Unified fractional kinetic equation and a fractional diffusion equation. *Astrophys. Space Sci.* **2004**, *209*, 299–310. [CrossRef]
18. Mittag-Leffler, G.M. Sur la nouvelle fonction $E_\alpha(x)$. *CR Acad. Sci. Paris* **1903**, *137*, 554–558.
19. Prabhakar, T.R. A singular integral equation with a generalized Mittag-Leffler function in kernel. *Yokohama Math. J.* **1971**, *19*, 7–15.
20. Kumar, D. Solution of fractional kinetic equation by a class of integral transform of pathway type. *J. Math. Phys.* **2013**, *54*. [CrossRef]
21. Diethelm, K. *The Analysis of Fractional Differential Equations*; Springer: Berlin, Germany, 2010.
22. Gorenflo, R.; Mainardi, F. Fractional calculus, integral and differential equations of fractional order. In *Fractals and Fractional Calculus in Continuum Mechanics*; Carpinteri, A., Mainardi, F., Eds.; Springer: Wien, Austria, 1997; pp. 223–276.

axioms

MDPI

Article

An Overview of Generalized Gamma Mittag–Leffler Model and Its Applications

Seema S. Nair

St. Gregorios College, Kottarakara, Kerala 691531, India; seema.cms@gmail.com; Tel.: +91-4692611138

Academic Editor: Hans J. Haubold
Received: 6 May 2015; Accepted: 27 July 2015; Published: 26 August 2015

Abstract: Recently, probability models with thicker or thinner tails have gained more importance among statisticians and physicists because of their vast applications in random walks, Lévi flights, financial modeling, *etc.* In this connection, we introduce here a new family of generalized probability distributions associated with the Mittag–Leffler function. This family gives an extension to the generalized gamma family, opens up a vast area of potential applications and establishes connections to the topics of fractional calculus, nonextensive statistical mechanics, Tsallis statistics, superstatistics, the Mittag–Leffler stochastic process, the Lévi process and time series. Apart from examining the properties, the matrix-variate analogue and the connection to fractional calculus are also explained. By using the pathway model of Mathai, the model is further generalized. Connections to Mittag–Leffler distributions and corresponding autoregressive processes are also discussed.

Keywords: generalized Mittag–Leffler density and process; gamma density; Lévi density; pathway model; Tsallis statistics; superstatistics

---

1. Introduction

In model building situations in physical, biological, social and engineering sciences, the usual procedure is to select a probability model from a parametric family of distributions. In many practical problems, it is often found that the selected model is not a good fit for the experimental data, because it requires a model with a thicker or thinner tail than the ones available from the parametric family of distributions. In order to make the tail thicker or thinner, a technique is introduced here by augmenting a series to the original density. Our first step is to construct the thicker or thinner tailed distribution associated with the Mittag–Leffler function, because this function is connected to fractional calculus, the Mittag–Leffler stochastic process, non-Gaussian time series, Lévi flights and in a limiting process to the topics of Tsallis statistics, superstatistics, as well as to statistical distribution theory.

In reaction rate theory, input-output type situations and reaction-diffusion problems in physics and chemistry, when the integer derivatives are replaced by fractional derivatives, the solutions automatically go in terms of Mittag–Leffler functions and their generalizations; see Haubold and Mathai (2000) [1]. The ordinary and generalized Mittag–Leffler functions interpolate between a purely exponential law and power-law-like behavior of phenomena governed by ordinary kinetic equations and their fractional counterparts; see Kilbas *et al.* (2004) [2], Kiryakova (2000) [3], Mathai (2010) [4] and Mathai *et al.* (2010) [5]. This paper examines a new family of statistical distributions associated with Mittag–Leffler functions, which gives an extension to the gamma family, which will then connect to fractional calculus and statistical distribution theory through the theory of special functions. The model investigated in this paper is useful in the study of life testing problems, reliability analysis, in physical situations to describe stable solutions, as well as unstable and chaotic neighborhoods, *etc.* We will start with the definition of the Mittag–Leffler function.

A two-parameter Mittag–Leffler function is defined as follows:

$$E_{\alpha,\beta}(x^{\alpha}) = \sum_{k=0}^{\infty} \frac{x^{\alpha k}}{\Gamma(\alpha k+\beta)}, \Re(\alpha) > 0, \Re(\beta) > 0, \tag{1}$$

where $\Re(\cdot)$ denotes the real part of $(\cdot)$. Observe that the Mittag–Leffler function is an extension of the exponential function. When $\alpha = 1, \beta = 1$, Equation (1) reduces to e^x. In statistical model building, usually, the parameters are real, but since the results to be discussed hold for complex parameters, as well, we will state the corresponding relevant conditions. Various properties, generalizations and applications of the Mittag–Leffler function can be seen from Kilbas *et al.* (2004) [2].

Consider a probability density of the form:

$$f(x) = \begin{cases} a^{\beta}[1+\frac{\delta}{a^{\alpha}}]x^{\beta-1}e^{-ax}E_{\alpha,\beta}(-\delta x^{\alpha}), \beta > 0, \alpha > 0, a > 0, x \geq 0 \\ 0, \text{elsewhere}, \end{cases} \tag{2}$$

where C is the normalizing constant and $E_{\alpha,\beta}(-\delta x^{\alpha})$ is the Mittag–Leffler function. Some interesting special cases of Equation (2) are the following: The density in Equation (2) includes two-parameter gamma, exponential, chi square, noncentral chi square and the like. When $\delta = 0$, Equation (2) reduces to the two-parameter gamma density:

$$f_1(x) = C_1 x^{\beta-1}e^{-ax}, x > 0. \tag{3}$$

For $\alpha = \beta = 1$ in Equation (2), we have the exponential density. For $\beta = \frac{n}{2}$ and $a = \frac{1}{2}$ in $f_1(x)$, we have the chi square density with n degrees of freedom. For $\beta = p, p = 2, 3, \cdots$ in Equation (3), we have the Erlang density. For $p = 1$ in Erlang density, we have the exponential density. For fixed values of a, β and for various values of δ, we can look at the graphs that give a suitable interpretation to the model in Equation (2).

The above figures show a comparison between gamma density and gamma Mittag–Leffler density for different values of δ. Observe that $\delta = 0$ corresponds to the gamma density. In Figure 1, as the value of δ decreases, the right tail of the new density becomes thicker and thicker compared to that of a gamma density. Similarly the peakedness of the curve slowly decreases. In Figure 2, also, $\delta = 0$ corresponds to the gamma density. When the values of δ increases from $\delta = 0$, the right tail of the new density becomes thinner and thinner compared to that of a gamma density. Similarly, the peakedness of the curve slowly increases. Hence, when we look for a model with a thicker or thinner tail while a gamma density is found to be more or less a proper fit, then a member from the new family of densities introduced here will become quite useful and handy. Observe that the new density is mathematically and computationally easily tractable, just like a gamma density.

The moment-generating function of Equation (2) is given by:

$$\begin{aligned} M_x(t) &= a^{\beta}[1+\tfrac{\delta}{a^{\alpha}}]\int_0^{\infty} x^{\beta-1}e^{-(a-t)x}\sum_0^{\infty}\tfrac{(-\delta x^{\alpha})^k}{\Gamma(\alpha_k+\beta)}dx \\ &= a^{\beta}[1+\tfrac{\delta}{a^{\alpha}}]\sum_0^{\infty}\tfrac{(-\delta)^k}{\Gamma(\alpha_k+\beta)}\int_0^{\infty} x^{\alpha k+\beta-1}e^{-(a-t)x}dx. \end{aligned}$$

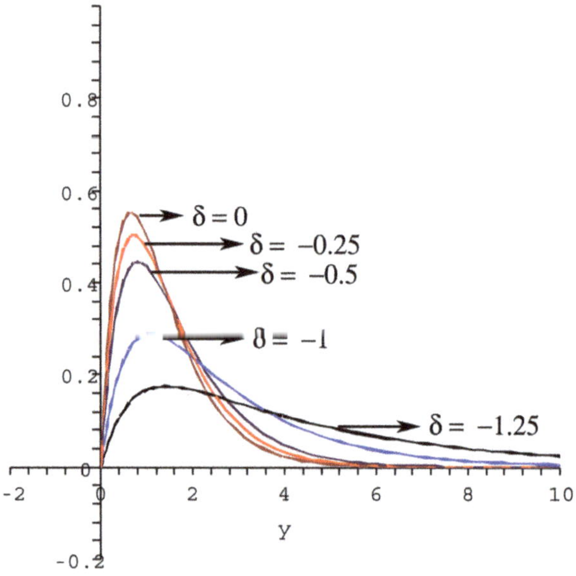

Figure 1. For $a = 1.5, \beta = 2$ and $\delta < 0$.

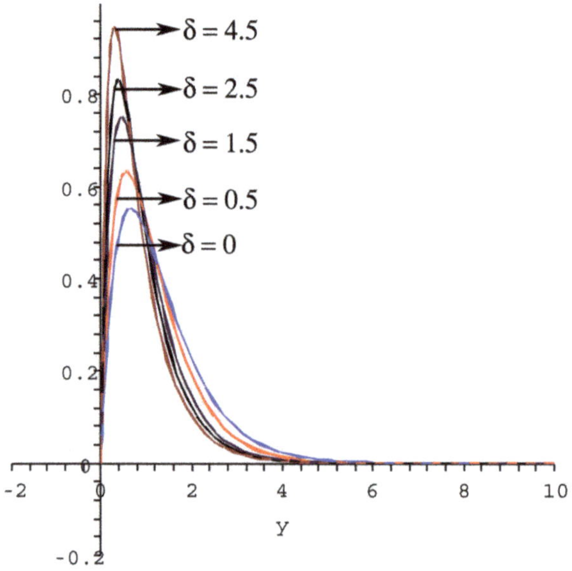

Figure 2. For $a = 1.5, \beta = 2$ and $\delta > 0$.

On simplification, we obtain mgfas:

$$M_x(t) = \frac{a^{\beta}}{(a-t)^{\beta}} \frac{(1+\frac{\delta}{a^{\alpha}})}{(1+\frac{\delta}{(a-t)^{\alpha}})}, \left|\frac{\delta}{(a-t)^{\alpha}}\right| < 1, (a-t) > 0. \quad (4)$$

The characteristic function can be obtained if we replace t by $it, i = \sqrt{-1}$. If we put $-t$ instead of t, then we will obtain the Laplace transform of the density. Using the Laplace transform or moment-generating function, we can easily obtain the integer moments by using the following formula:

$$\mu_r' = E(x^r) = (-1)^r \frac{\mathrm{d}^r}{\mathrm{d}t^r} L_f(t)|_{t=0}.$$

Thus, the mean value will be obtained as:

$$\mu_1' = \frac{\beta}{a} - \frac{\delta\alpha}{a(a^\alpha + \delta)}, \tag{5}$$

and:

$$\mu_2' = \frac{1}{a^2}\left\{2\alpha^2\left(\frac{\delta}{a^\alpha + \delta}\right)^2 - (\alpha^2 + \alpha(2\beta + 1))\left(\frac{\delta}{a^\alpha + \delta}\right) + \beta(\beta + 1)\right\}. \tag{6}$$

Variance:

$$\mu_2 = \frac{\beta}{a^2} - \frac{\delta\alpha[(\alpha + 1)a^\alpha + \delta]}{a^2(a^\alpha + \delta)^2}.$$

Arbitrary moments can be obtained in terms of the generalized Wright hypergeometric function. That is,

$$\mu_\gamma' = E(x^\gamma) = \int_0^\infty x^\gamma f(x)\mathrm{d}x$$

which is nothing but the Mellin transform of the function f with $\gamma = s - 1$.

$$\mu_\gamma' = \frac{(1 + \frac{\delta}{a^\alpha})}{a^\gamma} {}_2\Psi_1\left[{}^{(1,1),(\beta+\gamma,\alpha)}_{(\beta,\alpha)}\Big| -\frac{\delta}{a^\alpha}\right] \tag{7}$$

where ${}_p\Psi_q(z)$ is the generalized Wright's hypergeometric function defined for $z \in C$, complex $a_i, b_j \in C$ and $\alpha_i, \beta_j \in \Re_+ = (0, \infty), a_i, b_j \neq 0; i = 1, 2, \ldots, p; j = 1, 2, \ldots, q$ by the series:

$${}_p\Psi_q(z) \equiv {}_p\Psi_q\left[{}^{(a_i,\alpha_i)1,p}_{(b_j,\beta_j)1,q}\Big| z\right] = \sum_{k=0}^{\infty} \frac{\left\{\prod_{i=1}^{p} \Gamma(a_i + \alpha_i k)\right\} z^k}{\left\{\prod_{j=1}^{q} \Gamma(b_j + \beta_j k)\right\} k!}. \tag{8}$$

The function in Equation (8) was introduced by Wright and is called the generalized Wright's hypergeometric function. For convergence conditions, the existence of various contours and other properties, see Wright (1940) [6] or the theory of *H*-function to be discussed later. If we take γ as integers in Equation (7), then we will obtain integer moments. It may be observed that the distribution function is available in terms of a series of incomplete gamma functions.

2. Estimation of Parameters

In this section, we have given explicit forms of the estimators of the parameters using the method of moments. The motivation for the method of moments comes from the fact that the sample moments are consistent estimators for the corresponding population moments. To start with, let us consider the case for $a = 1, \beta = 1$ in Equation (2), then the model will become the Mittag–Leffler extension of the standard exponential distribution and has the density of the form:

$$g_1(x) = (1 + \delta)\mathrm{e}^{-x} E_\alpha(-\delta x^\alpha). \tag{9}$$

The moments can be obtained from Equations (5) and (6) with the parameter value $a = 1, \beta = 1$.

The moment estimators of δ and α are given by:

$$\hat{\alpha} = \frac{\sum(x_i - \overline{x})^2 - n\overline{x}^2}{n(\overline{x}-1)} + 1, \overline{x} \neq 1$$

and:

$$\hat{\delta} = \frac{n(\overline{x}-1)^2}{n\overline{x} - \sum(x_i-\overline{x})^2}, n\overline{x} - \sum(x_i - \overline{x})^2 \neq 0.$$

Now, consider the Mittag–Leffler extension of standard gamma density. For that, we take $a = 1$ in Equation (2); thus, the model has the following form:

$$g_2(x) = (1+\delta)x^{\beta-1}e^{-x}E_{\alpha,\beta}(-\delta x^{\alpha}). \tag{10}$$

Using the same procedure as above, one can obtain the estimators as:

$$\hat{\beta} = \frac{\sum(x_i-\overline{x})^3 - [\sum(x_i-\overline{x})^2][\sum x_i^2 + n\overline{x}] + 2n\overline{x}[n\overline{x}^3 - \overline{x}^2]}{\sum x_i^2[4n\overline{x} - 3n - 3] + n\overline{x}^2[8n\overline{x} + 13n] + n^2\overline{x}}$$

$$\hat{\alpha} = \frac{\sum(x_i-\overline{x})^2 - n(\overline{x}-\hat{\beta})^2 - n\overline{x}}{n(\overline{x}-\hat{\beta})}, n(\overline{x}-\hat{\beta}) \neq 0$$

and:

$$\hat{\delta} = \frac{(\hat{\beta}-\overline{x})}{\hat{\alpha}+\overline{x}-\hat{\beta}}.$$

If we retain all parameters α, β, δ and a, then the analytical solution is quite difficult, but the numerical solution can be obtained by using software like MATLAB and MAPLE.

3. The Q-Analogue of Generalized Gamma Mittag–Leffler Density

In this section, a generalized density of Equation (2) is considered. Mathai (2005) [7] introduced the pathway model, where the scalar version of the pathway density is given as follows:

$$\begin{aligned} f_1(x) &= c_1\,|x|^{\beta-1}[1+a(q-1)|x|^{\rho}]^{-\frac{1}{q-1}} \\ &\quad \text{for} a>0, \beta>0, q>1, -\infty<x<\infty \end{aligned} \tag{11}$$

where$c_1 = \frac{\rho[a(q-1)]^{\frac{\beta}{\rho}}\Gamma(\frac{1}{q-1})}{2\Gamma(\frac{\beta}{\rho})\Gamma(\frac{1}{q-1}-\frac{\beta}{\rho})}, \frac{1}{q-1} > \frac{\beta}{\rho}$.

For $q < 1$, writing $q - 1 = -(1-q)$, we have the form:

$$\begin{aligned} f_2(x) &= c_2\,|x|^{\beta-1}[1-a(1-q)|x|^{\rho}]^{\frac{1}{1-q}} \\ &\quad \text{for} a>0, \beta>0, q<1, 1-a(1-q)|x|^{\rho}>0 \end{aligned} \tag{12}$$

where$c_2 = \frac{\rho[a(1-q)]^{\frac{\beta}{\rho}}\Gamma(\frac{1}{1-q}+\frac{\beta}{\rho}+1)}{2\Gamma(\frac{\beta}{\rho})\Gamma(\frac{1}{1-q}+1)}$.

As $q \to 1$, $f_1(x)$ and $f_2(x)$ tend to $f_3(x)$, which we refer to as the extended symmetric generalized gamma distribution, where $f_3(x)$ is given by:

$$f_3(x) = \frac{\rho a^{\frac{\beta}{\rho}}}{2\Gamma(\frac{\beta}{\rho})}|x|^{\beta-1}e^{-a|x|^{\rho}}, -\infty<x<\infty, a, \beta, \rho > 0. \tag{13}$$

Now, generalize Equation (2) by using Equation (11). Since the normalizing constants obtained in this section are in terms of H-functions, a definition of the H-function is given here.

$$H_{p,q}^{m,n}\left[z\Big|_{(b_1,\beta_1),\ldots,(b_q,\beta_q)}^{(a_1,\alpha_1),\ldots,(a_p,\alpha_p)}\right] = \frac{1}{2\pi i}\int_L \phi(s)z^{-s}ds, \tag{14}$$

where:

$$\phi(s) = \frac{\left\{\prod_{j=1}^{m}\Gamma(b_j+\beta_j s)\right\}\left\{\prod_{j=1}^{n}\Gamma(1-a_j-\alpha_j s)\right\}}{\left\{\prod_{j=m+1}^{q}\Gamma(1-b_j-\beta_j s)\right\}\left\{\prod_{j=n+1}^{p}\Gamma(a_j+\alpha_j s)\right\}}$$

where $\alpha_j, j = 1,2,...,p$, $\beta_j, j = 1,2,...,q$ are real positive numbers, $a_j, j = 1,2,...,p$ and $b_j, j = 1,2,...,q$ are complex numbers and L is a contour separating the poles of $\Gamma(b_j+\beta_j s), j = 1,2,...,m$ from those of $\Gamma(1-a_j-\alpha_j s), j = 1,2,...,n$. The details of convergence conditions, various properties and applications of H-function are available in Mathai *et al.* (2010) [5].

Consider:

$$f_4(x) = C_4 x^{\beta-1}[1+a(q-1)x^{\rho}]^{-\frac{1}{q-1}}E_{\frac{\alpha}{\rho},\frac{\beta}{\rho}}(-\delta x^{\alpha}), \tag{15}$$
$$x \geq 0, q > 1, a > 0, \rho > 0, \alpha > 0, \beta > 0$$

where:

$$C_4^{-1} = \frac{1}{\rho\Gamma(\frac{1}{q-1})[a(q-1)^{\frac{\beta}{\rho}}]}H_{1,2}^{2,1}\left[\frac{\delta}{[a(q-1)]^{\frac{\alpha}{\rho}}}\Big|_{(0,1),(\frac{1}{q-1}-\frac{\beta}{\rho},\frac{\alpha}{\rho})}^{(0,1)}\right], q > 1. \tag{16}$$

The general density in Equation (15), for $q > 1$, includes the Mittag–Leffler extension of Type 2 beta density and F density. If $x > 0$ is replaced by $|x|, -\infty < x < \infty$, with the appropriate change in the normalizing constant, then we have the Student t density and Cauchy density as special cases, and many more. As $q \to 1$, $f_4(x)$ reduces to $f_5(x)$, where:

$$f_5(x) = C_5 x^{\beta-1}e^{-ax^{\rho}}E_{\frac{\alpha}{\rho},\frac{\beta}{\rho}}(-\delta x^{\alpha}), C_5 = \rho a^{\frac{\beta}{\rho}}(1+\frac{\delta}{a^{\frac{\alpha}{\rho}}}). \tag{17}$$

This includes the Mittag–Leffler extension of generalized gamma, Weibull, chi-square, Maxwell–Boltzmann and related densities. If $\delta = 0$, Equation (17) reduces to the generalized gamma density. When $\rho = 1$, Equation (17) reduces to Equation (2). In particular, if $\rho = \beta$, the Mittag–Leffler extension of Weibull density is obtained,

$$f_6(x) = C_6 x^{\beta-1}e^{-ax^{\beta}}E_{\frac{\alpha}{\rho},1}(-\delta x^{\alpha}). \tag{18}$$

For $x > 0$ and $q < 1$, $f_4(x)$ takes the following form:

$$f_7(x) = C_7 x^{\beta-1}[1-a(1-q)x^{\rho}]^{\frac{1}{1-q}}E_{\frac{\alpha}{\rho},\frac{\beta}{\rho}}(-\delta x^{\alpha}), \tag{19}$$
$$0 \leq x \leq \frac{1}{[a(1-q)]^{\frac{1}{\rho}}}, a, \alpha, \beta > 0, q < 1$$

where:

$$C_7^{-1} = \frac{\Gamma(1+\frac{1}{1-q})}{\rho[a(1-q)^{\frac{\beta}{\rho}}]}H_{1,2}^{1,1}\left[\frac{\delta}{[a(1-q)]^{\frac{\alpha}{\rho}}}\Big|_{(0,1),(-\frac{1}{1-q}-\frac{\beta}{\rho},\frac{\alpha}{\rho})}^{(0,1)}\right], q < 1. \tag{20}$$

This general density includes the Mittag–Leffler extension of Type 1 beta density, triangular density, uniform density and many more. As $q \to 1$, Equation (15) reduces to Equation (17). For $\rho = \beta$, then Equations (15) and (19) will give the two forms of extended q-Weibull density as given below.

$$f_8(x) = C_8 x^{\beta-1}[1+a(q-1)x^{\beta}]^{-\frac{1}{q-1}}E_{\frac{\alpha}{\beta},1}(-\delta x^{\alpha}), \tag{21}$$

$$x \geq 0, q > 1, a > 0, \alpha > 0, \beta > 0$$

$$f_9(x) = C_9 x^{\beta-1}[1 - a(1-q)x^{\beta}]^{\frac{1}{1-q}} E_{\frac{\alpha}{\beta},1}(-\delta x^{\alpha}), \tag{22}$$

$$0 \leq x \leq \frac{1}{[a(1-q)]^{\frac{1}{\beta}}}, a, \alpha, \beta > 0, q < 1$$

where the normalizing constants C_8 and C_9 are obtained from Equations (16) and (20), respectively. In particular, if $\delta = 0$, then $f_8(x)$ and $f_9(x)$ will coincide with q-Weibull density. In particular, for $\beta = 1$ and $\rho = 1$, the integral of Equation (19) can be treated as the pathway fractional integral transform of the Mittag–Leffler function with appropriate transformation of the variable. More details about pathway fractional integral transform can be found in Seema Nair ((2009) [8], (2011) [9]).

The model in Equation (15) can be connected to the superstatistics of Beck and Cohen (2003) [10] and Beck (2006) [11] for $q = 2, \delta = 0$. There is a vast literature on Beck–Cohen superstatistics in statistical mechanics with many physical interpretations; see, for example, Beck ((2010) [12], (2009) [13]). From a statistical point of view, it can be obtained as an unconditional density from Bayes' setup, by considering generalized gamma as a conditional density with the assumption that the scale parameter a has a prior gamma density; see Nair and Kattuveettil (2010) [14]. For $\beta = 1, \rho = 1, a = 1$ and $\delta = 1$, we have Tsallis statistics for $q > 1$ and $q < 1$, respectively, from the models in Equations (15) and (19). Thus, we can make a connection to nonextensive statistical mechanics; see, for example, Tsallis ((1988) [15], (2004) [16], (2009) [17]) and Mathai *et al.* (2010) [5]. (It is stated that over 3000 papers have been published on this topic during the past 20 years, and over 5000 people from around the globe are working in this area).

4. Connection to Mittag–Leffler Distributions and Autoregressive Processes

Especially when $a = 0$, $\beta = \alpha$, such that $0 < \alpha \leq 1$ in Equation (2), we have the Mittag–Leffler density, which has the probability density function:

$$g_1(x) = \delta x^{\alpha-1} E_{\alpha,\alpha}(-\delta x^{\alpha}), 0 < \alpha \leq 1, \delta > 0 \tag{23}$$

and has Laplace transform $\delta(s^{\alpha} + \delta)^{-1}$, a special case of the general class of Laplace transforms discussed by Mathai *et al.* (2006) [18]. Similarly, for $a = 0$ in (2) and replacing β by $\alpha\gamma$ and δ by $\frac{1}{\rho}$, one can arrive at the generalized Mittag–Leffler density, which has the probability density function:

$$g(x) = \frac{x^{\alpha\gamma-1}}{\rho^{\gamma}} E^{\gamma}_{\alpha,\alpha\gamma}(-\frac{x^{\alpha}}{\rho}), 0 < \alpha < 1, \gamma, \rho > 0, x \geq 0 \tag{24}$$

where $E^{\delta}_{\alpha,\beta}(\cdot)$ is the generalized Mittag–Leffler function defined as:

$$E^{\delta}_{\alpha,\beta}(z) = \sum_{k=0}^{\infty} \frac{(\delta)_k z^k}{\Gamma(\alpha k + \beta) k!}, \alpha > 0, \beta > 0, \gamma > 0,$$

where $(\delta)_k = \delta(\delta+1)(\delta+2)\cdots(\delta+k-1)$ and Equation (24) has the Laplace transform of the form:

$$L_g(t) = [1 + \rho t^{\alpha}]^{-\gamma}, |\rho t^{\alpha}| < 1. \tag{25}$$

The Laplace transform in Equation (25) is associated with infinitely-divisible and geometrically infinitely-divisible distributions, α-Laplace and Linnik distributions. The class of Laplace transforms relevant in geometrically infinitely-divisible and α-Laplace distributions is of the form $L(p) = \frac{1}{1+\eta(p)}$, where $\eta(p)$ satisfies the condition $\eta(bp) = b^{\alpha}\eta(p)$ and $\eta(p)$ is a periodic function for fixed α. In Mathai (2010) [4], it is noted that there is a structural representation in terms of the positive Lévi random variable by using Equation (25).

A positive Lévi random variable $u > 0$, with parameter α is such that the Laplace transform of the density of $u > 0$ is given by $e^{-t^{\alpha}}$. That is:

$$E[e^{-tu}] = e^{-t^{\alpha}}. \tag{26}$$

Theorem 4.1. *Let $y > 0$ be a Lévi random variable with Laplace transform as in (26) and $x \geq 0$ an exponential random variable with parameter δ, and let x and y be independently distributed. Then, $u = yx^{\frac{1}{\alpha}}$ is distributed as generalized Mittag–Leffler random variable with Laplace transform (see [4]):*

$$L_u(t) = [1 + \rho t^{\alpha}]^{-\gamma}$$

Proof: For proving this result, we will make use of the following lemma of the conditional argument.

Lemma 1.1. For two random variables u and v having a joint distribution,

$$E(u) = E[E(u|v)] \tag{27}$$

whenever all of the expected values exist, where the inside expectation is taken in the conditional space of u given v and the outside expectation is taken in the marginal space of v.

Now, by applying Equation (27), we have the following: Let the density of u be denoted by $g(u)$. Then, the conditional Laplace transform of g, given x, is given by:

$$E[e^{-(tx^{\frac{1}{\alpha}})y}|x] = e^{-t^{\alpha}x}. \tag{28}$$

However, the right side of Equation (28) is in the form of a Laplace transform of the density of x with parameter t^{α}. Hence, the expected value of the right side, with respect to x, is available by replacing the Laplace parameter t in $[1 + \rho t]^{-\gamma}$ (the Laplace transform of a gamma density with the scale parameter ρ and shape parameter γ) by t^{α},

$$L_g(t) = [1 + \rho t^{\alpha}]^{-\gamma}, \rho > 0, \gamma > 0 \tag{29}$$

which establishes the result. From Equation (27), one can observe that if we consider an arbitrary random variable y with the Laplace transform of the form:

$$L_g(t) = e^{-[\phi(t)]} \tag{30}$$

whenever the expected value exists, where $\phi(t)$ is such that:

$$\phi(tx^{\frac{1}{\alpha}}) = x\phi(t), \lim_{t \to 0} \phi(t) = 0, \tag{31}$$

then from Equation (28), we have:

$$E[e^{-(tx^{\frac{1}{\alpha}})y}|x] = e^{-x[\phi(t)]}. \tag{32}$$

Now, let x be an arbitrary positive random variable having Laplace transform, denoted by $L_x(t)$ where $L_x(t) = \psi(t)$. Then, from Equations (27) and (30), we have:

$$L_g(t) = \psi(\phi(t)).$$

In particular, when $\delta = 1$ in Equation (23), we have the standard Mittag–Leffler density introduced by Pillai (1990) [19]. For the properties of Mittag–Leffler distributions connected to the autoregressive process, see the papers of Pillai and Jayakumar (1995) [20] and Jayakumar (2003) [21].

4.1. Moments from Mellin–Barnes Integral Representation

It is shown in Mathai (2010) [4] that for handling problems connected to Mittag–Leffler densities, it is convenient to use the Mellin–Barnes representation of a Mittag–Leffler function and then proceed from there. The following gives the Mellin–Barnes integral representation of $g(x)$.

$$\begin{aligned} g(x) &= \frac{1}{\Gamma(\gamma)}\frac{1}{2\pi i}\int_{c-i\infty}^{c+i\infty}\frac{\Gamma(s)\Gamma(\gamma-s)}{\Gamma(\alpha\gamma-\alpha s)}x^{\alpha\gamma-1}x^{-\alpha s}ds, \Re(\gamma)>0 \\ &= \frac{1}{\alpha\Gamma(\gamma)}\frac{1}{2\pi i}\int_{c-i\infty}^{c+i\infty}\frac{\Gamma(\gamma-\frac{1}{\alpha}+\frac{s}{\alpha})\Gamma(\frac{1}{\alpha}-\frac{s}{\alpha})}{\Gamma(1-s)}x^{-s}ds, \end{aligned} \tag{33}$$

by taking $\alpha\gamma-1-\alpha s=-s_1$. One can also look upon the Mellin transform of the density of a positive random variable as the $(s-1)$-th moment, and therefore, from Equation (33), one can write the Mellin transform as an expected value. That is,

$$\begin{aligned} M_g(s) &= E(x^{s-1})ing \\ &= \frac{1}{\Gamma(\gamma)}\frac{\Gamma(\gamma-\frac{1}{\alpha}+\frac{s}{\alpha})\Gamma(\frac{1}{\alpha}-\frac{s}{\alpha})}{\alpha\Gamma(1-s)}, \\ & \text{for} 1-\alpha<\Re(s)<1,0,\alpha\le 1,\gamma>0 \\ &= \frac{1}{\Gamma(\gamma)}\frac{\Gamma(\gamma-\frac{1}{\alpha}+\frac{s}{\alpha})\Gamma(1+\frac{1}{\alpha}-\frac{s}{\alpha})}{\Gamma(2-s)}. \end{aligned} \tag{34}$$

If we replace $s-1$ by ρ, then one can obtain the ρ-th moment of the generalized Mittag–Leffler density as follows:

$$E(x^\rho)=\frac{\Gamma(\gamma+\frac{\rho}{\alpha})\Gamma(1-\frac{\rho}{\alpha})}{\Gamma(\gamma)\Gamma(1-\rho)}, -\alpha<\Re(\rho)<\alpha<1.$$

In particular, for $\gamma=1$, we can arrive at the ρ-th moment of the Mittag–Leffler density of Pillai (1990) [19], Pillai and Jayakumar (1995) [20]. For $\alpha\to 1$, Equation (34) reduces to:

$$M_g(t)=\frac{1}{\Gamma(\gamma)}\Gamma(\gamma-1+s).$$

Its inverse Mellin transform is then:

$$g_1(x)=\frac{1}{\Gamma(\gamma)}\frac{1}{2\pi i}\int_L\Gamma(\gamma-1+s)x^{-s}ds=\frac{1}{\Gamma(\gamma)}x^{\gamma-1}e^{-x}, x\ge 0,\gamma>0 \tag{35}$$

which is the one-parameter gamma density, and for $\gamma=1$, it reduces to the exponential density. Hence, the generalized Mittag–Leffler density g can be taken as an extension of a gamma density, such as the one in Equation (35), and the Mittag–Leffler density g_1 as an extension of exponential density for $\gamma=1$. It is shown that generalized Mittag–Leffler density $(GMLD(\alpha,\gamma,\rho))$ is infinitely and geometrically infinitely divisible for $0<\alpha\le 1, 0<\gamma\le 1$, and also, it belongs to the class L; see [21]. Jose *et al.* (2010) [22] discussed the first order autoregressive process with $GMLD(\alpha,\beta)$ marginals. In a similar manner, we shall construct a first order autoregressive process with $GMLD(\alpha,\gamma,\rho)$ marginals, which will be stated as a remark.

Remark: The first order autoregressive process $X_n=aX_{n-1}+\eta_n, a\in(0,1)$ is strictly stationary Markovian with $GMLD(\alpha,\gamma,\rho)$ if and only if the $\{\eta_n\}$ are distributed independently and identically as the γ-fold convolution of the random variable $\{V_n\}$ where:

$$V_n=\begin{cases} 0 \text{withprobability} a^\alpha \\ M_n \text{withprobability} 1-a^\alpha \end{cases} \tag{36}$$

where $\{M_n\}$ are independently and identically distributed Mittag–Leffler random variables, provided $X_0\stackrel{d}{=}GMLD(\alpha,\gamma,\rho)$ and independent of η_n.

5. Applications in Reaction-Diffusion Problems

We can look at the model Equation (2) in another way, as well; consider the total integral as:

$$1 = C \int_0^\infty x^{\beta-1} e^{-ax} E_{\alpha,\beta}(-\delta x^\alpha) dx, \tag{37}$$

which can be treated as the Laplace transform of the function $x^{\beta-1}E_{\alpha,\beta}(-\delta x^\alpha)$, and hence, $C^{-1} = \frac{a^{\alpha-\beta}}{a^\alpha+\delta}$, where C is the normalizing constant of Equation (2), is nothing but the Laplace transform of the given function. It is shown to be very relevant in fractional reaction-diffusion problems in physics, since it naturally occurs in the derivation of the inverse Laplace transform of the functions of the type $a^\alpha(d + ba^\beta)$, where a is the Laplace transform parameter and $d$andb are constants.

For more details, see the papers of [8,18,23]. Reaction-diffusion equations are modeling tools for the dynamics presented by a competition between two or more species, activators and inhibitors or production and destruction, that diffuse in a physical medium. As an example, [1] considered the evolution of a star like the Sun, which is governed by a second order system of differential equations, the kinetic equations, describing the rate of change of chemical composition of the star for each species in terms of the reaction rates for destruction and production of that species [24–26]. Methods for modeling processes of destruction and production have been developed for bio-chemical reactions and their unstable equilibrium states [27] and for chemical reaction networks with unstable states, oscillations and hysteresis [28].

Consider an arbitrary reaction characterized by a time-dependent quantity $N = N(t)$. It is possible to equate the rate of change $\frac{dN}{dt}$ to a balance between the destruction rate d and the production rate p of N, that is $\frac{dN}{dt} = -d + p$. In general, through feedback or other interaction mechanisms, destruction and production depend on the quantity N itself: $d = d(N)$ or $p = p(N)$. This dependence is complicated, since the destruction or production at time t depends not only on $N(t)$, but also on the past history $N(\tau), \tau < t$, of the variable N. This may be formally represented by:

$$\frac{dN}{dt} = -d(N_i) + p(N_i) \tag{38}$$

where N_i denotes the function defined by $N_i(t^*) = N(t - t^*), t^* > 0$. The production and destruction of species is described by kinetic equations governing the change of the number density N_i of species i over time, that is,

$$\frac{dN_i}{dt} = -\sum_j N_i N_j < \sigma v >_{ij} + \sum_{k,l \neq i} N_k N_l < \sigma v >_{kl}, \tag{39}$$

where $< \sigma v >_{mn}$ denotes the reaction probability for an interaction involving species mand n, and the summation is taken over all reactions, which either produce or destroy the species i [29]. The first sum in Equation (39) can also be written as:

$$-\sum_j N_i N_j < \sigma v >_{ij} = -N_i(\sum_j N_j < \sigma v >_{ij}) = N_i a_i, \tag{40}$$

where a_i is the statistically expected number of reactions per unit volume per unit time destroying the species i. It is also a measure of the speed at which the reaction proceeds. In the following, we are assuming that there are $N_j(j = 1, \ldots i, \ldots)$ species j per unit volume and that for a fixed N_i, the number of other reacting species that interact with the i-th species is constant in a unit volume. Following the same argument, for the second sum in Equation (39) accordingly,

$$+\sum_{k,l \neq i} N_k N_l < \sigma v >_{kl} = +N_i b_i, \tag{41}$$

where $N_i b_i$ is the statistically expected number of the i-th species produced per unit volume per unit time for a fixed N_i. Note that the number density of species i, $N_i = N_i(t)$, is a function of time, while the $< \sigma v >_{mn}$, containing the thermonuclear functions (see Haubold and Kumar (2008) [30]), are assumed to depend only on the temperature of the gas, but not on the time t and number densities N_i. Then, Equation (38) implies that:

$$\frac{\mathrm{d}N_i(t)}{\mathrm{d}t} = -(a_i - b_i)N_i(t). \tag{42}$$

For Equation (42), we have three distinct cases, $c_i = a_i - b_i > 0$, $c_i < 0$ and $c_i = 0$, of which the last case says that N_i does not vary over time, which means that the forward and reverse reactions involving species i are in equilibrium; such a value for N_i is called a fixed point and corresponds to a steady-state behavior. The first two cases exhibit that either the destruction $(c_i > 0)$ of species i or production $(c_i < 0)$ of species i dominates.

For the case $c_i > 0$, we have:

$$\frac{\mathrm{d}N_i(t)}{\mathrm{d}t} = -c_i N_i(t), \tag{43}$$

with the initial condition that $N_i(t = 0) = N_0$ is the number density of species i at time $t = 0$, and it follows that:

$$N_i(t)\mathrm{d}t = N_0 \mathrm{e}^{-c_i t}\mathrm{d}t. \tag{44}$$

The exponential function in Equation (44) represents the solution of the linear one-dimensional differential Equation (43) in which the rate of destruction of the variable is proportional to the value of the variable. Equation (43) does not exhibit instabilities, oscillations or chaotic dynamics, in striking contrast to its cousin, the logistic finite-difference equation [26,29]. A thorough discussion of Equation (43) and its standard solution in Equation (44) is given in [25].

Let us consider the standard fractional kinetic equation:

$$N(t) - N_0 = -c^{\nu}\, {}_0D_t^{-\nu}N(t),\ \nu > 0 \tag{45}$$

which is derived by [1]. ${}_0D_t^{-\nu}$ is the Riemann–Liouville fractional integral of order ν. It is very easy to find the solution of this equation. On applying Laplace transforms on both sides, with parameter p, one has:

$$\begin{aligned} L\{N(t)\} - N_0 L\{1\} &= -c^{\nu}\, L\{{}_0D_t^{-\nu}N(t)\} \\ L\{N(t)\} - \tfrac{N_0}{p} &= -c^{\nu}\, p^{-\nu}L\{N(t)\} \\ (1 + c^{\nu}\, p^{-\nu})L\{N(t)\} &= \tfrac{N_0}{p} \\ L\{N(t)\} &= \frac{N_0}{p[1+(\frac{p}{c})^{-\nu}]} = N_0\left[\sum_{k=0}^{\infty}(-1)^k c^{\nu k}p^{-(\nu k+1)}\right]. \end{aligned}$$

Now, applying inverse Laplace transform,

$$N(t) = N_0 \sum_{k=0}^{\infty}(-1)^k c^{\nu k}L^{-1}\left\{p^{-(\nu k+1)}\right\} = N_0\sum_{k=0}^{\infty}(-1)^k\frac{(ct)^{\nu k}}{\Gamma(\nu k+1)}.$$

This solution can be written in terms of the Mittag–Leffler function as:

$$N(t) = N_0\, E_{\nu}(-c^{\nu}t^{\nu}),\ \nu > 0. \tag{46}$$

Theorem 5.1. *For $\nu > 0$, let the coefficient of N_0 of the fractional integral equation be g(t) an arbitrary function, so that the fractional integral becomes:*

$$N(t) - N_0\, g(t) = -c^{\nu}\, {}_0D_t^{-\nu}N(t). \tag{47}$$

Then, the Laplace transforms of the fractional integral equation will be:

$$L\{N(t)\} = \frac{N_0\, G(p)}{1+(\frac{c}{p})^{\nu}}, \text{where} G(p) = L\{g(t)\} \tag{48}$$

Now, one can take various forms for $G(p)$; all of the special cases are considered in [23,31]. For example:

Let us consider $G(p) = p_1^{-\beta} F_1(\gamma_1, \beta_1, p^{-\alpha})$ for $\alpha > 0, \beta > 0, \nu > 0, c > 0$, then Equation (47) becomes:

$$\begin{aligned} L\{N(t)\} &= \frac{N_0\, p^{-\beta} \sum_{k=0}^{\infty} \frac{(\gamma_1)_k}{(\beta_1)_k} \frac{p^{-\alpha k}}{k!}}{\left[1+(\frac{c}{p})^{\nu}\right]} \\ &= N_0\, p^{-\beta} \sum_{k=0}^{\infty} \frac{(\gamma_1)_k}{(\beta_1)_k} \frac{p^{-\alpha k}}{k!} \sum_{n=0}^{\infty} (-1)^n (\frac{c}{p})^{\nu n} \\ &= N_0 \sum_{k=0}^{\infty} \frac{(\gamma_1)_k}{(\beta_1)_k k!} \sum_{n=0}^{\infty} (-1)^n c^{\nu n} p^{-(\nu n+\alpha k+\beta)} \end{aligned} \tag{49}$$

On applying inverse Laplace transform, we get:

$$\begin{aligned} N(t) &= N_0 \sum_{k=0}^{\infty} \frac{(\gamma_1)_k}{(\beta_1)_k k!} \sum_{n=0}^{\infty} (-1)^n c^{\nu n} \frac{t^{\nu n+\alpha k+\beta-1}}{\Gamma(\nu n+\alpha k+\beta)} \\ &= N_0 t^{\beta-1} \sum_{k=0}^{\infty} \frac{(\gamma_1)_k}{(\beta_1)_k k!} t^{\alpha k} E_{\nu,\alpha k+\beta}(-c^{\nu} t^{\nu}) \end{aligned} \tag{50}$$

Haubold and Mathai (2000) [1] generalized the standard kinetic Equation (43) to a standard fractional kinetic Equation (45), derived solutions of a fractional kinetic equation that contains the particle reaction rate (or thermonuclear function) as a time constant and provided the analytic technique to further investigate possible modifications of the reaction rate through a kinetic equation. The Riemann–Liouville operator in the fractional kinetic equation introduces a convolution integral with a slowly-decaying power-law kernel, which is typical for memory effects referred to in [32]. This technique may open an avenue to accommodate changes in standard solar model core physics as proposed by [33]. In the solution of the standard fractional kinetic Equation (45), given in Equation (46), the standard exponential decay is recovered for $\nu = 1$. However, the Equation (46), for $0 < \nu < 1$, shows a power-law behavior for $t \to \infty$ and is constant (initial value N_0) for $t \to 0$.

6. Application in Financial Modeling

In this section, we present an application of generalized gamma Mittag–Leffler density in modeling currency exchange rates. Here, we considered 1152 observations starting from 2004 to 2008, of U.S. dollar-Indian rupee foreign exchange rates, which is available at www.rbi.org. The log returns of the exchange rates are considered. The following Figure 3 is the graph of the log-transformed data embedded with the generalized gamma Mittag–Leffler density.

From the data, the summary statistics for transformed currency exchange rates obtained are as follows:

$$\begin{aligned} \text{Mean} &= 0.0110 \\ \text{Variance} &= 2.0805 \times 10^{-6} \\ \text{Coefficientofskewness} &= 0.2728 \end{aligned}$$

For convenience, assume that $\alpha = 1$; by using the method of moments, estimates of parameters are obtained and are given as:

$$\begin{aligned} \hat{\beta} &= 76.3057 \\ \hat{\delta} &= 6227.6449 \\ \hat{a} &= 7413.998 \end{aligned}$$

From the figures, we can observe that the generalized gamma Mittag–Leffler density fits well to the data with the above estimated values of the parameters. Here, we used the chi square statistic to measure the goodness of fit. The calculated chi square value is 2.44, and the corresponding tabled value is 15.507. Hence, we conclude that the model in Equation (2) is a good fit to the dataset considered.

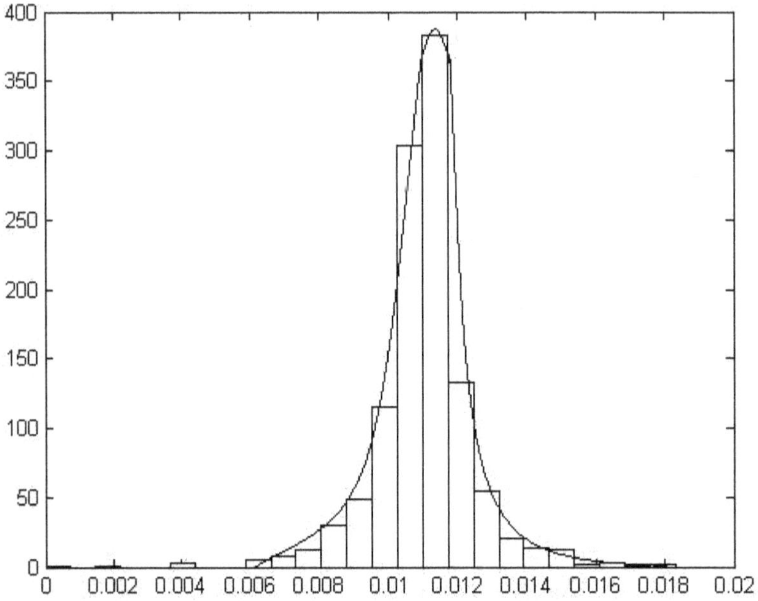

Figure 3. Generalized gamma Mittag–Leffler density fitted to the data on dollar-rupee exchange rate.

7. Multivariate Generalization

In this section, a multivariate analogue of the density in Equation (2) is considered. The multivariate Mittag–Leffler function will be defined as:

$$E_p[\alpha_1,\cdots,\alpha_p;\beta_1,\cdots,\beta_p;-\delta;x_1^{\alpha_1}\cdots x_p^{\alpha_p}]=\sum_{k=0}^{\infty}\frac{x_1^{\alpha_1 k}\cdots x_p^{\alpha_p k}(-\delta)^k}{\Gamma(\alpha_1 k+\beta_1)\cdots\Gamma(\alpha_p k+\beta_p)}, \quad (51)$$
$$\text{where}\Re(\alpha_j)>0,\Re(\beta_j>0),j=1,2,\ldots,p.$$

A function close to $E_p(\cdot)$ of Equation (37) is the multi-index Mittag–Leffler function of Kiryakova; see, for example, Kiryakova (2000) [3]. Now, the general density is defined as:

$$f_p(x_1,\ldots,x_p)=\begin{cases} C_p x_1^{\beta_1-1}\cdots x_p^{\beta_p-1}\mathrm{e}^{-a_1x_1-\cdots-a_px_p}E_p[\alpha_1,\cdots,\alpha_p;\beta_1,\cdots,\beta_p;-\delta;x_1^{\alpha_1}\cdots x_p^{\alpha_p}], \\ 0\le x_j<\infty,\Re(\alpha_j)>0,\Re(\beta_j>0),a_j>0,j=1,2,\ldots,p \\ 0,\text{elsewhere.}\end{cases}$$

The normalizing constant is available, by proceeding as before, as:

$$C_p = a_1^{\beta_1}\cdots a_p^{\beta_p}(1+\frac{\delta}{a_1^{\alpha_1}\cdots a_p^{\alpha_p}}),|\frac{\delta}{a_1^{\alpha_1}\cdots a_p^{\alpha_p}}|<1. \quad (52)$$

Various properties of one form of a multivariate gamma density can be seen from Griffiths (1984) [34]. The following sections reveal the joint Laplace transform and product moments of the density function.

The Joint Laplace transform is available from Equation (52) by replacing a_j by $a_j + t_j, j = 1, \ldots, p$ and then dividing by the normalizing constant. That is:

$$E(e^{-t_1x_1}\cdots e^{-t_px_p}) = \frac{a_1^{\beta_1}\cdots a_p^{\beta_p}}{(a_1+t_1)^{\beta_1}\cdots(a_p+t_p)^{\beta_p}} \frac{(1+\frac{\delta}{a_1^{\alpha_1}\cdots a_p^{\alpha_p}})}{(1+\frac{\delta}{(a_1+t_1)^{\alpha_1}\cdots(a_p+t_p)^{\alpha_p}})}, a_j + t_j > 0, \quad \left|\frac{\delta}{(a_1+t_1)^{\alpha_1}\cdots(a_p+t_p)^{\alpha_p}}\right| < 1, j = 1, \ldots, p. \tag{53}$$

The arbitrary product moments are obtained in terms of generalized Wright function and are given by:

$$\begin{aligned} E(x_1^{\gamma_1}\cdots x_p^{\gamma_p}) &= \int_0^\infty \cdots \int_0^\infty x_1^{\gamma_1}\cdots x_p^{\gamma_p} C_p x_1^{\beta_1-1}\cdots x_p^{\beta_p-1} e^{-a_1x_1}\cdots e^{-a_px_p} \\ &\quad \times E_p(\alpha_j;\beta_j, j = 1,\ldots,p; -\delta; x_1^{\alpha_1}\cdots x_p^{\alpha_p})dx_1\cdots dx_p \\ &= \frac{1}{a_1^{\gamma_1}\cdots a_p^{\gamma_p}}(1 + \frac{\delta}{a_1^{\alpha_1}\cdots a_p^{\alpha_p}}) \quad {}_{p+1}\Psi_p\left[\begin{matrix}(\beta_1+\gamma_1,\alpha_1),\cdots,(\beta_p+\gamma_p,\alpha_p),(1,1) \\ (\beta_1,\alpha_1),,(\beta_p,\alpha_p)\end{matrix}; -\frac{\delta}{a_1^{\alpha_1}\cdots a_p^{\alpha_p}}\right], \\ &\quad \Re(\beta_j + \gamma_j) > 0, j = 1, \ldots, p. \end{aligned} \tag{54}$$

Conditional density of x_1 given $x_2, \ldots, x_p$ is given by:

$$f(x_1|x_2,\ldots,x_p) = a_1^{\beta_1}x_1^{\beta_1-1}e^{-a_1x_1}\frac{E_p(\alpha_j;\beta_j, j = 1,\ldots,p; -\delta; x_1^{\alpha_1}\cdots x_p^{\alpha_p})}{E_{p-1}(\alpha_j;\beta_j, j = 2,\ldots,p; -\frac{\delta}{a_1^{\alpha_1}}; x_2^{\alpha_2}\cdots x_p^{\alpha_p})}. \tag{55}$$

Regression analysis is concerned with constructing the best predictor of one variable at the preassigned values of some other variables. Under the minimum mean square principle, the best predictors can be seen to be the conditional expectation of x_1 given $x_2, \cdots, x_p$. Hence, $E(x_1|x_2, \cdots, x_p)$ is defined as the regression of x_1 on $x_2, \cdots, x_p$. In our case:

$$\begin{aligned} E(x_1|x_2,\ldots,x_p) &= \int_0^\infty x_1 f(x_1|x_2,\ldots,x_p)dx_1 \\ &= \frac{1}{K(X)}a_1^{\beta_1}\int_0^\infty x_1^{\beta_1}e^{-a_1x_1}\sum_{k=0}^{\infty}\frac{(-\delta)^k x_1^{\alpha_1 k}\cdots x_p^{\alpha_p k}}{\Gamma(\alpha_1k+\beta_1)\cdots\Gamma(\alpha_pk+\beta_p)} \\ \text{where} K(X) &= E_{p-1}(\alpha_j;\beta_j, j = 2,\ldots,p; -\frac{\delta}{a_1^{\alpha_1}}; x_2^{\alpha_2}\cdots x_p^{\alpha_p}) \end{aligned}$$

$$E(x_1|x_2,\ldots,x_p) = \frac{\beta_1}{a_1} - \frac{\delta\alpha_1 x_2^{\alpha_2}\cdots x_p^{\alpha_p}}{a_1^{\alpha_1+1}}\frac{E_{p-1}^{(2)}(\alpha_j;\beta_j+\alpha_j, j = 2,\ldots,p; -\frac{\delta}{a_1^{\alpha_1}}; x_2^{\alpha_2}\cdots x_p^{\alpha_p})}{E_{p-1}(\alpha_j;\beta_j, j = 2,\ldots,p; -\frac{\delta}{a_1^{\alpha_1}}; x_2^{\alpha_2}\cdots x_p^{\alpha_p})}, \tag{56}$$

where $E_{p-1}^{(2)}$ is the same function appearing in Equation (37), but with an additional Pochhammer symbol $(2)_k$ sitting in the numerator. The conditional expectation, $E(x_1|x_2)$, is the best predictor, best in the sense of minimizing the expected squared error, also known as the regression of x_1 on x_2, and is given as:

$$E(x_1|x_2) = \frac{\beta_1}{a_1} - \frac{\delta\alpha_1 x_2^{\alpha_2}}{a_1^{\alpha_1+1}a_3^{\alpha_3}\cdots a_p^{\alpha_p}}\frac{E_1^{(2)}(\alpha_2;\beta_2+\alpha_2; -\frac{\delta}{a_1^{\alpha_1}a_3^{\alpha_3}\cdots a_p^{\alpha_p}}; x_2^{\alpha_2})}{E_1(\alpha_2;\beta_2; -\frac{\delta}{a_1^{\alpha_1}a_3^{\alpha_3}\cdots a_p^{\alpha_p}}; x_2^{\alpha_2})}. \tag{57}$$

In a similar manner, we can extend the results in Section 2 to matrix-variate cases, as well. Let $X = (x_{rs})$ be $m \times n$, where all x_{rs}'s are distinct, X of full rank and having a joint density $f(X)$, where $f(X)$ is a real-valued scalar function of X. Then, the characteristic function of $f(X)$, denoted by $\phi_X(T)$, is given by:

$$\phi_X(T) = E[e^{itr(XT)}], i = \sqrt{-1} \text{and} T = (t_{rs}) \tag{58}$$

is an $m \times n$ matrix of distinct parameters t_{rs}'s, and let T be of full rank. As an example, we can look at the real matrix-variate Gaussian distribution, given by the density geometric:

$$g(X) = \frac{|A|^{\frac{n}{2}}|B|^{\frac{m}{2}}}{\pi^{\frac{mn}{2}}} \mathrm{e}^{-\mathrm{tr}(AXBX')} \tag{59}$$

where X is $m \times n$, $A = A' > 0$, $B = B' > 0$ are $m \times m$ and $n \times n$ positive definite constant matrices, having Fourier transform as:

$$\phi_g(T) = \mathrm{e}^{-\frac{1}{4}\mathrm{tr}(B^{-\frac{1}{2}}TA^{-1}T'B^{-\frac{1}{2}})} \tag{60}$$

where T is the parameter matrix. Motivated by Equations (58) and (60), Mathai (2010) [4] defined matrix-variate Linnik density and gamma-Linnik density. The following result was given recently in Mathai (2010) [4], which will be needed for the discussion to follow.

Theorem 7.1. *Let y be a real scalar random variable, distributed independently of the $m \times n$ matrix X and having a gamma density with the Laplace transform:*

$$L_y(t) = (1 + \rho t)^{-\gamma}, \rho > 0, \gamma > 0$$

and let X be distributed as a real rectangular matrix-variate Linnik variable having characteristic function $e^{-[tr(T\Sigma_1 T'\Sigma_2)]^{\frac{\alpha}{2}}}, 0 < \alpha \leq 2$; then, the rectangular matrix $U = y^{\frac{1}{\alpha}}X$ has the characteristic function:

$$\phi_U(T) = [1 + \rho(tr(T\Sigma_1 T'\Sigma_2))^{\frac{\alpha}{2}}]^{-\beta} \tag{61}$$

The matrix variable U in Equation (61) will be called a real rectangular matrix-variate Gaussian Linnik variable when $\alpha = 2$ and gamma Linnik variable for the general α. By using the multi-index Mittag–Leffler function of Kiryakova (2000) [3] in Equation (37), one can also define a gamma-Kiryakova vector or matrix variable. Now, let us consider two independently-distributed real random variables y and X where y has a general Mittag–Leffler density as in Equation (24) and X has a real rectangular matrix-variate Gaussian density as in Equation (59). Then, we have the following results.

Theorem 7.2. *Let y follow $GMLD(\alpha, \gamma, \rho)$ and the real rectangular matrix X have the density in Equation (59), and let X and y be independently distributed. Then, $U_1 = y^{\frac{1}{\alpha}}X$ has the characteristic function:*

$$\phi_{U_1}(T) = [1 + \frac{\rho}{4^\alpha}(tr(A^{-1}T'B^{-1}T))^\alpha]^{-\gamma} \tag{62}$$

Theorem 7.3. *Let $U_2 = y^{\frac{1}{\beta}}X$, where y and X are independently distributed with y having the density in Equation (24), and X is a real rectangular matrix-variate Linnik variable with the parameters γ, A, B; then U_2 has the characteristic function:*

$$\phi_{U_2}(T) = [1 + \rho(tr(A^{-1}T'B^{-1}T)^{\frac{\alpha\beta}{2}})]^{-\gamma}, \rho, \gamma, \alpha, \beta > 0 \tag{63}$$

8. Conclusion

The model introduced in this paper will be useful for investigators in disciplines of physical sciences, particularly superstatistics of Beck ((2006) [11], (2009) [13]), nonextensive statistical mechanics Tsallis ((1988) [15], (2004) [16], (2009) [17]), reaction-diffusion problems in physics and Haubold (2000) [5], statistical distribution theory and model building. In many physical situations, when the integer derivatives are replaced by fractional derivatives, naturally, the solutions are obtained in terms of Mittag–Leffler functions and their generalizations. However, fractional integrals are shown to be connected to Laplace convolutions of positive random variables. Thus, one can connect the ideas of fractional calculus to statistical distribution theory via the Mittag–Leffler function. the Mittag–Leffler distribution can also be used as waiting time distributions, as well as first passage time distributions for certain renewal process.

Acknowledgments: The author acknowledges gratefully the encouragement given by Professor A.M. Mathai, Department of Mathematics and Statistics, McGill University, Montreal, Canada, in this work.

Conflicts of Interest: The author declares no conflict of interest.

References

1. Haubold, H.J.; Mathai, A.M. The fractional kinetic equation and thermonuclear functions. *Astrophys. Space Sci.* **2000**, *273*, 53–63. [CrossRef]
2. Kilbas, A.A.; Saigo, M.; Saxena, R.K. Generalized Mittag–Leffler function and generalized fractional calculus operators. *J. Integral Transforms Spec. Funct.* **2004**, *15*, 31–49. [CrossRef]
3. Kiryakova, V. Multi-index Mittag–Leffler function, related Gelfond-Leontiev operator and Laplace type transforms. *J. Comput. Appl. Math.* **2000**, *118*, 241–259. [CrossRef]
4. Mathai, A.M. Some properties of Mittag–Leffler functions and matrix-variate analogues: A statistical perspective. *Fract. Calc. Appl. Anal.* **2010**, *13*, 113–132.
5. Mathai, A.M.; Saxena, R.K.; Haubold, H.J. *The H Function: Theory and Applications*; Springer: New York, NY, USA, 2010.
6. Wright, E.M. The generalized Bessel functions of order greater than one. *Quart. J. Math. Oxf. Ser.* **1940**, *11*, 36–48. [CrossRef]
7. Mathai, A.M. A pathway to matrix-variate gamma and normal densities. *Linear Algebra Its Appl.* **2005**, *396*, 317–328. [CrossRef]
8. Seema Nair, S. Pathway fractional integration operator. *Fract. Calc. Appl. Anal.* **2009**, *12*, 237–252.
9. Seema Nair, S. Pathway fractional integral operator and matrix-variate functions. *Integral Transforms Spec. Funct.* **2011**, *22*, 233–244. [CrossRef]
10. Beck, C.; Cohen, E.G.D. Superstatistics. *Phys. A* **2003**, *322*, 267–275. [CrossRef]
11. Beck, C. Stretched exponentials from superstatistics. *Phys. A* **2006**, *365*, 96–101. [CrossRef]
12. Beck, C. Generalized statistical mechanics for superstatistical systems. arXiv.org/1007.0903. 2010.
13. Beck, C. Recent developments in superstatistics. *Braz. J. Phys.* **2009**, *39*, 357–363. [CrossRef]
14. Seema Nair, S.; Katuveettil, A. Some remarks on the paper "On the q-type distributions". In *Proceedings Astrophysics & Space Science*; Springer: Berlin, Germany, 2010; pp. 11–15.
15. Tsallis, C. Possible generalizations of Boltzmann-Gibbs statistics. *J. Stat. Phys.* **1988**, *52*, 479–487. [CrossRef]
16. Tsallis, C. What should a statistical mechanics satisfy to reflect nature? *Phys. D* **2004**, *193*, 3–34. [CrossRef]
17. Tsallis, C. *Introduction to Nonextensive Statistical Mechanics-Approaching a Complex World*; Springer: New York, NY, USA, 2009.
18. Mathai, A.M.; Saxena, R.K.; Haubold, H.J. A Certain Class of Laplace Transforms with Applications to Reaction and Reaction-Diffusion Equations. *Astrophys. Space Sci.* **2006**, *305*, 283–288. [CrossRef]
19. Pillai, R.N. On Mittag–Leffler functions and related distributions. *Ann. Inst. Statist. Math.* **1990**, *42*, 157–161. [CrossRef]
20. Pillai, R.N.; Jayakumar, K. Discrete Mittag–Leffler distributions. *Stat. Probab. Lett.* **1995**, *23*, 271–274. [CrossRef]
21. Jayakumar, K. On Mittag–Leffler process. *Math. Comput. Model.* **2003**, *37*, 1427–1434. [CrossRef]
22. Jose, K.K.; Uma, P.; Seethalekshmi, V.; Haubold, H.J. Generalized Mittag–Leffler distributions and process for applications in astrophysics and time series modeling. *Astrophys. Space Sci. Proc.* **2010**, *79*, 79–92. [CrossRef]
23. Saxena, R.K.; Mathai, A.M.; Haubold, H.J. Unified fractional kinetic equation and a fractional diffusion equation. *Astrophys. Space Sci.* **2004**, *209*, 299–310. [CrossRef]
24. Clayton, D.D. *Principles of Stellar Evolution and Nucleosynthesis*, 2nd ed.; The University of Chicago Press: Chicago, IL, USA, 1983.
25. Kourganoff, V. *Introduction to the Physics of Stellar Interiors*; D. Reidel Publishing Company: Dordrecht, The Netherlands, 1973.
26. Perdang, J. *Lecture Notes in Stellar Stability*; Instituto 13 di Astronomia: Padova, Italy, 1976.
27. Murray, J.D. *Mathematical Biology*; Springer-Verlag: Berlin, Germany, 1989.

28. Nicolis, G.; Prigogine, I. *Self-Organization in Nonequilibrium Systems-From Dissipative Structures to Order Through Fluctuations*; Wiley: New York, NY, USA, 1977.
29. Haubold, H.J.; Mathai, A.M. A heuristic remark on the periodic variation in the number of solar neutrinos detected on Earth. *Astrophys. Space Sci.* **1995**, *228*, 113–134. [CrossRef]
30. Haubold, H.J.; Kumar, D. Extension of thermonuclear functions through the pathway model including Maxwell-Boltzmann and Tsallis distributions. *Astropart. Phys.* **2008**, *29*, 70–76. [CrossRef]
31. Seema Nair, S.; Kattuveettil, A. Some aspects of Mittag–Leffler functions, their applications and some new insights. *STARS Int. J. Sci.* **2007**, *1*, 118–131.
32. Kaniadakis, G.; Lavagno, A.; Lissia, M.; Quarati, P. Anomalous diffusion modifies solar neutrino fluxes. *Phys. A* **1998**, *261*, 359–373. [CrossRef]
33. Schatzman, E. Role of gravity waves in the solar neutrino problem. *Phys. Rep.* **1999**, *311*, 143–150. [CrossRef]
34. Griffiths, R.C. Characterization of infinitely divisible multivariate gamma distributions. *J. Multivar. Anal.* **1984**, *15*, 13–20. [CrossRef]

axiom s

Article

Limiting Approach to Generalized Gamma Bessel Model via Fractional Calculus and Its Applications in Various Disciplines

Nicy Sebastian

Indian Statistical Institute(ISI), Chennai CIT Campus, Taramani, Chennai 600113, India; nicyseb@yahoo.com or nicycms@gmail.com; Tel.: +91-0487-242-0435

Academic Editor: Hans J. Haubold
Received: 15 July 2015; Accepted: 10 August 2015; Published: 26 August 2015

Abstract: The essentials of fractional calculus according to different approaches that can be useful for our applications in the theory of probability and stochastic processes are established. In addition to this, from this fractional integral, one can list out almost all of the extended densities for the pathway parameter $q < 1$ and $q \to 1$. Here, we bring out the idea of thicker- or thinner-tailed models associated with a gamma-type distribution as a limiting case of the pathway operator. Applications of this extended gamma model in statistical mechanics, input-output models, solar spectral irradiance modeling, *etc.*, are established.

Keywords: fractional integrals; statistical distributions; Bessel function; gamma model

MSC: MSC (2000) 26A33, 33C10, 33B15

1. Introduction

In recent years, considerable interest has been shown in so-called fractional calculus, which allows one to consider integration and differentiation of any order, not necessarily integer. Fractional calculus is a rapidly growing field both in theory and in applications to real-world problems. There is a revived interest in fractional integrals and fractional derivatives due to their recently-found applications in reaction, diffusion, reaction-diffusion problems, in solving certain partial differential equations, in input-output models and related areas; see, for example [1–6]. There are many books in the area, some of which are [7–11]. The classical left- and right-hand-sided Riemann–Liouville fractional integral operators of order $\alpha \in \mathbb{C}, \Re(\alpha) > 0$, are defined as:

$$ {}_0D_x^{-\alpha} f = (I_{0+}^{\alpha} f)(x) = \frac{1}{\Gamma(\alpha)} \int_0^x (x-t)^{\alpha-1} f(t) \mathrm{d}t, x > 0, \Re(\alpha) > 0 \quad (1) $$

$$ {}_xD_{\infty}^{-\alpha} f = (I_{-}^{\alpha} f)(x) = \frac{1}{\Gamma(\alpha)} \int_x^{\infty} (t-x)^{\alpha-1} f(t) \mathrm{d}t, x > 0, \Re(\alpha) > 0 \quad (2) $$

The traditional special functions are also related to the classical fractional calculus (FC) and later to the generalized fractional calculus and are shown to be representable as fractional order integration or differentiation operators of some basic elementary functions. Such relations provided some alternative definitions for the special functions by means of Poisson-type and Euler-type integral representations and Rodrigues-type differential formulas. An example of such a unified approach on special functions, based on a generalized fractional calculus, can be seen in [8]. The essentials of fractional calculus according to different approaches that can be useful for our applications in the theory of probability and stochastic processes are established with the help of the pathway idea in [12].

The pathway idea was originally proposed by Mathai in the 1970s in connection with population models and later rephrased and extended in [12] to cover scalar, as well as matrix cases, and it was made suitable for modeling data from statistical and physical situations. The main idea behind the derivation of this model is the switching properties of going from one family of functions to another and, yet, another family of functions. It is shown that through a parameter q, called the pathway parameter, one can connect a generalized Type 1 beta family of densities, a generalized Type 2 beta family of densities and generalized gamma family of densities, in the scalar, as well as in the matrix cases, also in the real and complex domains. It is shown that when the model is applied to physical situations, then the current hot topics of Tsallis statistics and superstatistics in statistical mechanics become special cases of the pathway model, and the model is capable of capturing many stable situations, as well as the unstable or chaotic neighborhoods of the stable situations and transitional stages. Mathai [12] deals mainly with rectangular matrix-variate distributions, and the scalar case is a particular case there. For the real scalar case, the pathway model is the following:

$$h_1(x) = k_1 x^{\gamma-1}[1-a(1-q)x^{\theta}]^{\frac{\eta}{1-q}}, 1-a(1-q)x^{\theta} > 0, a, \theta, \gamma, \eta > 0, q < 1 \tag{3}$$

where k_1 is the normalizing constant if a statistical density is needed. For $q < 1$, the model remains as a generalized Type 1 beta model in the real case. Other cases available are the regular Type 1 beta density, Pareto density, power function, triangular and related models. Observe that Equation (3) is a model with the right tail cut off. When $q > 1$, we may write $1-q = -(q-1), q > 1$, so that $h_2(x)$ assumes the form:

$$h_2(x) = k_2 x^{\gamma-1}\left[1+a(q-1)x^{\theta}\right]^{-\frac{\eta}{q-1}}, x \geq 0, a, \theta, \gamma, \eta > 0, q > 1 \tag{4}$$

which is a generalized Type 2 beta model for real x, and k_2 is the normalizing constant, if a statistical density is required. Beck and Cohen's superstatistics belongs to this case Equation (4), and dozens of published papers are available on the topic of superstatistics in astrophysics. For $\gamma = 1, a = 1, \theta = 1$, we have Tsallis statistics for $q > 1$ from Equation (4). Other standard distributions coming from this model are the regular Type 2 beta, the F-distribution, Lévi models and related models. When $q \to 1$, the forms in Equations (3) and (4) reduce to:

$$h_3(x) = k_3 x^{\gamma-1} e^{-bx^{\theta}}, x \geq 0, b = a\eta > 0, \gamma, \theta > 0 \tag{5}$$

where k_3 is the normalizing constant. This includes generalized gamma, gamma, exponential, chi-square, Weibull, Maxwell–Boltzmann, Rayleigh and related models; for more details, see [13,14]. If x is replaced by $|x|$ in Equation (3), then more families of distributions are covered in Equation (3). Note that q is the most important parameter here, which enables one to move from one family of functions to another family. The other parameters are the usual parameters within each family of functions.

The paper is organized as follows: In Section 2, the connections of fractional integral operators to statistical distribution theory and incomplete integrals are given. Section 3 covers the limiting approach to the generalized gamma model via the pathway operator. The application of the extended generalized gamma model in statistical mechanics is introduced in Section 4. Generalized Laplacian density and the stochastic process are introduced in Section 5. In Section 6, we consider the application of the generalized gamma model in solar spectral irradiance modeling.

2. Statistical Interpretations of Fractional Integrals

A general pathway fractional integral operator is discussed in [15], which generalizes the classical Riemann–Liouville fractional integration operator. The pathway fractional integral operator has found applications in reaction-diffusion problems, non-extensive statistical mechanics, non-linear waves, fractional differential equations, non-stable neighborhoods of physical system, *etc*. By means of the pathway model [12], the pathway fractional integral operator (pathway operator) is defined as follows: Let $f(x) \in L(a,b), \eta \in \mathbb{C}, \Re(\eta) > 0, a > 0$ and $q < 1$, then:

$$\left(P_{0+}^{(\eta,q)} f\right)(x) = x^{\eta-1} \int_0^{\frac{x}{a(1-q)}} \left[1 - \frac{a(1-q)t}{x}\right]^{\frac{\eta}{(1-q)}-1} f(t)\mathrm{d}t \tag{6}$$

where q is the pathway parameter and $f(t)$ is an arbitrary function.

When $q \to 1_-$, $\left[1 - \frac{a(1-q)t}{x}\right]^{\frac{\eta}{1-q}} \to \mathrm{e}^{-\frac{a\eta}{x}t}$. Thus, the operator will become:

$$P_{0+}^{\eta,1} = x^{\eta-1} \int_0^{\infty} \mathrm{e}^{\frac{-a\eta}{x}t} f(t)\mathrm{d}t = x^{\eta-1} L_f\left(\frac{a\eta}{x}\right)$$

the Laplace transform of f with parameter $\frac{a\eta}{x}$. When $q = 0, a = 1$ in Equation (6), the integral will become,

$$\int_0^x (x-t)^{\eta-1} f(t)\mathrm{d}t = \Gamma(\eta) I_{0+}^{\eta}$$

where I_{0+} is the left-sided Riemann–Liouville fractional integral operator.

Fractional integrals in the matrix-variate cases and their connection to statistical distributions are pointed out in [16,17]. Let $x > 0$ and $y > 0$ be statistically-independently-distributed positive real scalar random variables. Let the densities of x and y be $f_1(x)$ and $f_2(y)$, respectively. Then, the joint density of x and y is $f(x,y) = f_1(x)f_2(y)$. Let $u = x + y, t = y$. Then, the density of u, denoted by $g_1(u)$, is given by:

$$g_1(u) = \int_{t=0}^{u} f_1(u-t) f_2(t)\mathrm{d}t \tag{7}$$

Here, Equation (7) is in the same format of the Riemann–Liouville left-sided fractional integral for $f_1(x) = c_1 x^{\alpha-1}$ and $f_2(y) = c_2 f(y)$, where c_1 and c_2 are normalizing constants to create densities. Thus, a constant multiple of the left-sided Riemann–Liouville fractional integral can be interpreted as the density $g_1(u)$ of a sum of two independently-distributed real positive scalar random variables.

Now, let us look at $u = x - y$ with the additional assumption that $u = x - y > 0$. Then, the density of u, denoted by $g_2(u)$, will have the format:

$$g_2(u) = \int_{t=u}^{\infty} f_1(t) f_2(t-u)\mathrm{d}t \tag{8}$$

By taking $f_2(y) = c_2 y^{\alpha-1}$ and $f_1(x) = c_1 f(x)$, where c_1 and c_2 are some normalizing constants, Equation (8) agrees with the density of a structure $u = x - y$ with $x - y > 0, x > 0, y > 0$. Thus, the right-sided Riemann–Liouville fractional integral can be interpreted as the density of $u = x - y > 0, x > 0, y > 0$, where x and y are statistically-independently- distributed real scalar random variables.

Let us look into some examples from [16,18]. A real positive scalar random variable x is said to have a gamma density if its density function is of the form:

$$f(x) = \frac{m^{\alpha}}{\Gamma(\alpha)} x^{\alpha-1} \mathrm{e}^{-mx}, 0 \leq x < \infty, \alpha > 0, m > 0$$

and $f(x) = 0$ elsewhere. Here, $f(x) \geq 0$ for all x and $\int_{-\infty}^{\infty} f(x)\mathrm{d}x = 1$, so that $f(x)$ can be a statistical density. In this case:

$$1 = \frac{m^{\alpha}}{\Gamma(\alpha)} \int_0^{\infty} t^{\alpha-1} \mathrm{e}^{-mt}\mathrm{d}t$$

Let us take a fraction of this integral, such as e^{-ax} times this total integral one. That is,

$$\begin{aligned} e^{-mx}(1) &= e^{-mx}\frac{m^\alpha}{\Gamma(\alpha)}\int_0^\infty t^{\alpha-1}e^{-mt}dt \\ &= \frac{m^\alpha}{\Gamma(\alpha)}\int_0^\infty t^{\alpha-1}e^{-m(t+x)}dt, u=t+x \\ &= \frac{m^\alpha}{\Gamma(\alpha)}\int_{u=x}^\infty (u-x)^{\alpha-1}e^{-u}du \\ &= m^\alpha(I_-^\alpha f)(x) \text{with} f(u)=e^{-mu} \end{aligned}$$

Thus, the constant multiple of the right-sided Riemann–Liouville fractional integral when $f(u) = e^{-mu}$ can be interpreted as a fraction of the total integral coming from a gamma density. Let us examine a fraction of the Type 1 beta density. A real scalar random variable u is said to have a Type 1 beta density if the density function is given by:

$$f(u) = \frac{u^{\alpha-1}(1-u)^{\beta-1}}{B(\alpha,\beta)}, 0 \le u < 1, \alpha > 0, \beta > 0$$

and zero elsewhere, where $B(\alpha,\beta) = \Gamma(\alpha)\Gamma(\beta)/\Gamma(\alpha+\beta)$. The total probability in this case is given by:

$$1 = \int_0^1 \frac{u^{\alpha-1}(1-u)^{\beta-1}}{B(\alpha,\beta)}du$$

Let us consider a fraction of this total probability and consider $b^{\alpha+\beta-1}(1)$. That is,

$$\begin{aligned} b^{\alpha+\beta-1} &= b^{\alpha+\beta-1}\int_0^1 \frac{u^{\alpha-1}(1-u)^{\beta-1}}{B(\alpha,\beta)}du \\ &= \int_0^b \frac{(b-t)^{\alpha-1}t^{\beta-1}}{B(\alpha,\beta)}dt \\ &= \frac{\Gamma(\alpha+\beta)}{\Gamma(\beta)}(I_{0+}^\alpha f)(x), f(t)=t^{\beta-1} \end{aligned}$$

Thus, the left-sided Riemann–Liouville fractional integral when $f(t) = t^{\beta-1}$ can be interpreted as a fraction of the total integral coming from a beta density.

Similarly, a constant multiple of the left-sided pathway fractional integral can be interpreted as the density of a sum of two independently-distributed real positive scalar random variables; see [17]. Let $x > 0$ and $y > 0$ be statistically-independently-distributed positive real scalar random variables with densities $f_1(x)$ and $f_2(y)$, respectively. Let $u = x + a(1-q)y, t = y$. Then, the density of u is given by:

$$g_3(u) = \int_{t=0}^{\frac{u}{a(1-q)}} f_1(u - a(1-q))f_2(t)dt \quad (9)$$

This is in the same format of the left-sided pathway fractional integral for $f_1(x) = c_1\left(\frac{x}{u}\right)^{\frac{\eta}{(1-q)}-1}$ and $f_2(y) = c_2u^{\eta-1}f(y)$. That is:

$$\begin{aligned} g_3(u) &= c_1c_2x^{\eta-1}\int_0^{\frac{x}{a(1-q)}}\left[1-\frac{a(1-q)t}{x}\right]^{\frac{\eta}{(1-q)}-1}f(t)dt \\ &= c_1c_2P_{0+}^{(\eta,q)} \end{aligned}$$

Likewise, statistical interpretations can also be given for other fractional integrals. If we replace $f(t)$ by a non-negative integrable function, one can obtain a statistical density through this operator. In addition to this, from this fractional integral, one can list out almost all of the extended densities for the pathway parameter $q < 1$ and $q \to 1$; for more details, see [17].

3. Limiting Approach to the Generalized Gamma Bessel Model via the Pathway Operator

Here, we bring out the idea of thicker- or thinner-tailed models associated with a gamma-type distribution as a limiting case of the pathway operator. Let the integrand of Equation (6) be denoted by $I_{(\eta,q)}$.

$$I_{(\eta,q)} = \left[1 - \frac{a(1-q)t}{x}\right]^{\frac{\eta}{(1-q)}-1} f(t), \eta > 0 \tag{10}$$

If we consider any real-valued positive integrable scalar function of t instead of any arbitrary real-valued scalar function of t, one can bring out a statistical density from the pathway fractional integral operator. Thus, one can say that:

$$f_q(t) = CI_{(\eta,q)}(t)$$

is a statistical density. Hence, Equation (6) generalizes all of the left-sided standard fractional integrals and almost all of the extended densities for $q < 1$ and $q \to 1$. In Equation (6), when $q \to 1$, the integrand $I_{(\eta,q)}$ will become:

$$I_{(\eta,1)} = e^{-\frac{a\eta}{x}t} f(t)$$

In particular, if we take $f(t) = 1$ and $\frac{a\eta}{x} = b > 0$, then one has obtained the Gaussian or normal density. For $q \to 1$ and if $f(t)$ is replaced by t^{β}, we have the gamma density. Similarly, for the standard Type 1 beta density, the pathway model for $q < 1$, chi-square density, exponential density and many more can be obtained as a special case of the pathway integral operator. From Equation (10), one can obtain the generalized gamma Bessel density as a limiting case. When $q \to 1_-$ and replacing $f(t)$ by $t^{\beta-1}{}_0F_1(;\beta;\delta t)$, then $g(t)$ will be:

$$g(t) = \begin{cases} Ct^{\beta-1}e^{-bt}{}_0F_1(;\beta;\delta t); & t \geq 0, \beta, b > 0 \\ 0; & \text{otherwise} \end{cases} \tag{11}$$

Some of the special cases of Equation (11) are given in Table 1. For fixed values of β and b, we can look at the graphs for $\delta > 0$, as well as for $\delta < 0$. These graphs give a suitable interpretation, when tail areas are considered. In Figure 1a, note that $\delta = 0$ is the case of a gamma density. Thus, when δ increases from $\delta = 0$, the right tail of the density becomes thicker and thicker. Thus, when fitting a gamma-type model to given data and if it is found that a model with a thicker tail is needed, then one can select a member from this family for appropriate $\delta > 0$. In Figure 1b, observe that $\delta = 0$ is the case of gamma density. When δ decreases from $\delta = 0$, the right tail gets thinner and thinner. Thus, if we are looking for a gamma-type density, but with a thinner tail, then one from this family may be appropriate for $\delta < 0$. For more details of the model in Equation (11), see [19,20]. When $q \to 1_-, \eta = 1$ and replacing $f(t)$ by $t^{\beta-1}{}_0F_1(;\beta;\delta t)$ in the pathway fractional integral operator, then we are essentially dealing with distribution functions under a gamma Bessel-type model in a practical statistical problem, which provides a connection between statistical distribution theory and fractional calculus, so that one can make use of the rich results in statistical distribution theory for further development of fractional calculus and *vice versa*.

Table 1. Special cases of the generalized gamma model associated with the Bessel function.

$\delta = 0$	Two-parameter gamma density
$\delta = 0, a = 1$	One-parameter gamma density
$\delta = 0, \beta = 1$	Exponential density
$\delta = 0, a = \frac{1}{2}, \beta = \frac{n}{2}, n = 1, 2, \cdots$	Chi-square density
$\delta = \lambda, a = \frac{1}{2}, \beta = \frac{n}{2}, n = 1, 2, \cdots$	Noncentral chi-square density

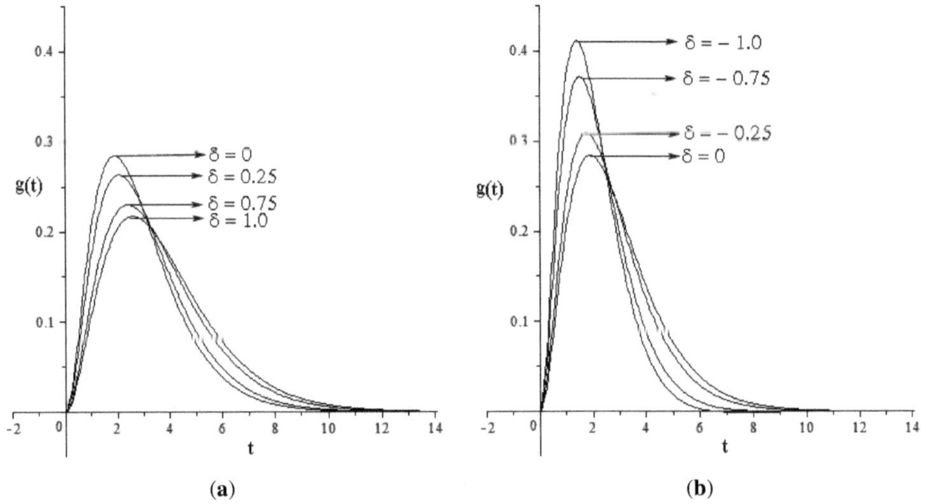

Figure 1. (**a**) Gamma Bessel model for $\delta > 0$; (**b**) gamma Bessel model for $\delta < 0$.

We can look at the model in another way also. Consider the total integral as:

$$1 = C \int_0^\infty t^{\beta-1} e^{-bt} {}_0F_1(;\beta;\delta t) dt$$

which can be treated as the Laplace transform of the function $t^{\beta-1}{}_0F_1(;\beta;\delta t)$, and hence, $C = \frac{b^\beta}{\Gamma(\beta) e^{\frac{\delta}{b}}}$, where C, the normalizing constant of Equation (11), is nothing but the Laplace transform of the given function. It is shown to be very relevant in fractional reaction-diffusion problems in physics. Similarly for $b = 0$, it will become the Mellin transform of the function ${}_0F_1(;\beta;\delta t)$.

The q-analogue of generalized gamma Bessel density can also be deduced from the pathway fractional integral operator, by putting $x = 1, \eta = 1$ and replacing $f(t)$ by $t^{\beta-1}{}_0F_1(;\beta;\delta t)$, then, $g_q(t)$ will be:

$$g_q(t) = \begin{cases} Kt^{\beta-1}[1 - b(1-q)t]^{\frac{1}{1-q}} {}_0F_1(;\beta;\delta t); & q < 1, 1 - b(1-q)t > 0, t > 0, \beta, b > 0 \\ 0; & \text{otherwise} \end{cases} \quad (12)$$

where K is the normalizing constant. For fixed values of b and β, we can look at the graphs for $\delta = -0.5$, $q < 1$, $\delta = 0.5, q < 1$, as well as for $\delta = 0, q < 1$. From Figures 2 and 3, we can see that when q moves from -1 to one, the curve becomes thicker tailed and less peaked. It is also observed that when $\delta > 0$, the right tail of the density becomes thicker and thicker. Similarly, when $\delta < 0$, the right tail gets thinner and thinner. Observe that for $q > 1$, writing $1 - q = -(q-1)$ in Equation (11) produces the extended Type 2 beta form, which is given by:

$$f_q(t) = \begin{cases} Pt^{\beta-1}[1 + b(q-1)t]^{-\frac{1}{q-1}} {}_0F_1(;\beta;\delta t); & q > 1, t > 0, \beta, b > 0 \\ 0; & \text{otherwise} \end{cases} \quad (13)$$

where P is the normalizing constant. From Figure 4, we can see that when q moves from one to ∞, the curve becomes less peaked. In this case, also, it is observed that when $\delta > 0$, the right tail of the density becomes thicker and thicker, and when $\delta < 0$, the right tail gets thinner and thinner.

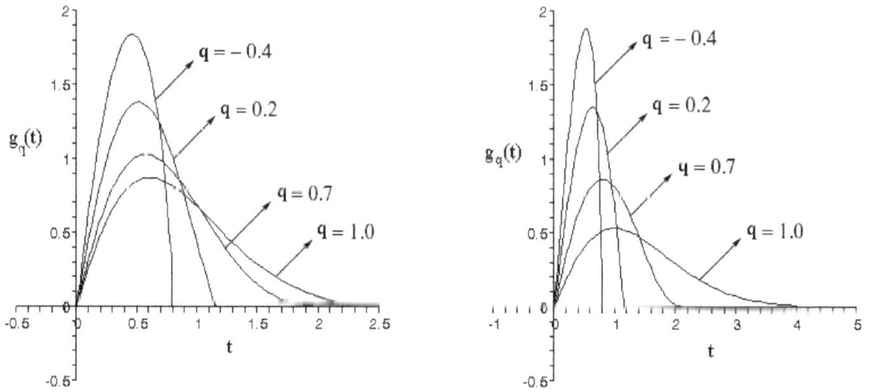

Figure 2. (**a**) The q-gamma Bessel model for $q < 1, \delta = -0.50$; (**b**) The q-gamma Bessel model for $q < 1, \delta = 0.5$.

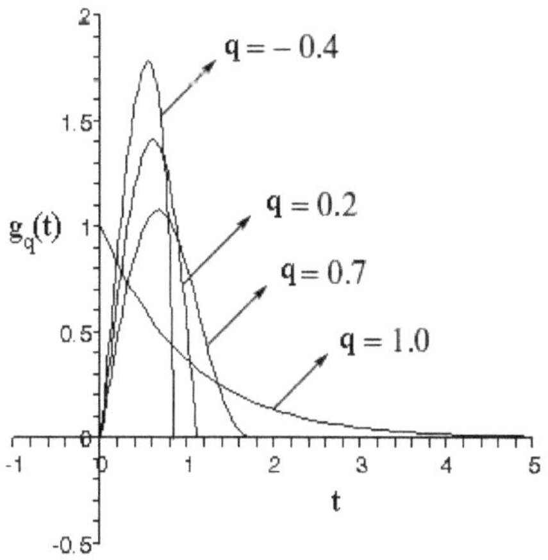

Figure 3. The q-gamma Bessel model for $q < 1, \delta = 0$.

Densities exhibiting thicker or thinner tails occur frequently in many different areas of science. For practical purposes of analyzing data from physical experiments and in building up models in statistics, we frequently select a member from a parametric family of distributions. However, it is often found that the model requires a distribution with a thicker or thinner tail than the ones available from the parametric family.

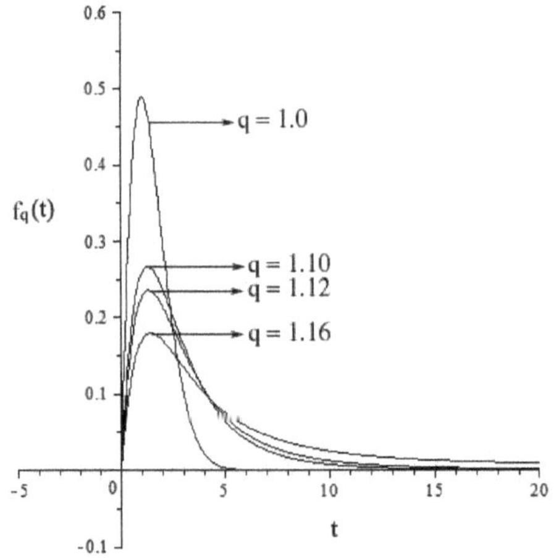

(a)

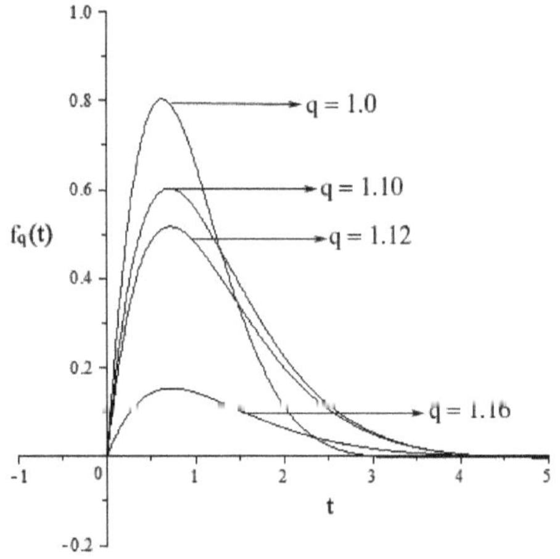

(b)

Figure 4. (**a**) q-gamma Bessel model for $\delta = 0.5, q > 1$; (**b**) q-gamma Bessel model for $\delta = -0.5, q > 1$.

4. Applications in Statistical Mechanics

Nonequilibrium complex systems often exhibit dynamics that can be decomposed into several dynamics on different time scales. As a simple example, consider a Brownian motion of a particle moving through a changing fluid environment, characterized by temperature variations on a large scale. In this case, two dynamics are relevant: one is a fast dynamics describing the local motion of the Brownian particle, and the other one is a slow one due to the large global variations of the environment with spatio-temporal inhomogeneities. These effects produce a superposition of two different statistics, which is referred to as superstatistics. The concept of superstatistics has been introduced by [21,22] after some preliminary considerations in [23,24]. The stationary distributions of superstatistical systems typically exhibit a non-Gaussian behavior with fat tails, which can decay, for example, as a power law, a stretched-exponential law or in an even more complicated way [25]. Essential for this approach is the existence of an intensive variable, say β, which fluctuates on a large spatio-temporal scale.

For the above-mentioned example of a superstatistical Brownian particle, β is the fluctuating inverse temperature of the environment. In general, however, β may also be an effective friction constant, a changing mass parameter, a variable noise strength, the fluctuating energy dissipation in turbulent flows, a fluctuating volatility in finance, an environmental parameter for biological systems, a local variance parameter extracted from a signal, and so on. Superstatistics offers a very general framework for treating nonequilibrium stationary states of such complex systems. After the original work in [21], much effort has been made for further theoretical elaboration; see [26,27]. At the same time, it has also been applied successfully to a variety of systems and phenomena, including hydrodynamic turbulence, pattern formation, cosmic rays, mathematical finance, random matrices and hydro-climatic fluctuations.

From a statistical point of view, the procedure is equivalent to starting with a conditional distribution of a gamma type for every given value of a parameter a. Then, a is assumed to have a prior known density of the gamma type. Then, the unconditional density is obtained by integrating out over the density of a. Let us consider the conditional density of the form:

$$f_{x|a}(x|a) = k_1 x^{\gamma-1} e^{-ax^{\rho}} {}_0F_1(;\frac{\gamma}{\rho};\delta x^{\rho}); 0 \le x < \infty, \rho, a, \gamma > 0 \tag{14}$$

and $f(x) = 0$ elsewhere, where k_1 is the normalizing constant. When $\delta = 0$, Equation (14) reduces to generalized gamma density. Note that this is the generalization of some standard statistical densities, such as gamma, Weibull, exponential, Maxwell–Boltzmann, Rayleigh and many more. When we put $\rho = 1$ in Equation (14), it reduces to Equation (11). When $\delta = 0, \rho = 2$, Equation (14) reduces to folded standard normal density.

Suppose that a has a gamma density given by:

$$f_a(a) = \frac{\lambda^{\eta} a^{\eta-1} e^{-\lambda a}}{\Gamma(\eta)}; 0 \leq a < \infty, \eta, \lambda > 0 \tag{15}$$

and $f_a(a) = 0$ elsewhere. In a physical problem, the residual rate of change may have small probabilities of it being too large or too small, and the maximum probability may be for a medium range of values for the residual rate of change a. This is a reasonable assumption. Then, the unconditional density of x is given by:

$$f_x(x) = \int_a f_{x|a}(x|a) f_a(a) \mathrm{d}a = \frac{\rho \lambda^{\eta} x^{\gamma-1}}{\Gamma(\frac{\gamma}{\rho})\Gamma(\eta)} {}_0F_1(;\frac{\gamma}{\rho};\delta x^{\rho}) I_{11} \tag{16}$$

where:

$$I_{11} = \int_0^{\infty} a^{\frac{\gamma}{\rho}+\eta-1} e^{-a(\lambda+x^{\rho})-\frac{\delta}{a}} \mathrm{d}a \tag{17}$$

Note that one form of the inverse Gaussian probability density function is given by:

$$h_1(x) = c x^{-\frac{3}{2}} e^{-\frac{\xi}{2}[\frac{x}{\nu^2}+\frac{1}{x}]}, \nu \neq 0, \xi > 0, x \geq 0$$

where c is the normalizing constant. Put $\frac{\gamma}{\rho} + \eta - 1 = -\frac{3}{2}, \lambda + x^{\rho} = \frac{\xi}{2\nu^2}, \delta = \frac{\xi}{2}$ in I_{11}; we can see that the inverse Gaussian density is the integrand in I_{11}. Hence, I_{11} can be used to evaluate the moments of inverse Gaussian density. Furthermore, I_{11} is the special case of the reaction rate probability integral in nuclear reaction rate theory, Krätzel integrals in applied analysis, *etc.* (see [28–31]). For the evaluation of this integral and for more details, see [19,20]. Hence, we have the unconditional density:

$$f_x(x) = \frac{\rho\lambda^{\eta}}{\Gamma(\frac{\gamma}{\rho})\Gamma(\eta)} \frac{x^{\gamma-1}}{(\lambda+x^{\rho})^{\frac{\gamma}{\rho}+\eta}} {}_0F_1(;\frac{\gamma}{\rho};\delta x^{\rho}) G^{2,0}_{0,2}[\delta(\lambda+x^{\rho})|_{0,\frac{\gamma}{\rho}+\eta}] \tag{18}$$

where the G-function is defined as the following Mellin–Barnes integral:

$$G^{m,n}_{p,q}\left[z|^{a_1,\ldots,a_p}_{b_1,\ldots,b_q}\right] = \frac{1}{2\pi i}\int_{\mathcal{L}} \Phi(s) z^{-s} \mathrm{d}s$$

where:

$$\Phi(s) = \frac{\left\{\prod_{j=1}^{m}\Gamma(b_j+s)\right\}\left\{\prod_{j=1}^{n}\Gamma(1-a_j-s)\right\}}{\left\{\prod_{j=m+1}^{q}\Gamma(1-b_j-s)\right\}\left\{\prod_{j=n+1}^{p}\Gamma(a_j+s)\right\}}$$

with $a_j, j = 1, \ldots, p$ and $b_j, j = 1, \ldots, q$ being complex numbers and $\mathcal{L}$ a contour separating the poles of $\Gamma(b_j+s), j = 1, \ldots, m$ from those of $\Gamma(1-a_j-s), j = 1, \ldots, n$. Convergence conditions, properties and applications of the G-function in various disciplines are available in the literature. For rexample, see [7]. Equation (18) is a superstatistics, in the sense of superimposing another distribution or the distribution of x with superimposed distribution of the parameter a. In a physical problem, the parameter may be something like temperature having its own distribution. Several physical interpretations of superstatistics are available from the papers of Beck and others.

We can easily obtain the series representation of the unconditional density Equation (18), given by:

$$
\begin{aligned}
f_x(x) &= \frac{\rho\lambda^{\eta}}{\Gamma(\frac{\gamma}{\rho})\Gamma(\eta)} \frac{x^{\gamma-1}}{(\lambda+x^{\rho})^{\frac{\gamma}{\rho}+\eta}} {}_0F_1(;\tfrac{\gamma}{\rho};\delta x^{\rho}) \sum_{k=0}^{\infty} \frac{\Gamma(\frac{\gamma}{\rho}+\eta-k)(-1)^k[\delta(\lambda+x^{\rho})]^k}{k!} \\
&= \frac{\rho\Gamma(\frac{\gamma}{\rho}+\eta)\lambda^{\eta}}{\Gamma(\frac{\gamma}{\rho})\Gamma(\eta)} \frac{x^{\gamma-1}}{(\lambda+x^{\rho})^{\frac{\gamma}{\rho}+\eta}} {}_0F_1(;\tfrac{\gamma}{\rho};\delta x^{\rho}) {}_0F_1(;1-\tfrac{\gamma}{\rho}-\eta;\delta(\lambda+x^{\rho})) \\
&\quad \lambda,\rho,\eta,\delta>0, \tfrac{\gamma}{\rho}>0, 1-\tfrac{\gamma}{\rho}-\eta \neq -\nu, \nu=0,1,\cdots, x\geq 0
\end{aligned} \tag{19}
$$

This series representation provides an extension of the Beck and Cohen statistic. Thus, Equation (19) gives a suitable interpretation, when tail areas are shifted. This model has wide potential applications in physical sciences, especially in statistical mechanics; see [19,20].

5. Applications in the Growth-Decay Mechanism

If x is replaced by $|x|$ in Equation (3) and when $q \to 1$, the real scalar case of the pathway model takes the form,

$$h_4(x) = c_4|x|^{\gamma-1}e^{-a|x|^{\theta}}, -\infty < x < \infty, a > 0 \tag{20}$$

The density in Equation (20) for $\gamma = 1, \theta = 1$ is the simple Laplace density. For $\gamma = 1$, we have the symmetric Laplace density. A general Laplace density is associated with the concept of the Laplacianness of quadratic and bilinear forms. For the concept of the Laplacianness of bilinear forms, corresponding to the chi-squaredness of quadratic forms, and for other details, see [14,32]. Laplace density is also connected to input-output-type models. Such models can describe many of the phenomena in nature. When two particles react with each other and energy is produced, part of it may be consumed or converted or lost, and what is usually measured is the residual effect. The water storage in a dam at a given instant is the residual effect of the water flowing into the dam minus the amount taken out of the dam. Grain storage in a silo is the input minus the grain taken out. Hence, it is of great importance in modeling this residual effect, and there are many studies on this concept. There are several input-output-type situations in economics, social sciences, industrial production, commercial activities, cosmological studies, and so on. It is shown in [33] that when we have independently-distributed gamma-type input and gamma-type output, the residual part $z = x - y, x =$ input variable, $y =$ output variable, then the special cases of the density of z is a Laplace density. In this case, one can also obtain the asymmetric Laplace and generalized Laplace densities, which are currently used very frequently in stochastic processes, as special cases of the input-output model.

The generalized gamma Bessel model in Equation (11) has the moment generating function:

$$M_x(t) = \frac{b^{\beta_1}}{e^{\frac{\delta_1}{b}}} \frac{e^{\frac{\delta_1}{b-t}}}{(a_1-t)^{\beta_1}}, b-t>0, \beta_1>0$$

Let x and y be two independently-distributed generalized gamma Bessel models having parameters $(\alpha_1, \beta_1, \delta_1)$ and $(\alpha_2, \beta_2, \delta_2)$, respectively, $\alpha_i > 0, \beta_i > 0, \delta_i, i = 1, 2$. Let $z = x - y$. Due to the independence of x and y, the moment-generating function of u is given by:

$$M_z(t) = \frac{\alpha_1{}^{\beta_1}}{e^{\frac{\delta_1}{\alpha_1}}} \frac{e^{\frac{\delta_1}{\alpha_1-t}}}{(\alpha_1-t)^{\beta_1}} \frac{\alpha_2{}^{\beta_2}}{e^{\frac{\delta_2}{\alpha_2}}} \frac{e^{\frac{\delta_2}{\alpha_2+t}}}{(\alpha_2+t)^{\beta_2}}, \alpha_1-t>0, \alpha_2+t>0$$

when $\alpha_1 = \alpha_2 = \alpha, \beta_1 = \beta_2 = \beta, \delta_1 = \delta_2 = \delta = 0$, then the above equation reduces to that of the generalized Laplacian model of Mathai.

6. Applications in Solar Spectral Irradiance Modeling

Any object with a temperature above absolute zero emits radiation. The Sun, our singular source of renewable energy, sits at the center of the solar system and emits energy as electromagnetic radiation at an extremely large and relatively constant rate, 24 h per day, 365 days of the year. With an effective temperature of approximately 6000 K, the Sun emits radiation over a wide range of wavelengths, commonly labeled from high energy shorter wavelengths to lower energy longer wavelengths as gamma ray, X-ray, ultraviolet, visible, infrared and radio waves. These are called spectral regions; see Figure 5. The rate at which solar energy reaches a unit area at the Earth is called the "solar irradiance" or "insolation". The units of measure for irradiance are watts per square meter (W/m^2). Solar irradiance is an instantaneous measure of rate and can vary over time. The units of measure for solar radiation are joules per square meter (J/m^2), but often watt-hours per square meter (Wh/m^2) are used. As will be described above, solar radiation is simply the integration or summation of solar irradiance over a time period. For more details, see [34,35].

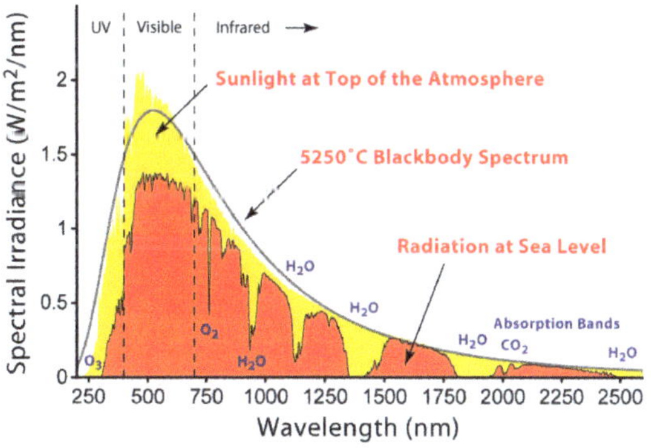

Figure 5. Solar irradiance spectrum above the atmosphere and at the surface prepared by Robert A. Rohde (used by copyright from http://www.globalwarmingart.com/wiki/File:Solar-Spectrum-png).

Good quality, reliable solar radiation data are becoming increasingly important in the field of renewable energy, with regard to both photovoltaic and thermal systems. It helps well-founded decision making on activities, such as research and development, production quality control, determination of optimum locations, monitoring the efficiency of installed systems and predicting the system output under various sky conditions. Especially with larger solar power plants, errors of a few percent can significantly impact the return on investment. Scientists studying climate change are interested in understanding the effects of variations in the total and spectral solar irradiance on Earth and its climate.

A recent set of typical meteorological year (TMY) datasets for the United States, called TMY2 datasets, has been derived from the 30-year historical National Solar Radiation Data Base. In 2000, the American Society for Testing and Materials developed an AM0 reference spectrum (ASTM E-490) for use by the aerospace community. That ASTM E490 Air Mass Zero solar spectral irradiance is based on data from satellites, space shuttle missions, high-altitude aircraft, rocket soundings, ground-based solar telescopes and modeled spectral irradiance. Our dataset consists of 1522 observations collected from http://rredc.nrel.gov/solar/spectra/am0/. Here, mathematical software MAPLE and MATLAB are used for the data analysis. The model considered here is the density function given in Equation (11). In many situations, the gamma model is used to model the spectral density. Figure 6 is the histogram of the data embedded with gamma and our new probability models. We have not specified any parameters here to plot the function. The same program generated the two different graphs as shown below. We calculated the Kolmogorov–Smirnov test statistic for the two different probability models.

For gamma density, the value of the statistic is obtained as 0.11139, and for our new probability model, the value is 0.10808. From the table, the value obtained is 0.410. We can see that the two different probability models are consistent with the data. However, the distance measure of the statistic of our new probability model is less than the other probability model, and hence, our model is better fit to the data than the other one.

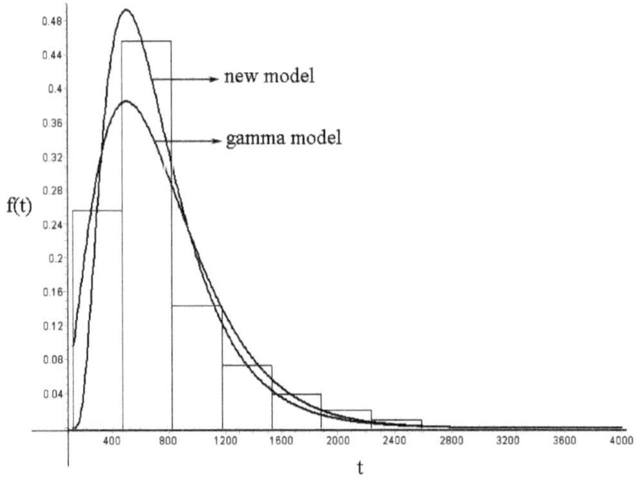

Figure 6. The graph of the histogram embedded with the probability models.

Acknowledgments: The author acknowledges gratefully the encouragement given by Arak M. Mathai, Department of Mathematics and Statistics, McGill University, Montreal H3A 2K6, Canada.

Conflicts of Interest: The authors declare no conflict of interest.

References

1. Haubold, H.J.; Mathai, A.M. The fractional kinetic equation and thermonuclear functions. *Astrophys. Space Sci.* **2000**, *273*, 53–63. [CrossRef]
2. Henry, B.I.; Wearne, S.L. Existence of Turing instabilities in a two-species fractional reaction-diffusion system. *SIAM J. Appl. Math.* **2002**, *62*, 870–887. [CrossRef]
3. Kilbas, A.A.; Saigo, M.; Saxena, R.K. Generalized Mittag-Leffler function and generalized fractional calculus operators. *Integral Transform. Special Funct.* **2004**, *15*, 31–49. [CrossRef]
4. Mainardi, F.; Luchko, Y.; Pagnini, G. The fundamental solution of the space-time fractional diffusion equation. *Fract. Calc. Appl. Anal.* **2001**, *4*, 153–192.
5. Mathai, A.M.; Haubold, H.J. Pathway model, superstatistics, Tsallis statistics and a generalized measure of entropy. *Phys. A* **2007**, *375*, 110–122. [CrossRef]
6. Saxena, R.K.; Mathai, A.M.; Haubold, H.J. Fractional reaction-diffusion equations. *Astrophys. Space Sci.* **2006**, *305*, 289–296. [CrossRef]
7. Kilbas, A.A.; Saigo, M. *H-Transforms: Theory and Applications*; CRC Press: New York, NY, USA, 2004.
8. Kiryakova, V. *Generalized Fractional Calculus and Applications*; CRC Press: New York, NY, USA, 1994.
9. Miller, K.S.; Ross, B. *An Introduction to the Fractional Calculus and Fractional Differential Equations*; Wiley: New York, NY, USA, 1993.
10. Podlubny, I. *Fractional Differential Equations*; Academic Press: San Diego, CA, USA, 1999.
11. Samko, S.G.; Kilbas, A.A.; Marichev, O.I. *Fractional Integrals and Derivatives: Theory and Applications*; Gordon and Breach: New York, NY, USA, 1990.
12. Mathai, A.M. A pathway to matrix-variate gamma and normal densities. *Linear Algebra Its Appl.* **2005**, *396*, 317–328. [CrossRef]

13. Honerkamp, J. *Stochastic Dynamical Systems: Concepts, Numerical Methods, Data Analysis*; VCH Publishers: New York, NY, USA, 1994.
14. Mathai, A.M. On non-central generalized Laplacianness of quadratic forms in normal variables. *J. Multivar. Anal.* **1993**, *45*, 239–246. [CrossRef]
15. Nair, S.S. Pathway fractinal integration operator. *Fract. Calc. Appl. Anal.* **2009**, *12*, 237–252.
16. Mathai, A.M. Fractional integrals in the matrix-variate cases and connection to statistical distributions. *Integral Transform. Special Funct.* **2009**, *20*, 871–882. [CrossRef]
17. Nair, S.S. Pathway fractional integral operator and matrix-variate functions. *Integral Transform. Special Funct.* **2011**, *22*, 233–244. [CrossRef]
18. Sebastian, N. Some statistical aspects of fractional calculus. *J. Kerala Stat. Assoc.* **2009**, *20*, 23–33.
19. Sebastian, N. A generalized gamma model associated with Bessel function and its applications in statistical mechanics. In Proceedings of AMADE-09, Proceedings of Institute of Mathematics, Natinal Academy of Sciences, Minsk, Belarus; 2009; pp. 114–119.
20. Sebastian, N. A generalized gamma model associated with a Bessel function. *Integral Transform. Special Funct.* **2011**, *22*, 631–645. [CrossRef]
21. Beck, C.; Cohen, E.G.D. Superstatistics. *Phys. A* **2003**, *322*, 267–275. [CrossRef]
22. Beck, C. Stretched exponentials from superstatistics. *Phys. A* **2006**, *365*, 96–101. [CrossRef]
23. Beck, C. Dynamical Foundations of Nonextensive Statistical Mechanics. *Phys. Rev. Lett.* **2001**, *87*, 180601. [CrossRef]
24. Beck, C. Non-additivity of Tsallis entropies and fluctuations of temperature. *Europhys. Lett.* **2002**, *57*, 329–333. [CrossRef]
25. Touchette, H. *Nonextensive Entropy-Interdisciplinary Applications*; Gell-Mann, M., Tsallis, C., Eds.; Oxford University Press: Oxford, UK, 2004.
26. Beck, C. Comment on critique of q-entropy for thermal statistics. *Phys. Rev. E* **2004**, *69*, 038101.
27. Chavanis, P.H. Coarse-grained distributions and superstatistics. *Phys. A* **2006**, *359*, 177–212. [CrossRef]
28. Joseph, D.P. Gamma distribution and extensions using pathway idea. *Stat. Pap.* **2011**, *52*, 309–325. [CrossRef]
29. Krätzel, E. Integral transformations of Bessel type. 1979; 148–155.
30. Mathai, A.M.; Haubold, H.J. *Mordern Problems in Nuclear and Neutrino Astrophysics*; Akademie-Verlag: Berlin, Germany, 1988.
31. Mathai, A.M. A Versatile Integral. CMS Project *SR*/*S4*/*MS*:287/05. *Preprint*. 2007; 9.
32. Mathai, A.M.; Provost, S.B.; Hayakawa, T. *Bilinear Forms and Zonal Polynomials*; Lecture notes in Statistics, No. 102; Springer-Verlag: New York, MY, USA, 1995.
33. Mathai, A.M. The residual effect of growth-decay mechanism and the distributions of covariance structure. *Can. J. Stat.* **1993**, *21*, 277–283. [CrossRef]
34. Gueymard, C. The sun's total and spectral irradiance for solar energy applications and solar radiation models. *Sol. Energy* **2004**, *76*, 423–453. [CrossRef]
35. Stoffel, T.; Renné, D.; Myers, D.; Wilcox, S.; Sengupta, M.; George, R.; Turchi, C. Concentrating Solar Power. Available online: http://www.nrel.gov/docs/fy10osti/47465.pdf (accessed on 13 August 2015).

axioms

MDPI

Article

Multivariate Extended Gamma Distribution

Dhannya P. Joseph

Department of Statistics, Kuriakose Elias College, Mannanam, Kottayam, Kerala 686561, India; dhannyapj@gmail.com; Tel.: +91-9400-733-065

Academic Editor: Hans J. Haubold

Received: 7 June 2016; Accepted: 13 April 2017; Published: 24 April 2017

Abstract: In this paper, I consider multivariate analogues of the extended gamma density, which will provide multivariate extensions to Tsallis statistics and superstatistics. By making use of the pathway parameter β, multivariate generalized gamma density can be obtained from the model considered here. Some of its special cases and limiting cases are also mentioned. Conditional density, best predictor function, regression theory, etc., connected with this model are also introduced.

Keywords: pathway model; multivariate extended gamma density; moments

1. Introduction

Consider the generalized gamma density of the form

$$g(x) = c_1 x^{\gamma} e^{-ax^{\delta}},\ x \geq 0,\ a > 0,\ \delta > 0,\ \gamma + 1 > 0, \tag{1}$$

where $c_1 = \frac{\delta a^{\frac{\gamma+1}{\delta}}}{\Gamma(\frac{\gamma+1}{\delta})}$, is the normalizing constant. Note that this is the generalization of some standard statistical densities such as gamma, Weibull, exponential, Maxwell-Boltzmann, Rayleigh and many more. We will extend the generalized gamma density by using pathway model of [1] and we get the extended function as

$$g_1(x) = c_2 x^{\gamma}[1 + a(\beta - 1)x^{\delta}]^{-\frac{1}{\beta-1}},\ x \geq 0,\ \beta > 1,\ a > 0,\ \delta > 0 \tag{2}$$

where $c_2 = \frac{\delta (a(\beta-1))^{\frac{\gamma+1}{\delta}} \Gamma(\frac{1}{\beta-1})}{\Gamma(\frac{\gamma+1}{\delta})\Gamma(\frac{1}{\beta-1} - \frac{\gamma+1}{\delta})}$, is the normalizing constant.

Note that $g_1(x)$ is a generalized type-2 beta model. Also $\lim_{\beta \to 1} g_1(x) = g(x)$, so that it can be considered to be an extended form of $g(x)$. For various values of the pathway parameter β a path is created so that one can see the movement of the function denoted by $g_1(x)$ above towards a generalized gamma density. From the Figure 1 we can see that, as β moves away from 1 the function $g_1(x)$ moves away from the origin and it becomes thicker tailed and less peaked. From the path created by β we note that we obtain densities with thicker or thinner tail compared to generalized gamma density. Observe that for $\beta < 1$, writing $\beta - 1 = -(1 - \beta)$ in Equation (2) produce generalized type-1 beta form, which is given by

$$g_2(x) = c_3 x^{\gamma}[1 - a(1 - \beta)x^{\delta}]^{\frac{1}{1-\beta}},\ 1 - a(1-\beta)x^{\delta} \geq 0,\ \beta < 1,\ a > 0,\ \delta > 0$$

where $c_3 = \frac{\delta(a(1-\beta))^{\frac{\gamma+1}{\delta}}\Gamma(\frac{1}{1-\beta}+1+\frac{\gamma+1}{\delta})}{\Gamma(\frac{\gamma+1}{\delta})\Gamma(\frac{1}{1-\beta}+1)}$, is the normalizing constant (see [2]).

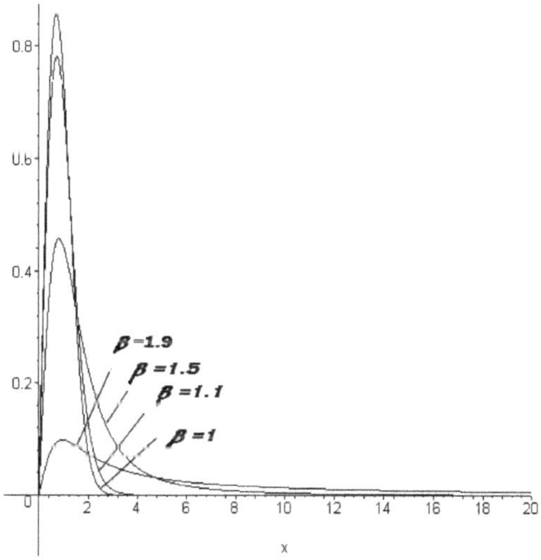

Figure 1. The graph of $g_1(x)$, for $\gamma = 1, a = 1,\ \delta = 2,\ \eta = 1$ and for various values of β.

From the above graph, one can see the movement of the extended gamma density denoted by $g_1(x)$ towards the generalized gamma density, for various values of the pathway parameter β. Beck and Cohen's superstatistics belong to the case (2) [3,4]. For $\gamma = 1,\ a = 1,\ \delta = 1$ we have Tsallis statistics [5,6] for $\beta > 1$ from (2).

Several multivariate extensions of the univariate gamma distributions exist in the literature [7–9]. In this paper we consider a multivariate analogue of the extended gamma density (2) and some of its properties.

2. Multivariate Extended Gamma

Various multivarite generalizatons of pathway model are discussed in the papers of Mathai [10,11]. Here we consider the multivariate case of the extended gamma density of the form (2). For $X_i \geq 0,\ i = 1,\ 2, \ldots, n$, let

$$\begin{aligned} f_\beta(x_1, x_2, \ldots, x_n) &= k_\beta x_1^{\gamma_1} x_2^{\gamma_2} \ldots x_n^{\gamma_n} [1 + (\beta - 1)(a_1 x_1^{\delta_1} + a_2 x_2^{\delta_2} + \ldots + a_n x_n^{\delta_n})]^{-\frac{\eta}{\beta-1}}, \\ &\quad \beta > 1,\ \eta > 0,\ \delta_i > 0,\ a_i > 0,\ i = 1,\ 2, \ldots, n, \end{aligned} \tag{3}$$

where k_β is the normalizing constant, which will be given later. This multivariate analogue can also produce multivariate extensions to Tsallis statistics [5,12] and superstatistics [3]. Here the variables are not independently distributed, but when $\beta \to 1$ we have a result that $X_1, X_2, \ldots, X_n$ will become independently distributed generalized gamma variables. That is,

$$\begin{aligned} \lim_{\beta \to 1} f_\beta(x_1, x_2, \ldots, x_n) &= f(x_1, x_2, \ldots, x_n) \\ &= k x_1^{\gamma_1} x_2^{\gamma_2} \ldots x_n^{\gamma_n} \mathrm{e}^{-b_1 x_1^{\delta_1} - \ldots - b_n x_n^{\delta_n}}, \\ &\quad x_i \geq 0,\ b_i = \eta a_i > 0,\ \delta_i > 0,\ i = 1, 2, \ldots, n, \end{aligned} \tag{4}$$

where $k = \prod_{i=1}^{n} \frac{\delta_i b_i^{\frac{\gamma_i+1}{\delta_i}}}{\Gamma(\frac{\gamma_i+1}{\delta_i})}$, $\gamma_i + 1 > 0$, $i = 1, 2, \ldots, n$.

The following are the graphs of 2-variate extended gamma with $\gamma_1 = 1$, $\gamma_2 = 1$, $a_1 = 1$, $a_2 = 1$, $\delta_1 = 2$, $\delta_2 = 2$ and for various values of the pathway parameter β. From the Figures 2–4, we can see the effect of the pathway parameter β in the model.

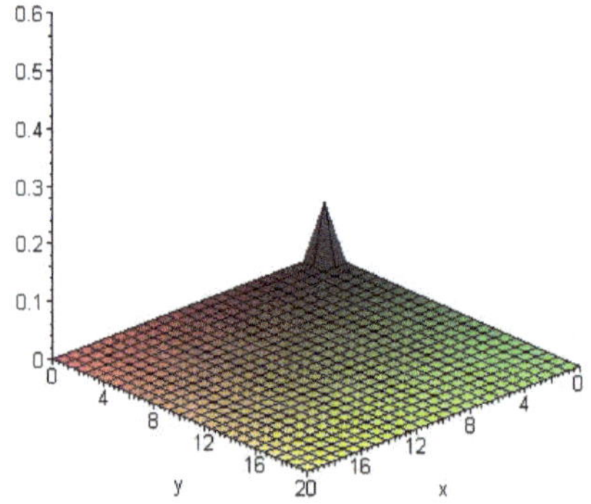

Figure 2. $\beta = 1.2$.

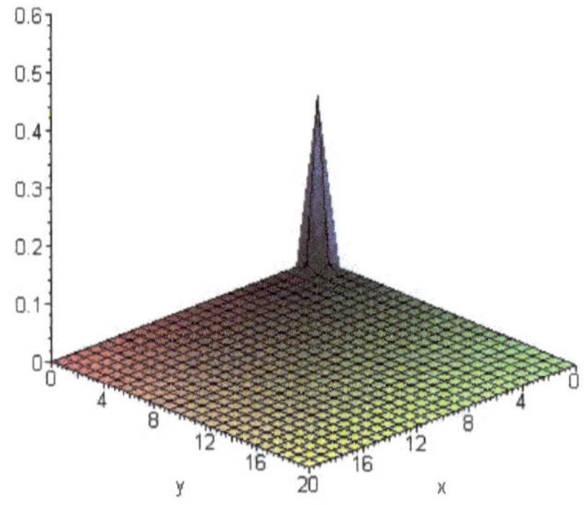

Figure 3. $\beta = 1.5$.

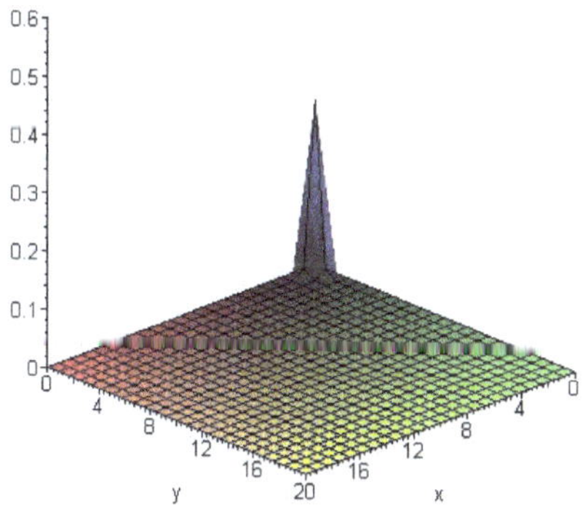

Figure 4. $\beta = 2$.

Special Cases and Limiting Cases

1. When $\beta \to 1$, (3) will become independently distributed generalized gamma variables. This includes multivariate analogue of gamma, exponential, chisquare, Weibull, Maxwell Boltzmann, Rayleigh, and related models.
2. If $n = 1$, $a_1 = 1$, $\delta_1 = 1$, $\beta = 2$, (3) is identical with type-2 beta density.
3. If $\beta = 2$, $a_1 = a_2 = \ldots = a_n = 1$, $\delta_1 = \delta_2 = \ldots = \delta_n = 1$ in (3), then (3) becomes the type-2 Dirichlet density,

$$D(x_1, x_2, \ldots, x_n) = dx_1^{\nu_1 - 1} x_2^{\nu_2 - 1} \ldots x_n^{\nu_n - 1} [1 + x_1 + x_2 + \ldots + x_n]^{-(\nu_1 + \ldots + \nu_{n+1})}, \; x_i \geq 0, \quad (5)$$

where $\nu_i = \gamma_i + 1$, $i = 1, 2, \ldots, n$, $\nu_{n+1} = \eta - (\nu_1 + \ldots + \nu_n)$ and d is the normalizing constant (see [13,14]).

A sample of the surface for $n = 2$ is given in the Figure 5.

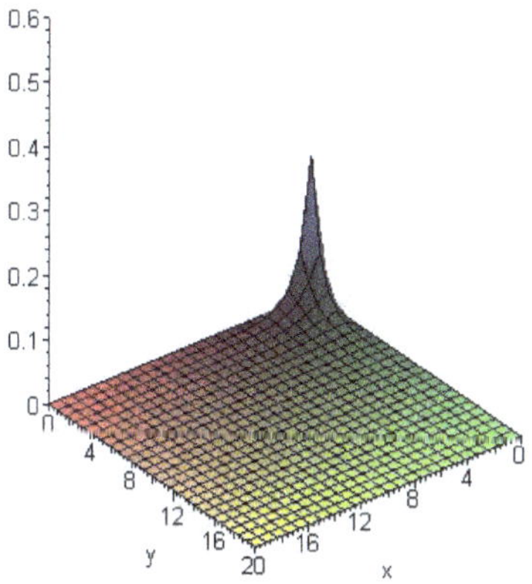

Figure 5. The graph of bivariate type-2 Dirichlet with $\gamma_1 = \gamma_2 = 1$, $\eta = 6$.

3. Marginal Density

We can find the marginal density of X_i, by integrating out $X_1,\ X_2, \ldots, X_{i-1},\ X_{i+1}, \ldots, X_{n-1},\ X_n$. First let us integrate out X_n, then the joint density of $X_1,\ X_2, \ldots, X_{n-1}$ denoted by f_1 is given by

$$\begin{aligned} f_1(x_1,\ x_2, \ldots, x_{n-1}) &= \int_{x_n>0} f_\beta(x_1, x_2, \ldots, x_n)\mathrm{d}x_n \\ &= k_\beta x_1^{\gamma_1} x_2^{\gamma_2} \ldots x_{n-1}^{\gamma_{n-1}}[1+(\beta-1)(a_1 x_1^{\delta_1}+\ldots+a_{n-1}x_{n-1}^{\delta_{n-1}})]^{-\frac{\eta}{\beta-1}} \\ &\quad \times \int_{x_n} x_n^{\gamma_n}[1+Cx_n^{\delta_n}]^{-\frac{\eta}{\beta-1}}\mathrm{d}x_n, \end{aligned} \tag{6}$$

where $C = \frac{(\beta-1)a_n}{[1+(\beta-1)(a_1 x_1^{\delta_1}+\ldots+a_{n-1}x_{n-1}^{\delta_{n-1}})]}$. By putting $y = Cx_n^{\delta_n}$ and integrating we get

$$\begin{aligned} f_1(x_1,\ x_2, \ldots, x_{n-1}) &= \frac{k_\beta \Gamma(\frac{\eta}{\beta-1} - \frac{\gamma_n+1}{\delta_n})\Gamma(\frac{\gamma_n+1}{\delta_n})}{\delta_n[a_n(\beta-1)]^{\frac{\gamma_n+1}{\delta_n}}\Gamma(\frac{\eta}{\beta-1})} x_1^{\gamma_1}x_2^{\gamma_2}\ldots x_{n-1}^{\gamma_{n-1}} \\ &\quad \times [1+(\beta-1)(a_1x_1^{\delta_1}+a_2x_2^{\delta_2}+\ldots+a_{n-1}x_{n-1}^{\delta_{n-1}})]^{-\left[\frac{\eta}{\beta-1}-\frac{\gamma_n+1}{\delta_n}\right]}, \end{aligned} \tag{7}$$

$x_i \geq 0,\ i = 1,2,\ldots,n-1,\ a_i > 0,\ \delta_i > 0,\ i = 1,2,\ldots,n,\ \beta > 1,\ \eta > 0,\ \frac{\eta}{\beta-1} - \frac{\gamma_n+1}{\delta_n} > 0,\ \gamma_n+1 > 0.$

In a similar way we can integrate out $X_1,\ X_2, \ldots, X_{i-1},\ X_{i+1}, \ldots, X_{n-1}$. Then the marginal density of X_i is denoted by f_2 and is given by

$$f_2(x_i) = k_2 x_i^{\gamma_i}[1+(\beta-1)a_i x_i^{\delta_i}]^{-\left[\frac{\eta}{\beta-1}-\frac{\gamma_n+1}{\delta_n}-\ldots-\frac{\gamma_{i-1}+1}{\delta_{i-1}}-\frac{\gamma_{i+1}+1}{\delta_{i+1}}-\ldots\frac{\gamma_1+1}{\delta_1}\right]}, \tag{8}$$

where $x_i \geq 0,\ \beta > 1,\ \delta_i > 0,\ \eta > 0,$

$$k_2 = \frac{\delta_i(a_i(\beta-1))^{\frac{\gamma_i+1}{\delta_i}}\Gamma(\frac{\eta}{\beta-1}-\frac{\gamma_n+1}{\delta_n}-\ldots-\frac{\gamma_{i-1}+1}{\delta_{i-1}}-\frac{\gamma_{i+1}+1}{\delta_{i+1}}-\ldots-\frac{\gamma_1+1}{\delta_1})}{\Gamma(\frac{\gamma_i+1}{\delta_i})\Gamma(\frac{\eta}{\beta-1}-\frac{\gamma_1+1}{\delta_1}-\ldots-\frac{\gamma_n+1}{\delta_n})},$$

$\gamma_i+1 > 0,\ \frac{\eta}{\beta-1} - \frac{\gamma_n+1}{\delta_n} - \ldots - \frac{\gamma_{i-1}+1}{\delta_{i-1}} - \frac{\gamma_{i+1}+1}{\delta_{i+1}} - \ldots - \frac{\gamma_1+1}{\delta_1} > 0,\ \frac{\eta}{\beta-1} - \frac{\gamma_1+1}{\delta_1} - \ldots - \frac{\gamma_n+1}{\delta_n} > 0.$

If we take any subset of $(X_1, \ldots, X_n)$, the marginal densities belong to the same family. In the limiting case they will also become independently distributed generalized gamma variables.

Normalizing Constant

Integrating out X_i from (8) and equating to 1, we will get the normalizing constant k_β as

$$k_\beta = \frac{\delta_1\delta_2\ldots\delta_n(a_1(\beta-1))^{\frac{\gamma_1+1}{\delta_1}}(a_2(\beta-1))^{\frac{\gamma_2+1}{\delta_2}}\ldots(a_n(\beta-1))^{\frac{\gamma_n+1}{\delta_n}}\Gamma(\frac{\eta}{\beta-1})}{\Gamma(\frac{\gamma_1+1}{\delta_1})\Gamma(\frac{\gamma_2+1}{\delta_2})\ldots\Gamma(\frac{\gamma_n+1}{\delta_n})\Gamma(\frac{\eta}{\beta-1}-\frac{\gamma_1+1}{\delta_1}-\ldots-\frac{\gamma_n+1}{\delta_n})}, \tag{9}$$

$\delta_i > 0,\ a_i > 0,\ \gamma_i+1 > 0,\ i = 1,2,\ldots,n,\ \beta > 1,\ \eta > 0,\ \frac{\eta}{\beta-1} - \frac{\gamma_1+1}{\delta_1} - \ldots - \frac{\gamma_n+1}{\delta_n} > 0.$

4. Joint Product Moment and Structural Representations

Let $(X_1, \ldots, X_n)$ have a multivariate extended gamma density (3). By observing the normalizing constant in (23), we can easily obtained the joint product moment for some arbitrary $(h_1, \ldots, h_n)$,

$$E(x_1^{h_1}x_2^{h_2}\dots x_n^{h_n}) = k_\beta \frac{\Gamma(\frac{\gamma_1+h_1+1}{\delta_1})\dots\Gamma(\frac{\gamma_n+h_n+1}{\delta_n})\Gamma(\frac{\eta}{\beta-1}-\frac{\gamma_1+h_1+1}{\delta_1}-\dots-\frac{\gamma_n+h_n+1}{\delta_n})}{\delta_1\delta_2\dots\delta_n(a_1(\beta-1))^{\frac{\gamma_1+h_1+1}{\delta_1}}\dots(a_n(\beta-1))^{\frac{\gamma_n+h_n+1}{\delta_n}}\Gamma(\frac{\eta}{\beta-1})}$$

$$= \frac{\Gamma(\frac{\eta}{\beta-1}-\frac{\gamma_1+h_1+1}{\delta_1}-\dots-\frac{\gamma_n+h_n+1}{\delta_n})\prod_{i=1}^{n}\Gamma(\frac{\gamma_i+h_i+1}{\delta_i})}{\Gamma(\frac{\eta}{\beta-1}-\frac{\gamma_1+1}{\delta_1}-\dots-\frac{\gamma_n+1}{\delta_n})\prod_{i=1}^{n}[a_i(\beta-1)]^{\frac{h_i}{\delta_i}}\Gamma(\frac{\gamma_i+1}{\delta_i})}, \tag{10}$$

$\frac{\eta}{\beta-1}-\sum_{i=1}^{n}\frac{\gamma_i+h_i+1}{\delta_i}>0,\ \gamma_i+h_i+1>0,\ \frac{\eta}{\beta-1}-\sum_{i=1}^{n}\frac{\gamma_i+1}{\delta_i}>0,\ \gamma_i+1>0,\ a_i>0,\ \beta>1,\ \delta_i>0,$ $i=1,2,\dots,n.$

Property 1. *The joint product moment of the multivariate extended gamma density can be written as*

$$E(x_1^{h_1}x_2^{h_2}\dots x_n^{h_n}) = \frac{\Gamma\left(\frac{\eta}{\beta-1}-\sum_{i=1}^{n}\frac{\gamma_i+h_i+1}{\delta_i}\right)}{\Gamma\left(\frac{\eta}{\beta-1}-\sum_{i=1}^{n}\frac{\gamma_i+1}{\delta_i}\right)}\prod_{i=1}^{n}E(y_i^{h_i}), \tag{11}$$

where Y_i' s are generalized gamma random variables having density function

$$f_y(y_i) = c_i y_i^{\gamma_i}e^{-[a_i(\beta-1)y_i]^{\delta_i}},\ y_i\geq 0,\ \beta>1,\ a_i>0,\ \delta_i>0, \tag{12}$$

where $c_i = \frac{\delta_i[a_i(\beta-1)]^{\frac{\gamma_i+1}{\delta_i}}}{\Gamma(\frac{\gamma_i+1}{\delta_i})}$, $\gamma_i+1>0$, $i=1,2,\dots,n$, *is the normalizing constant.*

Property 2. *Letting $h_2=\dots=h_n=0$, in (10), we get*

$$E(x_1^{h_1}) = \frac{\Gamma(\frac{\eta}{\beta-1}-\frac{\gamma_1+h_1+1}{\delta_1}-\frac{\gamma_2+1}{\delta_2}-\dots-\frac{\gamma_n+1}{\delta_n})\Gamma(\frac{\gamma_1+h_1+1}{\delta_1})}{\Gamma(\frac{\eta}{\beta-1}-\frac{\gamma_1+1}{\delta_1}-\dots-\frac{\gamma_n+1}{\delta_n})[a_1(\beta-1)]^{\frac{h_1}{\delta_1}}\Gamma(\frac{\gamma_1+1}{\delta_1})}, \tag{13}$$

$\frac{\eta}{\beta-1}-\frac{\gamma_1+h_1+1}{\delta_1}-\sum_{i=2}^{n}\frac{\gamma_i+1}{\delta_i}>0,\ \frac{\gamma_1+h_1+1}{\delta_1}>0,\ \frac{\eta}{\beta-1}-\sum_{i=1}^{n}\frac{\gamma_i+1}{\delta_i}>0,\ \gamma_1+1>0,\ a_1>0,$ $\beta>1,\ \delta_1>0$. *(13) is the h_1^{th} moment of a random variable with density function of the the form (8),*

$$f_3(x_1) = k_3 x_1^{\gamma_1}[1+(\beta-1)a_1x_1^{\delta_1}]^{-\left[\frac{\eta}{\beta-1}-\frac{\gamma_2+1}{\delta_2}-\dots-\frac{\gamma_n+1}{\delta_n}\right]}, \tag{14}$$

where k_3 is the normalizing constant. Then

$$E(x_1^{h_1}) = k_3\int_0^\infty x_1^{\gamma_1}[1+a_1(\beta-1)x_1^{\delta_1}]^{-[\frac{\eta}{\beta-1}-\frac{\gamma_2+1}{\delta_2}-\dots-\frac{\gamma_n+1}{\delta_n}]}dx_1 \tag{15}$$

Making the substitution $y=a_1(\beta-1)x_1^{\delta_1}$, then it will be in the form of a type-2 beta density and we can easily obtained the h_1^{th} moment as in (13).

Property 3. *Letting $h_3 = \ldots = h_n = 0$, in (10), we get*

$$E(x_1^{h_1} x_2^{h_2}) = \frac{\Gamma(\frac{\eta}{\beta-1} - \frac{\gamma_1+h_1+1}{\delta_1} - \frac{\gamma_2+h_2+1}{\delta_2} - \ldots - \frac{\gamma_n+1}{\delta_n})\Gamma(\frac{\gamma_1+h_1+1}{\delta_1})\Gamma(\frac{\gamma_2+h_2+1}{\delta_2})}{\Gamma(\frac{\eta}{\beta-1} - \frac{\gamma_1+1}{\delta_1} - \ldots - \frac{\gamma_n+1}{\delta_n})[a_1(\beta-1)]^{\frac{h_1}{\delta_1}}[a_2(\beta-1)]^{\frac{h_2}{\delta_2}}\Gamma(\frac{\gamma_1+1}{\delta_1})\Gamma(\frac{\gamma_2+1}{\delta_2})}, \quad (16)$$

$\frac{\eta}{\beta-1} - \frac{\gamma_1+h_1+1}{\delta_1} - \frac{\gamma_2+h_2+1}{\delta_2} - \sum_{i=3}^{n} \frac{\gamma_i+1}{\delta_i} > 0,\ \frac{\eta}{\beta-1} - \sum_{i=1}^{n} \frac{\gamma_i+1}{\delta_i} > 0,\ \beta > 1,\ \gamma_i + h_i + 1 > 0,$ $\gamma_i + 1 > 0,\ a_i > 0, \delta_i > 0,\ i = 1,2,$ *which is the joint product moment of a bivariate extended gamma density is denoted by f_4 and is given by*

$$f_4(x_1x_2) = k_4 x_1^{\gamma_1} x_2^{\gamma_2}[1 + (\beta-1)(a_1x_1^{\delta_1} + a_2x_2^{\delta_2})]^{-\left[\frac{\eta}{\beta-1} - \frac{\gamma_3+1}{\delta_3} - \ldots - \frac{\gamma_n+1}{\delta_n}\right]}, \quad (17)$$

where k_4 is the normalizing constant. (17) is obtained by integrating out $X_3, \ldots, X_n$ from (3). By putting $h_4 = \ldots = h_n = 0$, in (10), we get the joint product moment of trivariate extended gamma density and so on.

Theorem 1. *When $X_1, \ldots, X_n$ has density in (3), then*

$$\begin{aligned} & E\{x_1^{h_1} \ldots x_n^{h_n}[1 + (\beta-1)(a_1x_1^{\delta_1} + \ldots + a_nx_n^{\delta_n})]^{h'}\} \\ &= k_\beta \frac{\Gamma(\frac{\gamma_1+h_1+1}{\delta_1}) \ldots \Gamma(\frac{\gamma_n+h_n+1}{\delta_n})\Gamma(\frac{\eta}{\beta-1} - h' - \frac{\gamma_1+h_1+1}{\delta_1} - \ldots - \frac{\gamma_n+h_n+1}{\delta_n})}{\delta_1\delta_2\ldots\delta_n(a_1(\beta-1))^{\frac{\gamma_1+h_1+1}{\delta_1}} \ldots (a_n(\beta-1))^{\frac{\gamma_n+h_n+1}{\delta_n}}\Gamma(\frac{\eta}{\beta-1} - h')} \\ &= \frac{\Gamma(\frac{\eta}{\beta-1})\Gamma(\frac{\eta}{\beta-1} - h' - \frac{\gamma_1+h_1+1}{\delta_1} - \ldots - \frac{\gamma_n+h_n+1}{\delta_n})\prod_{i=1}^{n}\Gamma(\frac{\gamma_i+h_i+1}{\delta_i})}{\Gamma(\frac{\eta}{\beta-1} - \frac{\gamma_1+1}{\delta_1} - \ldots - \frac{\gamma_n+1}{\delta_n})\Gamma(\frac{\eta}{\beta-1} - h')\prod_{i=1}^{n}\left\{[a_i(\beta-1)]^{\frac{h_i}{\delta_i}}\Gamma(\frac{\gamma_i+1}{\delta_i})\right\}}, \end{aligned} \quad (18)$$

$\frac{\eta}{\beta-1} - h' - \sum_{i=1}^{n} \frac{\gamma_i+h_i+1}{\delta_i} > 0,\ \gamma_i + h_i + 1 > 0,\ \frac{\eta}{\beta-1} - \sum_{i=1}^{n} \frac{\gamma_i+1}{\delta_i} > 0,\ \gamma_i + 1 > 0,\ a_i > 0,\ \beta > 1,$ $\eta > 0, \delta_i > 0, i = 1,2,\ldots,n.$

Corollary 1. *When $X_1, \ldots, X_n$ has density in (3), then*

$$E\{[1 + (\beta-1)(a_1x_1^{\delta_1} + \ldots + a_nx_n^{\delta_n})]^{h'}\} = \frac{\Gamma(\frac{\eta}{\beta-1})\Gamma(\frac{\eta}{\beta-1} - h' - \frac{\gamma_1+h_1+1}{\delta_1} - \ldots - \frac{\gamma_n+h_n+1}{\delta_n})}{\Gamma(\frac{\eta}{\beta-1} - \frac{\gamma_1+1}{\delta_1} - \ldots - \frac{\gamma_n+1}{\delta_n})\Gamma(\frac{\eta}{\beta-1} - h')}, \quad (19)$$

$$\frac{\eta}{\beta-1} - h' - \sum_{i=1}^{n} \frac{\gamma_i+h_i+1}{\delta_i} > 0,\ \frac{\eta}{\beta-1} - h' > 0,\ \frac{\eta}{\beta-1} - \sum_{i=1}^{n} \frac{\gamma_i+1}{\delta_i} > 0,\ a_i > 0,\ \beta > 1, \eta > 0,$$
$$\delta_i > 0,\ i = 1,2,\ldots,n.$$

4.1. Variance-Covariance Matrix

Let X be a $n \times 1$ vector. Variance-covariance matrix is obtained by taking $E[(X - E(X))(X - E(X))']$. Then the elements will be of the form

$$E[(X-E(X))(X-E(X))'] = \begin{bmatrix} \mathrm{Var}(x_1) & \mathrm{Cov}(x_1,x_2) & \dots & \mathrm{Cov}(x_1,x_n) \\ \mathrm{Cov}(x_2,x_1) & \mathrm{Var}(x_2) & \dots & \mathrm{Cov}(x_2,x_n) \\ \vdots & & \ddots & \\ \mathrm{Cov}(x_n,x_1) & \mathrm{Cov}(x_n,x_2) & \dots & \mathrm{Var}(x_n) \end{bmatrix}$$

where

$$\mathrm{Cov}(x_i,x_j) = E(x_ix_j) - E(x_i)E(x_j),\ i,\ j = 1,2,\dots,n,\ i \neq j \tag{20}$$

and

$$\mathrm{Var}(x_i) = E(x_i^2) - [E(x_i)]^2,\ i = 1,2,\dots,n. \tag{21}$$

$E(x_ix_j)$'s are obtained from (10) by putting $h_i = h_j = 1$ and all other $h_k = 0,\ k = 1,2,\dots,n,$ $k \neq i,\ j$. $E(x_i)$'s and $E(x_i^2)$'s are respectively obtained from (10) by putting $h_i = 1$ and $h_i = 2$ and all other $h_k = 0,\ k = 1,2,\dots,n,\ k \neq i$. Where

$$E(x_1x_2) = \int_0^\infty \int_0^\infty x_1x_2f_2(x_1,x_2)\mathrm{d}x_1\mathrm{d}x_2. \tag{22}$$

4.2. Normalizing Constant

Integrate out x_i from (8) and equate with 1, we will get the normalizing constant K_α as

$$K_\alpha = \frac{\delta_1\delta_2\cdots\delta_n(a_1(\alpha-1))^{\frac{\gamma_1+1}{\delta_1}}(a_2(\alpha-1))^{\frac{\gamma_2+1}{\delta_2}}\cdots(a_n(\alpha-1))^{\frac{\gamma_n+1}{\delta_n}}\Gamma(\frac{\eta}{\alpha-1})}{\Gamma(\frac{\gamma_1+1}{\delta_1})\Gamma(\frac{\gamma_2+1}{\delta_2})\cdots\Gamma(\frac{\gamma_n+1}{\delta_n})\Gamma(\frac{\eta}{\alpha-1}-\frac{\gamma_1+1}{\delta_1}-\dots-\frac{\gamma_n+1}{\delta_n})}. \tag{23}$$

5. Regression Type Models and Limiting Approaches

The conditional density of X_i given $X_1,X_2,\dots,X_{i-1},X_{i+1},\dots,X_n$ is denoted by f_5 and is given by

$$\begin{aligned} f_5(x_i|x_1,x_2,\dots,x_{i-1},x_{i+1},\dots,x_n) &= \frac{f_\beta(x_1,x_2,\dots,x_n)}{f_6(x_1,x_2,\dots,x_{i-1},x_{i+1},\dots,x_n)} \\ &= \frac{\delta_i[a_i(\beta-1)]^{\frac{\gamma_i+1}{\delta_i}}\Gamma(\frac{\eta}{\beta-1})}{\Gamma(\frac{\gamma_i+1}{\delta_i})\Gamma(\frac{\eta}{\beta-1}-\frac{\gamma_i+1}{\delta_i})}x_i^{\gamma_i} \\ &\times [1+\frac{(\beta-1)a_ix_i^{\delta_i}}{1+(\beta-1)(a_1x_1^{\delta_1}+a_2x_2^{\delta_2}+\dots+a_{i-1}x_{i-1}^{\delta_{i-1}}+a_{i+1}x_{i+1}^{\delta_{i+1}}+\dots+a_nx_n^{\delta_n})}]^{-\frac{\eta}{\beta-1}} \\ &\times [1+(\beta-1)(a_1x_1^{\delta_1}+a_2x_2^{\delta_2}+\dots+a_{i-1}x_{i-1}^{\delta_{i-1}}+a_{i+1}x_{i+1}^{\delta_{i+1}}+\dots+a_nx_n^{\delta_n})]^{-\frac{\gamma_i+1}{\delta_i}}, \end{aligned} \tag{24}$$

where f_6 is the joint density of $X_1,X_2,\dots,X_{i-1},X_{i+1},\dots,X_n$. When we take the limit as $\beta \to 1$ in Equation (24), we can see that the conditional density will be in the form of a generalized gamma density and is given by

$$\lim_{\beta\to 1} f_5(x_i|x_1,x_2,\dots,,x_{i-1},x_{i+1},\dots,x_n) = \frac{\delta_i(\eta a_i)^{\frac{\gamma_i+1}{\delta_i}}}{\Gamma(\frac{\gamma_i+1}{\delta_i})}x_i^{\gamma_i}e^{-a_ix_i^{\delta_i}}, \tag{25}$$

$$x_i \geq 0,\ \delta_i > 0,\ \eta > 0,\ \gamma_i + 1 > 0.$$

Theorem 2. *Let $(X_1, X_2, \ldots, X_n)$ have a multivariate extended gamma density (3), then the limiting case of the conditional density $f_\beta(x_i|x_1, x_2, \ldots, x_{i-1}, x_{i+1}, \ldots, x_n)$ will be a generalized gamma density (25).*

Best Predictor

The conditional expectation, $E(x_n|x_1, \ldots, x_{n-1})$, is the best predictor, best in the sense of minimizing the expected squared error. Variables which are preassigned are usually called independent variables and the others are called dependent variables. In this context, X_n is the dependent variable or being predicted and $X_1, \ldots, X_{n-1}$ are the preassigned variables or independent variables. This 'best' predictor is defined as the regression function of X_n on $X_1, \ldots, X_{n-1}$.

$$\begin{aligned} E(x_n|x_1, \ldots, x_{n-1}) &= \int_{x_n=0}^{\infty} x_n f_7(x_n|x_1, \ldots, x_{n-1}) \mathrm{d}x_n \\ &= \frac{\delta_n [a_n(\beta-1)]^{\frac{\gamma_n+1}{\delta_n}} \Gamma(\frac{\eta}{\beta-1})}{\Gamma(\frac{\gamma_n+1}{\delta_n})\Gamma(\frac{\eta}{\beta-1} - \frac{\gamma_n+1}{\delta_n})} [1 + (\beta-1)(a_1 x_1^{\delta_1} + a_2 x_2^{\delta_2} + \ldots + a_{n-1} x_{n-1}^{\delta_{n-1}})]^{-\frac{\gamma_n+1}{\delta_n} + \frac{\eta}{\beta-1}} \\ &\quad \times \int_{x_n=0}^{\infty} x_n^{\gamma_n+1} [1 + (\beta-1)(a_1 x_1^{\delta_1} + a_2 x_2^{\delta_2} + \ldots + a_n x_n^{\delta_n})]^{-\frac{\eta}{\beta-1}} \mathrm{d}x_n. \end{aligned} \tag{26}$$

We can integrate the above integral as in the case of Equation (6). Then after simplification we will get the best predictor of X_n at preassigned values of $X_1, \ldots, X_{n-1}$ which is given by

$$\begin{aligned} E(x_n|x_1, \ldots, x_{n-1}) = & \frac{\delta_n [a_n(\beta-1)]^{-\frac{1}{\delta_n}} \Gamma(\frac{\eta}{\beta-1} - \frac{\gamma_n+2}{\delta_n}) \Gamma(\frac{\gamma_n+2}{\delta_n})}{\Gamma(\frac{\gamma_n+1}{\delta_n})\Gamma(\frac{\eta}{\beta-1} - \frac{\gamma_n+1}{\delta_n})} \\ & \times [1 + (\beta-1)(a_1 x_1^{\delta_1} + a_2 x_2^{\delta_2} + \ldots + a_{n-1} x_{n-1}^{\delta_{n-1}})]^{-\frac{1}{\delta_n}}, \end{aligned} \tag{27}$$

$\delta_n > 0$, $a_n > 0$, $\beta > 1$, $x_i > 0$, $i = 1, 2, \ldots, n-1$, $\frac{\eta}{\beta-1} - \frac{\gamma_n+2}{\delta_n} > 0$, $\gamma_n + 1 > 0$. We can take the limit $\beta \to 1$ in (27). For taking limit, let us apply Stirling's approximations for gamma functions, see for example [15]

$$\Gamma(z+a) \to (2\pi)^{\frac{1}{2}} z^{z+a-\frac{1}{2}} e^{-z}, \text{for } |z| \to \infty \text{ and } a \text{ is bounded} \tag{28}$$

to the gamma's in (27). Then we will get

$$\lim_{\beta \to 1} E(x_n|x_1, \ldots, x_{n-1}) = \frac{\delta_n \Gamma(\frac{\gamma_n+2}{\delta_n})}{(a_n \eta)^{\frac{1}{\delta_n}} \Gamma(\frac{\gamma_n+1}{\delta_n})} \tag{29}$$

which is the moment of a generalized gamma density as given in (25).

6. Multivariate Extended Gamma When $\beta < 1$

Consider the case when the pathway parameter β is less than 1, then the pathway model has the form

$$g(x) = K x^{\gamma} [1 - a(1-\beta) x^{\delta}]^{\frac{\eta}{1-\beta}}, \ \beta < 1, \ a > 0, \ \delta > 0, \ \eta > 0, \tag{30}$$

$1 - a(1-\beta)x^{\delta} \geq 0$, and K is the normalizing constant. $g(x)$ is the generalized type-1 beta model. Let us consider a multivariate case of the above model as

$$\begin{aligned} g_{\beta}(x_1, x_2, \ldots, x_n) = & K_{\beta} x_1^{\gamma_1} x_2^{\gamma_2} \ldots x_n^{\gamma_n} [1 - (1-\beta)(a_1 x_1^{\delta_1} + a_2 x_2^{\delta_2} + \ldots + a_n x_n^{\delta_n})]^{\frac{\eta}{1-\beta}}, \\ & \beta < 1,\ \eta > 0,\ \delta_i > 0,\ a_i > 0,\ i = 1, 2, \ldots, n, \\ & 1 - (1-\beta)(a_1 x_1^{\delta_1} + a_2 x_2^{\delta_2} + \ldots + a_n x_n^{\delta_n}) \geq 0. \end{aligned} \tag{31}$$

where K_{β} is the normalizing constant and it can be obtained by solving

$$K_{\beta} \int \ldots \int x_1^{\gamma_1} \ldots x_n^{\gamma_n} [1 - (1-\beta)(a_1 x_1^{\delta_1} + \ldots + a_n x_n^{\delta_n})]^{\frac{\eta}{1-\beta}} \mathrm{d}x_1 \ldots \mathrm{d}x_n = 1 \tag{32}$$

Integration over x_n yields the following,

$$K_{\beta}\ x_1^{\gamma_1} x_2^{\gamma_2} \ldots x_{n-1}^{\gamma_{n-1}} [1 - (1-\beta)(a_1 x_1^{\delta_1} + \ldots + a_{n-1} x_{n-1}^{\delta_{n-1}})]^{\frac{\eta}{1-\beta}} \int_0^u x_n^{\gamma_n} [1 + C_1 x_n^{\delta_n}]^{-\frac{\eta}{\beta-1}} \mathrm{d}x_n, \tag{33}$$

where $u = [\frac{1-(1-\beta)(a_1 x_1^{\delta_1} + \ldots + a_{n-1} x_{n-1}^{\delta_{n-1}})}{a_n(1-\beta)}]^{\frac{1}{\delta}}$ and $C_1 = \frac{(1-\beta)a_n}{[1-(1-\beta)(a_1 x_1^{\delta_1} + \ldots + a_{n-1} x_{n-1}^{\delta_{n-1}})]}$. Letting $y = C_1 x_n^{\delta_n}$, then the above integral becomes a type-1 Dirichlet integral and the normalizing constant can be obtained as

$$K_{\beta} = \frac{\prod_{j=1}^{n} [\delta_j((1-\beta)a_j)^{\frac{\gamma_j+1}{\delta_j}}] \Gamma(1 + \frac{\eta}{1-\beta} + \frac{\gamma_1+1}{\delta_1} + \ldots + \frac{\gamma_n+1}{\delta_n})}{\Gamma(\frac{\gamma_1+1}{\delta_1}) \ldots \Gamma(\frac{\gamma_n+1}{\delta_n}) \Gamma(1 + \frac{\eta}{1-\beta})} \tag{34}$$

When $\beta \to 1$, (31) will become the density of independently distributed generalized gamma variables. By observing the normalizing constant in (34), we can easily obtaine the joint product moment for some arbitrary $(h_1, \ldots, h_n)$,

$$\begin{aligned} E(x_1^{h_1} x_2^{h_2} \ldots x_n^{h_n}) = & K_{\beta} \frac{\Gamma(\frac{\gamma_1+h_1+1}{\delta_1}) \ldots \Gamma(\frac{\gamma_n+h_n+1}{\delta_n}) \Gamma(1 + \frac{\eta}{1-\beta})}{\prod_{j=1}^{n} [\delta_j((1-\beta)a_j)^{\frac{\gamma_j+h_j+1}{\delta_j}}] \Gamma(1 + \frac{\eta}{1-\beta} + \frac{\gamma_1+h_1+1}{\delta_1} + \ldots + \frac{\gamma_n+h_n+1}{\delta_n})} \\ = & \frac{\Gamma(1 + \frac{\eta}{1-\beta} + \frac{\gamma_1+1}{\delta_1} + \ldots + \frac{\gamma_n+1}{\delta_n}) \Gamma(\frac{\gamma_1+h_1+1}{\delta_1}) \ldots \Gamma(\frac{\gamma_n+h_n+1}{\delta_n})}{\prod_{j=1}^{n} [((1-\beta)a_j)^{\frac{h_j}{\delta_j}}] \Gamma(1 + \frac{\eta}{1-\beta} + \frac{\gamma_1+h_1+1}{\delta_1} + \ldots + \frac{\gamma_n+h_n+1}{\delta_n}) \Gamma(\frac{\gamma_1+1}{\delta_1}) \ldots \Gamma(\frac{\gamma_n+1}{\delta_n})}, \end{aligned} \tag{35}$$

$$\gamma_i + h_j + 1 > 0,\ \gamma_j + 1 > 0,\ a_j > 0,\ \beta < 1,\ \delta_j > 0,\ j = 1, 2, \ldots, n.$$

Letting $h_2 = \ldots = h_n = 0$, in (35), we get

$$E(x_1^{h_1}) = \frac{\Gamma(1 + \frac{\eta}{1-\beta} + \frac{\gamma_1+1}{\delta_1} + \ldots + \frac{\gamma_n+1}{\delta_n}) \Gamma(\frac{\gamma_1+h_1+1}{\delta_1})}{[(1-\beta)a_1]^{\frac{h_1}{\delta_1}}] \Gamma(1 + \frac{\eta}{1-\beta} + \frac{\gamma_1+h_1+1}{\delta_1} + \frac{\gamma_2+1}{\delta_2} \ldots + \frac{\gamma_n+1}{\delta_n}) \Gamma(\frac{\gamma_1+1}{\delta_1})}, \tag{36}$$

$$\gamma_1 + h_1 + 1 > 0,\ \gamma_j + 1 > 0,\ a_1 > 0,\ \beta < 1,\ \delta_j > 0,\ \eta > 0,\ j = 1, 2, \ldots, n.$$

(13) is the h_1^{th} moment of a random variable with density function,

$$g_1(x_1) = K_1 x_1^{\gamma_1}[1-(1-\beta)a_1 x_1^{\delta_1}]^{\frac{\eta}{1-\beta}+\frac{\gamma_2+1}{\delta_2}+\ldots+\frac{\gamma_n+1}{\delta_n}}, \tag{37}$$

where K_1 is the normalizing constant.

Letting $h_3 = \ldots = h_n = 0$, in (35), we get

$$E(x_1^{h_1}x_2^{h_2}) = \frac{\Gamma(1+\frac{\eta}{1-\beta}+\frac{\gamma_1+1}{\delta_1}+\ldots+\frac{\gamma_n+1}{\delta_n})\Gamma(\frac{\gamma_1+h_1+1}{\delta_1})\Gamma(\frac{\gamma_2+h_2+1}{\delta_2})}{\prod_{j=1}^{2}[((1-\beta)a_j)^{\frac{h_j}{\delta_j}}]\Gamma(1+\frac{\eta}{1-\beta}+\sum_{i=1}^{2}\frac{\gamma_i+h_i+1}{\delta_i}+\sum_{j=3}^{n}\frac{\gamma_j+1}{\delta_j})\Gamma(\frac{\gamma_1+1}{\delta_1})\Gamma(\frac{\gamma_2+1}{\delta_2})}, \tag{38}$$

$$\gamma_1+h_1+1)>0,\ (\gamma_2+h_2+1>0,\ \gamma_j+1>0,\ a_1>0,\ a_2>0\ \beta<1,\ \delta_j>0,\ \gamma_j+1>0,\ j=1,2,\ldots,n.$$

If we proceed in the similar way as in Section 4.1, here we can deduce the variance-covariance matrix of multivariate extended gamma for $\beta < 1$.

7. Conclusions

Multivariate counterparts of the extended generalized gamma density is considered and some properties are discussed. Here we considered the variables as not independently distributed, but when the pathway parameter $\beta \to 1$ we can see that $X_1, X_2, \ldots, X_n$ will become independently distributed generalized gamma variables. Joint product moment of the multivariate extended gamma is obtained and some of its properties are discussed. We can see that the limiting case of the conditional density of this multivariate extended gamma is a generalized gamma density. A graphical representation of the pathway is given in Figures 1–4.

Acknowledgments: The author acknowledges gratefully the encouragements given by Professor A. M. Mathai, Department of Mathematics and Statistics, McGill University, Montreal, QC, Canada in this work.

Conflicts of Interest: The author declares no conflict of interest.

References

1. Mathai, A.M. A Pathway to matrix-variate gamma and normal densities. *Linear Algebra Appl.* **2005**, *396*, 317–328.
2. Joseph, D.P. Gamma distribution and extensions by using pathway idea. *Stat. Pap. Ger.* **2011**, *52*, 309–325.
3. Beck, C.; Cohen, E.G.D. Superstatistics. *Physica A* **2003**, *322*, 267–275.
4. Beck, C. Stretched exponentials from superstatistics. *Physica A* **2006**, *365*, 96–101.
5. Tsallis, C. Possible generalizations of Boltzmann-Gibbs statistics. *J. Stat. Phys.* **1988**, *52*, 479–487.
6. Mathai, A.M.; Haubold, H.J. Pathway model, superstatistics Tsallis statistics and a generalized measure of entropy. *Physica A* **2007**, *375*, 110–122.
7. Kotz, S.; Balakrishman, N.; Johnson, N.L. *Continuous Multivariate Distributions*; John Wiley & Sons, Inc.: New York, NY, USA, 2000.
8. Mathai, A.M.; Moschopoulos, P.G. On a form of multivariate gamma distribution. *Ann. Inst. Stat. Math.* **1992**, *44*, 106.
9. Furman, E. On a multivariate gamma distribution. *Stat. Probab. Lett.* **2008**, *78*, 2353–2360.
10. Mathai, A.M.; Provost, S.B. On q-logistic and related models. *IEEE Trans. Reliab.* **2006**, *55*, 237–244.

11. Haubold, H.J.; Mathai, A.M.; Thomas, S. An entropic pathway to multivariate Gaussian density. *arXiv* **2007**, arXiv:0709.3820v.
12. Tsallis, C. Nonextensive Statistical Mechanics and Thermodynamics. *Braz. J. Phys.* **1999**, *29*, 1–35.
13. Thomas, S.; Jacob, J. A generalized Dirichlet model. *Stat. Probab. Lett.* **2006**, *76*, 1761–1767.
14. Thomas, S.; Thannippara, A.; Mathai, A.M. On a matrix-variate generalized type-2 Dirichlet density. *Adv. Appl. Stat.* **2008**, *8*, 37–56.
15. Mathai, A.M. *A Handbook of Generalized Special Functions for Statistical and Physical Sciences;* Oxford University Press: Oxford, UK, 1993; pp. 58–116.

axioms

MDPI

Editorial

Scientific Endeavors of A.M. Mathai: An Appraisal on the Occasion of his Eightieth Birthday, 28 April 2015

Hans J. Haubold [1,*] and Arak M. Mathai [2]

[1] Office for Outer Space Affairs, United Nations, Vienna International Centre, A-1400 Vienna, Austria
[2] Department of Mathematics and Statistics, McGill University, Montreal H3A 2K6, Canada; mathai@math.mcgill.ca
* Author to whom correspondence should be addressed; hans.haubold@gmail.com; Tel.: +43-676-4252050.

Academic Editor: Angel Garrido
Received: 6 May 2015; Accepted: 24 June 2015; Published: 3 July 2015

Abstract: A.M. Mathai is Emeritus Professor of Mathematics and Statistics at McGill University, Canada. He is currently the Director of the Centre for Mathematical and Statistical Sciences India. His research contributions cover a wide spectrum of topics in mathematics, statistics, physics, astronomy, and biology. He is a Fellow of the Institute of Mathematical Statistics, National Academy of Sciences of India, and a member of the International Statistical Institute. He is a founder of the Canadian Journal of Statistics and the Statistical Society of Canada. He was instrumental in the implementation of the United Nations Basic Space Science Initiative (1991–2012). This paper highlights research results of A.M. Mathai in the period of time from 1962 to 2015. He published over 300 research papers and over 25 books.

Keywords: special functions; fractional calculus; entropic functional; mathematical physics; applied analysis; statistical distributions; geometrical probabilities; multivariate analysis

---

1. Early Work in Design of Experiment and Related Problems

A.M. Mathai's first paper was in the area of Design of Experiment and Analysis of Variance in Statistics. This work was done after finishing M.A in Mathematics at University of Toronto and waiting to register for Ph.D, during July–August 1962. This was the first publication which appeared in the journal *Biometrics* in 1965, Mathai (1965) [*Biometrics*, **21**(1965), 376–385]. This problem was suggested by Professor Ralph Wormleighton of the University of Toronto. In two-way classification with multiple observations per cell the analysis becomes complicated due to lack of orthogonality in the design. If two factors, such as the amount of fertilizer used and planting methods in an agricultural experiment to study the yield of corn, are to be tried and if the experiment is planned to replicate n times, it may happen that some observations in some replicate may get lost and as a result, instead of n observations per cell one may have n_{ij} observations in the (i,j)th cell. When doing the analysis of the data, for estimating the effects of fertilizers, say, $\alpha_1, ..., \alpha_p$, one has to solve a singular system of linear equations of the type $(I - A)\alpha = G$ where G is known and $I - A$ is singular and the unknown quantity $\alpha' = (\alpha_1, ..., \alpha_p)$ is to be evaluated. Due to singularity, one cannot write $\alpha = (I - A)^{-1}G$. This $A = (a_{ij})$ is the incidence matrix and has the property that all elements are positive and $\sum_{j=1}^{p} a_{ij} = 1$ for each $i = 1, ..., p$. Mathai observed that this property means that a norm of A, namely $\parallel A \parallel = \max_i \sum_{j=1}^{p} |a_{ij}| = 1$ and further, since the design is taking care of a general effect, one can impose a condition on $\alpha_1, ..., \alpha_p$ such as $\alpha_1 + ... + \alpha_p = 0$. Now, consider A being rewritten as $A = A - C + C$ where C is a matrix where all the first row elements are equal to the median a_1 of the first row elements of A, all the second row elements are the median a_2 of the second row elements of A and so on. Now, by using the conditions on α_i's, $C\alpha = O$ (null). Then

$$(I - A)\alpha = G \Rightarrow (I - B)\alpha = G$$

where $B = (b_{ij}), b_{ij} = a_{ij} - a_i, j = 1, ..., p$ or $\sum_{j=1}^{p} |b_{ij}| = \sum_{j=1}^{p} |a_{ij} - a_i|$ = sum of absolute deviations from the median a_i, which is the least possible. Hence the norm $\| B \| = \max_i \sum_{j=1}^{p} |a_{ij} - a_i|$ is the least possible and evidently < 1. Then

$$(I - B)\alpha = G \Rightarrow \alpha = (I - B)^{-1}G = \left(I + B + B^2 + ...\right)G$$

Note that the convergence of the matrix series is made the fastest possible due to the fact that the mean absolute deviation is least when taken from the median. Thus, successive approximations are available from $BG, B^2G, ...$ but for all practical purposes of testing hypotheses it is found that the approximation $\alpha \approx BG$ is sufficient. This approximation avoids matrix inversion or other complicated operations except one matrix multiplication, namely BG. Encouraged by this work, the thesis was written on sampling distributions under missing values. A concept called "dispersion theory" was also developed in the thesis. It is shown that statistical decision making is nothing but a study of a properly defined measure of scatter or dispersion in random variables. Some dispersions are also defined in terms of some norms or metrics. Papers were published on the concept, in the journal *Metron*, **XXVII-34.1–2**(1968), 125–135.

2. Work on Generalized Distributions

This work started in 1965 when R.K. Saxena from Jodhpur, India, was visiting McGill University as a post-doctoral fellow of Charles Fox, the father of Fox's H-function. Mathai's group is responsible to call this function as Fox's H-function. Such Mellin-Barnes type representations were available from 1888 onwards but since Charles Fox revived the whole area and given a new life it was decided to call the function as Fox's H-function. Mathai translated some statistical problems in terms of special functions and Saxena immediately gave the solutions. Several general densities were introduced. General compatible structures for conditional densities and prior densities, so that the unconditional and posterior densities could be easily evaluated in Bayesian analysis problems, were investigated. The paper got published in the Annals [Mathai and Saxena, *Ann. Math. Statist.*, **40**(1969), 1439–1448].

Mathai decided to study the area of special functions and the result of this study is the book from Oxford University Press; Mathai (1993): *A Handbook of Generalized Special Functions for Statistical and Physical Sciences*, Oxford University Press 1993.

2.1. Early Work on Multivariate Analysis

Some problems from multivariate statistical analysis were posed by Mathai, and Saxena could give the solutions in terms of G and H-functions. These functions were not computable and hence difficult to utilize in statistics or mathematics. This prompted Mathai and Saxena to look into computable series forms for G and H-functions and Mathai developed an operator which could solve the difficulties and computable series forms could be obtained. The work of Mathai and Saxena in the area of special functions resulted in the following books : A.M. Mathai and R.K. Saxena (*Generalized Hypergeometric Functions with Applications in Statistics and Physical Sciences*, Springer-Verlag, Heidelberg and New York, Lecture Notes No.348, 1973; *The H-function with Applications in Statistics and Other Disciplines*, Wiley Eastern, New Delhi and Wiley Halsted, New York, 1978); A.M. Mathai and H.J. Haubold, *Special Functions for Applied Scientists*, Springer, New York, 2008, A.M. Mathai, R.K. Saxena and H.J. Haubold, *The H-function: Theory and Applications*, Springer, New York, 2010.

When exploring statistical distributions and their structural decompositions Mathai established several results in characterizations of densities, see Mathai (*Canadian Mathematical Bullettin*, **9**(1966),

95–102, **10**(1967), 239–245; *South African Journal of Statistics*, **1**(10)(1967), 43–48), Gordon and Mathai (*Annals of Mathematical Statistics*, **43**(1972), 205–229). These works and others' related results were put together and brought out a monograph on characterizations, see A.M. Mathai and G. Pederzoli, *Characterizations of the Normal Probability Law*, Wiley Eastern, New Delhi and Wiley Halsted, New York, 1977.

3. Work in Multivariate Analysis

Mathai had already noted the densities of several structures could be written in terms of G and H-functions. Consider $x_1, x_2, ..., x_r, x_{r+1}, ..., x_k$ mutually independently distributed positive random variables such as exponential variables, type-1 or type-2 beta variables or gamma variables or generalized gamma variables *etc.* Consider the structures

$$u = \frac{x_1...x_r}{x_{r+1}...x_k}, v = \frac{x_1^{\delta_1}...x_r^{\delta_r}}{x_{r+1}^{\delta_{r+1}}...x_k^{\delta_k}} \tag{1}$$

where $\delta_1, .., \delta_k$ are some arbitrary real powers. Then taking the Mellin transforms or the $(s-1)$th moments of u and v and then taking the inverse Mellin transform one can write the density of u as a G-function in most cases or as a H-function, and that of v as a H-function. Product of independently distributed type-1 beta random variables has the same structure of general moments of the likelihood ratio criterion or λ-criterion, or a one-to-one function of it, in many of testing hypotheses problems connected with one or more multivariate Gaussian populations and exponential populations. This showed that one could write the exact densities in the general cases as G-functions in most of the cases. Mathai was searching for computable representations in the general cases.

During one summer camp at Queen's University, Kingston, Ontario, Canada, Mathai met P.N. Rathie, a post-doctoral fellow of L.L. Campbell of Queen's University, again from Jodhpur, India, and also a student of R.K. Saxena. They started the collaboration in information theory and at the same time investigated ways and means of putting G-function in nice computable series form. First they developed an operator and later Mathai perfected it, see Mathai (*Annals of the Institute of Statistical Mathematics*, **23**(1971), 181–197). The operator is of the form

$$G_\nu = \left[\frac{\partial}{\partial s} + (-\ln x)\right]^\nu \tag{2}$$

which is an operator operating on the integrand in the Mellin-Barnes representation of the density functions when the densities are written in terms of G functions. By using this operator, general series expansions are obtained for G-functions of the types $G_{0,p}^{p,0}$, which is coming from product of independent gamma variables, $G_{p,p}^{p,0}$, which is coming from product of independent type-1 beta variables, $G_{p,p}^{p,p}$, which is coming from product of independent type-2 beta variables and the general $G_{p,q}^{m,n}$, see Mathai (*Metron*, **28**(1970), 122–146; *Mathematische Nachrichten*, **48**(1970), 129–139; *South African Journal of Statistics*, **5**(1971), 71–90), Mathai and Rathie (*Royal Belg. Akad. Class des Sci.*, **56**(1970), 1073–1084; *Sankhya Series A*, **33**(1983), 45–60), Mathai and Saxena (*Kyungpook Mathematics Journal*, **12**(1972), 61–68; Book: *Generalized Hypergeometric Functions with Applications in Statistics and Physical Sciences*, Springer Lecture Notes No. 348, Heidelberg and New York, 1973).

By using the same operator in Equation (2) the exact distributions of almost all λ-criteria associated with tests of hypotheses on the parameters of one or more Gaussian populations and exponential populations are worked out, see Mathai (*Publ. l'ISUP Paris*, **19**(1970), 1–15; *Journal of the Indian Statistical Association*, **8**(1970), 1–17; *Annals of the Institute of Statistical Mathematics*, **23**(1971), 181–197; *Trabajos de Estadistica*, **23**(1972), 67–83, 111–124; *Skand. Aktuar.*, **55**(1972), 193–198; *Sankhya Series A*, **34**(1972), 161–170; *Annals of the Institute of Statistical Mathematics*, **24**(1972), 53–65, **25**(1973), 557–566), Mathai and Rathie (*Journal of Statistical Research*, **4**(1970), 140–159; *Annals of the Institute of Statistical*

Mathematics, **22**(1970), 69–116; *Statistica*, **31**(1971), 673–688; *Sankhya Series A*, **33**(1971), 45–60; *Annals of Mathematical Statistics*, **42**(1971), 1010–1019). Mathai popularized Mellin transform techniques, and special function technique in general, in statistical distribution theory. Exact distributions of almost all λ-criteria, in the null and non-null cases, are given in explicit computable forms for the most general cases by Mathai and his co-researchers. The exact distributions in some non-null cases could not be obtained for the general cases. For example, in testing equality of covariance matrices or equality of populations in k multivariate normal populations are still open problems for $k \geq 3$, in the sense that some representations for the general case are not available.

3.1. Development of 11-digit Accurate Percentage Points for Multivariate Test Statistics

Even after giving the explicit computable series forms for the various exact distributions of test statistics in the null (when the hypothesis is true) and non-null (under the alternate hypothesis) for the general parameters, the series forms were complicated and exact percentage points could not be computed. When Mathai visited University of Campinas in Brazil he met the physicist R.S. Katiyar. After six months of joint work of simplifying the complicated gamma products, psi and zeta functions, Katiyar was able to come up with a computer program. The first paper in the series giving the exact percentage points up to 11-digit accurate was produced. This paper made all the complicated theory usable in practical situations of testing of hypotheses in multivariate statistical analysis. The paper appeared in Biometrika and other papers followed, see Mathai and Katiyar (*Biometrika*, **66**(1979), 353–356; *Annals of the Institute of Statistical Mathematics*, **31**(1979), 215–224; *Sankhya Series B*, **42**(1980), 333–341), Mathai (*Journal of Statistical Computation and Simulation*, **9**(1979), 169–182).

3.2. Development of a Computer Algorithm for Nonlinear Least Squares

After developing a computer program for computing exact 11-digit accurate percentage points from complicated series forms of the exact densities of λ-criteria for almost all multivariate test statistics, the problem of developing a computer program for non-linear least squares was re-examined. Starting with Marquardt's methods, there were a number of algorithms available in the literature but all these algorithms had deficiencies. There are a few (around 11) standard test problems to test the efficiency of a computer program. The efficiency of a computer program is measured by checking the following two items: In how many test functions the computer program fails and how many function evaluations are needed to come up to the final solution. These are the usual two criteria used in the field to test a new algorithm. A new algorithm for non-linear least squares was developed by Mathai and Katiyar which did not fail in any of the test functions and the number of function evaluations needed was least compared to all other algorithms available in the literature. The paper was published in a Russian journal, see Mathai and Katiyar (*Researches in Mathematical Statistics (Russian)*, **207(10)**(1993), 143–157). This paper was later translated into English by the American Mathematical Society.

3.3. Integer Programming Problem

The usual optimization problems such as optimizing a quadratic form or quadratic expression, subject to linear or quadratic constraints, optimizing a linear form subject to linear (linear programming problem) or quadratic constraints *etc.* deal with continuous variables. When the variables are continuous then these optimization problems can be handled by using calculus or related techniques. Suppose that the variables can only take integer values such as positive integers $1, 2, 3, \ldots$ then the problem becomes complicated. Many of the standard results available when the variables are continuous are no longer true when the variables are integer-valued. One such problem was brought to the attention of Mathai by S. Kounias. This was solved and a joint paper was published, see Kounias and Mathai (*Optimization*, **19**(1988), 123–131).

4. Work on Information Theory

When the exact distributions for the test statistics being worked out, side by side the work on information theory was also progressing. Characterizations of information and statistical concepts were the ones attempted as a joint venture by Mathai and Rathie. Several characterization theorems were established for various information measures and for statistical concepts such as covariance, variance, correlation *etc.*, see for example, Mathai and Rathie (*Sankhya Series A*, **34**(1972), 441–442; *Annals of the Institute of Statistical Mathematics*, **24**(1972), 473–483; in the book *Measures of Information and Their Applications*, IIT Bombay, pp. 1–10, 1974; in the book *Essays in Probability and Statistics*, Shinko Tsusho, Tokyo, pp. 607–633, 1976. This collaboration resulted in the first book in the area of characterizations of information measures, A.M. Mathai and P.N. Rathie*, *Basic Concepts in Information Theory and Statistics: Axiomatic Foundations and Applications*, Wiley Eastern, New Delhi and Wiley Halsted, New York, 1975. One of the measures discussed there is Havrda-Charvát α-generalized entropy

$$H = \frac{\int_{-\infty}^{\infty}[f(x)]^{\alpha}\mathrm{d}x - 1}{2^{1-\alpha} - 1} \tag{3}$$

where $f(x)$ is a density function. This is the continuous version. There is also a discrete analogue. The denominator is put into the form of the exponent of 2 for ready applications to binary systems. When $\alpha \to 1$ one has H in Equation (3) going to the Shannon entropy $S = -\int_{-\infty}^{\infty} f(x) \ln f(x)\mathrm{d}x$ and hence Equation (3) is called an α-generalized entropy. There are several α-generalized entropies in the literature, including the one given by Mathai. This Equation (3) in a modified form with the denominator replaced by $1 - \alpha$ is developed later by C. Tsallis, as the basis for the whole area of non-extensive statistical mechanics. The Mathai-Rathie (1975) book can be considered to be the first book on characterizations. As a side result, as an application of functional equations, Mathai and Rathie solved a problem in graph theory, see *Journal of Combinatorial Theory*, **13**(1972), 83–90. Other applications of information theory concepts in social sciences, population studies *etc.* may be seen from Kaufman and Mathai (*Journal of Multivariate Analysis*, **3**(1973), 236–242), Kaufman, Mathai, Rathie (*Sankhya Series A*(1972), 441–442), Mathai (*Transactions of the 7th Prague Conference on Information Theory*, pp. 353–357).

4.1. Applications to Real-Life Problems

Applications of the concepts of information measures, 'entropy' or the measure of 'uncertainty', directed divergence (a concept of pseudo-distance), 'affinity' or closeness between populations, concept of 'distance between social groups' *etc.* were applied to solve problems in social statistics, population studies *etc.* Mathai had developed a generalized measure of 'affinity' as well as 'distance between social groups'. On application side, dealing with applications of information theory type measures, see George and Mathai (*Canadian Studies in Population*, **2**(1975), 91–100, **7**(1980), 1–7; *Journal of Biosocial Sciences (UK)*, **6**(1975), 347–356; *The Manpower Journal*, **14**(1978), 69–78).

5. Work on Biological Modeling

During one of the visits of Mathai to the Indian Statistical Institute in Calcutta, India, he came across the biologist T.A. Davis. Davis had a number of problems for which he needed answers. He had a huge collection of data on the number of petals in certain flowers of one species of plant. He noted that the petals were usually 4 in each flower but sometimes the number of petals was 5. He wanted to know whether the occurrence of 5-petaled flowers showed any pattern. His data were insufficient to come up with any pattern. Patterns, if any, would be connected to genetical factors. Then he had a question about how various patterns come in nature, in the growth of leaves, flowers, arrangements of petals and seeds in flowers *etc.* and whether any mathematical theory could be developed to explain these. Then he brought in the observations on sunflower. When we look at flowers, certain flowers such as rose flower, sunflower *etc.* look more beautiful than other flowers. This appeal is due to the arrangements of petals, florets, and color combinations. When we look at a sunflower at the florets or at the seed formations, after the florets dry up, we see some patterns in the arrangements of these seeds on the flower disk called capitulum. The seeds look like arranged along some spirals coming from the periphery going to the center. Let us call these as radial spirals. If one marks a point on the periphery and then one looks to the left of the mark one sees one set of radial spirals and if one looks to the right one sees a different set of radial spirals going in the opposite direction. The numbers of these two sets are always two successive numbers from a Fibonacci sequence of numbers $1, 1, 2, 3, 5, 8, 13, 21, ...$ (the sum of two successive numbers is the next number). Another observation made is that if one looks along a radial spiral this spiral does not go to the center but it becomes fuzzy after a while. At that stage if one draws a concentric circle and then look into the inside of this circle then one will see that if one started with the pair $(13, 21)$, then this has shifted to $(8, 13)$ and then to $(5, 8)$ and so on. The same sort of arrangement can be seen in pineapple, in the arrangement of leaves on a coconut tree crown and at many other places. If one takes a coconut crown and project onto a circle then the positions of the leaves on the crown form a replica of the seed arrangement in a sunflower. In a coconut crown if the oldest leaf is in a certain direction, call it 0-th direction then the next older leaf is not the next one to the oldest, but it is about θ degrees either to the right or to the left and this θ is such that $\frac{\theta}{2\pi-\theta}$ = golden ratio $= \frac{\sqrt{5}-1}{2}$. This golden ratio also appears at many places in nature and the above $\theta \approx 137.5^o$. Davis wanted a mathematical explanations for these and related observations. These observations were made by biologists over centuries. Many theories were also available on the subject. All the theories were trying to explain the appearance of radial spirals. Mathematicians try with differential equations and others from other fields try with their own tools. Mathai figured out that the radial spirals that one sees may be aftermath of something else and radial spirals are not generated per se. Also the philosophy is that nature must be working on very simple principles. If one buys sunflower seeds from a shop or look at sunflower seeds on a capitulum the seeds are all of the same dimensions if one takes one from the periphery or from any other spot on the capitulum. Such a growth can happen if something is growing along an Archimedes' spiral, which has the equation in polar coordinates $r = k\theta$ after one leaves the center. Davis' artist was asked to mark points on an Archimedes' spiral, differentiating from point to point at $\theta =\approx 137.5^\circ$, something like a point moving along Archimedes' spiral at a constant speed so that when the first points reaches θ mark a second point starts, both move at the same speed whatever be the speed. When the second point comes to the mark θ a third point starts, and so on. After creating a certain number of points, may be 200 points, remove the Archimedes' spiral from the paper and fill up the space with any symmetrical object, such as circle, diamonds *etc.*, with those points being the centers. Then if one looks from the periphery the two types of radial spirals can be seen. No such spirals are there but it is one's vision that is creating the radial spirals. Thus a sunflower pattern was recreated from this theory and Mathai and Davis proposed a theory of growth and forms. Consider a capillary a very thin tube with built-in chambers. Consider a viscous fluid being continuously pumped in from the bottom. The liquid enters the first chamber. When a certain pressure is built up, an in-built valve opens and the fluid moves into the second chamber and so on. Suppose that the tube opens in the center part of a pan (with a hole at the center). If the pan is fully sealed so that the only force acting on the liquid is Earth's gravity. The flow of the liquid will be governed by the functional equation $f(\theta_1) + f(\theta_2) = f(\theta_1 + \theta_2)$ whose continuous solution is the linear function $f(\theta) = k\theta$. This is Archimedes' spiral.

The paper was sent for publication in the journal of Mathematical Biosciences the editor 'enthusiastically accepted for publication'. In this paper, Mathai and Davis*(*Mathematical Biosciences*, **20**(1974), 117–133), a theory of growth and form is proposed. This theory still stands and since then there were many papers in physics, chemistry and other areas supporting various aspects of the theory and none has disputed the theory so far. In 1976 the journal has taken Mathai-Davis sunflower head as the cover design for the journal and it is still the cover design.

5.1. *Work on Coconut Tree Crown*

The coconut crown was also examined from many mathematical points of view and found to be an ideal crown. This paper may be seen from Mathai and Davis (*Proceedings of the National Academy of Sciences, India*, **39**(1973), 160–169).

5.2. *Engineering Wonder of Bayya Bird's Nest and Other Biological Problems*

Further problems looked into by Mathai and Davis are the following: (1) The engineering aspect of the egg chamber of bayya bird's nest. The nest hangs from the tips of tree branches, the mother bird goes into the egg chamber through the tail opening of the nest, the nest oscillates violently during heavy winds or storms but no egg comes out of the egg chamber and fall through the tail opening. Naturally the tail opening is bigger than the diameter of the eggs because the mother bird goes through that opening. This shape, beng an engineering marvel, was examined by Mathai and Davis; (2) thermometer birds in Andaman Nicobar Islands; (3) transfer of Canadian Maple Syrup technology in the production of palm sugar and jaggery in Tamilnadu, India; (4) Nipa palms to prevent sea erosion along Kannyakumari sea coast; (5) rejuvenation of Western Ghats in Kannyakumari region; All these projects were undertaken jointly by the Centre for Mathematical Sciences, Trivandrum Campus (CMS) where A.M. Mathai was the Honorary Director and Haldane Research Institute of Nagarcoil, Tamilnadu (HRI) where T.A. Davis was the Director and A.M. Mathai was the Honorary Chairman. Earlier to these studies, George and Mathai had done work in population problems, especially in the study of inter-live-birth intervals, that is, the interval between two live births among women in child-bearing age group, see George and Mathai (*Sankhya Series B*, **37**(1975), 332–342; *Demography of India*, **5**(1976), 163–180; *The Manpower Journal*, **14**(1978), 69–78). Here, Mathai had introduced the concepts of affinity and distance between social groups.

5.3. *Introducing the Phrase 'Statistical Sciences'*

By 1970 Mathai was working to establish a Canadian statistical society and a Canadian journal of statistics. The phrase 'statistical sciences' was framed and defined it as a systematic and scientific study of random phenomena so that the theoretical developments of probability and statistics and applications in all branches of knowledge will come under the heading 'statistical sciences', and random variables as an extension of mathematical variables or mathematical variables as degenerate random variables. After launching Statistical Science Association of Canada, the term 'statistical science' became a standard phrase. Journals and organizations started using the name 'statistical science'. Mathai was responsible to introduce these terms into scientific literature.

When G.P.H. Styan, a colleague of Mathai, was editing the news bulletin of the Institute of Mathematical Statistics he posed the question whether the phrase 'statistical science' was ever used before launching statistical science association of Canada. There was a response from a Japanese scientist claiming that he had used the term 'statistical science' before. Incidentally, later the Institute of Mathematical Statistics changed the name of Annals of Mathematical Statistics to Annals of Statistics and hence that name was no longer available when statistical science association of Canada changed its name back to the original proposed name Statistical Society of Canada.

6. Work on Probability and Geometrical Probabilities

Work in mathematical statistics and special functions continued. As a continuation of the investigation of structural properties of densities, Mathai came across the distributions of lengths, areas and volume contents of random geometrical configurations such as random distance, random area, random volume and random hyper-volume. All the theories of G and H-functions, products and ratios of positive random variables *etc.* could be used in examining the distributional aspects of volume of random parallelotopes and simplices. By analyzing the structure of general moments, Mathai noted that these could be generated by products of independently (1) gamma distributed points; (2) uniformly distributed points; (3) type-1 beta distributed points; (4) type-2 beta distributed points. Out of these, (1) fell into the category of $G^{p,0}_{0,p}$, the second and third fell into $G^{p,0}_{p,p}$ category and (4) fell into $G^{p,p}_{p,p}$ category, for all of which the necessary theory was already developed by Mathai and his team. Papers were published on the distributional aspects, see Mathai (*Sankhya Series A.* **45**(1983), 313–323; Mathai and Tracy (*Communications in Statistics A,* **12(15)**(1983), 625–632, 1727–1736; Mathai (*Proceedings of ISPS VI Annual Conference*, pp. 3–8, 1987; *International Journal of Mathematical and Statistical Sciences*, **3(1)**(1994), 79–109, **7(1)** (1998), 77–96; *Rendiconti del Circolo Matematico di Palermo, Serie II, Suppl.*, **65**(2000), 219–232), Mathai and Pederzoli (*American Journal of Mathematical and Management Sciences* **9**(1989), 113–139; *Rendiconti del Circolo Matematico di Palermo, Serie II, Suppl.*, **50**(1997), 235–258).

6.1. A Conjecture in Geometric Probabilities

Then Mathai came across a conjecture posed by an Australian scientist R.E. Miles, regarding the asymptotic normality of a certain random volume coming from uniformly distributed random points. This was proved to be true by H. Ruben. In fact Ruben brought this area to the attention of Mathai. The structure of the random geometric configuration was known to Mathai and that it was a G-function of the type $G^{p,0}_{p,p}$ and Mathai realized that a very simple proof of the conjecture could be given by using the asymptotic formula, or Stirling's formula which is the first approximation there, for gamma functions. This was worked out and shown that the conjecture could be proved very easily. This paper appeared in the journal in probability, see Mathai (*Annals of Probability*, **10**(1982), 247–251). Incidentally, there is a mistake there. Final representation is given in terms of a confluent hypergeometric function $_1F_1$ there but it should be a Gauss hypergeometric function $_2F_1$, one parameter is missed there in writing the final form. Then Mathai noted that the same conjecture can be formulated in terms of type-1 beta distributed random points and similar conjectures could be formulated for type-2 beta distributed random points and gamma distributed random points. These conjectures were formulated and solved, see Mathai (*Sankhya Series A*, **45**(1983), 313–323; *American Journal of Mathematical and Management Sciences*, **9**(1989), 113–139); Mathai and Tracy (*Communications in Statistics A*, **12(15)**(1983), 1727–1736; *Metron*, **44**(1986), 101–110).

6.2. Random Volumes and Jacobians of Matrix Transformations

Side by side Mathai was developing functions of matrix argument. The work in this area will be given later but its connection to geometrical probabilities will be mentioned here. The area of stochastic geometry or geometrical probabilities is a fusion of geometry and measure theory. When measure theory is mixed with geometry the standard axiomatic definition for probability measure is not sufficient. It is quite evident to see that an additional property of invariance is needed because a geometrical object can be moved around in a plane or in space and the probability statements must remain the same. The famous Betrand's paradoxes or Russell's paradoxes come from lack of invariance conditions there. The details are discussed in the book, A.M. Mathai, *Introduction to Geometrical Probability: Distributional Aspects and Applications*, Gordon and Breach, New York, 1999. Consider a circle of radius r. Take two points A and B at random and independently on the circumference of this circle. Here, 'at random' could mean that the probability of finding a point, such as A, in an interval of length δ is $\frac{\delta}{2\pi r}$. Consider the chord AB. Then AB is a random chord. Let P be the mid point of this chord and O the center of the circle. Then OP is fixed when AB is fixed and OP is perpendicular to AB. Consider another situation of selecting a point P at random inside the circle. This can be done by assigning probability of finding P in a region R inside the circle is $\frac{R}{\pi r^2}$. If P is fixed and if P is the midpoint of a chord then the chord is automatically fixed. In many ways one can geometrically uniquely determine a chord. The chord can be made 'random' by assigning probabilities in many ways. Two ways are described above. If one asks a question, what is the probability that the length of this random chord is less than a specified number? The answer will be different for different ways of assigning probabilities. This is the paradox. Note that all steps in the derivations of the answers will be correct and valid steps as per the usual axioms of probability.

In stochastic probability area the methods used are the methods from differential and integral geometry and usually very difficult. Even if one wishes to talk about the distribution of random volume of a parallelotope through differential or integral geometry the process is very involved. Mathai noted that such problems could be easily answered through Jacobians of matrix transformations. A paper was published in advances in applied probability, see Mathai *Advances in Applied Probability*, **31(2)**(1999), 487–506). More papers were published, see Mathai (*Rendiconti del Circolo Matematico di Palermo, Serie II, Suppl.*, **65**(2000), 219–232; in the book *Probability and Statistical Models with Applications*, pp. 293–316, Chapman and Hall, 2001, *Rendiconti del Circolo Matematico di Palermo* **XLVIII**(1999), 487–506); Mathai and Moschopoulos (*Statistica*, **LIX(1)**(1999), 61–81; *Rendiconti del Circolo Matematico di Palermo*, **XLVIII**(1999), 163–190).

6.3. Applications in Transportation Problems

As an application of geometrical probability problems Mathai explored the travel distance from the suburb to city core for circular and rectangular grid cities. Many of the European cities are designed with a city center and circular and radial streets from the center whereas in North America most of the cities are designed in rectangular grids. Travel distances, time taken and associated expenses are random quantities and related to the nature of city design. Some problems of this type were analyzed by Mathai (*Environmetrics*, **9**(1998), 617–628); Mathai and Moschopoulos (*Environmetrics*, **10**(1999), 791–802).

7. Work in Astrophysics

After publishing the two books on generalized hypergeometric functions in 1973 and H-function in 1978, physicists were interested to use those results in their works. A number of people from different parts of Germany were using these results. The German group working in astrophysics problems were trying to solve some problems connected with reaction rate theory. Then H.J. Haubold, came to McGill University with open problems where help from special function theory was needed. After converting their problems in terms of integral equations, Mathai noted that the basic integral to be evaluated was of the following form:

$$I(\gamma,a,b)=\int_0^\infty x^\gamma \mathrm{e}^{-ax-bx^{-\frac{1}{2}}}\mathrm{d}x \tag{4}$$

and generalizations of this integral. Note that if a or b is zero then the integral can be evaluated by using a gamma integral. Mathematically, if the nonlinear exponent is of the form $x^{-\frac{1}{2}}$ or of the form $x^{-\rho}, \rho>0$ it would not make any difference. Mathai could not find any such integrals in any of the books of tables of integrals. He noted that the integrand consisted of integrable functions and therefore one could make statistical densities out of them. For example, $f_1(x)=c_1x^\gamma\mathrm{e}^{-ax}, 0\le x<\infty$ is a density where c_1 is the normalizing constant. Similarly $f_2(x)=c_2\mathrm{e}^{-x^\rho}, \rho>0, 0\le x<\infty$ is a density where c_2 is the normalizing constant. Then the structure in Equation (4) can be written as follows:

$$g(u)=\int_v \frac{1}{v}f_1(v)f_2\left(\frac{u}{v}\right)\mathrm{d}v \tag{5}$$

where $g(u)$ can represent the density of $u=x_1x_2$ where x_1 and x_2 are independently distributed positive real scalar random variables with the densities $f_1(x_1)$ and $f_2(x_2)$ respectively. Once the structure in Equation (4) is identified as that in Equation (5) then, since the density being unique, it is only a matter of finding the density $g(u)$ by using some other means. We can easily use the properties of arbitrary moments. For example

$$E(u)^{s-1}=E(x_1x_2)^{s-1}=E\left(x_1^{s-1}\right)E\left(x_2^{s-1}\right)$$

due to statistical independence of x_1 and x_2, where E denotes the expected value. Note that $E\left(x_1^{s-1}\right)$ is available from $f_1(x_1)$ and $E\left(x_2^{s-1}\right)$ from $f_2(x_2)$. Then $g(u)$ is available from the inverse, that is,

$$g(u) = \frac{1}{2\pi i}\int_{c-i\infty}^{c+i\infty} E\left(u^{s-1}\right)u^{-s}\mathrm{d}s \tag{6}$$

where $i = \sqrt{-1}$ and c is determined from the poles of $E(u^{s-1})$. Thus, by using statistical techniques the integral in Equation (4) was evaluated. After working out many results it was realized that one could also use Mellin convolution of a product to solve integrals of the type in Equation (4). This was not seen when the method through statistical distribution theory was devised. Various types of thermonuclear reactions, resonant, non resonant, depleted case, high energy cut off case *etc.* were investigated. The work also went into exploring exact analytic solar models, gravitational instability problems, solar neutrino problems, reaction-rates, nuclear energy generation *etc.* The work until 1988 was summarized in the monograph Mathai and Haubold (*Modern Problems in Nuclear and Neutrino Astrophysics*, Akademie-Verlag, Berlin, 1988). Since then a lot of work was done, some of them are the following: Haubold and Mathai (*Annalen der Physik*, **44**(1987), 103–116; *Astronomische Nachrichten*, **308(5)**(1987), 313–318; *Journal of Mathematical Physics*, **29(9)**(1988), 2069–2077; *Astronomy and Astrophysics*, **203**(1988), 211–216; *Astronomische Nachrichten*, **312(1)**(1991), 1–6; *Astrophysics and Space Science*, **176**(1991), 51–60, **197**(1992), 153–161,**214**, 49–70,139–149, **228**(1995), 77–86, **258**(1988), 185–199; *American Institute of Physics, Conference Proceedings*, **320**(1994), 102-116, **320**(1994), 89–101; *SIAM Review*, **40(4)**(1998), 995–997). The collaboration also resulted in two encyclopedia articles, see Haubold and Mathai (Sun, *Enclyclopaedia of Planetary Sciences*, pp. 786–794, 1997, Structure of the Universe, *Encyclopedia of Applied Physics*, **23**(1998), pp. 47–51).

7.1. New Results in Mathematics Through Statistical Techniques

After evaluating the basic integrals in physics problems by using statistical techniques, it was realized that such statistical techniques could be used to obtain results in mathematics. Some summation formulae, computable series representations, extensions of several mathematical identities *etc.* were obtained through statistical techniques, see Mathai and Tracy (*Metron*, **XLII-N1-2**(1985), 117–126), Mathai and Pederzoli (*Metron*, **XLIII-N3-4**(1985), 157–166, Mathai and Provost (*Statistical Methods*, **4(2)**(2002), 75–98).

8. Work on Differential Equations

One of the problems investigated in connection with problems in astrophysics was the gravitational instability problem. The problem was brought to the attention of Mathai by Haubold. Papers by Russian researchers were there on the problem of mixing two types of cosmic dusts. Mathai looked at it and found that by making a transformation in the dependent variable and by changing the operator to $t\frac{\mathrm{d}}{\mathrm{d}t}$ instead of the integer order differential operator $D = \frac{\mathrm{d}}{\mathrm{d}t}$ one could identify the differential equation as a particular case of the differential equation satisfied by a G-function. Then G-function theory could be used to solve the problem of mixing k different cosmic dusts. Thus the first paper in integer order differential equation was written and published in the MIT journal, see Mathai (*Studies in Applied Mathematics*, **80**(1989), 75–93). Two follow-up papers were written developing the differential equation and applying to physics problems, see Haubold and Mathai (*Astronomische Nachrichten*, **312(1)**(1991), 1–6; *Astrophysics and Space Science*, **214(1&2)**(1994), 139–149).

9. The Idea of Laplacianness of Bilinear Forms and Work on Quadratic and Bilinear Forms

In the 1980's two students of Mathai, S.B. Provost and D. Morin-Wahhab, finished their Ph.Ds in the area of quadratic form. Mathai has also published a number of papers on quadratic and bilinear forms by this time. Then it was decided to bring out a book on quadratic forms in random variables. On the mathematical side, there were books on quadratic forms but there was none in the area of quadratic forms in random variables. Only real random variables and samples coming from Gaussian population were considered. Later in 2005 Mathai extended the theory to cover very general classes of populations. This aspect will be considered later when pathway models are discussed. Only when I. Olkin pointed out to Mathai about the many applications of complex Gaussian case in communication theory, after the book appeared in print, Mathai and Provost realized that an equal amount of material was missed out: A.M. Mathai and S.B. Provost, *Quadratic Forms in Random Variables: Theory and Applications*, Marcel Dekker, New York, 1992. Work on quadratic forms and related topics may be seen from Mathai(*Communications in Statistics A*, **20(10)**(1991) 3159–3174; *International Journal of Mathematical and Statistical Sciences*, **1(1)**(1992), 5–20; *Journal of Multivariate Analysis*, **41(2)**(1992), 178–193; *Annals of the Institute of Statistical Mathematics*, **44**(1992), 769–779; *Journal of Applied Statistical Sciences*, **1(2)**(1993), 169–178; *The Canadian Journal of Statistics*, **21(3)**(1993), 277–283; *Journal of Multivariate Analysis*, **45**(1993), 239–246; *Journal of Statistical Research*, **27(1&2)**(1993), 57–80).

9.1. Chisquaredness of Quadratic Forms and Laplacianness of Bilinear Forms

In the 1980's two students of Mathai, S.B. Provost and D. Morin-Wahhab, finished their Ph.Ds in the area of quadratic form. Mathai has also published a number of papers on quadratic and bilinear forms by this time. Then it was decided to bring out a book on quadratic forms in random variables. On the mathematical side, there were books on quadratic forms but there was none in the area of quadratic forms in random variables. Only real random variables and samples coming from Gaussian population were considered. Later in 2005 Mathai extended the theory to cover very general classes of populations. This aspect will be considered later when pathway models are discussed. Only when I. Olkin pointed out to Mathai about the many applications of complex Gaussian case in communication theory, after the book appeared in print, Mathai and Provost realized that an equal amount of material was missed out: A.M. Mathai and S.B. Provost, *Quadratic Forms in Random Variables: Theory and Applications*, Marcel Dekker, New York, 1992. Work on quadratic forms and related topics may be seen from Mathai(*Communications in Statistics A*, **20(10)**(1991) 3159–3174; *International Journal of Mathematical and Statistical Sciences*, **1(1)**(1992), 5–20; *Journal of Multivariate Analysis*, **41(2)**(1992), 178–193; *Annals of the Institute of Statistical Mathematics*, **44**(1992), 769–779; *Journal of Applied Statistical Sciences*, **1(2)**(1993), 169–178; *The Canadian Journal of Statistics*, **21(3)**(1993), 277–283; *Journal of Multivariate Analysis*, **45**(1993), 239–246; *Journal of Statistical Research*, **27(1&2)**(1993), 57–80).

9.2. Bilinear Form Book

After publishing the quadratic form book in 1992, a lot of work had been done on bilinear forms. Even though a bilinear form can be written as a quadratic form, there are many properties enjoyed by bilinear form and not enjoyed by quadratic forms. Quadratic forms do not have covariance structures. Then T. Hayakawa of Japan contacted Mathai asking why not bring out a book on bilinear form, parallel to the one on quadratic form including chapters on zonal polynomials. This book on bilinear forms and zonal polynomials was brought out in 1995: A.M. Mathai, S.B. Provost and T. Hayakawa, *Bilinear Forms and Zonal Polynomials*, Springer, New York, 1995, in the lecture notes series. Additional papers may be seen from Mathai and Pederzoli (*Journal of the Indian Statistical Society*, **3**(1995), 345–356; *Statistica*, **LVI(4)**(1996), 4-7-41).

10. Functions of Matrix Argument

Meanwhile Mathai's work on functions of matrix argument was progressing. These are real-valued scalar functions where the argument is a real or complex matrix. The theory is well developed when the argument matrix is real positive definite or hermitian positive definite. Note that when A is a square or rectangular matrix we do not have a concept corresponding to the square root of a scalar quantity uniquely defined. But if the matrix A is real positive definite or hermitian positive definite, written as $A > O$, operations such as square root can be uniquely defined. Hence the theory is developed basically for real positive definite or hermitian positive definite matrices. Gordon and

Mathai tried to develop a matrix series and a pseudo analytic function involving general matrices, the attempt was not fully successful but some characterization theorems for multivariate normal population could be established, see Gordon and Mathai (*Annals of Mathematical Statistics*, **43**(1972), 205–229). Gordon has two more papers in the area, one in the *Annals of Statistics* and the other in the *Annals of the Institute of Statistical mathematics*. Hence the theory of real-valued scalar functions of matrix argument is developed when the matrix is real or hermitian positive definite. There are three approaches available in the literature. One is through matrix-variate Laplace transform and inverse Laplace transform developed by C. Herz and others, see for example, Herz (*Annals of Mathematics*, **61(3)**(1955), 474–523). Here one basic assumption is functional commutativity $f(AB) = f(BA)$ even if $AB \neq BA$, where A and B are $p \times p$ matrices. Under functional commutativity we have the following result, observing that when A is symmetric there exists and orthonormal matrix $P, PP' = I, P'P = I$ such that $P'AP = D$ where D is a diagonal matrix with the diagonal elements being the eigenvalue of A. Then

$$f(A) = f(AI) = f(APP') = F(P'AP) = f(D)$$

Thus, the original function of $p(p+1)/2$ real scalar variables, can be reduced to a function of p variables, the eigenvalues of A. Another approach is through zonal polynomials, developed by Constantine, James and others, see for example James (*Annals of Mathematics*, **74**(1961), 456–469) and Constantine (*Annals of Mathematical Statistics*, **34**(1963), 1270–1285). In this definition a general hypergeometric function with r upper parameters and s lower parameters is defined as follows:

$$rF_s(X) = {}_rF_s(a_1, ..., a_r; b_1, ..., b_s; X) = \sum_{k=0}^{\infty}\sum_{K} \frac{(a_1)_K...(a_r)_K}{(b_1)_K...(b_s)_K}\frac{C_K(X)}{k!} \tag{7}$$

where $C_K(X)$ is zonal polynomial of order k, $K = (k_1, ..., k_p), k_1 + ... + k_p = k$, and for example,

$$(a)_K = \prod_{j=1}^{p}(a - \frac{(j-1)}{2})_{k_j}, (b)_k = b(b+1)...(b+k-1), (b)_0 = 1, b \neq 0 \tag{8}$$

Here also functional commutativity is assumed. They claim uniqueness for the above series by claiming that Equation (7) satisfies both the integral equations defining matrix-variate function through the definition of Laplace and inverse Laplace pair. The third approach is due to Mathai and it is defined in terms of a general matrix transform or M-transform. The M-transform of $f(-X)$ defined by the equation

$$g(\rho) = \int_{X>O} |X|^{\rho - \frac{p+1}{2}} f(-X)\mathrm{d}X, \Re(\rho) > \frac{p-1}{2} \tag{9}$$

where $\Re(\cdot)$ means the real part of $(\cdot)$. Under functional commutativity, $f(-X)$ in Equation (9) reduces to a function of p variables, the eigenvalues of X. But, still the left side of Equation (9) is a function of only one variable ρ. Hence unique determination of f through $g(\rho)$ need not be expected. It is conjectured that f is unique when f is analytic in the cone of positive definite matrices. Right now, $f(-X)$ in Equation (9) remains as a class of functions satisfying the integral equation Equation (9). In this definition, a general hypergeometric function with r upper and s lower parameters will be defined as that class of functions for which the M-transform is the following:

$$g(\rho) = \frac{\Gamma_p(a_1 - \rho)...\Gamma_p(a_r - \rho)}{\Gamma_p(b_1 - \rho)...\Gamma_p(b_s - \rho)}, \Re(\rho) > \frac{p-1}{2} \tag{10}$$

where $\Gamma_p(a)$ is the real matrix-variate gamma given by

$$\Gamma_p(a) = \pi^{\frac{p(p-1)}{4}}\Gamma(a)\Gamma(a - \frac{1}{2})...\Gamma(a - \frac{(p-1)}{2}), \Re(a) > \frac{p-1}{2} \tag{11}$$

Then that class of function $f(-X)$ is given by the equation Equation (11). It is seen that M-transform technique is the most powerful in extending univariate results to matrix-variate cases. Some of the results may be seen from Mathai (*Mathematische Nachrichten* **84**(1978), 171–177; *Communications in Statistics A*, **A8(1)**(1979), 47–55, **A9(8)**(1980), 795–801;*Annals of the Institute of Statistical Mathematics*, **33**(1981), 35–43, **34**(1982), 591–597; *Sankhya Series A*, **45**(1983), 313–323; *Proceedings of the VI ISPS Conference*, pp. 3–8, 1987; *Indian Journal of Pure Applied Mathematics*, **22(11)**(1991), 887–903; *Journal of Multivariate Analysis*, **41(2)**(1992), 178–193; *Proceedings of the National Academy of Sciences*, **LXV(II)**(1995), 121–142, **LXV(III)**(1995), 227–251, **LXVI(IV)**(1995), 367–393, **LXVI(AI)**(1996), 1–22; *Indian Journal of Pure and Applied Mathematics*, **24(9)**(1993), 513–531; *Advances in Applied Probability*, **31(2)**(1999), 343–354; *Rendiconti del Circolo Matematico di Palermo, Series II, Suppl.*, **65**(2000), 219–232; *Linear Algebra and Its Applications*, **183**(1993), 202–221; in *Probability and Statistical Methods with Applications*, pp. 293–316, Chapman and Hall, 2001), Mathai and Saxena (*Journal de Matematica e Estatistica*, **1**(1979), 91–108), Mathai and Rathie (*Statistica*, **XL**(1980), 93–99; *Sankhya Series A*, **42**(1980), 78–87;), Mathai and Tracy (*Communications in Statistics A*, **12(15)**(1983), 1727–1736; *Metron*, **44**(1986), 11–110), Mathai and Pederzoli (*Metron*, **LI(3-4)**(1993), 3–24; *Indian Journal of Pure Applied Mathematics*, **27(3)**(1996), 7–32; *Linear Algebra and Its Applications*, **253**(1997), 209–226, **269**(1998), 91–103). The important publication in this area is the book on Jacobians of matrix transformation: A.M. Mathai, *Jacobians of Matrix Transformations and Functions of Matrix Argument*, World Scientific Publishing, New York, 1997. The work on functions of matrix argument is continuing in the form of applications in pathway models, fractional calculus and so on. These will be mentioned later.

In connection with matrix-variate integrals it is a very often asked question that whether matrix-variate integrals can be evaluated by treating them as multiple integrals and by using standard techniques in calculus. Mathai explored the possibility of explicitly evaluating matrix-variate gamma and beta integrals as multiple integrals in calculus. The basic matrix-variate integrals are the gamma integral and beta integrals, where X is a $p \times p$ real positive definite matrix or hermitian positive definite matrix. For example, when X is real and $X > O$ (positive definite) the gamma integral is

$$\int_{X>O} |X|^{\alpha-\frac{p+1}{2}} \mathrm{e}^{-\mathrm{tr}(X)} \mathrm{d}X, \Re(\alpha) > \frac{p-1}{2}$$

and the beta integral is

$$\int_{O<X<I} |X|^{\alpha-\frac{p+1}{2}} |I-X|^{\beta-\frac{p+1}{2}} \mathrm{d}X, \Re(\alpha) > \frac{p-1}{2}, \Re(\beta) > \frac{p-1}{2}$$

The corresponding integrals are there in the complex-variate case also. It is shown that this can be done explicitly for $p = 2$ and a recurrence relation can be obtained so that step by step they can be evaluated but for $p > 2$ this method of treating as multiple integrals is not a feasible proposition. See Mathai (*Journal of the Indian Mathematical Society*, **81(3–4)**(2014), 259–271; *Applied Mathematics and Computation*, **247**(2014), 312–318.)

11. Multivariate Gamma and Beta Models

Corresponding to a univariate model there is nothing called a unique multivariate analogue. Explorations of some convenient multivariate models corresponding to univariate gamma, type-1 beta, type-2 beta, Dirichlet models *etc.* were conducted in a series of papers. See, for example, Mathai (In *Time Series Methods in Hydrosciences,*, pp. 27–36, Elsevier, 1982), Mathai and Moschopoulos (*Journal of Multivariate Analysis*, **39**(1991), 135–153; *Annals of the Institute of Statistical Mathematics*,**44(1)**(1992), 97–106; *Statistica*, **LVII(2)** (1992), 189–197, **LIII(2)**(1993), 231–21). These were some of the works on the multivariate analogues of gamma and beta densities. Dirichlet models themselves are multivariate extensions of type-1 and type-2 beta integrals or beta densities. When working on order statistics from logistic populations, Mathai came across the need for a generalized form of type-1 Dirichlet model, see Mathai (*IEEE Trans. Reliability*, **52(2)**(2003), 20–206; in *Statistical Methods and Practice: Recent Advances*, pp. 57–67, Narosa Publishing, India, 2003; *Proceedings of the 7th Conference of the Society for Special Functions and Their Applications*, **7**(2006), pp. 131–142,). Various types of generalizations of type-1 and type-2 Dirichlet densities were considered, see for example, Jacob, Jose and Mathai (*Journal of the Indian Academy of Mathematics*, **26(1)**(2004), 175–189); Kurian, Kurian and Mathai (*Proceedings of the National Academy of Sciences*, **74(A)II**(2004), 1–10), Jacob, George and Mathai (*Proceedings of the National Academy of Sciences*, **15(3)**(2005, 1–9), Thomas and Mathai (*Advances in Applied Statistics*, **8(1)**(2008), 37–56; *Sankhya Series A*, **71(1)**(2009), 49–63), Thomas, Thannippara and Mathai (*Journal of Probability and Statistical Science*, **6(2)**(2008), 187–200).

11.1. Power Transformation and Exponentiation

Another problem explored is to see the nature of models available by power transformations and exponentiation of standard probability models. Such a study is useful when looking for an appropriate model for a given data. These explorations are done in Mathai (*Journal of the Society for Probability and Statistics (ISPS)*, **13**(2012), 1–19).

11.2. Symmetric and Assymetric Models

A symmetric model, symmetric at $x = a$ where a could be zero also, means that for $x < a$ the behavior of the function or the shape of the function is the same as its behavior for $x > a$. In many practical situations, symmetry may not be there. The behavior for $x < a$ may be different from that for $x > a$. Many authors have considered asymmetric models where asymmetry is introduced by giving different weighting factors for $x < a$ and for $x > a$ so that the total probability under the curve will be 1. But the shape of the curve itself may change for $x < a$ and for $x > a$. A method is proposed in the paper referred to in 11.1 above (Mathai 2012) where asymmetry is introduced through a scaling parameter so that the shape itself will be different for $x < a$ and $x > a$ cases but the total probability remaining as 1, which may have more practical relevance.

12. The Pathway Model

The basic idea was there in a paper of 1970's in the area of population studies where it was shown that by a limiting process one can go from one class of functions to another class of functions, the property is basically coming from the theory of hypergeometric functions from the aspect of getting rid off a numerator or a denominator parameter. This idea was revived and written as a paper on functions of matrix argument where the variable matrix is a rectangular one, see Mathai (*Linear Algebra and Its Applications*, **396**(2005), 317–328). Let X be a real $m \times n$ matrix, $m \leq n$ and of rank m be a matrix variable. Let A be $m \times m$ and B be $n \times n$ constant nonsingular matrices. Consider the function

$$f(X) = C|AXBX'|^{\gamma}|I - (1-\alpha)AXBX'|^{\frac{\eta}{1-\alpha}}, \eta > 0 \tag{12}$$

where α, η, C be scalar constants. This C can act as a normalizing constant if we wish to create statistical density out of Equation (12). Consider the case when $m = 1, n = 1$ and $x > 0$. Then one can also take powers for x and the model in Equation (12) can be written as

$$f_1(x) = c_1 x^{\gamma}\left[1 - a(1-\alpha)x^{\delta}\right]^{\frac{\eta}{1-\alpha}} \tag{13}$$

where $a > 0, \delta > 0, \eta > 0, x \geq 0$. In the matrix-variate case in Equation (12) arbitrary powers for matrices is not feasible even though $AXBX'$ is positive definite because even for a positive definite matrix, Y, arbitrary power such as Y^{δ} may not be uniquely defined. Even when uniquely defined transformation such as $Z = Y^{\delta}$ will create problems when computing the Jacobians. The types of difficulties that can arise may be seen for the case $\delta = 2$ described in the book, A.M. Mathai, *Jacobians of Matrix Transformations and Functions of Matrix Argument*, World Scientific Publishing, New York 1997. Hence for the matrix case we consider only when $\delta = 1$. Consider case $-\infty < \alpha < 1$. Then

Equation (13) remains as it is given in Equation (13) which is a generalized type-1 beta function. But if $\alpha > 1$ then writing $1 - \alpha = -(\alpha - 1)$ the form in Equation (13) changes to the following:

$$f(x) = c_2 x^{\gamma} \left[1 + a(\alpha - 1)x^{\delta}\right]^{-\frac{\eta}{\alpha-1}} \tag{14}$$

for $a > 0, \alpha > 1, \eta > 0, \delta > 0, x \geq 0$. This model is a generalized type-2 beta model. When $\alpha \to 1$ in Equation (13) and Equation (14), $f_1(x)$ and $f_2(x)$ reduce to the the form

$$f_3(x) = c_3 x^{\gamma} \mathrm{e}^{-a\eta x^{\delta}}, a > 0, \eta > o, x \geq 0 \tag{15}$$

This is a generalized gamma model. Thus three functional forms $f_1(x), f_2(x), f_3(x)$ are available for $\alpha < 1, \alpha > 1, \alpha \to 1$. This parameter α is called the pathway parameter, a pathway showing three different families of functions.

The practical utility of the model is that if Equation (15) is the stable or ideal situation in a physical system then the unstable neighborhoods or functions leading to Equation (15) are given in Equation (13) and Equation (14). In a model building situation, if the underlying data show a gamma-type behavior then a best-fitting model can be constructed for some values of the parameters or for some value of α the ideal model can be determined. Most of the statistical models in practical use in the areas of statistics, physics and engineering fields can be seen to be a member or products of members from f_1, f_2, f_3 above. Note that for $\alpha > 1$ and $\alpha \to 1$ situations we can take $\delta > 0$ or $\delta < 0$ and both these situations can create statistical densities. Note that f_1 is a family of finite range models whereas f_2 and f_3 are families of infinite range models. Extended models are available by replacing x by $|x|$ so that the whole real line will be covered. In this case the nonzero part of model Equation (13) will be in the range $\pm[a(1-\alpha)]^{-\frac{1}{\delta}}$ and for others $-\infty < x < \infty$. Note that in Equation (12) all individual variables x_{ij}'s are allowed to vary over the whole real line subject to the condition $I - (1-\alpha)AXBX' > O$ (positive definite). This model is also extended to complex rectangular matrix-variate case, see Mathai and Provost (*Linear Algebra and Its Applications*, **410**(2005), 198–216).

Note that Equation (13) for $\gamma = 0, \delta = 1, a = 1, \eta = 1$ is Tsallis statistics in nonextensive statistical mechanics. The function, without the normalizing constant c_1 will then be

$$g(x) = [1 - (1-\alpha)x]^{\frac{1}{1-\alpha}} \tag{16}$$

which is Tsallis statistics. This can be generated by optimizing Tsallis entropy or Havrda-Charvát entropy with the denominator factor $1 - \alpha$ instead of $2^{1-\alpha} - 1$, subject the constraint that the first moment is fixed and this condition can be connected to the principle of the total energy being conserved. Note that Equation (16) is also a power function model.

$$\frac{\mathrm{d}}{\mathrm{d}x} g(x) = -[g(x)]^{\alpha}$$

Also Equation (14) for $a = 1, \delta = 1, \eta = 1$ is superstatistics in nonextensive statistical mechanics.

Mathai's students have introduced a pathway fractional integral operator based on Equation (13) and a pathway transform based on Equation (13) and Equation (14). Equation (13) and Equation (14) can also be obtained by optimizing Mathai's entropy

$$M_{\alpha}(f) = \frac{\int_{-\infty}^{\infty} [f(x)]^{2-\alpha} \mathrm{d}x - 1}{\alpha - 1}, \alpha \neq 1, \alpha < 2$$

subject to two moment type constraints and also the pathway parameter α can be derived in terms of moments of $f_1(x)$ or $f_2(x)$. Thus, in terms of entropies one can establish a entropic pathway, in terms of distributions as explained above one can create a distributional pathway, one can also look into the corresponding differential equations and create a differential pathway, covering the three sets of functions belonging to generalized and extended type-1 beta family, type-2 beta family and gamma family. The theory of quadratic and bilinear forms in random variables is extended to cover pathway populations, instead of Gaussian population. Note that Gaussian population is a special case of the extended pathway population or pathway model, see Mathai* (*Linear Algebra and Its Applications*, **425**(2007), 162–170). Applications and advancement of theory of pathway model by Mathai and his associates may be seen from the following: Mathai and Haubold (*Physica A*, **375**(2007), 110–122, **387**(2007), 2462–2470; *Physics Letters A*, **372**(2008), 2109–2113; *Integral Transforms and Special Functions*, **21(11)**(2011), 867–875; *Applied Mathematics and Computations*, **218**(2011), 799–804; *Mathematica Aeterna*, **2(1)**,(2012), 51–61; *Sun and Geosphere*, **8(2)**(2013), 63–70, *UN Proceedings (2013)*; *Entropy*, **15**(2013), 4011–4025), Mathai and Provost (*IEEE Transactions on Reliability*, **55(2)**(2006), 237–244; *Journal of Probability and Statistical Science*, **9(1)**(2011), 1–20; *Physica A*, **392(4)**(2013), 545–551).

12.1. Input-Output Models

Many practical situations are input-output situations where what is observed is really the residual effect. Energy may be produced and consumed and what is observed is the net result or the residual effect. Water flows into a dam, which is the input variable, and water is taken out of the dam, which is the output variable and the storage at any instant is the residual effect of the input minus the output. In any production-consumption, creation-destruction, growth-decay situation what is observed is $z = x - y$ where x is the input variable and y is the output variable and z is the residual effect. Mathai explored a number of situations where x and y are independently distributed real scalar random variables or matrix random variables. Observations as widely different as solar neutrinos and the amount of melatonin present in human body are all residual observations. Some works in this direction may be seen from Mathai (*Annals of the Institute of Statistical Mathematics*, **34**(1982), 591–597; In *Time Series Methods in Hydrosciences*, pp. 27–36, Elsevier, Amsterdam, 1982; *Canadian Journal of Statistics*, **21(3)**(1993), 277–283; *Journal of Statistical Research*, **27(1–2)**(1993), 57–80; *Integral Transforms and Special Functions*, **20(12)**(2009), 49–63), Haubold and Mathai (*Astrophysics and Space Science*, **228**(1995), 113–134; *Astrophysics and Space Science*, **273**(2000), 53–63), Saxena, Mathai and Haubold (*Astrophysics and Space Science*, **290**(2004), 299–310) and a number of papers on fractional reaction-diffusion equations.

13. Work on Mittag-Leffler Functions and Mittag-Leffler Densities

On Mittag-Leffler functions and their generalizations an overview paper is written, see Haubold, Mathai and Saxena (*Journal of Applied Mathematics*, **ID 298628**(2011), 51 pages). Mittag-Leffler function comes in naturally when looking for solutions of fractional differential equations. This aspect will be considered later. Three standard forms of Mittag-Leffler functions in current use are the following:

$$E_\alpha(x)=\sum_{k=0}^{\infty}\frac{x^k}{\Gamma(1+\alpha k)}, \Re(\alpha)>0$$

$$E_{\alpha,\beta}(x)=\sum_{k=0}^{\infty}\frac{x^k}{\Gamma(\beta+\alpha k)}, \Re(\alpha)>0, \Re(\beta)>0$$

$$E_{\alpha,\beta}^{\gamma}(x)=\sum_{k=0}^{\infty}\frac{(\gamma)_k x^k}{k!\Gamma(\beta+\alpha k)}, \Re(\alpha)>0, \Re(\beta)>0$$

There is no condition on the parameter γ. If these are to be written in terms of H-functions then α and γ have to be real and positive. A generalization can be made by introducing a general hypergeometric type function, which may be written as

$$E_{\alpha,\beta,b_1,\ldots,b_s}^{a_1,\ldots,a_r}\left(x^\delta\right)=\sum_{k=0}^{\infty}\frac{(a_1)_k\ldots(a_r)_k\left(x^\delta\right)^k}{k!\Gamma(\beta+\alpha k)(b_1)_k\ldots(b_s)_k}$$

where the notation $(a_j)_k$ and $(b_j)_k$ are Pochhammer symbols. Convergence conditions can be worked out for this general form.

A problem of interest in this case is a general Mittag-Leffler density because such a density is needed in non-Gaussian stochastic processes and time series areas. Such a density was introduced based on $E^{\gamma}_{\alpha,\beta}(x^{\delta})$ and it is shown that such a model is connected to fat-tailed models, Lévy, Linnik models. Structural properties and asymptotic behavior are also studied and it is shown that such models are not attracted to Gaussian models, see Mathai (*Fractional Calculus & Applied Analysis*, **13(1)** (2010), 113–132), Mathai and Haubold (*Integral Transforms and Special Functions*, **21(11)**(2011), 867–875).

14. Work on Krätzel Function and Krätzel Densities

Another area explored is Krätzel function, Krätzel transform and Krätzel densities. Since Krätzel transform is important in applied analysis area, a general density is introduced based on Krätzel integral. The basic Krätzel integral is of the form

$$g_1(x) = \int_0^{\infty} x^{\gamma} e^{-ax-\frac{y}{x}} dx, a > 0, y > 0 \tag{17}$$

which can be generalized to the form

$$g_2(x) = \int_0^{\infty} x^{\gamma} e^{-ax^{\alpha}-\frac{y}{x^{\beta}}} dx \tag{18}$$

for $a > 0, y > 0, \alpha > 0, \beta > 0$ or $\beta < 0$. The integrand in Eqaution (17), normalized, is the inverse Gaussian density. The integral itself can be interpreted as Mellin convolution of a product, the marginal density in a bivariate case *etc*. The integral in Eqaution (18) is connected the general reaction-rate probability integral in reaction-rate theory ($\beta = \frac{1}{2}, \alpha = 1$ is the basic integral in reaction-rate theory) , unconditional densities in Bayesian analysis, marginal densities in a bivariate set up, and so on. Different problems in a large number of areas can be connected to Eqaution (18). Note that $x^{\gamma}e^{-ax^{\alpha}}$, normalized can act as a marginal density of a real scalar random variable $x > 0$ and $e^{-\frac{y}{x^{\beta}}}$, normalized, can act as the conditional density of y, given x. In this case Eqaution (18) has the structure of unconditional density of y in a Bayesian analysis situation. One can also look at Eqaution (18), normalized, as the joint density of two real scalar positive random variables and in this case Eqaution (18) integral represents the marginal density of y. For $\beta > 0$, Eqaution (18) can act as the Mellin convolution of a product and for $\beta < 0$ it can represent the Mellin convolution of a ratio. In this case, one can connect it to the Laplace transform of a generalized gamma density for $\alpha = 1$. Many types of such properties are studied in Mathai (*International Journal of Mathematical Analysis*, **6(51)**(2012), 2501–2510; In *Frontiers of Statistics and its Applications*, Bonfring Publications, Germany, 2013; *Proceedings of the 10th and 11th Annual Conference of SSFA*, **10-11**(2011–2012), pp. 11–20). Mathai has also considered the matrix-variate version of Eqaution (17).

15. Work on Fractional Calculus

Mathai may be credited with making a connection of fractional integrals to statistical distribution theory, extending fractional calculus to matrix-variate cases, to complex matrix-variate cases, to many scalar variable (multiple) cases, to many matrix variable cases. Recently Mathai has given a geometrical interpretation of fractional integrals in a simplex as fractions of certain total integral in n-dimensional cube. Mathai has also given a new definition to the area of fractional integrals, and thereby fractional derivatives, as Mellin convolutions of products and ratios in the real scalar case and as M-convolutions of products and ratios in the matrix-variate case, where one function is of type-1 beta form, see Mathai (*Integral Transforms and Special Functions*, **20(12)**(2009), 871–882; *Linear Algebra and Its Applications*, **439**(2013), 2901–2913, **446**(2014), 196–215), Mathai and Haubold (*Fractional*

Calculus & Applied Analysis, **14(1)**(2011), 138–155; *Cornell University arXiv*, I-IV(2012) 4 papers; *Fractional Calculus & Applied Analysis*, **16(2)**(2013), 469–478). Papers are published solving various types of fractional reaction, diffusion, reaction-diffusion differential equations, see Haubold, Mathai, Saxena (*Bulletin of the Astronomical Society of India*, **35**(2007), 681–689; *Journal of Computational and Applied Mathematics*, **235**(2011), 1311–1316; *Journal of Mathematical Physics*, **51**(2010), 103506-8), Saxena, Mathai and Haubold (*Astrophysics and Space Science*, **305**(2006), 289–296, 297–303, 305–313; *Astrophysics and Space Science Proceedings*, (2010), pp. 35–40, 55–62; *Axiom*, **3(3)** (2014), 320–334; *Journal of Mathematical Physics*, **55**(2014), 083519, doi:10.1063/1.4891922.

MDPI AG
St. Alban-Anlage 66
4052 Basel, Switzerland
Tel. +41 61 683 77 34
Fax +41 61 302 89 18
http://www.mdpi.com

Axioms Editorial Office
E-mail: axioms@mdpi.com
http://www.mdpi.com/journal/axioms

www.ingramcontent.com/pod-product-compliance
Lightning Source LLC
Chambersburg PA
CBHW051717210326
41597CB00032B/5516

* 9 7 8 3 0 3 8 4 2 6 6 5 3 *